BIOCHAR
A Regional Supply Chain Approach in View of Climate Change Mitigation

Climate change poses a fundamental threat to humanity, and thus solutions for both mitigation and adaptation strategies are becoming increasingly necessary. Biochar can offer a range of environmental services, such as reclamation of degraded land, improvement of soil fertility and carbon sequestration. However, it also raises questions, regarding sustainable feedstock provision, biomass pyrolysis and soil amendment. These questions, among various others, are addressed in this state-of-the-art compendium.

Covering a broad geographical range, with regional assessments from North America, Europe, the Near East and Southeast Asia, this interdisciplinary volume focuses on the entire biochar supply chain, from the availability and economics of biomass resources, to pyrolysis, and ultimately to the impacts on soil properties.

The combination of theory with practical examples makes this a valuable book for researchers, policymakers and graduate students alike, in fields such as soil science, sustainable development, climate change mitigation, biomass and bioenergy, forestry, and environmental engineering.

Dr Viktor J. Bruckman is a postdoctoral researcher at the Commission for Interdisciplinary Ecological Studies, of the Austrian Academy of Sciences, Austria. He also leads the Task Force 'Sustainable Forest Biomass Network (SFBN)' of the International Union of Forest Research Organizations (IUFRO). Dr Bruckman served as principal coordinator of the FOREBIOM project, looking at climate change mitigation potentials of biochar.

Professor Esin Apaydın Varol is an Assistant Professor in the Department of Chemical Engineering at Anadolu University, Turkey. She is currently working on the utilization of biomass for energy and carbonaceous products, and her research interests include biomass composition, its thermal degradation, and product characterization via spectroscopic and chromatographic methods.

Professor Başak B. Uzun (1975–2015) was a Professor in the Department of Chemical Engineering at Anadolu University, Turkey, and a member of the Carbon Materials Processing Group. Her research attempted to determine the effect of pyrolysis conditions on product yields and quality, using different types of pyrolysis reactors for bio-oil and biochar production.

Professor Jay Liu is Head of Department and leader of the Intelligent Systems Laboratory (ISL) at Pukyong National University (South Korea). He is actively involved in research on process systems engineering, with a focus on energy issues. ISL's research covers process design and simulation, and the application of chemometrics, particularly for BTL (biomass-to-liquid) processes.

BIOCHAR

A Regional Supply Chain Approach in View of Climate Change Mitigation

Edited by

VIKTOR J. BRUCKMAN
Austrian Academy of Sciences

ESIN APAYDIN VAROL
Anadolu University

BAŞAK B. UZUN
Anadolu University

and

JAY LIU
Pukyong National University

CAMBRIDGE
UNIVERSITY PRESS

University Printing House, Cambridge CB2 8BS, United Kingdom

Cambridge University Press is part of the University of Cambridge.

It furthers the University's mission by disseminating knowledge in the pursuit of education, learning and research at the highest international levels of excellence.

www.cambridge.org
Information on this title: www.cambridge.org/9781107117099

© Cambridge University Press 2016

This publication is in copyright. Subject to statutory exception
and to the provisions of relevant collective licensing agreements,
no reproduction of any part may take place without the written
permission of Cambridge University Press.

First published 2016

Printed in the United Kingdom by TJ International Ltd. Padstow Cornwall

A catalogue record for this publication is available from the British Library

Library of Congress Cataloguing-in-Publication data
Names: Bruckman, Viktor J., editor.
Title: Biochar: A Regional Supply Chain Approach in View of Climate Change
Mitigation / edited by Viktor J. Bruckman, Esin Apaydin Varol,
Başak B. Uzun and Jay Liu.
Description: New York: Cambridge University Press, 2016. |
Includes bibliographical references and index.
Identifiers: LCCN 2016026756 | ISBN 9781107117099 (hard back)
Subjects: LCSH: Biochar. | Greenhouse gas mitigation. |
Climate change mitigation.
Classification: LCC TP248.B55 B537 2016 | DDC 577.2/2–dc23
LC record available at https://lccn.loc.gov/2016026756

ISBN 978-1-107-11709-9 Hardback

Cambridge University Press has no responsibility for the persistence or accuracy of URLs for external or third-party internet websites referred to in this publication, and does not guarantee that any content on such websites is, or will remain, accurate or appropriate.

We would like to dedicate this book, as a token of our esteem, to our dear colleague and true friend Professor Dr Başak Burcu Uzun Akınlar, who tragically and unexpectedly passed away during the period of editing this book. Her contributions to this book as an editor, to science in general, her friendly and kind way of interacting with colleagues and students and her heartwarming smile will always be remembered.

Contents

List of contributors		*page* x
Preface		xv
1.	Biochar in the View of Climate Change Mitigation: the FOREBIOM Experience	1
	Viktor J. Bruckman, Michaela Klinglmüller and Milutin Milenković	
Part I	**The Interdisciplinary Approach**	23
2.	A Supply Chain Approach to Biochar Systems	25
	Nathaniel M. Anderson, Richard D. Bergman and Deborah S. Page-Dumroese	
3.	Life Cycle Analysis of Biochar	46
	Richard D. Bergman, Hongmei Gu, Deborah S. Page-Dumroese and Nathaniel M. Anderson	
4.	Systems Integration for Biochar in European Forestry: Drivers and Strategies	70
	Saran P. Sohi and Tom Kuppens	
5.	Biochar as an Integrated and Decentralised Environmental Management Tool in the Botanic Garden Berlin-Dahlem	96
	Robert Wagner, René Schatten, Kathrin Rössler, Ines Vogel and Konstantin Terytze	
Part II	**Sustainable Biomass Resources**	121
6.	An Integrated Approach to Assess Sustainable Forest Biomass Potentials at Country Level	123
	Michael Englisch, Thomas Gschwantner, Thomas Ledermann and Klaus Katzensteiner	

7.	Sustainable Biomass Potentials from Coppice Forests for Pyrolysis: Chances and Limitations	139
	Valeriu-Norocel Nicolescu, Eduard Hochbichler and Viktor J. Bruckman	
8.	Towards Environmental and Economic Sustainability via the Biomass Industry: the Malaysian Case Study	162
	Kok Mun Tang, Wan Asma Ibrahim and Wan Rashidah Kadir	
9.	Carbon Sequestration Potential of Forest Biomass in Turkey	184
	Betül Uygur and Yusuf Serengil	

Part III	**Biochar Production**	**197**
10.	Biochar Production	199
	Frederik Ronsse	
11.	Biomass Pyrolysis for Biochar Production: Kinetics, Energetics and Economics	227
	Byungho Song	
12.	Pyrolysis: a Sustainable Way From Biomass to Biofuels and Biochar	239
	Başak B. Uzun, Esin Apaydın Varol and Ersan Pütün	
13.	The Role of Biochar Production in Sustainable Development in Thailand, Lao PDR and Cambodia	266
	Maliwan Haruthaithanasan, Orracha Sae-Tun, Natthaphol Lichaikul, Soktha Ma, Sithong Thongmanivong and Houngphet Chanthavong	

Part IV	**Biochar Application as a Soil Amendment**	**289**
14.	Biochar Applications to Agricultural Soils in Temperate Climates – More Than Carbon Sequestration?	291
	Gerhard Soja, Elena Anders, Jannis Bücker, Sonja Feichtmair, Stefan Gunczy, Jasmin Karer, Barbara Kitzler, Michaela Klinglmüller, Stefanie Kloss, Maximilian Lauer, Volker Liedtke, Franziska Rempt, Andrea Watzinger, Bernhard Wimmer, Sophie Zechmeister-Boltenstern and Franz Zehetner	
15.	Opportunities and Uses of Biochar on Forest Sites in North America	315
	Deborah S. Page-Dumroese, Mark D. Coleman and Sean C. Thomas	
16.	The Role of Mycorrhizae and Biochar in Plant Growth and Soil Quality	336
	İbrahim Ortaş	
17.	The Use of Stable Isotopes in Understanding the Impact of Biochar on the Nitrogen Cycle	351
	Rebecca Hood-Nowotny	

18. Biochar Amendment Experiments in Thailand: Practical Examples 368
   Thavivongse Sriburi and Saowanee Wijitkosum

Index 391
Colour plate section to be found between pages 238 and 239.

Contributors

Elena Anders
Austrian Institute of Technology GmbH, Health & Environment Department, Tulln, Austria, Elenaanders@gmx.at

Nathaniel M. Anderson
US Department of Agriculture, US Forest Service, Rocky Mountain Research Station, Missoula MT, United States, nathanielmanderson@fs.fed.us

Esin Apaydın Varol
Anadolu University, Department of Chemical Engineering, Eskişehir, Turkey, eapaydin@anadolu.edu.tr

Richard D. Bergman
US Department of Agriculture, US Forest Service, Forest Products Laboratory, Madison WI, United States, rbergman@fs.fed.us

Viktor J. Bruckman
Austrian Academy of Sciences, Commission for Interdisciplinary Ecological Studies, Vienna, Austria, viktor.bruckman@oeaw.ac.at

Jannis Bücker
Groundwater Research Institute GmbH, Dresden, Germany, jannis.buecker@googlemail.com

Houngphet Chanthavong
National University of Laos, Faculty of Forestry, Vientiane, Lao PDR, houngphet@hotmail.com

Mark D. Coleman
University of Idaho, Department of Forest, Rangeland, and Fire Sciences, Moscow ID, United States, mcoleman@uidaho.edu

Michael Englisch
Federal Research and Training Centre for Forests, Natural Hazards and Landscape, Department of Forest Ecology and Soils, Vienna, Austria, michael.englisch@bfw.gv.at

Sonja Feichtmair
Austrian Institute of Technology GmbH, Health & Environment Department, Tulln, Austria, Sonja.Feichtmair@gmx.at

Thomas Gschwantner
Federal Research and Training Centre for Forests, Natural Hazards and Landscape, Department of Forest Inventory, Vienna, Austria, thomas.gschwantner@bfw.gv.at

Hongmei Gu
US Department of Agriculture, US Forest Service, Forest Products Laboratory, Madison WI, United States, hongmeigu@fs.fed.us

Stefan Gunczy
Joanneum Research Forschungsgesellschaft mbH, Graz, Austria, stefan.gunczy@joanneum.at

Maliwan Haruthaithanasan
Kasetsart University, Biomass and Bioenergy Technology Department, Kasetsart Agriculture and Agro-industrial Product Improvement Institute, Bangkok, Thailand, aapmwt@ku.ac.th

Eduard Hochbichler
University of Natural Resources and Life Sciences, Department of Forest- and Soil Sciences, Institute of Silviculture, Vienna, Austria, eduard.hochbichler@boku.ac.at

Rebecca Hood-Nowotny
Austrian Institute of Technology GmbH, Health & Environment Department, Tulln, Austria, Rebecca.Hood@ait.ac.at

Wan Asma Ibrahim
Forest Research Institute Malaysia, Biomass Technology Programme, Kuala Lumpur, Malaysia, asma@frim.gov.my

Wan Rashidah Kadir
Forest Research Institute Malaysia, Soil Management Branch, Kuala Lumpur, Malaysia, rashidah@frim.gov.my

Jasmin Karer
Austrian Institute of Technology GmbH, Health & Environment Department, Tulln, Austria, jasmin.karer@gmx.at

Klaus Katzensteiner
University of Natural Resources and Life Sciences Vienna, Department of Forest and Soil Sciences, Institute of Forest Ecology, Vienna, Austria, klaus.katzensteiner@boku.ac.at

Barbara Kitzler
Federal Research and Training Centre for Forests, Natural Hazards and Landscape, Department of Forest Ecology and Soils, Vienna, Austria, barbara.kitzler@bfw.gv.at

Michaela Klinglmüller
University of Natural Resources and Life Sciences (BOKU) Vienna, Department of Economics and Social Sciences, Institute for Sustainable Economic Development, Vienna, Austria, michaela.klinglmueller@gmail.com

Stefanie Kloss
University of Natural Resources and Life Sciences Vienna, Department of Forest and Soil Sciences, Institute of Soil Research, Vienna, Austria, stefanie.kloss@boku.ac.at

Tom Kuppens
Hasselt University, Centre for Environmental Sciences, Research Group Environmental Economics, Diepenbeek, Belgium, tom.kuppens@uhasselt.be

Maximilian Lauer
Joanneum Research Forschungsgesellschaft mbH, Institut für Energieforschung, Graz, Austria, RESSekretariat@joanneum.at

Thomas Ledermann
Federal Research and Training Centre for Forests, Natural Hazards and Landscape, Department of Forest Growth and Silviculture, Vienna, Austria, thomas.ledermann@bfw.gv.at

Natthaphol Lichaikul
Kasetsart University, Biomass and Bioenergy Technology Department, Kasetsart Agriculture and Agro-industrial Product Improvement Institute, Bangkok, Thailand, radnha_np@hotmail.com

Volker Liedtke
Aerospace & Advanced Composites GmbH, Department of Inorganic Composites, Wiener Neustadt, Austria, volker.liedtke@aac-research.at

Soktha Ma
Ministry of Agriculture, Forestry and Fisheries, Forestry Administration, Department of Forest Plantation and Private Forest Developments, Phnom Penh, Cambodia, masoktha79@gmail.com

Milutin Milenković
Vienna University of Technology, Department of Geodesy and Geoinformation, Vienna, Austria, Milutin.Milenkovic@geo.tuwien.ac.at

Valeriu-Norocel Nicolescu
Transylvania University of Brasov, Faculty of Silviculture and Forest Engineering, Brasov, Romania, nvnicolescu@unitbv.ro

İbrahim Ortaş
Cukurova University, Department of Soil Science and Plant Nutrition, Faculty of Agriculture, Adana, Turkey, iortas@cu.edu.tr

Deborah S. Page-Dumroese
US Department of Agriculture, US Forest Service, Rocky Mountain Research Station, Moscow ID, United States, ddumroese@fs.fed.us

Ersan Pütün
Anadolu University, Department of Materials Science and Engineering, Faculty of Engineering, Eskişehir, Turkey, eputun@anadolu.edu.tr

Franziska Rempt
Austrian Institute of Technology GmbH, Health & Environment Department, Tulln, Austria, franziska.rempt@gmail.com

Frederik Ronsse
Ghent University, Department of Biosystems Engineering, Faculty of Bioscience Engineering, Ghent, Belgium, Frederik.Ronsse@UGent.be

Kathrin Rössler
Freie Universität Berlin, Department of Earth Science, Institute of Geographical Sciences, Berlin, Germany, kathrin.roessler@fu-berlin.de

Orracha Sae-Tun
Kasetsart University, Biomass and Bioenergy Technology Department, Kasetsart Agriculture and Agro-industrial Product Improvement Institute, Bangkok, Thailand, orrachs@gmail.com

René Schatten
Freie Universität Berlin, Department of Earth Science, Institute of Geographical Sciences, Berlin, Germany, rene.schatten@fu-berlin.de

Yusuf Serengil
Istanbul University, Watershed Management Department, Faculty of Forestry, Istanbul, Turkey, serengil@istanbul.edu.tr

Saran P. Sohi
University of Edinburgh, School of Geosciences, Edinburgh, Scotland, saran.sohi@ed.ac.uk

Gerhard Soja
Austrian Institute of Technology GmbH, Health & Environment Department, Tulln, Austria, gerhard.soja@ait.ac.at

Byungho Song
Kunsan National University, Department of Chemical Engineering, Gunsan, South Korea, bhsong@kunsan.ac.kr

Thavivongse Sriburi
Chulalongkorn University, Chula Unisearch, Bangkok, Thailand, Thavivongse.S@chula.ac.th

Kok Mun Tang
Malaysia Biomass Industry Confederation, Putrajaya, Malaysia, tang.rapid@gmail.com

Konstantin Terytze
Freie Universität Berlin, Department of Earth Science, Institute for Geographical Sciences, Berlin, Germany, konstantin.terytze@fu-berlin.de

Sean C. Thomas
University of Toronto, Faculty of Forestry, Toronto, Canada, sc.thomas@utoronto.ca

Sithong Thongmanivong
National University of Laos, Faculty of Forestry, Vientiane, Laos, sithong@nuol.edu.la

Betül Uygur
Istanbul University, Watershed Management Department, Faculty of Forestry, Istanbul, Turkey, uygurb@istanbul.edu.tr

Başak B. Uzun*
Anadolu University, Department of Chemical Engineering, Eskişehir, Turkey

Ines Vogel
Freie Universität Berlin, Department of Earth Science, Institute for Geographical Sciences, Berlin, Germany, vogeline@zedat.fu-berlin.de

Robert Wagner
Freie Universität Berlin, Department of Earth Science, Institute for Geographical Sciences, Berlin, Germany, rowagner@zedat.fu-berlin.de

Andrea Watzinger
Austrian Institute of Technology GmbH, Health & Environment Department, Tulln, Austria, andrea.watzinger@ait.ac.at

Saowanee Wijitkosum
Chulalongkorn University, Environmental Research Institute, Bangkok, Thailand, w.m.saowanee@gmail.com

Bernhard Wimmer
Austrian Institute of Technology GmbH, Health & Environment Department, Tulln, Austria, bernhard.wimmer@ait.ac.at

Sophie Zechmeister-Boltenstern
University of Natural Resources and Life Sciences Vienna, Department of Forest and Soil Sciences, Institute of Soil Research, Vienna, Austria, sophie.zechmeister@boku.ac.at

Franz Zehetner
University of Natural Resources and Life Sciences Vienna, Department of Forest and Soil Sciences, Institute of Soil Research, Vienna, Austria, franz.zehetner@boku.ac.at

* Deceased

Preface

The world is currently facing major challenges such as climate change, a rising demand of biomass for food, feed, raw materials and energy, environmental degradation and pollution, as well as a considerable loss of biodiversity. Soils are at centre stage in all of these challenges and there are examples from the past where it was shown that the success of civilizations is tightly dependent on soil fertility and productivity. We begin to understand that the prosperity of the entire world population builds on soils, which are a non-renewable resource in human timescales and threatened on global scales by unsustainable management practices, climate change and other anthropogenic influences. Soils are not just fine grains of weathered rock, but can be described as living bioreactors that provide the basis for biomass production. They are regulators of the world's climate and represent the largest terrestrial carbon stock.

Scientists are now celebrating the discovery of something that was already there before humans actively managed and used land. Charred organic matter is an important component of soils in many ecosystems, especially those in Mediterranean regions, which are usually well-adapted to wildfires. The relative recalcitrance of charred organic matter makes the difference where things become interesting from an environmental engineering point of view. Why not bury charcoal in our soils where it decomposes slowly and hence sequesters carbon and ideally also improves soil properties such as nutrient and water retention? Indeed there is evidence that this is a promising strategy, as very fertile soils were discovered in tropical South America, where one would only expect heavy weathered clay minerals with poor fertility. It turned out that charred organic matter plays a role in improving and conserving fertility in at least some of these soils, which are called 'Terra Preta de Indio'. However, this might quickly lead to misunderstandings and misconceptions as it is by no means as simple as adding charcoal to poor soil to end up with 'Terra Preta'. There are many interactions and interdependencies between different types of carbon in the soil, microorganisms, moisture and the way land is being managed. We are currently only at the beginning of understanding the entire system and although we learn more with every single effort to study biochar and its application in the environment, there are still many questions, and some of them need time to find a satisfying answer.

One of these recent approaches was made by an international collaboration project between Austria, South Korea and Turkey, with the aim of studying the potential for greenhouse gas (GHG) mitigation using biochar in the respective consortium member countries. This KORANET (Korean scientific cooperation network with the European Research Area) collaboration provided the basis to prepare this book, with contributions selected based on discussions with participants of the project workshops and in some cases also external expertise. The aim of this book is not to replicate existing knowledge and recently published books on this topic, but the discussions showed clearly that the biochar topic must be addressed in an interdisciplinary way by using a systems approach. Therefore, we purposely tried to include chapters on the entire supply chain, from biomass availability and provision to the actual conversion process, pyrolysis, to the final application.

The first part of the book (Chapters 2–5) provides a more integrative overview and describes the entire supply chain from different points of view. Part 2 (Chapters 6–9) is focused on the feedstock potentials and implications for biomass markets and regional trade scales. Chapters 10–13 comprise Part 3, which is devoted to the production of biochar from a technological point of view, but also considers byproducts and tradeoffs. Finally, in Part 4, (Chapters 14–18) we focus on biochar–soil interactions and the potential benefits (including co-benefits) of biochar amendment in soils.

We focus in this book on woody biomass and biomass resources from forests as well as forest plantations as these are currently not well covered in the existing literature. Likewise, we present potential uses of biochar in forest ecosystems as well. A key strategy of this book is to combine theoretical examples and considerations with practical examples, and therefore we include at least one practical chapter in each section with original data from field experiments or demonstration sites.

The inclusion of expertise from different climatic and geographic regions of the world highlights that if biochar is to be considered as a tool for environmental or geo-engineering, one may need to expect different (regional to site-specific) challenges but also opportunities. As the scenarios of biochar amendment are infinite in terms of expectations, biochar properties and environmental responses, it needs efforts to better understand and characterize the mechanisms behind it and to employ robust standards. This would also allow and facilitate a market for biochar and its safe use. Nevertheless, it will still be necessary to decide at project level if and under which circumstances the use of biochar delivers the expected benefits. Success finally also depends on economics, and this is currently one of the major drawbacks. But still, biochar may be helpful to restore soil functions and improve soil fertility, and the wide range of feedstock materials as well as pyrolysis conditions could allow the production of specific biochars with distinct properties, triggering specific environmental responses. A profound understanding of the entire supply chain and interdisciplinary approaches are needed to address this issue, and the aim of this book is to provide a good insight into different steps of the supply chain, under different circumstances. Ultimately it should help to understand the use of biochar as a tool to tackle the current challenges, without fuelling exaggerated expectations, in the most efficient way. Even though it may seem to be impossible to employ biochar for the sake of a single function, it might be a viable solution when considering byproducts or co-benefits. The potential pathways of biochar utilization are endless, but science has to provide the basis for a sustainable and safe application and we hope that this book is of help for experts and students as well as engineers and land managers.

We would like to express our sincere gratitude to the institutions that funded the KORANET Project 'FOREBIOM', which allowed us to put this book together. These are the 7th Framework Programme for Research and Technological Development of the European Commission (FP7, Ref.: KORANET), The Austrian Federal Ministry of Science, Research and Economy (BMWFW, Ref.: BMWF-308.299/0023-II/6/2012), The National Research Foundation of South Korea (NRF) funded by the Ministry of Science, ICT and Future Planning (Ref.: 2012K1A3A7A03052140), and the Scientific and Technological Research Council of Turkey (TÜBİTAK, Ref.: 112M662). Mr Gerald Dunst, CEO of Sonnenerde, kindly provided the pyrolysis of woodchips for our field experiment.

We thank all the internal and officially appointed referees of the manuscripts in several development stages for their time and efforts to significantly improve the quality of the final chapters. We would also like to thank our colleagues for their advice in various discussions regarding the book project. Specifically, we would like to thank Hakkı Alma, Nele Ameloot, Nuray Çiçek Atikmen, Erland Bååth, Gero Becker, Torsten Berger, Paul Blanz, Anthony V. Bridgwater, Martin H. Gerzabek, Robert Jandl, Claudia Kammann, Rattan Lal, Sung-Rin Lim, Franz Oberwinkler, İbrahim Ortaş, Jun-Hyung Ryu, Inge Stupak, Mohd Nazip Suratman, Håkan Wallander, Michael Weiß, Karin Windsteig, Seung-Han Woo, Sophie Zechmeister-Boltenstern, Ronald Zirbs and a number of anonymous referees, including those reviewing the initial proposal for this book. We are thankful to members of the Task Force 'Sustainable Forest Biomass Network (SFBN)' and RG 7.01 (Impacts of air pollution and climate change on forest ecosystems) of the International Union of Forest Research Organizations (IUFRO) for continuous advice in several stages of this project. We would like to express our sincere gratitude to Emma Kiddle, Cassi Roberts and Zoe Pruce as well as their colleagues at Cambridge University Press (CUP) for guiding us through the entire publication process in an extraordinarily professional and dedicated way. Our friends, colleagues and most of all family members deserve our greatest gratitude for their support and patience during the compilation of this book.

The Editors
Viktor J. Bruckman, Vienna, Austria
Esin Apaydın Varol, Eskişehir, Turkey
Başak B. Uzun Akınlar, Eskişehir, Turkey
Jay Liu, Busan, South Korea
January, 2016

1

Biochar in the View of Climate Change Mitigation: the FOREBIOM Experience

VIKTOR J. BRUCKMAN, MICHAELA KLINGLMÜLLER
AND MILUTIN MILENKOVIĆ

Abstract

Biochar is currently one of the dominant topics in soil research, despite the fact that it is not a new discovery. It has the potential to address some of the most pressing questions humanity is currently facing, that is climate change, food security, energy security and environmental pollution. However, a soil system is very complex and together with the multitude of biochar production settings and nearly infinite number of potential feedstock resources it becomes evident that there is no single solution for these challenges available. This is specifically an issue when addressing the potential of biochar for climate change mitigation via reduction of greenhouse gases (GHG). Systems approaches are needed, covering the entire supply chain and backed up with life cycle assessments to ensure a positive impact by using biochar as a tool for environmental management.

This chapter provides a summary and brief introduction of the subsequent chapters of this book, with a focus on biochar for climate change mitigation, including an economic assessment of GHG abatement costs. The FOREBIOM project will be briefly introduced and results on biochar erosion after amendment of a forest floor are presented.

1.1 Introduction

Biochar, 'a solid material obtained from thermochemical conversion of biomass in an oxygen-limited environment' (IBI, 2013), has become one of the dominant research topics within the scientific soil community during the past years. Scientific conferences and meetings focusing on soil issues feature a considerable number of biochar related contributions. Unlike the case in a number of hypes in science, the biochar topic seems to be attractive also to the public, at least in some countries. The results of a simple web search on biochar application are just overwhelming, starting from private small-scale initiatives all the way through to industrial-scale production. One can easily retrieve reports, photos and even videos on

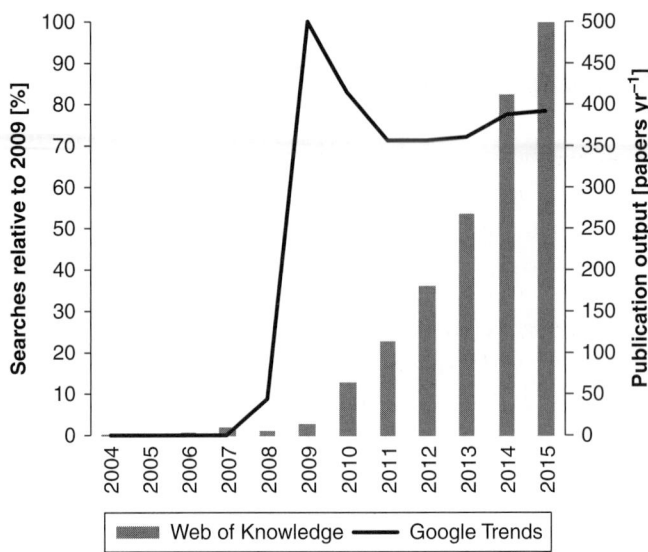

Figure 1.1. A comparison between scientific and public interest in biochar since the beginning of the recent boom at around 2006. Grey bars show the scientific output of literature (scientific articles and reviews with the term "biochar" in the title) according to the Institute for Scientific Information (ISI) Web of Knowledge (now Web of Science). The solid black line represents the relative share of searches based on the year 2009, where most searches occurred.

YouTube, explaining how to build your own biochar reactor and how to use biochar in your home vegetable garden. Biochar science is currently in a phase of rapid growth and the annual publication output is still increasing since beginning in around 2006. It is interesting to note, however, that the public interest in biochar, expressed as the number of web searches on Google, seems to have already peaked in 2009 and remains on a relative constant high level since then (Figure 1.1). It is assumed that Google Trends represents public interest in biochar, although a share of the searches likely also originates from experts. However, we hypothesize that they would use more complex search terms or combine biochar with other attributes. A closer analysis in terms of regional interests reveals that until January 2016, the leading country in terms of search volume was Australia, followed by Canada, the United States and the United Kingdom. India is the leading developing country, with 24% of the search volume, which is comparable to the figures in Germany or Italy. The reason for the rapid growth in popularity might be that it has been claimed that using biochar as a soil amendment may be helpful in addressing some of the global key challenges, such as population growth and increasing demands for food and feed supplies, climate change, and a recent boom in renewable energy. Indeed, the potentials seem enormous, but it is also clear that biochar alone is not the key for all these issues. A very comprehensive introduction and presentation of current knowledge of biochar is offered in the recently published second edition of *Biochar for Environmental Management: Science, Technology and Implementation* (Lehmann and Joseph, 2015a).

Biochar represents the solid residual when heating biomass to above 250°C in an oxygen-free or -depleted atmosphere, with the specific aim to use it as a soil amendment or growth medium. The production process is generally called pyrolysis and is explained in detail later in this book (Chapters 10 and 12). During pyrolysis, volatiles and water are driven off and carbon is re-organized in fused aromatic ring structures that are more recalcitrant to decomposition as compared to the carbon compounds of the original biomass. As the pyrolysis temperature (highest heating temperature, HHT) is increased, a large number of microcracks are formed, increasing the surface area and thus the potential chemical reactivity. Although the chemical characteristics change is remarkable, including a relative enrichment in carbon (C), calcium (Ca), phosphorus (P), and potassium (K) among others and typically a depletion of hydrogen (H) and oxygen (O), as well as the reorganization of C, there are only minor morphological impacts on the visible macro-scale, despite the black colour.

The material itself can be found in many forms in a large number of ecosystems worldwide, resulting from wildfires, where it is called char, or purposely made in kilns, where it is called charcoal. Charcoal is traditionally used as carrier of thermal energy. The use of both char and charcoal has a long history and is closely related to the development of human culture. The Chauvert cave in Southern France presents some of the oldest char drawings with an age of around 30,000 years BP (before present) according to radiocarbon dating (Pettitt, 2008). Much later in history, it was the most important source of thermal energy for smelting ores in developing cultures, until its peak use when it fuelled the industrial revolution from about the middle of the eighteenth century until fossil resources finally replaced charcoal. Woodlands, especially in relatively close proximity to urban areas, still show features such as abandoned forest railways, even plateaus in rugged terrain or simply distinct field names that indicate the significance of charcoal production in past times. Charcoal has clear advantages over fuelwood, especially in urban areas where the consumption of energy is high. As volatile matter and water is removed during the pyrolysis process, its weight is reduced and therefore transportation is easier and the emissions are lower during combustion, making it even suitable for indoor use if air circulation is ensured. Therefore, it is still of major importance in a number of regions worldwide, especially in Africa, South America and Southeast Asia, where charcoal is still produced using traditional methods (Chapter 13). However, as biochar is made for a specific purpose other than burning and harvesting thermal energy, it may be produced from different biogenic feedstock materials. Consequently, not all kinds of biochar are suitable for combustion and it might overlap with charcoal only under certain conditions (feedstock material and production characteristics etc.).

This diversity of biochar offers a great opportunity for producing materials that deliver dedicated environmental functions based on the properties of the soils on site scale and the desired environmental response after biochar amendment. Specific biochar properties may be achieved by the selection of feedstock resources or a blend of different materials, the pyrolysis conditions (especially HHT), mixing of different types of biochar, or post treatment, such as composting of biochar with other organic materials. There are a number of different terms in the literature for this targeted biochar production, such as 'designer biochar', 'bespoke biochar' and 'fit for purpose biochar', among others. Bespoke biochar may fulfil a specific and predictable function in a complex environment that includes various biotic and abiotic factors on a number of spatial scales. The basis of a targeted use is therefore a sound understanding of the site conditions, as well as the processes and mechanisms involved, especially in relation to soil microbiology and soil chemistry. In addition, it is necessary to formulate standards and guidelines to facilitate and promote biochar production and use in a safe and efficient way. The International Biochar Initiative (IBI) as well as the European Biochar Research Network (EBRN), among other institutions, developed biochar standards and certificates, with the aim of setting industry standards that ensure a certain level of quality and safety. In particular, heavy metals, which are part of the initial biomass in low doses, might be accumulated in the biochar after pyrolysis and therefore pose environmental risks. The secondary formation of potential hazardous contaminants is controlled by the feedstock material and the pyrolysis conditions. There are a range of potentially harmful contaminants that can be formed during pyrolysis, such as polycyclic aromatic hydrocarbons (PAHs) or dioxins.

A more direct and immediate use of biochar can be achieved by reducing the complexity of a biochar system. On local scales, and for distinct spatially confined scales, there are a number of possibilities to utilize biochar. This is especially the case in urban areas and in combination with new approaches of modern architecture. A worldwide trend of urbanization and a consequent city compaction threatens urban green space and, hence, a sustainable urban development with careful planning and new approaches is needed (Haaland and van den Bosch, 2015). The advantage of the low weight of biochar, in combination with a large surface area, makes it an ideal component for biochar-soil mixtures used on roofs or other elements of a building. One such example can be found in the 1 Utama shopping centre in Kuala Lumpur, the capital of Malaysia (Chapter 8). It is one of the largest shopping centres worldwide and features a rain forest as well as a rooftop garden, which are both well integrated in the building structure (Figure 1.2). Biochar plays an essential role as a growth medium in this particular example.

In fact, research already steps beyond green space when considering tight material cycles to reduce the environmental footprint. Such material cycles can even equal the boundaries of a single building, which makes them to a certain degree autonomous. For instance, biochar can be used for sustainable sanitation (Schuetze and Santiago-Fandiño, 2014). Likewise, Chapter 5 demonstrates the use of biochar in a system with clearly defined boundaries, where it also plays a role in sanitation to recover essential nutrients, such as nitrogen, phosphate and potassium, in the Berlin-Dahlem Botanical Garden.

1.2 Summary of the Contents of This Book

The initial idea of the underlying FOREBIOM project (potentials for realizing negative carbon emissions using forest biomass and subsequent biochar recycling) was to foster collaboration among South Korea and European countries towards the potentials of biochar for climate change mitigation. The scope of the current book is widened to

Biochar in the View of Climate Change Mitigation

Figure 1.2. The "Rain forest" (a) and the "Secret garden" (b) in the 1 Utama shopping centre in Kuala Lumpur both use a mixture of biochar, coconut fibre and clayey subsoil as a growth medium. The ratio of the three compartments was adjusted according to the actual vegetation. Charcoal briquettes from sawdust, produced for barbecue purposes, were used to provide biochar. Photographs by V.J. Bruckman. (A black and white version of this figure will appear in some formats. For the colour version, please refer to the plate section.)

ensure a holistic approach and a broader geographical perspective is included. A focus of this book is biochar production and use from a forestry perspective, as forests represent a potentially large source of biomass, and biochar application in forests may be of interest in certain circumstances (Chapters 4 and 15). The book consists of four major sections, divided into a more cross-disciplinary section, and three key steps involved in the biochar supply chain, each representing a section of four to five chapters (Figure 1.3). The key steps were identified and discussed during the FOREBIOM workshops and it was agreed that these should be considered in a comprehensive assessment of biochar potentials. Consequently, we introduce each step separately (1.2.1–1.2.3) and present links to the relevant chapters. On the other hand, we refer to the cross-disciplinary chapters (2–5) throughout this introductory chapter. The examples of integrating biochar as a functional element in Berlin and Kuala Lumpur demonstrate that the potential uses are indeed vast. Despite the fact that practical use of biochar does not play a major role at this time for a number of reasons, it requires a holistic and systems approach to ensure that biochar can be part of a sustainable

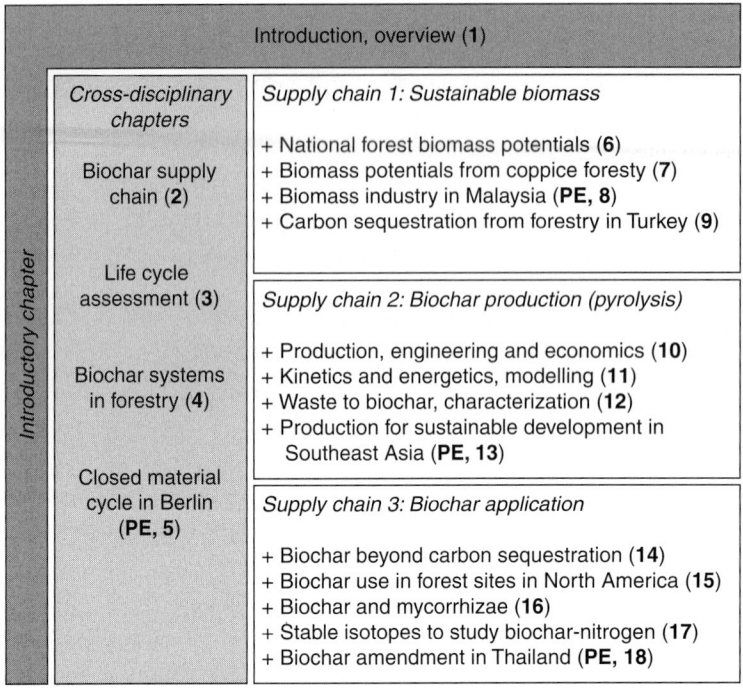

Figure 1.3. Summary of the contents of this book, divided into the introductory chapter and four major sections. The chapters are categorized according to their main focus and their numbers are given in brackets. Chapters describing practical examples (PE) or case studies are indicated. The three main elements of the supply chain are described in detail in Sections 1.2.1–1.2.3.

development. Suitable tools in order to account for the net carbon impact of a biochar system, such as the life cycle analysis (LCA), can provide essential information on the sustainability in terms of C balance (Chapter 3).

1.2.1 Sustainable Biomass

The production of biochar requires biomass. The choice of the appropriate biomass feedstock may be determined by the desired properties of the final product, biochar, but often a more practical constraint dictates the type of biomass used. The availability of inexpensive biomass is currently (and probably will always be) the main factor influencing the potential of biochar production and to some extent the properties of the final product. This makes all kinds of waste biomass ideal candidates as feedstock for pyrolysis. This approach can also help to close material cycles and reduce the GHG footprint by replacing external purchases of fertilizers and growth media (Chapter 5). Biochar derived from woody plant biomass has the advantage of an inherent porous structure and a high C content, while the potential availability in many regions is high. However, biomass from forestry and agriculture, which together represent the most important potential source of plant biomass for pyrolysis, is demanded by a whole range of industries. This became specifically an issue in recent years, when the term 'bioeconomy' was coined and new technologies were developed to use biomass as a source for energy or other industrial feedstocks. Indeed the consumption of biomass is increasing and there is a need to clearly define sustainable potentials of biomass to support a meaningful green economy. One such approach can be based on an integrated assessment, where a theoretical potential is firstly calculated and then further decreased by a subset of constraints, considering for instance accessibility, environmental services, market price, nutrient balance and land tenure (Chapter 6). Even if this approach provides rough numbers for the entire forestry sector, it is also possible to determine potential waste streams (i.e. from sawmills or the wood processing industry) that represent potentials for biochar production. We show that some of these are traditionally used for charcoal production in Southeast Asia (Chapter 13). It was demonstrated that policy regulations can have a strong impact on the GHG balance of the forestry sector. Forests in Turkey represent a net sink of C and reforestation activities ensure that this trend remains unchanged. Currently there are no considerable incentives to increase biomass yields, although there is a gap between supply and demand. Consequently, agricultural biomass is being used extensively and there is strong competition in the biomass market (Chapter 9), making the production of biochar a challenge. At the same time, forests sequester large amounts of C, which can be a promising alternative strategy, compared to biochar.

Traditional forestry usually aims at the production of high-quality timber and the potential biomass waste is limited and also demanded by other industries (i.e. woodchip or pellet production). Biomass plantations can therefore produce biomass that is dedicated for use in energy or raw material production at much higher efficiencies when using fast-growing species in combination with short rotations. Such systems demand high standards of the soils, and fertilization is commonly necessary to compensate for the high nutrient losses. Biochar can be a potential tool to help to sustain the soil functions in such biomass plantations or to help remediate problematic sites (Chapter 4) while enhancing the efficiency of fertilizer use and the minimisation of GHG emissions such as nitrous oxide (N_2O) (Chapter 14). Short-rotation forestry was once a major source of biomass for energy and other products that require small stem diameters. Unlike the case in plantations, forests consisting of endemic species were coppiced, after which woodlands regenerate from resprouting from stumps or root suckers (Bruckman et al., 2011). This specific type of silviculture is seen as a potential source of biomass as it is sustainable and may even help to increase biodiversity (Chapter 7).

Woody biomass can also be provided in large quantities from the agriculture sector. In Malaysia, for instance, huge areas are cultivated with oil palms and rubber trees. Both crops provide large amounts of lignocellulosic biomass, suitable for a range of applications, including pyrolysis (Chapter 8). Innovative examples of biochar utilization may stimulate further research and development and prove that a successful application is possible under certain circumstances, as demonstrated in the 1 Utama shopping centre (Figure 1.2).

1.2.2 Biochar Production (Pyrolysis)

Virtually any organic material with a substantial amount of C can be used for producing biochar. The process of pyrolysis can be described as a method of thermal decomposition, as complex chemical structures are decomposed into a number of chemical compounds that differ significantly from each other. Moreover, the resulting products can be a mix of solid, liquid or gas and the ratio of these products is largely dependent on the pyrolysis conditions, especially the HHT and process duration. The pyrolysis conditions, together with the feedstock type, have a strong impact on the properties of biochar (Zhao et al., 2013; Kloss et al., 2012). C retention and biochar recalcitrance are two of the important properties with regard to immediate C sequestration, which can be greatly influenced by the process settings. In fact, a range of different pyrolysis reactors were developed, where biochar represents either the main product or in some cases a byproduct of an industrial process (Chapter 10). The products that can be derived from pyrolysis range from solid (e.g. charcoal, activated carbon, carbon fibres, ash, soot, fertilizers) to liquid (e.g. bio-oil, pyroligneous acid (Chapter 13), functional chemicals) to gaseous compounds (e.g. methane, ethane, propane, hydrogen) and all represent valuable bio-products (Chapter 12). Some of these products may be used directly (e.g. biochar for soil amendment), as a feedstock for further industrial processes or can be converted into energy. While the utilization of the gaseous products represents standard technology that is already available on the market, it is somewhat problematic with the liquid product of bio-oil. Its chemical composition is very complex, with a high water content, and it is corrosive. Hence, it needs a range of purifying steps to be ready for use in a variety of stationary energy applications (Chapter 10).

The production of biochar is not a new technology, as there is a long history in producing charcoal. Traditional methods are still widely used despite the low efficiency, uncertain process conditions and therefore varying qualities and high emissions of potentially hazardous components during pyrolysis. However, recent developments aim at improving these technologies and making them a viable source of income for local communities (Chapter 13). Although the main commodity will remain charcoal as there is a high domestic and international demand, residues from charcoal production may be used as biochar after comprehensive analysis and characterization of this material to avoid any negative consequences as a result of contamination. The practical questions for potential operators of pyrolysis reactors are whether there is a market that can sustain biochar production and whether a valorization of byproducts is possible. A careful economic assessment is important and it has to include feedstock costs, the development of biochar reactors, potential subsidies and additional revenues from bio-energy co-production, and a potential impact from carbon trading schemes and markets (Chapter 10). The combined production of bio-energy and biochar offers an interesting approach in terms of climate change mitigation, as a certain share of the original carbon in biomass is used to substitute fossil fuels and the residual solid biochar may contribute to a long-term C pool in soils, where it ideally improves soil properties. This can lead to a secondary effect through increased biomass productivity as a consequence of higher fertility. Modern pyrolysis reactors can produce biochar continuously and the process is self-sustaining once the pyrolysis process is started (Lehmann and Joseph, 2015b). The ratio between harvestable energy and other products can be predicted using specific kinetic models that describe the thermal decomposition of a biomass feedstock by explaining the chemical reaction rates in relation to the process conditions (Chapter 11). Such approaches are of major importance as reactors can be described by a set of mathematical equations, and hence provide a basis for the design and virtual testing of new pyrolysis approaches that include the production of biochar as well as other products, such as bio-energy, simultaneously. Different scenarios can be computed and optimal settings defined before a potential plant is built, which reduces the costs and efforts for system testing and evaluation.

1.2.3 Biochar Application as a Soil Amendment

In the final step of the supply chain, biochar is being incorporated into the soil, mixed with soil and/or other compartments for creating growth media or distributed on the soil surface. The soil system is very complex, with numerous interactions, and the introduction of biochar as a soil amendment may trigger certain responses that impact both soil physics and soil chemistry. Without any consideration of biochar and soil properties, the effect may be also negative, even in terms of C sequestration. Therefore, a sound understanding of the processes and mechanisms involved is of key relevance and there is no standard solution for how to amend a soil with biochar to achieve the desired response in the most efficient way. However, the current trend in biochar research is based on a range of expectations for biochar as a tool for environmental management. Key functions that biochar can provide are the improvement of nutrient availability and water retention, reduction of GHG emissions (Chapter 14), sorption of pollutants and growth-inhibiting substances such as heavy metals and salts, storage of recalcitrant C in the soil profile, improvement of soil physical properties and providing a suitable habitat for soil biota (Chapter 16). The efficient and targeted provision of one or more of these desired functions does not only depend on the feedstock and pyrolysis conditions, but to a large extent on the actual soil properties, the climate conditions, the type of vegetation and the resulting quality and quantity of existing soil organic matter (SOM). The above mentioned functions addressing GHG are to a great extent coupled with the nitrogen (N) cycle. Biochar may significantly influence soil N cycles and processes by altering nitrification, adsorption of ammonia, which has shown to be bioavailable (Taghizadeh-Toosi et al., 2011), and by increasing the storage capacity of ammonium, resulting in effects on plant nutrition, GHG emission from soils, and leaching of nitrate (Clough and Condron, 2010). Especially in temperate regions, reducing GHG emissions by using biochar to influence the N cycle turns out to be a promising strategy in reducing N_2O emissions (Hüppi et al., 2015) (Chapter 14). N_2O outcompetes CO_2 by a factor of 265 in terms of radiative forcing over a period of 100 years (Myhre et al., 2013) and is hence a very effective GHG. There has been significant progress in recent years in assessing the mechanisms and processes behind biochar–nitrogen relations.

Studying stable isotopes has been shown to be a viable method with a number of recent discoveries, and it currently represents a highly dynamic research field (Chapter 17).

Soil fertility is to a large extent determined by the activity and abundance of soil biota, and it was shown that biochar can have a strong impact on it (Thies et al., 2015). High activity in soil microbial communities implies higher nutrient mineralization rates and therefore higher potential biomass productivity, but on the other hand it could lead to elevated soil GHG emissions. Many productive species in agriculture and forestry depend on symbiotic mycorrhizae, helping to extend the accessibility of mineral nutrient and water pools that are otherwise inaccessible for fine roots. Biochar can provide a suitable habitat for mycorrhizae and it is suggested that a combination of biochar amendment and inoculation of specific mycorrhizae shows the highest increase in biomass productivity (Chapter 16). This is of special concern in areas suffering from drought and it may be a successful strategy to reduce the amounts of fertilizer used via increasing the efficiency of their uptake. This would be a positive impact on the C budget as the production of mineral fertilizer requires large amounts of energy and the rates of GHG emissions and leaching may be lower. Small-scale local solutions of biochar production and subsequent soil amendment can contribute to food security and reduced pressure on forest land, especially in tropical and subtropical climates where biochar can significantly improve soil properties (Chapter 18). Such local biochar systems must be evaluated and tested from the point of view of sustainability, as promising examples may have a strong demonstrating function for regional development.

Biochar amendment is usually demonstrated in agricultural systems, and studies within forest ecosystems are rare. Therefore, we were explicitly looking at potential pathways to utilize biochar in forest ecosystems, which makes sense where a closed material cycle is anticipated or in a circular biochar system (Chapter 4). One of the key challenges in forest ecosystems is that it is in most instances, especially in traditional forestry, impossible to integrate biochar into the soil horizon and, hence, it needs to be spread on the surface (Bruckman et al., 2015b). This implies temporal dynamics in the effects of biochar amendment; immediate effects are likely to influence plant nutrition, while further effects are subject to integration of biochar in the organic and subsequently mineral soil horizon (Sackett et al., 2015). The pyrolysis of harvesting residuals, with subsequent amendment of the resulting biochar, may help in restoring degraded forest sites with a high impact from logging. Key areas of targeted biochar use can be skid trails or log landings, while utilizing fuelwood that otherwise poses a risk in regions threatened by wildfires (Chapter 15). Leaching of nutrients at the time of harvesting may be reduced, and therefore biochar can play an important role in protecting water quality in rivers and streams and help to minimize losses of mineral nutrients from the system.

1.3 Biochar and GHG Mitigation

The negotiations at the 2015 Paris Climate Conference (COP21) were undoubtedly a necessary step forward to a legally binding, multinational universal agreement on climate with the aim to counteract climate change. As anthropogenic GHG emissions are responsible for the drastic changes in the world's climate, solutions for mitigation and adaptation are in urgent need. Even the recent global record-breaking temperatures were reported to be very likely a consequence of anthropogenic climate forcing (Mann et al., 2016). Biochar can offer some interesting approaches in terms of climate change mitigation (e.g. C sequestration, reduction of GHG emissions), but also adaptation (increased water retention). Up to 12% of the current annual net C equivalent emissions (1.8 petagrams [Pg]) may be reduced by using biochar on a global scale (Woolf et al., 2010). The successful use of biochar for GHG mitigation needs a holistic and systems approach, where the entire supply chain is considered. Models can help to understand the interactions and allow an interactive comparison of multiple scenarios of a given biochar system. In order to represent an effective pathway of altering soil properties, such effects must be long-lasting and therefore the material needs to be resistant against degradation.

1.3.1 The Recalcitrance of Biochar

The first and immediate thought might be focused on the recalcitrance of charred organic matter and thus its potential to store C for a long period of time in soils prior to oxidation. The aim is to sequester C by transferring atmospheric CO_2 into stable pools in soils (Lal, 2004b). The theory behind this approach can be easily justified as we know from a number of examples that carbon in the form of char or charcoal may be stored over centuries at least in soils (see also Figure 16.4). Archaeologists use this property to study the evolution and social behaviour of early civilizations. However, the problem with this approach is that the effect cannot be easily quantified: we do not know the properties and amount of the original biomass, nor do we know the conditions under which the biomass was pyrolysed. Moreover, we do not even know how much of the original char is left in the soil and hence its rate of decay. We only know that a certain share of char must have been resistant to microbial decomposition and oxidation for a relatively long period of time. Pyrolysis produces a certain amount of stable C and also a labile fraction of C that will be immediately oxidized after the amendment with soil or application on the soil surface (Bruckman et al., 2015b), followed by a steady but slow decomposition of the remaining amount. In terms of C sequestration, the effect of locking C in the soil depends on the amount of C that can be stored and the duration until it is released back into the atmosphere. While some of these questions can be addressed fairly accurately at this time, using a combined approach of studying historical charcoal production sites or wild-fire ecosystems and current efforts in studying biochar recalcitrance (Wang et al., 2015; Cross and Sohi, 2013; Budai et al., 2013), long-term experiments are still missing as a consequence of a relative recent evolvement of this scientific topic. In addition to that, one has to acknowledge that soils are dynamic and may react to biochar amendment in an unexpected way. There is evidence that biochar amendment can shift soil microorganism communities towards families which can degrade more recalcitrant C compounds (Anderson et al., 2011), and this may reduce the positive effect on the C balance. It is also unclear whether a sudden change in environmental conditions, for example, triggered by climate change itself, has an unexpected and negative impact on biochar stability.

1.3.2 Systems Approach

The improvement of the C balance with biochar in the soil ultimately leads to the question of how the biomass used for pyrolysis would have been used (or not used) otherwise. Carbon locked up in biomass that will be pyrolysed (where a significant share of it is oxidized and hence released to the atmosphere), is being transported from its origin (A) to the site of amendment after pyrolysis (B). While it generates a benefit in terms of carbon sequestration in A, it creates a deficit in B, caused by the removal of biomass. On the other hand, pyrolysis byproducts may be used to produce energy, thereby offsetting emissions from fossil fuels. Biochar itself can also substitute C-intensive growth media, such as peat (Chapter 5). The net sequestration impact always relies on the relative difference between potential other uses of biomass and the anticipated biochar use expressed in C equivalents. If, for instance, biomass is used for wooden furniture that remains in a household for at least 100 years, the cumulative sequestration benefit in this period would be likely positive over a scenario where the biomass is used for producing biochar with subsequent soil amendment. Biochar amendment has to create at least a positive effect on plant growth (increase in biomass productivity) or reduce the emission of GHGs in order to create a net benefit in terms of C balance (Woolf et al., 2010). In addition, this effect has to be long-lasting, implying the need for a certain stability of biochar against degradation.

Agricultural production can be seen as critical because of direct impacts on the production of food and other raw materials. It is important to protect or even increase SOM to sustain the production of agricultural goods (Lal, 2004b). On the contrary, a number of agricultural practices lead to a net loss of C to the atmosphere, such as ploughing, fertilizers, pesticides and irrigation (Lal, 2004a). From a system perspective, biochar may address several of these C-intensive activities by increasing water and nutrient retention, or contributing to increased crop health (e.g. as a consequence of providing a habitat for soil biota) and therefore reducing amounts of pesticides. Increased soil fertility and changes in soil physical properties (i.e. increased aeration or water permeability of heavy soils) may also provide conditions suitable for no-till farming. The agricultural revolution that increased the efficient production of agricultural commodities was to a large extent possible due to the extensive use of mineral fertilizers. Despite the fact that the C footprint of mineral fertilizers is high, strong efforts to increase efficient use are necessary as some crucial nutrient elements, such as phosphorus (P), are becoming scarce (Cordell and White, 2015). Here biochar may help to increase fertilizer efficiency and to reduce the system losses due to its inherent reactive surfaces or due to secondary effects that improve microbial activity and turnover as well as the propagation of symbiotic organisms, such as those that form mycorrhizae. The effect is site- and climate-dependent but has the potential to support a range of environmental services (Stavi and Lal, 2012).

Pyrolysis conditions can be tuned to maximize a specific desired function of biochar, but not all of them can be maximized as varying pyrolysis conditions imply a trade-off between specific functions (Crombie et al., 2015). At the same time, a biochar system might need to generate co-benefits from pyrolysis, such as the production of energy and other valuable materials, in order to become economically viable. In certain scenarios, biochar might represent a byproduct, for example from gasification, with distinct properties that are determined by the requirements of the main product and not by the specific requirements for soil amendment. In such a case post-processing of biochar (e.g. mixing with manure, compost or other types of biochar, sometimes called "biochar activation") may be necessary. While the amount of C that can be harvested in terms of solid biochar might be lower than the theoretical maximum in a byproduct scenario, the energy produced offsets a certain amount of fossil fuels that would otherwise be used to produce the main energy product. Currently, most of the biochar produced in Europe is not used directly as a soil amendment but in animal farming, where it is used as a catalyst for manure or directly as a feed additive (Gerlach and Schmidt, 2014). While emissions of GHGs, such as methane, can be reduced it was also reported that it reduced bad odour in some cases. A positive side effect is that biochar would already be mixed with manure when using it subsequently as fertilizer.

1.3.3 The Economic Valuation of GHG Mitigation via Abatement Costs

Due to the diversity of biochar systems, the understanding of conditions under which biochar application achieves favourable GHG budgets is still weak (Gurwick et al., 2013). Economic research on biochar focuses on benefits, costs and risks of biochar application to quantify the economic GHG mitigation potential. Within the FOREBIOM project, we assessed the cost-effective mitigation potential of biochar application on cropland in Austria. Therefore, we used results from field experiments, literature reviews and revenue and cost estimates of domestic biochar production. The cost-effectiveness of biochar as a climate change mitigation strategy is compared with other C removal strategies by calculating average GHG abatement costs of biochar application.

The cost-effectiveness analysis combines feedstock-specific (straw and woody biomass) and technology-specific (slow and fast pyrolysis) production costs with the results of GHG measurements from a pot and field trial within a project applying biochar to different arable soils in Austria between 2010 and 2013 (see Chapter 14 for details). Reported values for both costs and the mitigation potential vary considerably depending on the underlying assumptions. We considered different cost and mitigation scenarios based on reported values (see Table 1.1) to cover the full range of GHG abatement costs.

The simulation results show GHG abatement costs ranging from EUR 150 to EUR 200 per ton of carbon dioxide equivalent (CO_2e) for slow pyrolysis and from EUR 380 to EUR 830 per ton CO_2e for fast pyrolysis depending on the underlying cost and mitigation scenario and plant capacity. Slow pyrolysis produces biochar at lower costs than fast pyrolysis and is more suitable for maximizing the biochar output.

The results of slow pyrolysis correspond well with the results of Shackley et al. (2011), who calculated GHG abatement costs for different feedstocks in the UK. They calculated abatement costs of about EUR 178 for barley straw and EUR 187 per ton CO_2e for forestry

Table 1.1. *Assumptions for calculating average GHG abatement costs*

Feedstock costs

Feedstock costs range between EUR 60 and EUR 120 per ton due to different qualities (e.g. wood waste compared to wood chips).

Transportation costs

Transportation costs rise as a function of required feedstock quantities. For distances less than 100 km, trucks with trailers are the cheapest means of transport and specific transportation costs including an unloaded return range from EUR 0.20 to EUR 0.60 per ton and kilometre (Kappler, 2008).

Capital and operation costs

Pyrolysis technology and plant size (we consider small-, medium- and large-scale pyrolysis) determine the biochar production costs. Investment cost data from commercial facilities are highly uncertain. Therefore, our calculation is based on the costs of bioenergy plants assuming that capital costs contribute about 27–31% to total production costs depending on system scale (Shackley et al., 2011). The operation costs were estimated as a fixed fraction of the investment costs of 5% (BMVBS, 2010).

Costs for soil application

Biochar application costs are estimated using data for applying agricultural fertilizers to soil (Bauernzeitung, 2012). We consider variable costs to apply 30 tons of biochar per hectare using a fertilizer spreader, including the transport by tractor over a distance of two kilometres and required personal costs.

Byproduct revenues

We assume slow pyrolysis yields of 35% biochar, 30% bio-oil and 35% syngas and of 15%, 70% and 15% respectively for fast pyrolysis (Bridgwater et al., 2002; Brown et al., 2011; McCarl et al., 2009). When using syngas for electricity generation, this yields 0.31 Megawatt hours (MWh) per ton feedstock (McCarl et al., 2009). The returns from energy sales are estimated assuming costs for electricity generation of EUR 7.89 (slow pyrolysis) and 31.58 per ton feedstock (fast pyrolysis) and sales prices between EUR 50 to 70 per MWh (EXAA, 2013). Bio-oil is used to provide process energy for feedstock drying as its use for biofuel production is not feasible at the moment (Shackley et al., 2011).

GHG mitigation potential

The potential of biochar to offset GHG emissions depends on the amount of sequestered carbon in soil. It amounts to about two tons CO_2e per ton biochar applied to soil (assuming a stable carbon content of 55%). Additionally, the following aspects are relevant: effects of biochar on other GHG fluxes from soils such as methane (CH_4) and nitrous oxide (N_2O); GHG emissions resulting from the current biomass use; and fossil fuel offsets from energy generation by pyrolysis. Due to the complex interactions it is not possible to cover all climate-relevant factors and secondary effects (e.g. higher productivity due to increased soil fertility). Fossil fuel substitution by syngas is assumed to avoid 0.3 tons CO_2e per ton biochar and the suppression of N_2O emissions from fertilized soil 0.1 tons CO_2e per ton biochar. Emissions from feedstock and biochar transport as well as from applying biochar to soil lower the GHG mitigation potential insignificantly.

Note: All mass units refer to dry matter.
Source: Bruckman and Klinglmüller, 2014.

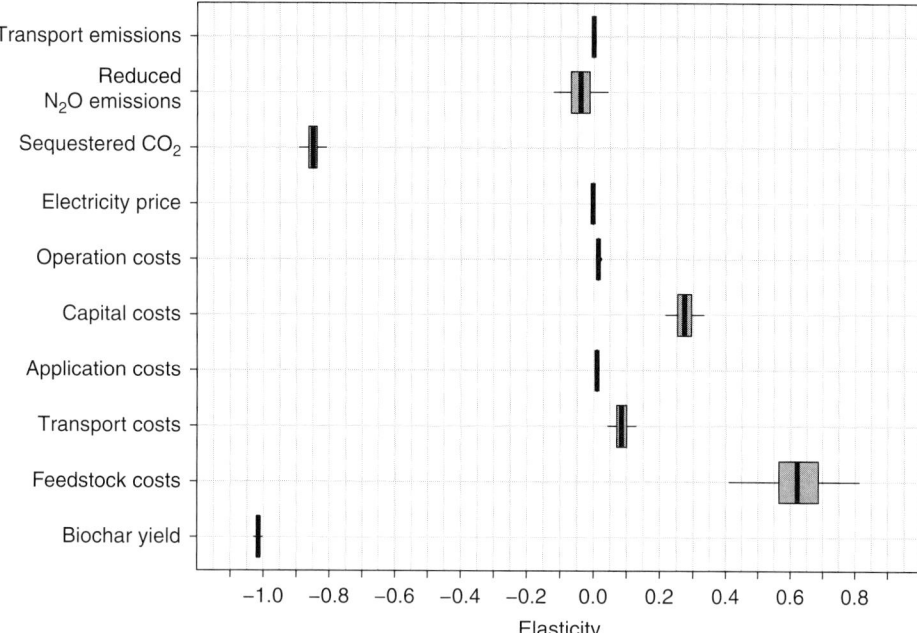

Figure 1.4 Tukey boxplot showing the elasticities between simulation input parameters and GHG abatement costs for a medium-scale slow pyrolysis plant. Boxes represent values between the 25th and 75th percentile, while the length of whiskers represents 1.5* IQR (interquartile range).

residues produced by small-scale, slow pyrolysis. Roberts et al. (2010) report even lower CO_2e abatement costs. Small-scale biochar production from forest residues in Tasmania may be feasible when achieving a minimum market price of EUR 300 per ton of biochar (Wrobel-Tobiszewska et al., 2015).

To assess the influence of uncertainties within our input data on abatement costs of biochar, we performed a Monte-Carlo simulation combined with a regression model (Schmidt et al., 2010; Höltinger et al., 2014) to estimate elasticities for all input parameters (see Figure 1.4).

The elasticity distribution reveals that the biochar yield from pyrolysis has the strongest influence on GHG abatement costs. An elasticity of -1 for the biochar yield implies that a 1% higher biochar output reduces abatement costs by 1%. Other important factors are the C sequestration rate of biochar and the feedstock costs. Fifty per cent higher feedstock costs, which are within the estimated uncertainty range, increase GHG abatement costs by about 30%. Therefore, the availability of cheap biomass is essential.

Our results confirm that biochar application to agricultural soils may lead to a reasonable GHG mitigation potential. However, from an economic point of view, biochar application to soils for GHG mitigation is not feasible under current market conditions. Compared to biochar, other carbon sequestration measures such as carbon capture and storage (CCS), reforestation and afforestation, or changes in agricultural practices such as grass planting or conservation tillage are cheaper GHG mitigation options. The costs for these measures range from EUR 15 to 80 per ton CO_2e (Finkenrath, 2011; Nabuurs et al., 2007).

1.4 Biochar Particle Behaviour on a Forest Floor – Results From the FOREBIOM Project

One of the main targets of the field study established within the FOREBIOM project was to assess the biochar particle behaviour on a forest floor until incorporation into the vegetation/litter layer. We therefore briefly summarise the approach and the results in this section. The motivation to evenly distribute biochar on the forest floor can result from various reasons as mentioned above. A potential aim is to consider the liming effect of biochar, compared to conventional liming in some managed forests. Another potential aim might be related to nutrient retention, especially in forests with repeating drastic events (e.g. clear-cuts, or windthrow-prone sites). This might also be important in terms of groundwater quality and impacts on hydrology (water retention in stands suffering from drought).

We used biochar from woodchips of *Picea abies* with a particle size of up to 5 mm and a moisture content of 35% during application to avoid dust. Biochar was evenly applied at 10 tons per hectare (ha^{-1}) in a *Picea abies* dominated stand prone to windthrow events due to topography. The herbaceous layer consists of grasses, shrubs and mosses and the stand is currently in its regeneration phase after a massive windthrow event in 2002. Further details about the experimental set-up can be obtained from Bruckman et al. (2015b). When biochar is applied to the forest floor, it is intended to remain in place until it is incorporated into the topsoil layer where it is protected from erosion. However, due to environmental forces such as gravity, precipitation and wind, it can be expected that lightweight biochar elements will move. Therefore, the first aim was to test whether biochar particles are prone to erosion, and if so, when, where and to what extent this occurs. Understanding these aspects of biochar behaviour in forest ecosystems would then allow *a priori* identification (or even prediction) of potential erosive conditions, which, in turn, could be used for better planning in further biochar amendments. A subplot of one of the biochar-amended experimental plots with an area of approximately 4 m^2 was selected to analyse the biochar movement in high resolution. The selected subplot has a sloped microtopography with a local depression at its bottom part, which was considered representative of the entire area. The relatively dense ground vegetation caused a portion of biochar to be submerged, while the remaining part was clearly exposed on the surface, filling local depressions around the moss or resting on and around the ground vegetation.

Several sensors were installed to observe the biochar movement and collect the axillary data. A time-lapse camera was set up at an edge of the subplot on a fixed tripod. A Sony DSC-R1 camera with a fixed focal length of 14.30 mm and a 10 megapixel sensor size observed the entire subplot continuously at 3-minute intervals over a period of 4-months, that is from 12 July 2013 until 24 November 2013. Simultaneously, environmental data (wind speed and precipitation) were continuously monitored by a weather station next to the plot. The recorded time-lapse images were used to observe and map biochar movements, while the acquired weather data were additionally linked to this information. The microtopography of the subplot was assessed by terrestrial laser scanning (TLS). This remote sensing technique utilizes laser light detection and ranging (LiDAR) principles to provide high-resolution, 3D geometrical information of the scanned object at a millimetre scale (Shan and Toth, 2008). TLS data, i.e. individual scans, were first pre-processed and then co-registered and merged into a single point cloud. The pre-processing was done as in Milenković et al. (2015). The microtopography was derived using the software OPALS (Pfeifer et al., 2014) and from the merged point cloud which contained all pre-processed scans. For our experiment, the point with the lowest elevation in each cell of a 1 cm grid was kept, and in a subsequent step, these local minimum points were interpolated using the moving planes interpolation to derive a regularly spaced digital elevation model (DEM), representing the microtopography of the subplot. The TLS measurements were further used to reconstruct the spatial position and orientation of the time-lapse camera, and thus to co-register the DEM with the observed time-lapse data. Finally, a simple water flow algorithm was used to predict the biochar particle movement under the given site conditions (surface properties and environmental conditions).

1.4.1 Results

The results show that the predicted particle flows represent the observed flows well; these were compiled in a single multimedia file that is available online (Bruckman et al., 2015a). The vast majority of the biochar particles were submerged in the herbaceous layer after the observation period and it was obvious that biochar particles are completely secured after just one vegetation period. While both precipitation and wind contribute to biochar erosion, the latter becomes less relevant after roughly a month. Wind erosion typically occurred after longer periods of drought when particles dried up and consequently lost weight. Strong precipitation events, typically occurring during summer storms, may trigger considerable biochar particle displacement until up to 2 months after application. After this period, movements can still be observed, but they do not lead to considerable erosion. Although valid for a specific case, these findings can be of practical relevance in terms of planning a potential surface application in forests. We would recommend spring after snowmelt or autumn before snowfall for a potential application to avoid erosion induced by heavy precipitation or wind after periods of drought. However, negative consequences due to soil compaction caused by heavy machinery should be considered, especially in spring.

1.5 Conclusions

Especially when considering biochar for GHG mitigation alone, it must be carefully assessed whether biochar is the most efficient solution. Besides agriculture, biochar production and amendment can also be an attractive option in a forest environment, ranging from traditional forestry to plantations of fast-growing woody species. A systems approach is necessary to be able to assess the potentials of biochar to mitigate GHG. Due to the high flexibility of a biochar system in terms of production and application, it is possible to provide specific solutions that have the potential to tackle several issues at the same time. Environmental and also societal circumstances should be considered before employing biochar systems (Abiven et al., 2014) to ensure environmental, social and economic sustainability. Especially under tropical conditions, biochar can help to secure food production on small scales while generating a number of co-benefits. A profound understanding of the effects of feedstock choice and production conditions is critical in order to develop specifically engineered biochar for different uses. The site-specific outcome of biochar application must be predictable for land managers planning to use biochar. Therefore, a solid classification system, based on existing work, has to be developed. This will also ensure safe application and help to establish a market. For a widespread application of biochar as a climate change mitigation option, a number of concerns have to be considered and sustainability needs to be addressed on site level. In most cases, it is expected that GHG mitigation will be successful only when taking into consideration effects beyond C sequestration.

The FOREBIOM project provided a good demonstration of the positive impact of transboundary collaboration and the exchange of ideas and experiences from a number of examples around the world. A clear message is that biochar systems are highly specific and the supply chain greatly varies among cases and geographic locations. The discussions revealed that there are still a number of research deficits, which also vary regionally (Bruckman and Klinglmüller, 2014; Uzun and Apaydın-Varol, 2015), but the current boom may help to widen our knowledge in this field. The functionality of biochar in soil and the response of soils to biochar inputs are still poorly understood and effects will most likely vary depending on soil type, biochar characteristics and environmental conditions. Regarding economic aspects, diversification, byproducts of pyrolysis, waste management, cascade utilization and purpose-designed biochar for specific cases are key strategies, while co-benefits can help to make a biochar system economically viable. However, the economic valuation of these co-benefits is difficult and complicates assessments of the overall profitability of biochar application.

Biochar can be a tool to address some of the greatest challenges humanity is currently facing, but certainly it is not the only one we have. Despite the fact that knowledge is still growing in this field, it is necessary to start using this tool in safe system boundaries of operation to demonstrate and study the potentials. Sustainability must be ensured and a holistic view supported by scientific evidence may help to avoid negative consequences.

References

Abiven, S., Schmidt, M. W. I. and Lehmann, J. (2014). Biochar by design. *Nature Geoscience*, 7, pp. 326–327.

Anderson, C. R., Condron, L. M., Clough, T. J., Fiers, M., Stewart, A., Hill, R. A. and Sherlock, R. R. (2011). Biochar induced soil microbial community change: implications for biogeochemical cycling of carbon, nitrogen and phosphorus. *Pedobiologia*, 54, pp. 309–320.

Bauernzeitung (2012). *Wie viel Wirtschaftsdünger wert sind*. [online] Available at: www.bauernzeitung.at/?id=2500,1014410 [Accessed 15 October 2015]

BMVBS Bundesministeriums für Verkehr, Ban und Stadtentwicklung (2010). Globale und regionale räumliche Verteilung von Biomassepotenzialen: Status Quo und Möglichkeit der Präzisierung. *BMVBS-Online-Publikation, Nr. 27/2010*. [online]. Available at: www.bbsr.bund.de/BBSR/DE/Veroeffentlichungen/BMVBS/Online/2010/DL_ON272010.pdf [Accessed 8 January 2016]

Bridgwater, A. V., Toft, A. J. and Brammer, J. G. (2002). A techno-economic comparison of power production by biomass fast pyrolysis with gasification and combustion. *Renewable and Sustainable Energy Reviews*, 6, pp. 181–246.

Brown, T. R., Wright, M. M. and Brown, R. C. (2011). Estimating profitability of two biochar production scenarios: slow pyrolysis vs fast pyrolysis. *Biofuels, Bioproducts and Biorefining*, 5, pp. 54–68.

Bruckman, V. J. and Klinglmüller, M. (2014). Potentials to mitigate climate change using biochar – the Austrian perspective. In: Bruckman, V. J., Liu, J., Uzun, B. B. and Apaydın-Varol, E. (eds.) *Potentials to Mitigate Climate Change Using Biochar – the Austrian Perspective*. IUFRO Occasional Papers, 27. Vienna.

Bruckman, V. J., Liu, J., Uzun, B. B. and Apaydın-Varol, E. (2015a). FOREBIOM time-lapse biochar. [online] Available at: http://dx.doi.org/10.1553/forebiom [Accessed 18 January 2016]

Bruckman, V. J., Terada, T., Uzun, B. B., Apaydın-Varol, E. and Liu, J. (2015b). Biochar for climate change mitigation: tracing the in-situ priming effect on a forest site. *Energy Procedia*, 76, pp. 381–387.

Bruckman, V. J., Yan, S., Hochbichler, E. and Glatzel, G. (2011). Carbon pools and temporal dynamics along a rotation period in Quercus dominated high forest and coppice with standards stands. *Forest Ecology and Management*, 262, pp. 1853–1862.

Budai, A., Zimmerman, A. R., Cowie, A. L., Webber, J. B. W., Singh, B. P., Glaser, B., Masiello, C. A., Andersson, D., Shields, F., Lehmann, J., Camps Arbestain, M., Williams, M., Sohi, S. and Joseph, S. (2013). Biochar carbon stability test method: an assessment of methods to determine biochar carbon stability. International Biochar Initiative document, Version: September 20, 2013.

Clough, T. J. and Condron, L. M. (2010). Biochar and the nitrogen cycle: introduction. *Journal of Environmental Quality*, 39, pp. 1218–1223.

Cordell, D. and White, S. (2015). Tracking phosphorus security: indicators of phosphorus vulnerability in the global food system. *Food Security*, 7, pp. 337–350.

Crombie, K., Mašek, O., Cross, A. and Sohi, S. P. (2015). Biochar – synergies and trade-offs between soil enhancing properties and C sequestration potential. *GCB Bioenergy*, 7, pp. 1161–1175.

Cross, A. and Sohi, S. P. (2013). A method for screening the relative long-term stability of biochar. *GCB Bioenergy*, 5, pp. 215–220.

EXAA (2013). EXAA Energy Exchange Austria. [online] Available at: www.exaa.at/en/marketdata/historical-data [Accessed 15 June 2015]

Finkenrath, M. (2011). *Cost and Performance of Carbon Dioxide Capture from Power Generation. IEA Working Paper*. Paris: International Energy Agency.

Gerlach, A. and Schmidt, H. (2014). The use of biochar in cattle farming. *The Biochar Journal*. [online] Available at: www.biochar-journal.org/en/ct/9 [Accessed 18 January 2016]

Gurwick, N. P., Moore, L. A., Kelly, C. and Elias, P. (2013). A systematic review of biochar research, with a focus on its stability *in situ* and its promise as a climate mitigation strategy. *PLoS ONE*, 8, p. e75932.

Haaland, C. and Van Den Bosch, C. K. (2015). Challenges and strategies for urban greenspace planning in cities undergoing densification: a review. *Urban Forestry & Urban Greening*, 14, pp. 760–771.

Höltinger, S., Schmidt, J., Schönhart, M. and Schmid, E. (2014). A spatially explicit techno-economic assessment of green biorefinery concepts. *Biofuels, Bioproducts and Biorefining*, 8, pp. 325–341.

Hüppi, R., Felber, R., Neftel, A., Six, J. and Leifeld, J. (2015). Effect of biochar and liming on soil nitrous oxide emissions from a temperate maize cropping system. *Soil*, 1, pp. 707–717.

IBI (2013). *Standardized product definition and product testing guidelines for biochar that is used in soil (Document version code IBI-STD-2.1, 23. November 2015).* [online]: International Biochar Initiative. Available at: www.biochar-international.org/characterizationstandard [Accessed 25 January 2016]

Kappler, G.O. (2008). Systemanalytische Untersuchung zum Aufkommen und zur Bereitstellung von energetisch nutzbarem Reststroh und Waldrestholz in Baden-Württemberg: eine auf das Karlsruher bioliq-Konzept ausgerichtete Standortanalyse. Dissertation, University of Freiburg.

Kloss, S., Zehetner, F., Dellantonio, A., Hamid, R., Ottner, F., Liedtke, V., Schwanninger, M., Gerzabek, M. H. and Soja, G. (2012). Characterization of slow pyrolysis biochars: effects of feedstocks and pyrolysis temperature on biochar properties. *Journal of Environmental Quality*, 41, pp. 990–1000.

Lal, R. (2004a). Carbon emission from farm operations. *Environment International*, 30, pp. 981–990.

Lal, R. (2004b). Soil carbon sequestration impacts on global climate change and food security. *Science*, 304, pp. 1623–1627.

Lehmann, J. and Joseph, S. (eds.) (2015a). *Biochar for Environmental Management: Science, Technology and Implementation.* 2nd Edition. Oxford, New York: Routledge.

Lehmann, J. and Joseph, S. (2015b). Biochar for environmental management: an introduction. In: Lehmann, J. and Joseph, S. (eds.) *Biochar for Environmental Management: Science, Technology and Implementation.* 2nd Edition. Oxford, New York: Routledge, pp. 1–12.

Mann, M. E., Rahmstorf, S., Steinman, B. A., Tingley, M. and Miller, S. K. (2016). The likelihood of recent record warmth. *Scientific Reports*, 6, document No. 19831.

McCarl, B. A., Peacocke, C., Chrisman, R., Kung, C.-C. and Sands, R. D. (2009). Economics of biochar production, utilization and greenhouse gas offsets. In: Lehmann, J. and Joseph, S. (eds.) *Biochar for Environmental Management – Science and Technology.* London, Washington, DC: Earthscan, pp. 341–358.

Milenković, M., Pfeifer, N. and Glira, P. (2015). Applying terrestrial laser scanning for soil surface roughness assessment. *Remote Sensing*, 7, pp. 2007–2045.

Myhre, G., Shindell, D., BréOn, F.-M., Collins, W., Fuglestvedt, J., Huang, J., Koch, D., Lamarque, J.-F., Lee, D., Mendoza, B., Nakajima, T., Robock, A., Stephens, G., Takemura, T. and Zhang, H. (2013). Anthropogenic and natural radiative forcing. In: Stocker, T. F., Qin, D., Plattner, G.-K., Tignor, M., Allen, S. K., Boschung, J., Nauels, A., Xia, Y., Bex, V. and Midgley, P. M. (eds.) *Climate Change 2013: The Physical Science Basis. Contribution of Working Group I to the Fifth Assessment Report of the Intergovernmental Panel on Climate Change.* Cambridge, UK and New York: Cambridge University Press, pp. 659–740.

Nabuurs, G.J., Masera, K., Andrasko, K., Benitez-Ponce, P., Boer, R., Dutschke, M., Elsiddig, J., Ford-Robertson, J., Frumhoff, P., Karjalainen, T., Krankina, O., Kurz, W. A., Matsumoto, M., Oyhantcabal, W., Ravindranath, N. H., Sanz Sanchez, M. J. and Zhang, X. (2007). Forestry. In: Metz, B., Davidson, O. R., Bosch, P. R., Dave, R. and Meyer, L. A. (eds.) *Climate Change 2007: Mitigation. Contribution of Working Group III to the Fourth Assessment Report of the Intergovernmental Panel on Climate Change*. Cambridge, UK and New York: Cambridge University Press, Chapter 9.

Pettitt, P. (2008). Art and the Middle-to-Upper Paleolithic transition in Europe: comments on the archaeological arguments for an early Upper Paleolithic antiquity of the Grotte Chauvet art. *Journal of Human Evolution*, 55, pp. 908–917.

Pfeifer, N., Mandlburger, G., Otepka, J. and Karel, W. (2014). OPALS – a framework for Airborne Laser Scanning data analysis. *Computers, Environment and Urban Systems*, 45, pp. 125–136.

Roberts, K. G., Gloy, B. A., Joseph, S., Scott, N. R. and Lehmann, J. (2010). Life cycle assessment of biochar systems: estimating the energetic, economic, and climate change potential. *Environmental Science & Technology*, 44, pp. 827–833.

Sackett, T. E., Basiliko, N., Noyce, G. L., Winsborough, C., Schurman, J., Ikeda, C. and Thomas, S. C. (2015). Soil and greenhouse gas responses to biochar additions in a temperate hardwood forest. *GCB Bioenergy*, 7, pp. 1062–1074.

Schmidt, J., Leduc, S., Dotzauer, E., Kindermann, G. and Schmid, E. (2010). Cost-effective CO_2 emission reduction through heat, power and biofuel production from woody biomass: a spatially explicit comparison of conversion technologies. *Applied Energy*, 87, pp. 2128–2141.

Schuetze, T. and Santiago-Fandiño, V. (2014). Terra Preta sanitation: a key component for sustainability in the urban environment. *Sustainability*, 6, pp. 7725–7750.

Shackley, S., Hammond, J., Gaunt, J. and Ibarrola, R. (2011). The feasibility and costs of biochar deployment in the UK. *Carbon Management*, 2, pp. 335–356.

Shan, J. and Toth, C. (eds) (2008). *Topographic Laser Ranging and Scanning: Principles and Processing*. Boca Raton, London, New York: CRC Press, Taylor & Francis Group.

Stavi, I. and Lal, R. (2012). Agroforestry and biochar to offset climate change: a review. *Agronomy for Sustainable Development*, 33, pp. 81–96.

Taghizadeh-Toosi, A., Clough, T. J., Sherlock, R. R. and Condron, L. M. (2011). Biochar adsorbed ammonia is bioavailable. *Plant and Soil*, 350, pp. 57–69.

Thies, J. E., Rillig, M. C. and Graber, E. R. (2015). Biochar effects on the abundance, activity and diversity of the soil biota. In: Lehmann, J. and Joseph, S. (eds.) *Biochar for Environmental Management: Science, Technology and Implementation*. 2nd Edition. Oxford, New York: Routledge, Chapter 13.

Uzun, B. B. and Apaydın-Varol, E. (2015). Potentials to mitigate climate change using biochar – Turkey's perspective. In: Bruckman, V. J., Liu, J., Uzun, B. B. and Apaydın-Varol, E. (eds.) *Potentials to Mitigate Climate Change Using Biochar – the Austrian Perspective*. IUFRO Occasional Papers, 27, Vienna.

Wang, J., Xiong, Z. and Kuzyakov, Y. (2015). Biochar stability in soil: meta-analysis of decomposition and priming effects. *GCB Bioenergy*, 8, pp. 512–523.

Woolf, D., Amonette, J. E., Street-Perrott, F. A., Lehmann, J. and Joseph, S. (2010). Sustainable biochar to mitigate global climate change. *Nature Communications*, 1, article No. 56.

Wrobel-Tobiszewska, A., Boersma, M., Sargison, J., Adams, P. and Jarick, S. (2015). An economic analysis of biochar production using residues from Eucalypt plantations. *Biomass and Bioenergy*, 81, pp. 177–182.

Zhao, L., Cao, X., Mašek, O. and Zimmerman, A. (2013). Heterogeneity of biochar properties as a function of feedstock sources and production temperatures. *Journal of Hazardous Materials*, 256–257, pp. 1–9.

Part I
The Interdisciplinary Approach

Part I

The Investigative Process

2

A Supply Chain Approach to Biochar Systems

NATHANIEL M. ANDERSON, RICHARD D. BERGMAN
AND DEBORAH S. PAGE-DUMROESE

Abstract

Biochar systems are designed to meet four related primary objectives: improve soils, manage waste, generate renewable energy, and mitigate climate change. Supply chain models provide a holistic framework for examining biochar systems with an emphasis on product life cycle and end use. Drawing on concepts in supply chain management and engineering, this chapter presents biochar as a manufactured product with a wide range of feedstocks, production technologies, and end use options. Supply chain segments are discussed in detail using diverse examples from agriculture, forestry and other sectors that cut across different scales of production and socioeconomic environments. Particular attention is focused on the environmental impacts of different production and logistics functions, and the relationship between supply chain management and life cycle assessment. The connections between biochar supply chains and those of various co-products, substitute products, and final products are examined from economic and environmental perspectives. For individuals, organizations, and broad associations connected by biochar supply and demand, achieving biochar's potential benefits efficiently will hinge on understanding, organizing, and managing information, resources and materials across the supply chain, moving biochar from a nascent to an established industry.

2.1 Biochar in a Supply Chain Context

Biochar production and application as a commercial enterprise connects a diverse constellation of organizations with varying capabilities, expertise, and objectives. From an industrial perspective, these organizations are bound together by a single goal: efficiently manufacture and deliver a product that effectively meets the needs of end users. This network of organizations is collectively known as a supply chain. Using the customer and organization focused framework of supply chain management (SCM), this chapter examines

Figure 2.1. Charcoal produced from loblolly pine (*Pinus taeda*) sawmill planer shavings using a high temperature (800–1100°C) pyrolysis system. Based on its feedstock and conversion process, this product will be classified as biochar if it is used as a soil amendment. Photograph by Nate Anderson.

biochar as a manufactured product used to meet soil improvement and climate change mitigation objectives, as well as waste management and energy needs.

In some ways, biochar supply chains are millennia in the making, dating back to the anthropogenic Terra Preta soils of the Amazon Basin, but as a component of modern economies biochar supply chains are new and rapidly evolving. Unlike many major agricultural and forest commodities, supply chains for biochar products are currently characterized by growing spot markets for diverse uses that are often in the early stages of development, with little or no historical information or market data to guide pioneering entrepreneurs. Varied raw material options, emerging conversion technologies, and intermittent distribution channels complicate this landscape. Furthermore, the needs of end users can be narrow, such as replacing mineral vermiculite in nursery potting media with a suitable organic alternative (Dumroese et al., 2011), or multifaceted, such as simultaneously managing crop residues, improving crop yields, and sequestering carbon in the soil to generate carbon credits (Roberts et al., 2010).

Biochar is well suited to examination in a supply chain context because its classification is closely bound to its end use. Charcoal is the carbon-rich solid product of thermal decomposition of biomass in the absence of oxygen (i.e. pyrolysis), and is used in a wide range of products, including solid fuels, industrial chemicals, sorbents, and consumer products like rubber, plastic, paints, inks and pigments. Charcoal that is used to improve the properties

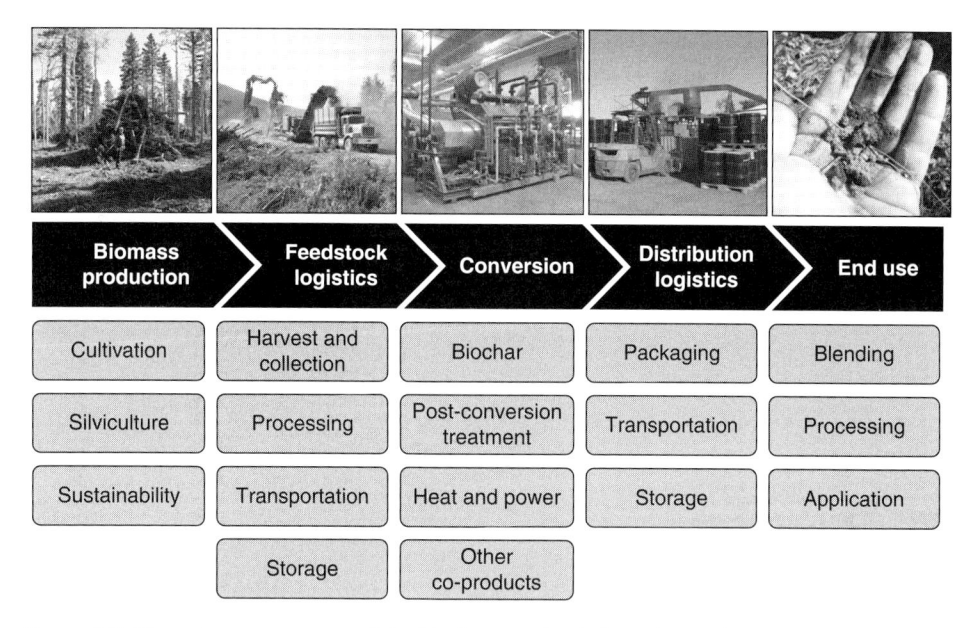

Figure 2.2. The primary segments of the biochar supply chain and associated activities related to material production, logistics, conversion, and end use. Photographs by Nate Anderson.

of soil, especially productivity, carbon storage and water holding capacity, is known as biochar (Lehmann and Joseph, 2009). This means that the charcoal shown in Figure 2.1, which was produced from loblolly pine (*Pinus taeda*) sawmill planer shavings using an advanced high temperature pyrolysis system, may or may not become biochar depending on how it is eventually used. If it is pelletized and used as solid fuel for co-firing with coal in a power plant, it remains charcoal, but if it is used as a soil amendment it becomes biochar. Similarly, though activated carbon (AC) shares many physical and chemical properties with biochar, AC used as a soil amendment for remediation of organic pollutants (as in Vasilyeva et al., 2006, for example) would not be classified as biochar if it is manufactured from fossil coal rather than biomass. In practice, classification of biochar based on end use as well as its parent material and production process links biochar to multiple co-products, substitute products, and end uses in complex and dynamic supply chains, but all biochar supply chains follow the same general supply chain model.

2.2 A Model Biochar Supply Chain

The biochar supply chain can be divided into five segments: biomass production, feedstock logistics, conversion, distribution logistics and end use (Figure 2.2). In manufacturing supply chains, each segment includes a variety of activities related to material production, logistics, conversion and end use functions. Material flows downstream from the site of harvest to the end user along the supply chain, with each activity adding value. Material is procured and transformed into intermediate and finished products, which are moved down the chain by logistics systems that include handling, transportation and storage (Goetschalckx, 2011). In addition to material flows, two other flows are critical to efficient and effective supply chains. Information flows back upstream from end users along the chain and can be used to coordinate activities, improve products, advance technologies, increase productivity and reduce costs. Financial transactions between the organizations involved in these activities underpin material and information flows in commercial supply chains. Regardless of the final product, material, information and financial flows are organized and managed to meet the needs of end users.

In the forest biomass example illustrated by the photographs in Figure 2.2, which represents one of many possible supply chain configurations, woody biomass is generated by silvicultural treatments prescribed by forest managers to harvest timber and reduce fire risk in a dry mixed conifer forest. Biomass is field dried in piles, and then collected and ground into a smaller, more uniform material with higher bulk density that can be efficiently delivered to a bioenergy facility by truck. Raw biomass from the forest becomes feedstock when it is processed – in this case biomass is ground into feedstock using a horizontal grinder. At the facility, the feedstock is further reduced in size, screened, dried, and then converted into biochar in a high temperature industrial pyrolysis system that also produces energy gas to fuel a generator providing power to the electrical grid. The biochar is packaged for distribution in 200 liter metal drums and delivered to an abandoned mine site, where it is finally used as a soil amendment for remediation and mine reclamation. Technical details associated with supply chains like this one, including the operations pictured, can be found in case studies throughout this book and also in Anderson et al. (2012, 2013), Keefe et al. (2014), and Kim et al. (2015).

Material flows are a useful way to characterize supply chains, but as described in Section 2.1, a supply chain is best thought of as a network of organizations engaged in activities to meet the needs of end users. Industrial manufacturing supply chains are generally dominated by private firms, businesses and corporations meeting the needs of consumer end users. Because of biochar's close connections to agriculture, forestry and climate change mitigation, and because it can be applied in various socioeconomic contexts around the world, in this case it is important to recognize a broad definition of organization, which includes public agencies, institutions, non-governmental organizations (NGO), family units and other groups.

In the example from Figure 2.2, the end user is a public National Forest in need of biochar for mine reclamation activities. A different public National Forest is the biomass producer, a private logging company is contracted to harvest and grind biomass on site and deliver it to the bioenergy facility, the bioenergy facility further processes the feedstock and carries out conversion, an independent co-located business packages and markets the biochar, and a common carrier freight company delivers the packaged biochar to an environmental engineering firm that has been contracted by the end user to remediate abandoned mine sites on public land. The material, information and financial flows in this example span seven different organizations, two of which are public agencies. It is also important to point out that the supply chains of other products and end uses are part of this network, and include logs that leave the site to be used in products like paper and solid wood products, as well as the co-product of electricity that is delivered from the bioenergy facility to customers over a grid that includes private companies, co-operative business, and public utilities.

Of course, it is possible for a single organization to carry out all of the functions and activities of the biochar supply chain. In fact, many authors have described a simple model of biochar production and application for small-scale agriculture in which farmers process waste biomass from crop residues in on-site small batch conversion systems like charcoal kilns to produce biochar for application to their fields (e.g. Sparrevik et al., 2013). Though simple in structure, in the context of climate change mitigation such supply chains may not be isolated from global markets. For example, Leach et al. (2012) examined the interplay between traditional biochar production and application for small-scale agriculture and biochar supply chains to meet global carbon management objectives through carbon markets.

The process of expanding the operations of an organization upstream or downstream along the supply chain is known as vertical integration, and a farmer who carries out all of the supply chain functions is fully vertically integrated with regards to biochar production and use. Horizontal integration occurs when an organization adds functions that are in a different sector or industry, such as a factory that manufactures charcoal briquettes for retail sale as cooking fuel expanding to develop and market a proprietary biochar soil amendment for home gardening applications. The costs and benefits of integration versus specialization vary widely by industry, but the general purposes of integration are to fill unmet consumer needs, capture value from new operations, and reduce the market leverage of suppliers and distributors. Most supply chains are made up of multiple organizations with varying levels of integration and specialization.

Much of the remainder of this book is devoted to case studies of existing real-world biochar systems and applications, but is it useful here to examine a network of hypothetical organizations. Figure 2.3 is a schematic of two hypothetical biochar supply chains (organizations #1 and #2 together in one chain, and #3 through #17 as another) showing the connections between 17 different organizations engaged in four general functions: biomass production, logistics, conversion and end use. These organizations are connected by material flows for raw biomass, intermediate products and biochar. Organization #1 is an almond (*Amygdalus communis*) orchard producing significant biomass residues in the form of branch trimmings, shells, cull trees and other byproducts. Organization #2 is a co-located company that produces biochar from these residues, as well as processing heat for orchard operations, including a greenhouse. Both companies have logistics capacity to transport, process and store materials using trucks, loaders, chippers, hammer mills, conveyors, bins, dryers, and other equipment. The orchard is the primary end user of the biochar produced, which it uses to improve soils in its orchards and reduce its carbon footprint through carbon sequestration. Like the example of a fully integrated farming operation, structurally this is a very simple supply chain made up of only two organizations, but it includes all of the major supply chain functions.

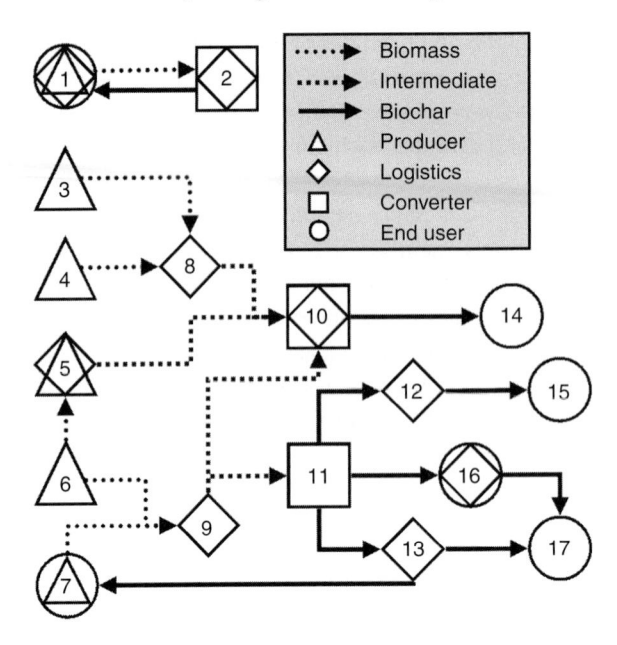

Figure 2.3. A schematic illustrating the interconnectedness of organizations and their functions in a biochar supply chain.

The connections between the remaining 15 organizations in Figure 2.3 are more complex and less vertically integrated. A forest park (#3) and tree plantation (#4) contract with loggers and trucking companies (#8) to chip and deliver woody biomass from forest management operations to a biochar company (#10) that produces, packages and delivers biochar to an organic strawberry grower (#14). A sawmill (#5) not only grinds, screens and delivers a portion of its mill residues to the biochar producer (#10), it also processes and delivers residues from a private industrial forest (#6) that also sells biomass to an equipment operator specializing in biomass harvesting operations (#9), who delivers ground and screened material to two different biochar producers (#10 and #11). One of the customers for this biochar is a forest reserve owned and operated by an environmental NGO (#7) that also provides biomass to both biochar producers (#10 and #11), through the biomass operator (#9). One of the end users (#16) uses biochar in potting mixes for its commercial greenhouse operations, but also sells the proprietary mix to retail customers (#17). Looking at the connections between #6 through #15 in this model, a linear material flow between specialized organizations nicely fits the metaphor of a "chain" for a simple product and single end user. However, as Christopher (2011), Stock and Lambert (2001), and others have pointed out, in reality even relatively simple products actually require complex flows of material, information, and capital in networks of organizations that integrate the supply chains of many different products and end uses.

What do models like these tell us about biochar systems? Over the last decade biochar has experienced a rapid expansion of awareness and interest closely tied to applications in agriculture, forestry, mining and climate change mitigation, some of which have been advocated for by scientists and various government agencies and non-profit initiatives, committees, centers, and other organizations. Compared to activity by these groups and in contrast to other industries in the agricultural and forest sectors, commercial enterprises devoted to the manufacture, marketing and use of biochar and biochar production equipment remain less common. As biochar evolves as a consumer product to meet various needs in diverse markets, we can expect the industry to move toward higher levels of complexity with varying degrees of integration and product differentiation across local, regional and global scales. This is true across a range of economic systems, including informal economies in developing countries for which biochar has been proposed as an accessible alternative to resource-intensive industrial agricultural inputs (Duku et al., 2011). For individuals, organizations and broad associations connected by biochar supply and demand, achieving biochar's potential benefits efficiently will hinge on understanding, organizing, and managing information, resources and materials across the supply chain, moving biochar from a nascent to an established industry.

2.3 Biochar Sustainability, SCM, and LCA

A supply chain framework is also a useful way to organize and analyze the various aspects of biochar production, effectiveness, economics, and environmental impacts that are discussed in detail in subsequent chapters of this book. To a larger extent than most manufactured products, biochar is fundamentally bound to sustainability (Part 2 of this book). This can be traced directly to its feedstocks, end uses, and intended benefits, which leverage environmental benefits from reduced greenhouse gas emissions, better waste disposal, and substitution for more environmentally damaging products. As a result, sustainability must be ingrained in biochar SCM (Section 2.5 of this chapter). Though sustainability is most commonly equated with environmental impacts, it also includes various socioeconomic aspects of production, such as land tenure, indigenous rights, labor rights, safety, legal standards, economic obligations and cultural protections. Most of these are formalized in various sustainability standards and certifications, including those for agriculture, forestry and even biochar specifically (e.g. FSC, 2010; Leonardo Academy, 2012; IBI, 2014).

With regards to quantifying and evaluating environmental impacts, the supply chain model closely parallels the life cycle assessment (LCA) method of evaluating environmental impacts (Chapter 3). The boundary of the biochar system defined and examined in LCA encompasses all of the supply chain functions illustrated in Figure 2.2, from raw material extraction ("the cradle") to end use and disposal ("the grave"), though each stage of the process may be segmented differently in LCA. For example, biomass harvest, collection, and processing may be attributed to raw material extraction rather than biomass production. In addition, the biochar system defined in LCA includes energy offsets and avoided emissions, with detailed accounting for emissions, effluents and waste from the system. As with supply chains, material conversions and logistics feature prominently in LCA, as do flows of materials, energy and capital. Though integration of LCA with SCM has not been without challenges (Hagelaar and van der Vorst, 2002), businesses increasingly view environmental impact, especially carbon footprint, as a key indicator of supply chain performance and value. In fact, formal integration of LCA is becoming a cornerstone of the relatively new field of environmental supply chain management (ESCM), also known as "green" SCM. In this book, LCA and SCM models provide a holistic framework to examine current biochar research with an emphasis on product life cycle and end use.

2.4 Biochar Supply Chains and End Use

As discussed in Section 2.1, biochar supply chains are focused on meeting four general and often overlapping needs of end use consumers (Lehman and Joseph, 2009): soil improvement, waste management, energy production and climate change mitigation. Much of the research on biochar is focused on understanding and quantifying biochar's effects on soil chemical, physical, and biological properties, particularly its impacts on soil chemistry, nutrient cycling, water availability, soil biota, and the nitrogen cycle (see subsequent chapters). In agricultural and forestry settings, beneficial changes to soil properties can be linked to increased productivity with lower inputs of nutrients and water, depending on the specific biochar used and a wide range of site-specific variables, especially soil texture, moisture regime and plant species. Whether for business or subsistence or both, increased productivity and more efficient water and nutrient use translate directly to higher yields at lower cost. These gains can be quantitatively measured against other options that might achieve similar outcomes, such as alternative soil amendments, chemical fertilizers, new irrigation technologies and genetically modified plants. Appropriate metrics for comparison typically include various market and non-market costs and benefits related to alternative financial, social and environmental outcomes.

As with biochar used to improve soil properties and enhance plant growth, the costs and benefits of using pyrolysis to process biomass waste like logging slash, stover, bagasse, nut shells, straw and other materials can be compared to alternative disposal options, including open burning, controlled combustion (i.e. incineration), biochemical conversion (i.e. decomposition or digestion) or burial. Similarly, biochar production systems often produce useable energy co-products in the form of heat, liquid fuels and energy gases (see 2.5.3 and Part 3 of this book), which can be compared to various substitute energy products, including both renewable and non-renewable options. For example, modern combustion and gasification systems that produce heat or combined heat and power (CHP) from biomass are technologically similar to some pyrolysis-based thermochemical conversion systems that produce biochar, and are similarly marketed for a broad range of waste-to-energy applications (Anderson et al., 2013).

Here it is important to distinguish between end use of biochar and its associated coproducts from the consumption of final goods. In economics, demand for biochar systems to meet soil, waste and energy needs is derived from demand for various final goods. For example, an integrated forest products company may use wood residues from sawmill

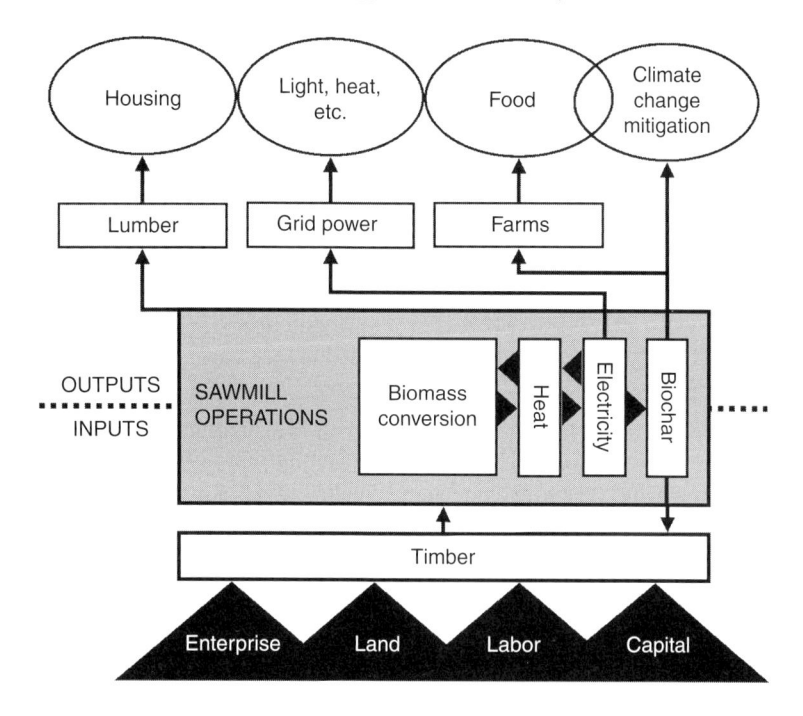

Figure 2.4. For a sawmill using mill residues to produce biochar, heat, and electricity, the demand for these intermediate goods is derived from demand for final goods like housing and food.

operations to produce biochar, heat, and power for on-site processes, and electricity to the grid using a distributed-scale biomass conversion system (Figure 2.4). The biochar may be marketed to local farms and also used to rehabilitate forest roads on company timberlands that have experienced soil compaction and erosion. However, in this example, demand for soil improvement, waste wood disposal, and energy are clearly derived from demand for other final goods, such as houses, home heating and lighting, and food (Figure 2.4). Recognizing that biochar is an intermediate good rather than a final good is important because its supply chains are subject to competition from alternative products that can be substituted for pyrolysis technologies and biochar to meet the same needs. However, climate change mitigation as an end use objective for biochar systems offers another level complexity.

In Figure 2.4, biochar can be used as an input to improve crop and timber production in ways previously described, connecting biochar to food and housing as final goods, for example. However, biochar systems can also be used primarily as a tool to meet climate change mitigation objectives, and in this application biochar can be considered a final good in itself. Section 2.5.5 discusses biochar used for climate change mitigation in more detail, but the direct connection between biochar production and climate change mitigation is closely tied to long-term sequestration of relatively stable carbon in the soil. This characteristic not only imparts potential carbon negative status on biochar and its co-products, depending on the details of the supply chain (Mathews, 2008), but has also distinguished biochar production and application as a potential means for geoengineering global-scale reductions in atmospheric carbon, independent of applications in agriculture, waste management and energy (Downie et al., 2012).

Though each of the four general end uses associated with biochar systems can be pursued independently to some degree, they are obviously bound together. For example, consider a subsistence farmer using a traditional charcoal kiln without energy capture or emissions controls to process crop residues into biochar for her fields. This simple supply chain incorporates soil improvement, waste disposal and carbon sequestration, though increased agricultural productivity from soil improvement is likely to be the main driver of use in this case. However, it should be clear at this point that the greatest net benefits from biochar are likely to occur when all four needs are met simultaneously in supply chains that include multiple products and market substitution for more carbon intensive products and practices. To what extent such benefits are realized rests squarely on the details of a specific supply chain.

2.5 A Closer Look at the Biochar Supply Chain

The generalized biochar supply chain segments and activities shown in Figure 2.2 cut across a wide range of specific feedstocks, logistics, conversion technologies, and end uses. Furthermore, given its close ties to agriculture and forestry, biochar has potential for production and use in diverse settings around the globe at many different scales within all types of economies. Section 2.5 takes a closer look at the range of materials, practices, activities, and technologies associated with each segment of the biochar supply chain, and examines the relationships between organizations that typically carry out critical functions at each stage of production. Subsequent chapters examine the technical details of specific cases, with an emphasis on biomass sustainability, innovative conversion technology, and end uses for soil improvement and climate change mitigation.

2.5.1 Biomass Production

By definition, biochar must be manufactured from biomass. Though pyrolysis of various petroleum and fossil coal products and derivatives can result in char that has similar chemical and physical properties (e.g. Ariyadejwanich et al., 2003), feedstock for biochar production must be derived from live or recently living organisms. Biochar is most often produced from herbaceous and woody plant materials, also known as cellulosic biomass, but it can also be made from algae, food waste, manure, and animal tissue. Though high in biomass content, mixed organic waste streams such as sewage and municipal solid waste (MSW) are generally not seen as viable feedstocks for biochar production because they can contain hazardous materials that contaminate soils (IBI, 2014). The emphasis here is on production of cellulosic biomass.

Biomass used as feedstock for pyrolysis can be a waste product (e.g. manure), a byproduct (e.g. bark), a co-product (e.g. wood chips), or a primary output of a dedicated feedstock production operation (e.g. *Miscanthus* cultivation). Primary products include crops and trees purposely grown as biomass feedstocks, such as switchgrass (*Panicum virgatum*), willow (*Salix* spp.), and hybrid poplar (*Populus* spp.). The difference between a waste, byproduct and co-product is variable by discipline, but SCM provides a relatively clean definition grounded in economics: waste products have disposal costs, byproducts have marginal costs and marginal value relative to primary products and co-products are manufactured jointly, have similar value, and use joint product costing in accounting. The complication with this definition is that the same material can be a waste, a byproduct, or a co-product depending on its value and costs, but it is useful to draw a clear line between waste as a material with net costs, especially for disposal, and production outputs that have market value and the potential to generate revenue.

In both theory and practice, biochar supply chains heavily favor the use of waste biomass as feedstock for several reasons. First, waste materials have disposal costs, generally making them a low cost raw material to procure. Poultry litter, which is a mix of waste bedding, feathers, feed, and excrement, falls into this category. Byproducts typically have some positive market value, but much of the cost of production is borne by some other higher value primary product. This makes them potentially less costly as feedstock, depending on other uses and markets. For example, wood chips and sawdust from lumber manufacturing traditionally have strong markets in areas with demand from pulp mills and wood panel manufacturers, but in areas distant from such facilities these may be good target feedstocks for pyrolysis. Second, waste materials often have disposal options with more damaging environmental impacts than processing via controlled thermochemical conversion. For example, open burning of agricultural and logging residues for disposal is widely practiced throughout the world, and has negative impacts from particulate and greenhouse gas (GHG) emissions (Loeffler and Anderson, 2014). In addition, logging residues are often burned in piles, which can result in long-term damage to the soil, invasion of non-native species, and loss of soil organic matter. Third, the use of waste biomass for biochar is unlikely to directly and negatively affect land use with regards to both conversion of forest to agriculture and transition from food crops to energy crops. Fourth, manufacturing facilities that generate waste and byproduct biomass in large quantities often need heat and power for production processes, which are co-products of some conversion systems.

Unlike waste and byproduct biomass, biomass purposely grown for bioenergy and bioproduct applications using agricultural, coppice, and plantation production systems must bear the full costs of production and feedstock logistics. In general, the trade-off here is between higher cost of production and higher productivity, which may result in lower per unit production costs. As with agricultural crops, this is often expressed as annual production per unit land area. Productivity for energy crops ranges from less than 2.0 megagram (Mg) ha^{-1} yr^{-1} (wheat straw) to 44.0 Mg ha^{-1} yr^{-1} (*Miscanthus*), with economically efficient and environmentally sustainable production systems generally characterized by easily established perennial crops rather than annual crops with high fertilizer, herbicide and pesticide inputs

(Laser and Lynd, 2014). In addition to increasing productivity, dedicated energy crops may reduce transportation and storage logistics costs when production and conversion are co-located and can hedge against feedstock price volatility, especially for vertically integrated firms. Co-locating production and conversion may also provide greater control over feedstock flow and quality, especially moisture content and homogeneity of feedstock physical and chemical properties. This is especially important for conversion systems that use catalysts to produce liquid fuels and chemicals.

The biomass production segment of the supply chain (Figure 2.2) is focused on the cultivation of crops and the silviculture of forests and woodlands. For most cellulosic feedstocks, this applies to biomass from dedicated energy crops, plantations, natural forests, and waste and byproduct biomass in both traditional and industrialized settings. Cultivation and silviculture as components of the supply chain include all aspects of site preparation, establishment, and tending. In agriculture, cultivation may include burning, tilling, fertilization, planting, pest and weed control, crop rotation, irrigation, and greenhouse and nursery operations. Silviculture may additionally include various practices for mechanical soil scarification, thinning, pruning and protection of forest health, such as sanitation cuttings to remove trees infected by insects and disease. The choice to develop and use different varieties of plants, including genetically modified organisms, hybrids and clones, is also included in feedstock production.

Even if the biomass used as feedstock is a waste or byproduct, the biochar supply chain appropriately begins in the field or forest, not with a pile of rice hulls or coconut shells at a processing plant. This has important implications for sustainability, which is the third component of feedstock production. A core concept of sustainability in agriculture and forestry is that sustainable practices do not degrade the long-term potential and productivity of the land, especially with regards to water, soil, and biodiversity. More recently, categorizing and quantifying GHG emissions have become central to assessing the sustainability of manufactured products. Though all segments of the supply chain have environmental impacts, sustainability features most prominently in biomass production because of the high potential for environmental damage due to deforestation, erosion, nutrient runoff, emissions and pollution from poor practices (Part 2 of this book).

2.5.2 Feedstock Logistics

Feedstock logistics includes activities to harvest, handle, collect, process, transport and store biomass from the field or forest to the conversion site. In industrial supply chains, these functions are often facilitated by specialized equipment (Figure 2.5). Waste and byproduct biomass is typically concentrated at the site of processing for primary products, such as a processing plant (e.g. nut shells, hulls, husks and bagasse), a concentrated animal feeding operation (e.g. manure), or at log landings and mills (e.g. logging and mill residues). These materials can also be left behind on field and forest sites in dispersed patterns, as in the case of corn stover, straw, orchard prunings, and some logging residues.

A Supply Chain Approach to Biochar Systems 37

Figure 2.5. Examples of industrial equipment used in woody biomass feedstock logistics, including: (a) a loader and horizontal grinder, (b) excavator and container truck, (c) self-unloading trailer, (d) rotary dryer, (e) feedstock conveyors and (f) a storage tent. Photographs by Nate Anderson. (A black and white version of this figure will appear in some formats. For the colour version, please refer to the plate section.)

Figure 2.6. Six different woody biomass feedstocks produced at a single sawmill (starting at 12 o'clock and running clockwise from left): dry planer shavings, ground wood fuel (also known as "hog fuel"), screened chips, sawdust, pulp chips and screened bark mulch. Photograph by Nate Anderson.

Forest biomass is particularly diverse with regards to concentration, ranging from widely dispersed tops, limbs, and foliage (i.e. "slash") left behind after cut-to-length logging operations, to piles of slash and unmerchantable logs resulting from road-side processing, to large volumes of homogenous sawdust, shavings, and wood chips concentrated at mill facilities (Keefe et al., 2014). Even when woody materials are concentrated as byproducts, the options for feedstock for use in biochar production can be highly variable at a single site (Figure 2.6). Sometimes waste and byproduct biomass can be procured at very low or even zero purchase price, but this should not be confused with the cost of logistics. For example, dispersed logging slash may have a very low purchase price per tonne, but the cost of logistics to harvest, process, and deliver this material to the conversion facility can be quite high – often more than the value of the feedstock once it is delivered (i.e. the "gate price").

Obviously, feedstock concentration is a good thing from a logistics perspective because dispersed feedstocks incur higher costs for collection and transportation, which translates to higher emissions from logistics in LCA. As a result, co-location of conversion systems with biomass production reduces logistics costs and associated emissions. This is true for waste and byproduct feedstocks, as well as dedicated biomass crops and plantations. Co-location can be achieved through integration of on-site biomass conversion, or by locating an independent biomass user at the site of biomass concentration. For example, many large forest industry operations use biomass-fueled combustion boilers for process heat and CHP, and biomass power plants tend to be located near biomass sources. Sometimes feedstock logistics systems feature intermediate concentration sites, such as feedstock silos and concentration yards. Such sites can improve transportation efficiency, product sorting and differentiation and processing (e.g. field drying).

Most conversion systems require some biomass processing prior to pyrolysis. The purpose of processing in feedstock logistics is to make the feedstock more suitable for conversion and more homogenous, which improves mechanized handling and reduces variability in solid, liquid, and gaseous conversion outputs. Specific needs for processing depend on technical specifications for feedstock moisture, particle size, ash content, and other characteristics. Typical processing functions include separation (e.g. debarking), drying, screening, and comminution by grinding, chipping, or hammering. Screening serves not only to narrow particle size distribution, but also to remove contaminants that may have detrimental effects on conversion, such as mineral soil and inorganic debris, like metal fragments from equipment and refuse.

Storage as a component of logistics is also important because it decouples conversion from feedstock production and delivery, allowing conversion to take place independently of feedstock production. This is especially critical when biomass is subject to seasonal availability or disruptions in supply due to weather or market conditions, which is the case for many agricultural and forest biomass resources. For large biomass operations, it is also important to consider systems for managing feedstock degradation, fugitive dust emissions and spontaneous combustion risk, which are all hazards in biomass storage and handling.

2.5.3 Conversion

The conversion segment of the supply chain includes three categories of activities: the chemical and physical transformation of biomass feedstock into biochar via thermochemical conversion, post-conversion treatments to enhance biochar effectiveness for specific end uses, and production of any co-products, including heat, power, energy gas, liquid fuels, and chemicals. Part 3 of this book examines pyrolysis conversion of biomass in detail, but several aspects of conversion are worth highlighting here. More than any other component of the biochar supply chain, conversion hinges on technology. The most striking aspect of biomass conversion from a supply chain standpoint is the diversity of technologies and scales that can be used to transform biomass into biochar. On one end of the spectrum, small traditional charcoal kilns and more modern small batch systems (Odesola and Owoseni, 2010) can be employed by farmers, gardeners, and horticulturalists to process residues into biochar for relatively small-scale, on-site applications, similar to the fully integrated production scenario described in Figure 2.2. On the other end of the spectrum, biochar can be a co-product of biofuel production by large, integrated biorefineries deploying cutting-edge conversion technologies at large scales (Rocke, 2014). In this context, biochar supply chains take on widely differing characteristics depending on the conversion technology employed, with biochar itself being variously a waste, byproduct, co-product or sole primary product, depending on the operation. Common co-products of pyrolysis include heat, bio-oil and gas that can be used as fuel for combustion (e.g. renewable natural gas) or as a raw material in the production of liquid fuels and chemicals via catalysis (e.g. synthesis gas).

Even among relatively comparable technologies, supply chains can be quite variable. For example, mobile and distributed-scale thermochemical conversion systems have received significant attention in recent years, mostly due to their relatively low capital investment and ability to be deployed in forward operations close to feedstocks, thereby producing dense, value-added products from waste biomass and reducing logistics costs (Anderson et al., 2013). Though many of these systems are similar in terms of size and configuration (Figure 2.7), they have different feedstock specifications and their different outputs necessitate significantly different downstream logistics. For example, fast pyrolysis systems that produce bio-oil as a co-product must include systems for liquid fuel handling, storage, transportation and safety, and biochar production cannot be decoupled from bio-oil production, regardless of independent market demand for the two products. Similarly, conversion technologies that capture and use gas for heat and power must include not only gas storage and handling systems, but also be well balanced with on-site energy demand.

Biochar can be used in its raw form to improve soils. However, in many cases, its performance as a soil amendment can be enhanced by post-conversion treatments. Such treatments include inoculation with desirable microbes, treatments to change pH or other chemical characteristics, granularization or pelletization to improve material handling and performance, composting or blending with chemical fertilizers and organics such as

Figure 2.7. Examples of mobile and distributed-scale pyrolysis conversion systems producing co-products with biochar: (a) biochar and heat, (b) biochar with low-energy gas and bio-oil, (c) biochar with low-energy gas and bio-oil and (d) biochar with medium-energy gas. Photographs by Nate Anderson. (A black and white version of this figure will appear in some formats. For the colour version, please refer to the plate section.)

manure, and activation by chemical or physical means to increase surface area and promote ion exchange. In addition to improving product performance, such treatments provide biochar producers with critical opportunities to both diversify their products to better meet the needs of different end users and also differentiate their products from other manufacturers marketing to the same customers.

2.5.4 Distribution Logistics

Distribution logistics includes activities to package, transport, and store biochar from the site of conversion to the site of end use. Depending on the feedstock and conversion method, biochar resulting from pyrolysis can be variously characterized as a fine powder or a coarse charcoal, hydrophobic or hydrophilic, physically stable or friable, and homogenous or heterogeneous in particle size and shape. Biochar may be dry or wet, depending on the cooling method used in production, and has various levels of performance in pneumatic and conveyor handling systems. These characteristics have important implications for distribution. Fine powders can be both difficult and dangerous to store and handle due to combustion risk and risk to health from aspiration of dust particles. Methods of pelletizing biochar to improve handling have proven effective (Reza et al. 2014), but come with added financial costs and energy requirements.

For large-scale applications and wholesale markets, biochar can be transported in bulk by rail or truck in specially designed rail cars and trailers. More commonly, raw biochar and biochar downstream products are packaged for delivery in forklift-able bulk containers, large polyethylene bulk bags (i.e. totes or "super sacks," which are common in agriculture), metal and plastic drums, large multi-ply paper bags, and low-volume plastic bag and bucket packaging for small-scale consumer applications. From a logistics standpoint, bulk packaging can be efficient for producers, but may not meet the needs of end users, especially if specialized equipment such as hydraulic lifts and rolling forklifts are needed for unloading. More broadly, distribution logistics must be well matched to both transportation modes and the capabilities of end users to handle and store the biochar before use.

2.5.5 End Use

In addition to agricultural and forest applications focused on improving soil productivity, several other biochar uses have gained prominence, including uses for mitigation and reclamation of mining sites, seed coating, potting media, storm water filtration, and restoration of soils on burned sites (Dumroese et al., 2011; Fellet et al., 2011; Delaney, 2015). Part 4 and other chapters of this book examine specific end uses of biochar for a variety of case studies.

The end use segment of the supply chain includes not only the application of biochar to soils, but any blending or pre-application processing that may occur at the site of end use. Biochar can be blended mechanically or by hand with soil and other soil additives, such as seeds, manure, compost or chemical fertilizers. Processing can include further grinding or screening, or additions of water or surfactants to improve handling during application. Application can be done by hand, but it is often performed by specialized agricultural and forestry equipment (Figure 2.8). Application generally relies on broadcasting by hand or application using planters, tillers, seeders, and spreaders at various scales and levels of mechanization. Biochar can also be applied using hydroseeding systems that spread a pressurized aqueous slurry of biochar, typically mixed with other additives, such as compost, mulch fertilizer, and tackifying agents to reduce loss of biochar in storm runoff.

As described in Section 2.1, biochar is used as a soil amendment, and biochar systems can meet a broad range of soil improvement, waste management, energy, and climate change mitigation needs. However, the same charcoal classified as biochar in soil applications has potential for use as a fuel and raw material in other applications. Alternative uses include fuel pellets and briquettes, chemicals, feedstock for gasification, gunpowder, pigments and dyes, industrial sorbents, and a precursor in the manufacture of activated carbon (Azargohar and Dalai, 2006; Anderson et al., 2013). In a supply chain context, biochar producers should be aware of alternative uses of charcoal for two reasons. Alternative uses

Figure 2.8. A six-wheeled forwarder, normally configured to carry logs, here mounted with a modified pellet spreader to apply biochar pellets on forested sites developed by the Missoula Technology Development Center, Missoula, MT. Photo by Han-Sup Han. (A black and white version of this figure will appear in some formats. For the colour version, please refer to the plate section.)

provide opportunities to diversify product lines and enter new, complementary markets. They also present the threat of competition from horizontal integration of biochar production and marketing by organizations that are already using biomass to manufacture carbon and charcoal products like solid fuels and activated carbon.

SCM considerations for biochar used primarily to meet climate change mitigation objectives (Section 2.4 and Figure 2.4) can be more complicated than the end uses discussed thus far. Gaunt and Cowie (2009) identified six specific characteristics of biochar that can result in net reductions of GHG emissions attributable to biochar systems: 1) sequestration of relatively stable carbon in the soil; 2) avoided emissions of methane and nitrogen oxides related to alternative disposal methods such as biomass decomposition and combustion; 3) avoided emissions of methane and nitrogen oxides related to changes in soil processes; 4) displacement of carbon intensive agricultural inputs through both direct substitution and increased efficiency; 5) carbon sequestration resulting from higher productivity leading to greater soil carbon; and 6) displacement of fossil fuels from biochar co-products. Only one of these, carbon sequestration in the soil, is a direct effect. The other benefits, though supported by research, are indirect and rely on assumptions about the fate of waste biomass, changes in soil processes and characteristics, and market substitutions for fertilizer, fossil fuels, and other carbon intensive inputs.

In a commercial context, monetizing climate change mitigation effects can turn these benefits from a desirable non-market secondary characteristic of biochar used primarily to improve productivity into a viable end use with potential to generate revenue. Biochar producers and end users may be able to capture value related to carbon sequestration through effective marketing and product differentiation, especially in the context of certification schemes and robust LCA (Section 2.3 and Chapter 3). When produced as a co-product of biofuels, biochar can be critical in meeting renewable fuel standards and capturing value from associated financial incentives (e.g. Wang et al., 2014). Monetizing climate benefits may also be possible through various international, national, regional, and independent frameworks that establish mechanisms to compensate, sell, and exchange net carbon offsets through markets and payments for ecosystem services (Jack et al., 2008; Gaunt and Cowie, 2009). Though these opportunities are closely tied to public policy, they can be incorporated explicitly into biochar SCM, and climate change mitigation can be considered a viable end use for biochar when conditions are favorable.

2.6 Conclusions

A supply chain approach to biochar systems is focused on meeting the needs of end users and emphasizes the interconnectedness of organizations involved in various stages of production, logistics, conversion and end use. It is an effective framework for dissecting and evaluating the economic, social, and environmental dimensions of biochar as a manufactured product used to meet diverse objectives, including improving soils, managing waste, producing renewable energy, and mitigating climate change. As the biochar industry evolves, SCM can be used to organize, coordinate and manage the material, information, and financial flows of the biochar supply chain, allowing organizations to more effectively and efficiently deliver the many potential benefits of biochar systems.

Acknowledgements

Funding for much of the research and analysis described in this chapter was provided to the authors by the Rocky Mountain Research Station of the US Forest Service and by the Biomass Research and Development Initiative of the US Department of Agriculture (USDA) National Institute of Food and Agriculture.

References

Anderson, N., Chung, W., Loeffler, D. and Jones, J. G. (2012). A productivity and cost comparison of two systems for producing biomass fuel from roadside forest treatment residues. *Forest Products Journal*, 62, pp. 223–233.

Anderson, N., Jones, J.G., Page-Dumroese, D., et al. (2013). A comparison of producer gas, biochar, and activated carbon from two distributed scale thermochemical conversion systems used to process forest biomass. *Energies*, 6, pp. 164–183.

Ariyadejwanich, P., Tanthapanichakoon, W., Nakagawa, K., Mukai, S. R. and Tamon, H. (2003). Preparation and characterization of mesoporous activated carbon from waste tires. *Carbon*, 41, pp. 57–164.

Azargohar, R. and Dalai, A. K. (2006). Biochar as a precursor of activated carbon. *Applied Biochemistry and Biotechnology*, 129–132, pp. 762–773.

Christopher, M. (2011). *Logistics and Supply Chain Management*. 4th Edition. Harlow, Essex: Pearson.

Deleney, M. (2015). *Northwest Biochar Commercialization Strategy Paper.* [online] Available at: http://nwbiochar.org/sites/default/files/sites/default/files/attached/nw_biochar_strategy_02-24-15.pdf [Accessed 16 March 2015].

Downie, A., Munroe, P., Cowie, A., Van Zwieten, L. and Lau, D. (2012). Biochar as a geoengineering climate solution: hazard identification and risk management. *Critical Reviews in Environmental Science and Technology*, 42, pp. 225–250.

Duku, M. H., Gu, S. and Hagan, E. B. (2011). Biochar production potential in Ghana – a review. *Renewable and Sustainable Energy Reviews*, 15, pp. 3539–3551.

Dumroese, K., Heiskanen, J., Englund, K. and Tervahauta, A. (2011). Pelleted biochar: chemical and physical properties show potential use as a substrate in container nurseries. *Biomass and Bioenergy*, 35, pp. 2018–2027.

Fellet, G., Marchiol, L., Delle Vedove, G. and Peressotti, A. (2011). Application of biochar on mine tailings: effects and perspectives for land reclamation. *Chemosphere*, 83, pp. 1262–1267.

Forest Stewardship Council (FSC, 2010). *FSC-US Forest Management Standard Version 1.0.* [online] Available at: https://ic.fsc.org/national-standards.247.htm [Accessed 16 March 2015].

Gaunt, J. L. and Cowie, A. (2009). Biochar, greenhouse gas accounting, and emissions trading. Chapter 18. In: Lehman, J. and Joseph, S. (eds.) *Biochar for Environmental Management: Science and Technology*. London: Earthscan.

Goetschalckx, M. (2011). *Supply Chain Engineering*. New York: Springer.

Hagelaar, G. and van der Vorst, J. (2002). Environmental supply chain management: using life cycle assessment to structure supply chains. *International Food and Agribusiness Management Review*, 4, pp. 399–412.

International Biochar Initiative (IBI, 2014). *Standardized Product Definition and Product Testing Guidelines for Biochar That is Used in Soil*. [online] Available at: www.biochar-international.org/characterizationstandard [Accessed 16 March 2015].

Jack, B. K., Kousky, C. and Sims, K. (2008). Designing payments for ecosystem services: lessons from previous experience with incentive-based mechanisms. *Proceedings of the National Academy of Sciences*, 105, pp. 9465–9470.

Keefe, R., Anderson, N., Hogland, J. and Muhlenfeld, K. (2014). Woody biomass logistics. Chapter 14. In: Karlen, D. (ed.) *Cellulosic Energy Cropping Systems*. Chichester, West Sussex: John Wiley and Sons.

Kim, D., Anderson, N. and Chung, W. (2015). Financial performance of a mobile pyrolysis system used to produce biochar from sawmill residues. *Forest Products Journal*, 65, pp. 189–197.

Laser, M. and Lynd, L. (2014). Introduction to cellulosic energy crops. Chapter 1. In: Karlen, D. (ed.) *Cellulosic Energy Cropping Systems*. Chichester, West Sussex: John Wiley and Sons.

Leach, M., Fairhead, J. and Fraser, J. (2012). Green grabs and biochar: revaluing African soils and farming in the new carbon economy. *Journal of Peasant Studies*, 39, pp. 285–307.

Lehmann, J. and Joseph, S. (2009). Biochar for environmental management: an introduction. Chapter 1. In: Lehman, J. and Joseph, S. (eds.) *Biochar for Environmental Management: Science and Technology*. London: Earthscan.

Leonardo Academy (2012). *National Sustainable Agriculture Standard, LEO-4000*. Madison, WI: Leonardo Academy.

Loeffler, D. and Anderson, N. (2014). Emissions tradeoffs associated with cofiring forest biomass with coal: a case study in Colorado, USA. *Applied Energy*, 113, 67–77.

Mathews, J. A. (2008). Carbon-negative biofuels. *Energy Policy*, 36, pp. 940–945.

Odesola, I. F. and Owoseni, T. A. (2010). Development of local technology for a small-scale biochar production processes from agricultural wastes. *Journal of Emerging Trends in Engineering and Applied Sciences*, 1, 205–208.

Reza, M. T., Uddin, M. H., Lynam, J. and Coronella, C. (2014). Engineered pellets from dry torrefied and HTC biochar blends. *Biomass and Bioenergy*, 63, 229–238.

Roberts, K., Gloy, B., Joseph, S., Scott, N. and Lehmann, J. (2010). Life cycle assessment of biochar systems: estimating the energetic, economic, and climate change potential. *Environmental Science and Technology*, 44, 827–833.

Rocke, M. (2014). *Cool Planet starts construction on first commercial facility: Louisiana facility to produce green fuels and biochar from sustainable wood residues*. [online] Available at: www.bloomberg.com/bb/newsarchive/aB3nqCdei4c0.html [Accessed 16 March 2015].

Sparrevik, M., Field, J. L., Martinsen, V., Breedveld, G. D. and Cornelissen, G. (2013). Life cycle assessment to evaluate the environmental impact of biochar implementation in conservation agriculture in Zambia. *Environmental Science and Technology*, 47, pp. 1206–1215.

Stock, J. R. and Lambert, D. M. (2001). *Strategic Logistics Management*. 4th Edition. New York: McGraw-Hill.

Vasilyeva, G. K., Strijakova, E. R. and Shea, P. J. (2006). Use of activated carbon for soil remediation, pp. 309–322. In: Twardowska, I., Allen, H. E., Haggblom, M. M. and Stefaniak, S. (eds.) *Soil and Water Pollution Monitoring, Protection and Remediation*. New York: Springer.

Wang, Z., Dunn, J. B., Han, J. and Wang, M. Q. (2014). Effects of co-produced biochar on life cycle greenhouse gas emissions of pyrolysis-derived renewable fuels. *Biofuels, Bioproducts and Biorefining*, 8, pp. 189–204.

3

Life Cycle Analysis of Biochar

RICHARD D. BERGMAN, HONGMEI GU,
DEBORAH S. PAGE-DUMROESE AND NATHANIEL M. ANDERSON

Abstract

All products, including bioproducts, have an impact on the environment by consuming resources and releasing emissions during their production. Biochar, a bioproduct, has received considerable attention because of its potential to sequester carbon in soil while enhancing productivity, thus aiding sustainable supply chain development. In this chapter, the environmental impacts of producing biochar using a holistic method called life-cycle assessment (LCA) or more generally life-cycle analysis are discussed. LCA is an internationally accepted method that can calculate greenhouse gas (GHG) and other emissions for part or all of a product life cycle. The present chapter will show how LCA can assess environmental impacts of the entire supply chain associated with all steps of the biochar system, from biomass harvesting through biochar production to soil amendment, with a focus on the production stage. Exploring a biochar system from a forestry LCA perspective, a new thermochemical conversion technology developed in the United States and used to process waste woody biomass, will be described. In particular, the conversion unit's environmental performance based on the LCA research conducted so far will be described. Although this chapter will present LCA mostly from a forestry perspective, non-forestry agricultural activities will also be discussed.

3.1 Introduction

Biomass as a sustainable feedstock for producing bioproducts has raised substantial attention (Guo et al., 2007). Biomass-derived fuels and products are one approach to reduce the need for oil and gasoline imports while supporting the growth of agriculture, forestry, and rural economies (Roberts et al., 2010; McKechnie et al., 2011). In particular, biochar as a bioproduct has received considerable attention because of its carbon (C) sequestration potential and ability to enhance soil productivity (Lehmann et al., 2006; Lorenz and Lal, 2014). Thus, biochar as a byproduct of bioenergy production from biomass, including production of heat, energy gas, and bio-oil, has the potential to reduce net greenhouse gas (GHG) emissions, improve local economies and energy security (Homagain et al., 2014), and may increase overall site productivity when added back to the soil. One approach to measure biochar's sustainability in the context of the above mentioned features is by conducting a life-cycle assessment (LCA).

LCA can be used to evaluate alternative scenarios for their GHG emissions. Categorizing GHG emissions has become crucial to assessing the sustainability of manufactured products. Scenarios include using wood residues such as logging slash or mill residues for feedstock to make biochar. However, there are alternative forest management practices for disposing of logging slash instead of collecting it for use as raw material for fuel, a low-value product. These include leaving the residues to decompose in the forest, thereby releasing GHG emissions or, worse yet from an emission standpoint, burning logging slash along the ground or in piles to either dispose of waste biomass or reduce impacts of potential wildfires. These practices tend to have worse emissions impacts because prescribed burning not only consumes the logging slash but also much of the down and dead wood on the forest floor, which releases unchecked GHG emissions and particulate matter in the form of smoke. Furthermore, incomplete combustion associated with open burning produces higher levels of methane and nitrogen oxides (NOx), which have higher global warming potentials (US EPA, 1995; NETL, 2013; Loeffler and Anderson, 2014; Pierobon et al., 2014). In the USA, wildfire-prevention policy objectives exist to drive the use of prescribed burning to reduce fuel loads, but open burning is also widely practiced in silviculture to open growing space for regeneration and in agriculture to dispose of crop residues and prepare fields for planting. LCA, as a widely accepted scientific method, can be used to capture these climate change impact differences for the various uses of wood residues and thus enable practitioners and policymakers to make sound decisions based on science. LCA can be thought of as an approach similar to financial accounting but instead, accounting for environmental costs and benefits to show which approach would cause the least negative impact.

3.1.1 Four Phases of Life Cycle Assessment

LCA measures the holistic environmental impacts of a product, including resources consumed and emissions released along with the associated environmental impacts. An LCA can cover the life of a product from extraction of raw materials to product production point (i.e. "cradle-to-gate") or through distribution, use, and to its final disposal point (i.e. "cradle-to-grave") (Figure 3.1) (ISO, 2006a, 2006b; ICLD, 2010). This approach is nicely aligned with the supply chain management (SCM) used in manufacturing (Chapter 2), but includes more detailed treatment of emissions, effluents, and waste.

LCAs are comprised of four phases (components) as defined by the International Organization for Standardization (ISO): (1) goal and scope definition; (2) life cycle inventory (LCI) analysis; (3) life cycle impact assessment (LCIA); and (4) interpretation (Figure 3.2). An LCA study includes all phases, but an LCI study does not include stage 3.

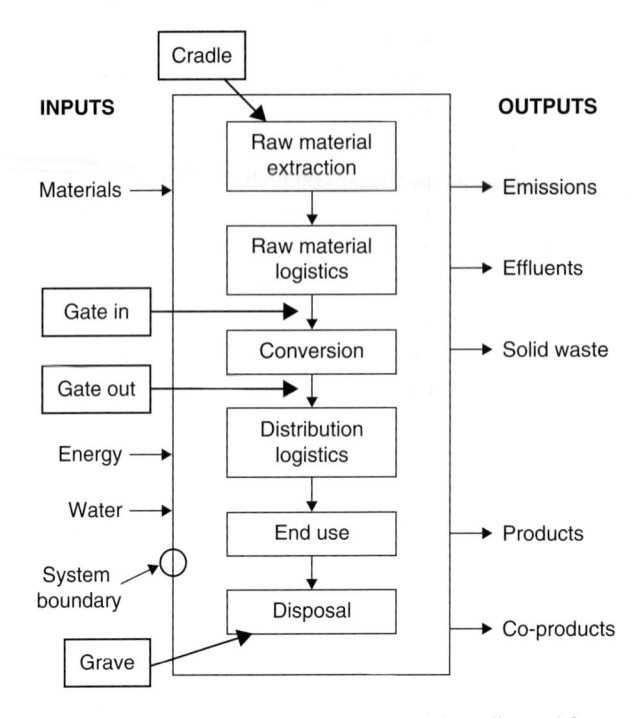

Figure 3.1. Complete life cycle from extraction of raw materials to disposal for product production.

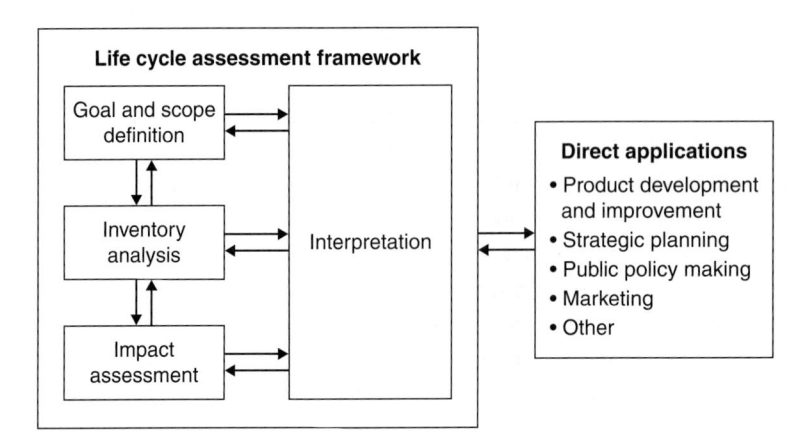

Figure 3.2. Four phases of life cycle assessment.
Source: This excerpt is adapted from ISO 14040:2006, Figure 1 on page 8, with the permission of ANSI on behalf of ISO. © ISO 2015 – All rights reserved.

An LCI measures all raw material and energy inputs and the associated environmental outputs to manufacture a particular product, process, or service on a per unit (functional) basis within carefully defined system boundaries. LCIAs as part of an LCA study can use LCI flows to calculate impacts in four areas: human health, social health, resource depletion, and ecosystem function. In the interpretation stage, alternative actions to reduce impacts are systematically evaluated after environmental "hotspots" have been identified (ISO, 2006a, 2006b; ILCD, 2010). Environmental hotspots can be identified as stages along the life cycle from the sourcing of raw materials through materials processing, manufacture, distribution, use, and disposal or recycling that generate higher environmental impacts than other stages. Some impact categories related to energy and material consumption are easier to calculate than others. The following sub-sections will discuss the four phases of LCA in more detail.

3.1.1.1 Goal and Scope

The goal and scope definition provides the study framework and explains how, and to whom, results are to be communicated. There are several important items to address during this phase. First, the functional unit is defined for the product system to provide a way to allocate raw material consumption, air emissions, water effluent, and solid waste generated during product production and to enable product comparison. The functional unit is similar to a production unit and can be defined as the quantity of a product serving a particular function for a set time. An example of a functional unit is one square meter of installed flooring with a service life of 100 years. This functional unit for the installed flooring can consist of renewable products such as wood or bamboo or non-renewable products like vinyl, and enable a product comparison on their environmental performance. Several refurbishments or replacements are typically necessary for most flooring products to reach the set time of 100 years (Bergman and Bowe, 2011). Second, a system boundary for the product is selected by setting what unit processes will be included in the analysis. The system boundary tracks the environmental inputs and environmental outputs crossing the boundary, as shown in Figure 3.1. The system boundary may cover the whole life cycle of a product or just a single part of the life cycle from gate to gate. Third, to address the most relevant life cycle stages, cut-off criteria are determined. In a practical sense, the cut-off criteria enable the LCA practitioner to complete the project in reasonable time by omitting inconsequential life cycle stages or life cycle stages typically omitted. Last, a protocol is described on how the collected primary data will be validated. Primary data are measured and collected in person and on-site for the study. For product LCAs, a mass balance is typically performed to aid in this endeavor. In addition, a common practice is to calculate the process energy consumption on a production unit basis (e.g. a cubic meter of dry sawn lumber) and then compare the results to a similar product or products found within secondary data sources such as peer-reviewed literature (ISO, 2006a, 2006b; ILCD, 2010).

3.1.1.2 Life Cycle Inventory Analysis

The life cycle inventory phase is the most time- and data-intensive part of conducting an LCA, primarily because primary data must be collected to develop LCI data or flows for the product system being evaluated. Data collection can occur at any stage of the life cycle such as during extraction of raw materials, product production, or use phase depending on the project goal and scope. As for data quality, certain requirements must be met, and the outcome reliability from LCA studies (i.e. LCIAs) highly depends on the degree to which these data quality requirements are met.

Once the primary data are collected, the data are validated and related to the functional unit to produce the aggregation of results (i.e. LCI flows or results). For industry products, a typical aggregation is by the production of the individual company, where data collected from the largest company carry the most weight in reporting. LCI flows include the raw material consumed, emissions to air and water, and solid waste generated per functional unit. An intricate step in this calculation process is the allocation of LCI flows, for example, releases to air and water. Complications arise because many existing product systems yield multiple products. As discussed in Chapter 2, the difference between a waste, byproduct, and co-product is variable by discipline, but LCA provides definitions based on assigning environmental impacts: waste products have disposal costs, byproducts have marginal costs and marginal value relative to primary products, and co-products are manufactured jointly and use joint product costing in accounting. The complication with this definition is that the same material can be a waste, a byproduct, or a co-product depending on its value and costs, but it is useful to draw a clear line between waste as a material with net costs, especially for disposal, and production outputs that have market value and the potential to generate revenue.

For example, sawmills not only produce sawn lumber as a product (i.e. the final product) but also produce chips, sawdust, bark, and shavings as co-products. As mentioned in Chapter 2 (Section 2.5.1), these "co-products" in the context of SCM would be considered byproducts and not co-products as they are in the LCA context because they have some economic value, although little in some circumstances. Therefore, the environmental outputs must often be allocated (i.e. assigned) to the different products and co-products. Waste products like boiler ash are considered the same in the LCA and SCM contexts. The following is recommended for allocation in order of preference: (1) wherever possible, allocation should be avoided by using system expansion; (2) where allocation is not avoidable, environmental inputs and environmental outputs should be partitioned between different functions or products in a way that corresponds to the underlying physical relationships between them, such as mass and energy; (3) if (1) or (2) are not viable, allocation should be carried out based on other existing relationships (e.g. in proportion to the revenue of the various products and co-products) (ISO, 2006a, 2006b; ILCD, 2010).

3.1.1.3 Life Cycle Impact Assessment

Life cycle impact assessment aims to show the potential environmental impacts by using LCI flows found in phase 2. The ISO14040 suggests an LCIA includes the following mandatory elements. The first is a selection of impact categories, category indicators, and characterization models. The second is classification, which is the assignment of individual inventory factors to impact categories. For example, CO_2 and N_2O are assigned to the global warming (GW) impact category. Other common impact categories are photo-oxidant formation, eutrophication, ozone depletion and acidification. The third mandatory element is characterization, which is the conversion of LCI flows to common units within each impact category, so the LCI flows can be aggregated into category indicator outputs. For example, CO_2 and N_2O are commonly emitted from burning of fossil fuels during transportation. However, though CO_2 is emitted at far greater levels than N_2O, CO_2 has less impact on climate change on a mass basis than N_2O. In addition, another complicating factor is that each GHG decays at a different rate in the atmosphere. Therefore, each emission must be considered separately for the quantity emitted along with its impact on the individual category indicator output being estimated. Overall, an LCIA provides a systematic approach for sorting and characterizing environmental impacts (ISO, 2006a, 2006b; ILCD, 2010). In the United States, a midpoint-oriented LCIA method referred to as the Tool for the Reduction and Assessment of Chemical and other environmental Impacts (TRACI) was developed by the US Environmental Protection Agency specifically using input parameters consistent with US locations (Bare, 2011). Limits and assumptions of the LCA study are listed to enable reproducibility of the results.

3.1.1.4 Interpretation

The object of the interpretation phase is to reach conclusions and recommendations in line with the defined goal and scope of the study. Results from the LCI and LCIA are combined together and reported to give a comprehensive, transparent, and unbiased account of the study. The interpretation is to be made iteratively with the other three phases.

The life cycle interpretation of an LCA or an LCI comprises three main elements: (1) identification of significant problems (i.e. environmental hotspots) based on the outcomes of the LCI and LCIA phases of an LCA; (2) evaluation of outcomes, which considers completeness, sensitivity, and consistency checks; and (3) conclusions and recommendations (ISO, 2006a, 2006b; ILCD, 2010).

3.1.2 Types of LCA

There are two main types of LCA along with various hybrid, dynamic, and streamlined methods that typically incorporate components of the two main types. We will only discuss the two basic types here.

3.1.2.1 Attributional

Attributional LCA (ALCA) uses a process-modeling method to find the critical environmental impacts for a particular product, referred to as "cradle-to-grave" (raw material extraction to waste disposal) analysis. This is the method that was discussed above. ALCA is a linear approach. Therefore, the magnitude of the functional unit (kg or tonne of biochar applied, for example) does not affect the LCIA outputs (Pennington et al., 2004). For example, one could state that for the GW impact a value of 10 kg CO_2-eq/kg or 10,000 kg CO_2-eq/tonne of biochar would be equal. Using the LCIA results, an ALCA can locate environmental "hot spots" for a given product system (cradle-to-gate) to provide information for manufacturers (decision-makers) regarding process improvements and design (Thomassen et al., 2008; Gaudreault et al., 2010). It is common for ALCA to use other allocation methods besides the system expansion listed in Section 3.1.1.2 if the LCA practitioner is unable to divide unit processes sufficiently to track impacts. These allocation methods assign environmental burdens to products and co-products. Common allocation methods include mass, energy and revenue allocations (ISO, 2006a, 2006b).

3.1.2.2 Consequential

Consequential LCA (CLCA) is similar to ALCA in that it is a process-modeling method but is used to describe the (indirect) consequences of a particular decision. CLCA estimates system-wide changes in (material and energy) resource flows and environmental burdens that result from different production levels of the functional unit based on a decision. It is the decision that alters the technology activity (Ekvall and Weidema, 2004; Ekvall and Andrae, 2006). CLCA studies use system expansion to describe the consequences instead of allocation by mass, energy, or revenue. This method examines the effects on marginal electricity consumption for a change in production, whereas ACLA evaluates environmental impacts based on modeling average technologies to create a "composite" technology. However, CLCA is not capable of locating "hot spots" as an ALCA is (Pennington et al., 2004; Thomassen et al., 2008; Gaudreault et al., 2010). Additionally, conducting a CLCA versus an ALCA usually results in greater uncertainty to an individual study, reducing its usefulness. Even so, some of the benefits of biochar production are indirect, such as substitution for non-renewable products yielding emissions offsets, making it a relevant method for biochar LCA. Other benefits that could be captured indirectly are: displacement of carbon-intensive agricultural inputs through both direct substitution and increased efficiency, and carbon sequestration resulting from higher productivity leading to greater soil carbon stocks.

3.1.2.3 Differences

An ALCA stays within carefully defined boundaries whereas a CLCA does not. CLCA activities may fall outside the original system boundary. For example, a sawmill produces sawn lumber as its final product while producing co-products such as sawdust (Bergman and Bowe, 2012). The sawdust is burned for fuel on-site to generate thermal energy for drying the sawn lumber or it is sold off-site. In an ALCA, material is not tracked once it crosses the system boundary and leaves the system. In the case of Bergman and Bowe (2012), sawdust from sawn lumber is not tracked beyond the system boundary, which is the sawmill gate. Its use as either fuel or as raw material for manufacturing wood panels by another mill does not impact the ALCA. An ALCA looks at a moment of time or a "snapshot" whereas the basis for a CLCA could be to evaluate a market decision. For example, a sawmill may decide to sell sawdust to wood panel manufacturers rather than use it as fuel,

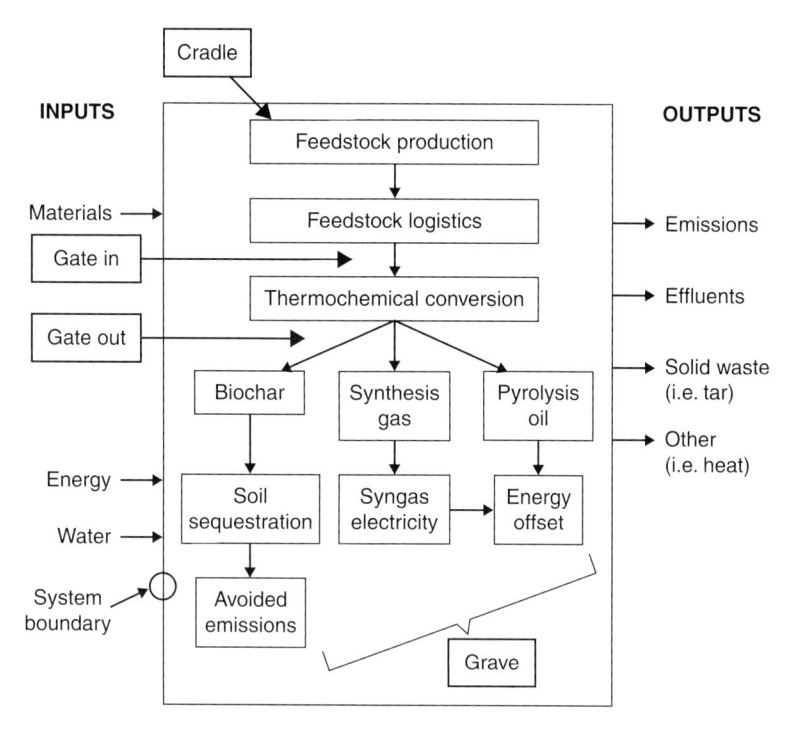

Figure 3.3. Whole life cycle of biochar production from cradle to grave.

and use natural gas to fire the boilers to dry sawn lumber because sawdust has more value as a feedstock for another than as a fuel. CLCA attempts to capture the potential environmental effects of selling the sawdust instead of burning it on-site at the sawmill for thermal energy. Depending on the goal and scope of the LCA study, both attributional and consequential methods are useful.

3.2 Life Cycle Stages for Biochar

As mentioned previously, LCAs can address environmental performance of biochar, including categorizing GHG emissions along the entire life cycle of a product such as carbon sequestration of the biochar when applied to the soil (Figure 3.3). The life cycle stages include: (1) raw material extraction (i.e. feedstock production); (2) raw material (feedstock) logistics; (3) thermochemical conversion; (4) biochar logistics; and (5) product end uses including soil carbon sequestration. However, for a more complete description of the supply chain, the previous chapter (Chapter 2) provides a detailed view of biochar systems in a supply chain context. This section will deal primarily with the thermochemical conversion process from an LCA perspective. Section 3.3 will describe a new thermochemical conversion system, an advanced pyrolysis using the life cycle stages defined here and in Section 3.3.

3.2.1 Raw Material Extraction

The raw material extraction stage for biochar involves interaction with agricultural and natural systems. In the case of biochar, the raw material is biomass, most often biomass from herbaceous and woody plants. Raw material extraction may include forest or agriculture activities involving cultivation, harvesting, collection, handling and processing including in-woods grinding and chipping, and screening. In-woods grinding and chipping are dominated by diesel fuel use. Inputs can include diesel, fertilizer, pesticides and herbicides and outputs can include fossil CO_2 and N_2O air emissions along with possible nitrogen fertilizer run-off. In the United States and other parts of the world, industrial timberlands tend to have greater inputs of nursery seedlings, herbicides, pesticides and fertilizers than naturally regenerating forests managed by non-industrial landowners. In this case, it is clear that management practices can have direct impacts on product attributes and corresponding LCA.

3.2.2 Raw Material Logistics

The second life cycle stage is raw material (i.e. feedstock) logistics. For biochar, feedstock transportation typically includes a diesel tractor trailer hauling the feedstock generated at the harvesting site from the landing to the thermochemical conversion facility. Inputs include diesel and outputs include fossil CO_2, volatile organic compounds, and particulate emissions. Raw material logistics may also include multi-stage, multi-mode transportation that includes intermediate facilities to store, concentrate or process biomass. From an LCA perspective, dispersed feedstocks incur higher costs for collection and transportation, which translates to higher emissions from the logistics stage.

3.2.3 Thermochemical conversion

The thermochemical conversion life cycle stage involves the production of biochar from biomass via gasification and pyrolysis or some similar process. These thermochemical conversion technologies are similar to traditional charcoal kilns but under much tighter control to prevent the release of N_2O, CH_4 (methane), and particulate emissions associated with the older technology (Woolf et al., 2010). These systems produce biochar, synthesis gas (syngas), and pyrolysis oil in different percentages. For pyrolysis systems, these systems always produce some biochar (Gaunt and Lehmann, 2008). The intent is to convert the incoming dry feedstock under a controlled environment while preventing the introduction of air (i.e. oxygen) into the system. Typically, the product production life cycle stage consumes the most energy and materials and thus has the highest environmental impact (Bergman and Gu, 2014; Dutta and Raghavan, 2014). Therefore, finding a mass balance and energy consumption at this stage is of utmost significance to accurately quantify LCI flows and the subsequent LCIA outputs. In addition, incoming feedstock with high moisture content (i.e. "green" feedstock) and large, heterogeneous particles may have to be dried, reduced and screened before thermochemical conversion, which can have large environmental impacts.

Feedstock preparation can be determined separately or be part of the thermochemical conversion life cycle stage, but its impacts must be captured. There are several reasons for these large impacts associated with the incoming green feedstock. First, mechanical size reduction by chipping, hammering and grinding and the subsequent screening required to ensure uniform size are energetically intensive activities. Second, feedstock drying to the appropriate moisture for the selected technology also has high energy demands. The sizing and moisture specification are highly dependent on the thermochemical conversion technology selected to optimize production. This processing ensures the feedstock is properly prepared before thermochemical conversion, but it comes at a price in terms of the energy consumed and its associated environmental impacts. Energy for drying feedstock can come from renewable or non-renewable sources, while the electricity for on-site grinding and chipping and handling comes primarily from grid power, which is dominated by fossil fuels in many locations. If woody biomass is burned as fuel for drying (as is common practice in the forest products industry), the drying process emits biogenic CO_2 emissions directly. However, boiler systems, although burning woody biomass as fuel, still consume grid power to operate motors and thus emit fossil CO_2 emissions indirectly.

It is noted that in addition to the direct effects of burning fuel for energy on-site and grid electricity captured within an LCA, the indirect effects of its cradle-to-gate production are also considered. We can use a common fossil fuel, natural gas, as an example to illustrate indirect effects. When natural gas is consumed on-site to provide thermal (process) energy for the thermochemical conversion system, the emissions are direct emissions. Contrarily, when natural gas is consumed at a power plant to generate electricity which releases combustion exhaust emissions, the emissions are indirect because they happen off-site. In addition to the indirect emissions released at the power plant, the production of the natural gas along its own supply chain releases emissions. All of these "indirect" emissions are categorized within the LCI from tracking consumption of natural gas and electricity at the thermochemical conversion site. One approach is inputting the consumption of natural gas and electricity into commercial LCA modelling software that has libraries containing LCI data, including LCI data for electricity produced from natural gas and production of natural gas such as the US LCI Database (NREL, 2012) and the US version of the European database Ecoinvent (www.ecoinvent.ch) referred to as US-EI. Geographical location of the biochar production plant has a substantial effect on the environmental impacts, especially if the energy source for generating electricity has a high portion of fossil fuels such as coal and natural gas, which is common in the eastern United States (US EPA, 2015). Inputs include biomass, electricity and fossil fuels and outputs include CO_2 and particulate emissions.

3.2.4 Biochar Logistics

Once the biochar has been produced, it can be packaged and transported to the application site by a tractor trailer and applied to the soil in several different ways, including manually, by logging equipment, or by modified agricultural equipment. Application sequesters black carbon (biochar) on or within the soil, depending on application method. Inputs include diesel and outputs include fossil CO_2, volatile organic compounds, and particulate emissions.

3.2.5 Soil Carbon Sequestration

Soil carbon sequestration is the process of transferring CO_2 from the atmosphere into the soil through agricultural crop (i.e. corn or wheat stover) or forest (i.e. logging slash) residues, and other organic solids, including biochar (Lal, 2004). These systems can provide GHG mitigation by storing atmospheric CO_2 in live biomass, organic matter and in the mineral soil (DeLuca and Aplet, 2008; McKechnie et al., 2011). In addition, biomass-derived black carbon (biochar), which is produced from pyrolysis, offers a large and long-term C sink when applied to soils (Lehmann et al., 2006). Although large-scale application of biochar to soils in agricultural and forest systems is still in its infancy, the potential exists to provide environmental services that improve non-productive or degraded soils and sequester C (Ippolito et al., 2012). Although some biochars contain bioavailable C, it is generally more stable in soil than the C in the original biomass (Ippolito et al., 2012). While biochars will vary, those produced under moderate to high temperatures have stable C that will likely persist for hundreds of years (Ippolito et al., 2012). Stable C can be considered permanently sequestered after 100 years (Wang et al., 2014). There are limited studies on the impacts of biochar on GHG emissions, and field and lab studies often reach different conclusions. In a large laboratory trial testing 16 different biochars on three different soil types, Spokas and Reicosky (2009) suggest that the impacts of biochar additions on GHG emissions are both soil and biochar specific. Generally, most biochars reduced the rate of net CH_4 oxidation in the soil, decreased CH_4 production, and decreased N_2O production activity. In addition, several studies have shown that biochar-amended soil CO_2 losses are inversely related to the pyrolysis temperature (Brewer et al., 2012; Kammann et al., 2012; Yoo and Kang, 2012). Overall, biochar type, pyrolysis conditions, and environmental factors (e.g. temperature, soil moisture, soil organic matter content) all play a role in the changes in GHG emissions from biochar-amended soils (Ippolito et al., 2012). Laboratory screening of biochar for its potential to reduce GHG emissions should be followed by field testing to ascertain specific soil responses. Feedstocks used to produce biochar influence the physical, chemical, and biological characteristics of biochar and therefore, care must be taken to optimize feedstock selection and pyrolysis production techniques and conditions (Spokas et al., 2012). Biochar can have positive, negative, or neutral effects on plant growth. For example, hardwood biochar applied once to a desert soil in the western United States produced no changes in corn growth one year following application, but a 36% yield decline was noted in year two (Lentz and Ippolito, 2012). In a forest stand in central Ontario the short-term impact of adding biochar was an increased availability of calcium and phosphorus, and long-term impacts are expected to be achieved when the biochar becomes incorporated into the mineral soil (Sackett et al., 2014). As illustrated in Chapter 15, in the western USA, tree growth after biochar additions can also be positive or neutral, but to date no detectable negative effects have been noted.

Life Cycle Analysis of Biochar 57

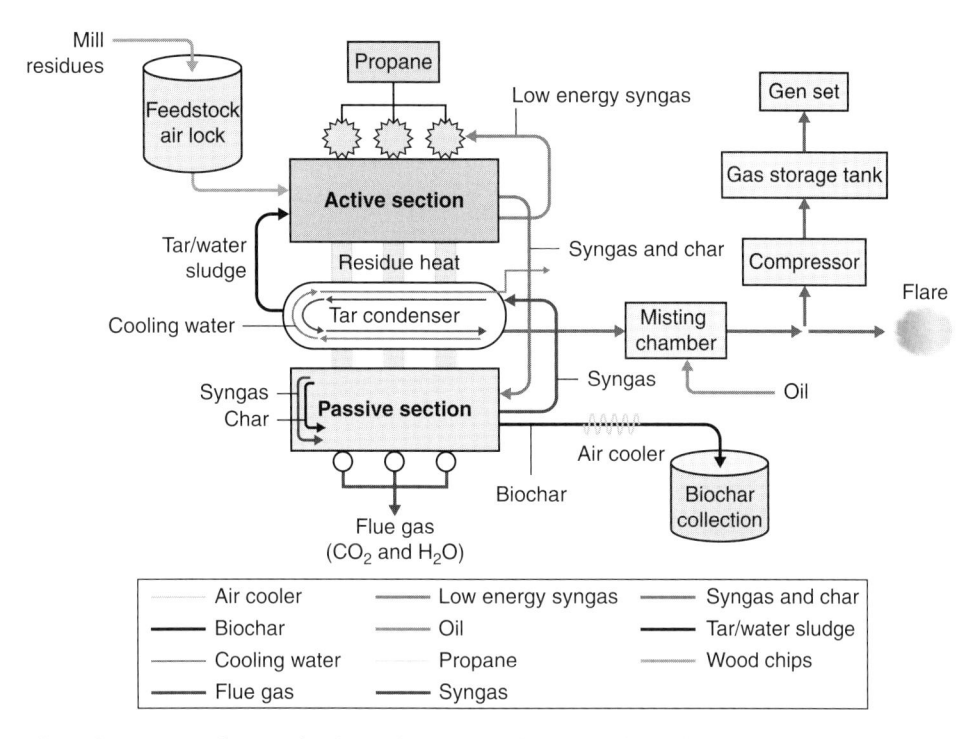

Figure 3.4. Process diagram for the Tucker Renewable Natural Gas unit. (A black and white version of this figure will appear in some formats. For the colour version, please refer to the plate section.)

3.3 LCA of an Advanced Pyrolysis System

A new thermochemical conversion technology, an advanced high-temperature pyrolysis system called the Tucker Renewable Natural Gas (RNG) thermal conversion unit, is under development by Tucker Engineering Associates (TEA), North Carolina, USA. The unit is designed to produce high yields of medium-energy syngas that can be used in heat and power applications, or be converted to liquid fuels by catalysis. The system produces a biochar co-product at 10–20% yield by dry input weight. This biochar can be used in its raw form, or activated by steam or chemicals to make activated carbon (AC) for liquid and gas filtering applications. In some uses, renewable bio-based AC would substitute for AC made from fossil coal. Figure 3.4 shows the system process for the Tucker RNG unit.

3.3.1 Operation

3.3.1.1 Feedstock Production and Logistics

Logs harvested in Montana, USA were processed into wood chips at a western Montana sawmill. The chips are a co-product of the mill's lumber production operations or produced directly by chipping poor-quality whole trees. An 812 kW$_e$ chipper was used for whole-tree chipping, while a 108 kW$_e$ screener operated in conjunction to produce the specified size.

Table 3.1. *Energy inputs for feedstock processing of 1.0 ovendry kg wood chips*

Source	Quantity	Unit
Electricity, chipping	0.154	kWh
Electricity, screening	0.0205	kWh
Electricity, chipping	0.0224	kWh
Thermal energy, drying[1]	3.74	MJ

[1] From wood fuel, 20.9 megajoules (MJ) per oven-dry kg.

These chips were then dried in a sawdust dryer to a moisture content of about 10% to meet the Tucker RNG unit system requirements. The sawdust dryer was fueled by a bark and wood fuel mixture during the drying operation, which released biogenic CO_2 emissions. Forest harvesting activities and log transportation were included in the analysis. Primary data for the whole-tree chipping, screening and drying processes were collected directly from the mill to help develop the LCI flows (Table 3.1). The analysis assumed co-location of the feedstock (chip) supplier and the advanced pyrolysis unit as this is the intent of the overall project.

3.3.1.2 Advanced Pyrolysis

The Tucker RNG unit is an advanced pyrolysis system comprised of active and passive sections (i.e. chambers). Feedstock logistics is embedded within the thermochemical conversion stage. The unit is engineered to maximize syngas output in a very low-oxygen reaction chamber at a high temperature between 760°C and 870°C. At these temperatures, the system is endothermic, requiring net inputs of energy (propane) to maintain the reaction (Table 3.2). Three propane burners provide continuous active heating for the reaction. The residence time for biomass feedstock in the Tucker RNG unit is estimated at three minutes for the complete reaction, with equal 1.5 minutes residence time in each section. Wood chip feedstock is sent through an air-locked auger system into the active section for high temperature heating. After passing through the active section, the partially converted biochar and hot syngas are transferred in an enclosed auger to the passive section, which uses the residual heat transferred from the active section through a vent system for additional pyrolyzing. After transferring heat from the combustion exhaust gases to the passive section, the exhaust gases from burning propane are released directly to the air. The biochar moves through augers inside the passive section of the Tucker unit, whereby additional conversion from higher molecular gases into methane occurs. The temperatures measured at the passive heating section are between 510°C and 760°C.

Syngas leaving the passive section is cooled in a tar condenser to help remove impurities (i.e. tars). The tar condenser has a mechanism to remove buildup of tar from the condensing of tars caused by the cooling of the syngas. After cooling, the medium-energy syngas goes through a misting chamber that removes oil and tars before leaving the Tucker RNG unit and going into an outside storage tank. The two primary products from the system – biochar and

Table 3.2. *Energy inputs for the advanced pyrolysis unit per 1.0 kg wood chips (8.19% water by weight)*

Source	Quantity	Unit
Propane, reaction	0.127	kg
Electricity, motors	0.010	kWh

medium-energy syngas – are collected at separated outlets. The syngas is intended to be combusted for electricity on-site. In this system, biochar is intended to be activated with steam to make AC, but can also be used in its raw form as a soil amendment or a coal replacement. Pyrolysis often produces residual tars, which can be a useful output or an undesirable waste product depending on production objectives. In the Tucker RNG system, the tar can be retorted back to the active heating chamber to produce a low-energy syngas for use as a propane substitute at about 30% of heating demand as shown in Figure 3.4. However, in this study, the tar/water mixture was considered a waste in the analysis and was collected. Therefore, no low-energy energy syngas was generated.

3.3.1.3 Two Product Components

Synthesis gas. The advanced pyrolysis system generates syngas, a medium-energy type. The medium-energy syngas will be burned to generate electricity for the grid. Medium-energy syngas will be referred to as syngas for the remainder of the chapter. The density of syngas is calculated at 1.08 kg/m^3. The higher heating value (HHV) was measured at 19.5 MJ/m^3 and the lower heating value (LHV) at 18.0 MJ/m^3. Electricity is intended to be produced from burning the Tucker RNG syngas in a commercial 1.6 MW$_e$ Caterpillar generator derated to 1.2 MW$_e$ because of the syngas's relatively low energy density compared to natural gas. Currently, the Tucker RNG unit will need to produce about two times the amount of syngas to generate the same electricity as natural gas does, since the HHV of the produced syngas is one half of the natural gas HHV, 38.3 MJ/m^3. The main components by mass of the syngas are CO (48%), CO_2 (11%), and CH_4 (15%).

Biochar. The pyrolysis unit also generates a solid product, biochar, but at a much smaller portion. Biochar on a dry basis has the following properties: (1) a fixed carbon content of about 89% and (2) an energy content of 32.1 and 31.9 MJ/kg for HHV and LHV, respectively. The energy content for biochar is about 50% higher on a dry basis than wood (Ince, 1979; FPL, 2004).

3.3.2 Four Phases of Tucker RNG Unit Life Cycle Assessment

3.3.2.1 Goal and Scope

The goal was to evaluate the critical environmental impacts of the bioenergy (syngas electricity) and bio-product (AC) converted from forest or mill residues using an advanced pyrolysis system (Figure 3.4). The scope of the study was to cover the cradle-to-grave life cycle of generating syngas electricity and AC by the advanced pyrolysis system and make comparison with fossil fuel alternatives. As previously noted, biochar is a precursor to making AC. However, the focus of the present analysis only covered biochar production and not AC production. The functional unit was 1.0 ovendry (OD) kg of incoming wood chips. OD units do not indicate that the feedstock was dried to 0% moisture content, but rather are used as a standardized unit that facilitates comparisons between feedstocks with different moisture contents.

3.3.2.2 Life Cycle Inventory

Mass balance. A mass balance was performed and verified data quality provided during a production run (Table 3.3). Thermochemical conversion turned the feedstock into syngas (65.4%), biochar (13.9%) and tar/water mixture (20.7%) by mass. The tar/water mixture is primarily water. Although the pyrolysis unit currently produces the tar/water mixture that could be converted to a low-energy syngas, this gas was not used as a propane substitute in the present analysis. Therefore, the low-energy syngas via residual tars is considered a waste under this study's LCA framework. Thus, the only products that have environmental inputs and environmental outputs (i.e. LCI flows) assigned to them are the syngas and biochar. These allocations can occur either by mass or energy. Allocations are 82.5% and 17.5% by mass and 72.8% and 27.2% by energy for the medium-energy gas and the biochar, respectively.

Cumulative energy consumption. Evaluating products for their cumulative energy consumption can be conducted through an LCA. Table 3.4 shows the cradle-to-gate cumulative energy of 16.6 MJ consumed from pyrolyzing 1.0 OD kg of incoming wood chips to produce syngas and biochar. In addition, Table 3.4 shows the various fuels that contribute to this 16.6 MJ value. Propane was the major contributor at 44.1%, and wood was second at 22.7%. Propane was burned to maintain the high temperatures during pyrolysis, while wood was burned to generate thermal energy to dry the incoming green feedstock.

Emissions to air and water. Table 3.5 shows some of the cradle-to-gate environmental outputs (e.g. emissions to the air and water) from wood pyrolysis. Fossil CO_2 emissions of 542 kg CO_2/OD kg of incoming wood chip came mostly from propane burning to maintain the endothermic reaction. Biogenic CO_2 emissions of 330 kg CO_2/OD kg came from burning wood residues as the heating source for the boiler used to dry the wood chips (i.e. green incoming feedstock). The total emission of each item is allocated to the two primary products based on the mass ratio of the two. Note that the total environmental outputs for the system listed in the last column in Table 3.5 will not change regardless of the allocation procedure used.

3.3.3 Life Cycle Impact Assessment for Syngas Electricity

Syngas produced from the Tucker RNG unit is intended to fuel an internal combustion generator to provide electricity to the power grid. Based on generating 0.732 kg (0.676 m^3) of syngas from 1.0 OD kg of incoming wood chips, 1.26 kilowatt hours (kWh) of electricity was generated.

Table 3.3. *Mass and energy balance for pyrolization of 1.0 kg of wood chips (8.19% water by weight)*

Source	Dried wood[1]	Synthesis gas[2]	Biochar[3]	Tar sludge[3]
Dry matter (kg)	0.918	0.654	0.139	<.01
Water (kg)	0.082	–	–	0.207
Mass allocation (%)	100	82.5	17.5	0.0
Higher heating value (MJ/kg)	18.41	17.96	31.74	10.54
Total energy (MJ)	18.41	11.75	4.40	2.18
Energy content (%)	100	64.1	24.0	11.9
Energy allocation (%)	100	72.8	27.2	0.0

[1] As measured from wood chips with 8.19% moisture (wt).
[2] Syngas energy value obtained from gas chromatography per ASTM D3588 (2011)/D1945 (2014) standards.
[3] Energy values for biochar and tar oil/water were obtained from the proximate analysis.

Table 3.4. *Cradle-to-gate cumulative energy consumption from pyrolyzing 1.0 ovendry kg wood chips*

Fuel	Unit	Quantity	Higher heating values (MJ/m^3)	Higher heating values (MJ/kg)	Energy (MJ)	Energy (%)
Natural gas (proxy for propane)	m^3	0.1898	38.4		7.288	44.1
Wood residue (ovendried)	kg	0.180		20.9	3.759	22.7
Natural gas	m^3	0.054	38.4		2.068	12.5
Crude oil	kg	0.04		45.5	1.811	11.0
Coal	kg	0.055		26.4	1.461	8.83
Nuclear	kg	3.66E-07		332000	0.121	0.73
Biomass	MJ	0.021			0.021	0.13
Hydro	MJ	0.014			0.014	0.08
Wind	MJ	0.0008			0.0008	0.005
Total					16.6	100

Table 3.5. *Cradle-to-gate environmental outputs from pyrolyzing 1.0 ovendry kg wood chips, mass allocation*

Substance	Unit	Syngas	Biochar	Total
Air emission				
Carbon dioxide, fossil	g	447	94.9	542
Carbon dioxide, biogenic	g	272	57.8	330
Sulphur dioxide	g	3.82	0.81	4.64
Methane	g	1.77	0.38	2.15
Nitrogen oxides	g	1.18	0.25	1.43
Carbon monoxide	g	0.83	0.18	1.01
Particulates, > 2.5 µm, and < 10 µm	g	0.73	0.16	0.89
Carbon monoxide, fossil	g	0.61	0.13	0.74
Methane, fossil	g	0.34	0.07	0.41
VOC, volatile organic compounds	g	0.13	0.03	0.15
Water effluent				
Suspended solids, unspecified	g	26.93	5.72	32.65
Chloride	g	21.50	4.57	26.07
Sodium	g	6.069	1.29	7.35
BOD5, biological oxygen demand	g	2.81	0.60	3.41
Calcium	g	1.91	0.41	2.32
Lithium	g	0.614	0.13	0.74
COD, chemical oxygen demand	g	0.17	0.04	0.21
Industrial waste				
Bark	g	1.19	0.253	1.44
Tar	g	35.1	7.5	42.6

Table 3.6. *Cradle-to-gate life cycle impact assessment of generating 1 kWh of syngas electricity*

Impact category	Unit	Quantity
Global warming	kg CO2 eq	0.525
Ozone depletion	kg CFC-11 eq	5.09E-08
Smog	kg O3 eq	0.081
Acidification	kg SO2 eq	0.006
Eutrophication	kg N eq	3.64E-04
Carcinogenics	CTUh	7.05E-08
Non-carcinogenics	CTUh	3.88E-08
Respiratory effects	kg PM2.5 eq	4.30E-04
Ecotoxicity	CTUe	0.699
Fossil fuel depletion	MJ surplus	1.158

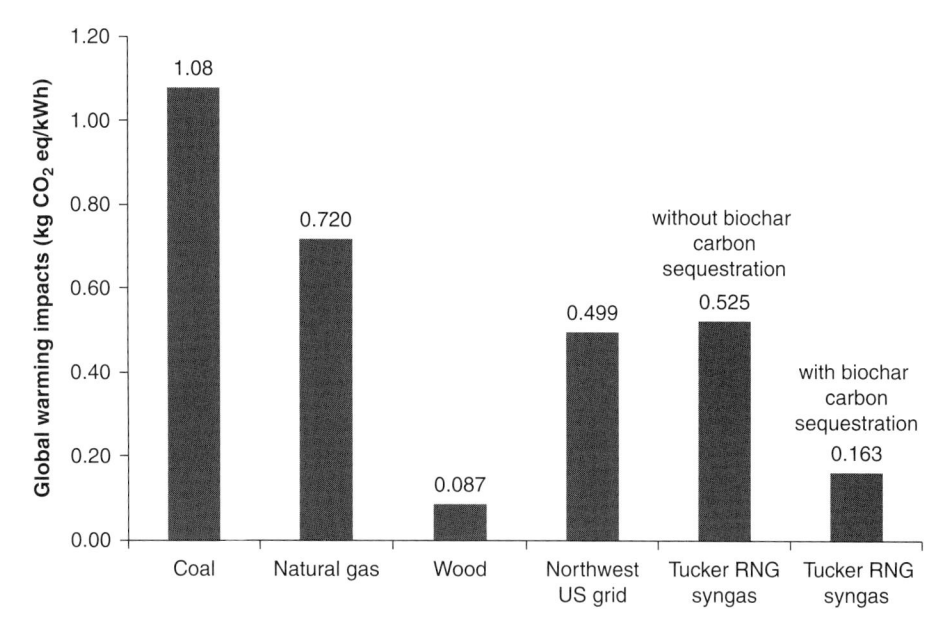

Figure 3.5. Global warming impacts of producing cradle-to-gate electricity in the United States.

For comparison, a wood power plant burning logging slash generates 1.14 kWh/OD kg (Bergman et al., 2013). Table 3.6 shows the values for the ten impact categories to produce 1 kWh of syngas electricity. A global warming (GW) impact of 0.525 kg CO$_2$-e/kWh of syngas electricity was estimated without biochar carbon sequestration being considered. In the context of the LCA for this section, biochar is considered a byproduct and thus carries no environmental burdens (Gu and Bergman, 2015).

In Table 3.6, note all environmental impacts were applied to the syngas electricity and none to the biochar. This was because the biochar will be applied to the soil for carbon sequestration. This means all the impacts tied to feedstock production and logistics and thermochemical conversion life cycle stages were assigned to the syngas electricity. Furthermore, the system boundary stopped at the gate of the conversion facility so the analysis did not include the impacts of transporting or applying the biochar in the field or in the forest. Of course, because of the additional fuel needed to transport and apply the biochar, adding these impacts would increase the quantity of GHG emissions released.

To calculate the permanent carbon sequestration benefit, the stable C portion of biochar was estimated at 80% at the end of 100 years. In Figure 3.5, GW impacts for the various electricity sources were calculated using LCA modeling software. The GW impact from the cradle-to-gate production of syngas electricity showed a notably lower value (0.163 kg eq CO$_2$/kWh) compared to electricity generated from bituminous coal (1.08 kg eq CO$_2$/kWh) and conventional natural gas (0.720 kg eq CO$_2$/kWh) when including carbon sequestration from biochar. Regarding the outcomes reported here, it is important to note the potential for change in future analysis because LCI databases are being constantly updated.

3.3.4 Interpretation

The LCA on the Tucker RNG unit provides insight into its environmental performance as well as that of other pyrolysis systems. There are several notable items described in the next three paragraphs.

First is the high fossil CO_2 emissions related to advanced pyrolysis (i.e. thermochemical conversion). Quantifying GW showed both the carbon benefits (e.g. low GHG emissions) and the carbon "hotspots" such as from burning propane to maintain the endothermic reaction in the Tucker RNG unit. If reducing or substituting propane usage in the Tucker RNG unit is possible, the GW impact could be further reduced. During the pyrolysis conversion in the Tucker RNG system, low-energy (waste) syngas was produced without being collected for use. We anticipate collecting and using this low-energy (waste) syngas to supplement propane usage would further reduce GHG emissions (i.e. fossil CO_2) associated with syngas electricity. Therefore, we conducted a scenario analysis with 30% propane reduction with the substitute of now-unused low-energy syngas produced from the Tucker RNG unit. The GW impact improved by 20% in total for the cradle-to-grave syngas electricity (Gu and Bergman, 2015).

Second, the interplay between primary products and co-products is important. As discussed previously, the difference between a waste, byproduct, and co-product is variable by discipline, but LCA provides a definition based on assigning environmental impacts: waste products have disposal costs, byproducts have marginal costs and marginal value relative to primary products, and co-products are manufactured jointly and use joint product costing in accounting. The complication with this definition is that the same material can be a waste, a byproduct, or a co-product depending on its value and costs, but it is useful to draw a clear line between waste as a material with net costs, especially for disposal, and production outputs that have market value and the potential to generate revenue. Regardless, all emissions have to be accounted for allocation.

Third, this system, which was engineered to produce high-quality biochar from a broad range of waste feedstocks, has relatively high environmental burdens for electricity compared to wood combustion, as noted in Figure 3.5, even though GW from syngas electricity was substantially lower than the other forms of electricity.

Furthermore, if producing high-quality biochar for field application is the main objective, there are many types of thermochemical conversion technologies that use less energy to create this form of biochar, many of which are exothermic and do not require energy inputs to maintain pyrolysis. However, in this case additional processing was performed for the Tucker RNG unit because the biochar produced by this system is meant to be used as a precursor for producing AC, thus more energy was required to meet the specific processing requirements of the AC. Perhaps most important, using biochar products as a soil amendment can significantly improve the GW profile of bioenergy technologies (Figure 3.5). Sequestering the biochar co-product in the soil as a GHG sink definitely lowers system impacts on climate change compared to other options, such as using it as a fuel (Gaunt and Lehmann, 2008; Roberts et al., 2010).

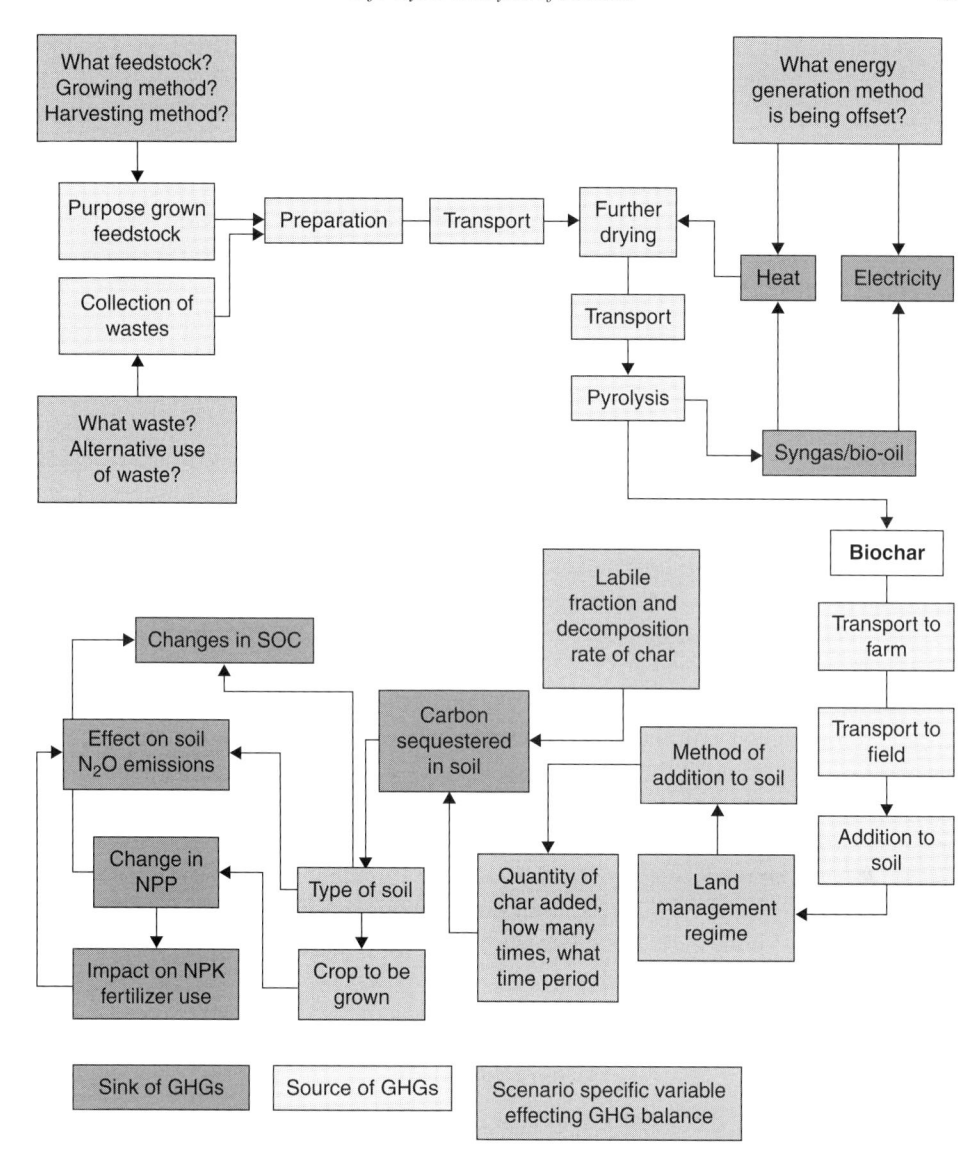

Figure 3.6. Diagram showing all factors and processes to be considered in a pyrolysis life cycle assessment. *Source*: Hammond et al. (2011): permission to use granted by Elsevier through Copyright Clearance Center's Rightslink service. (A black and white version of this figure will appear in some formats. For the colour version, please refer to the plate section.)

To fully analyze the environmental impacts of the biochar to be used as a soil amendment, a more detailed analysis across multiple potential use scenarios needs to be performed. Figure 3.6 from Hammond et al. (2011) provides an excellent framework for exploring a more detailed LCA. For example, the Tucker RNG unit LCIA results included only the direct carbon sequestration effect of applying biochar to the soil, whereas there are several indirect effects, as noted earlier, that should be considered, such as changes in net primary productivity and soil organic carbon, soil N_2O emission suppression, and fertilizer utilization. Indirect effects attributable to efficiency gains and various product substitutions, especially fossil fuel, can then be incorporated into the LCA, as described in more detail below.

As with SCM considerations, LCA considerations for biochar used primarily to meet climate change mitigation objectives can be more complicated than the other end uses discussed in Chapter 2. Gaunt and Cowie (2009) identified six specific characteristics of biochar application that can result in net reductions of GHG emissions attributable to biochar systems: (1) sequestration of moderately stable carbon in the soil; (2) avoided emissions of carbon dioxide and methane related to alternative disposal methods such as biomass combustion and decomposition; (3) suppression of methane and nitrous oxide emissions related to changes in soil processes especially for intensively fertilized, irrigated cropland; (4) displacement of carbon-intensive agricultural inputs through direct substitution and increased plant efficiency; (5) carbon sequestration resulting from higher productivity leading to greater soil carbon accumulation; and (6) displacement of fossil fuels from biochar co-products. Only the first one, carbon sequestration in the soil, is a direct effect. The other benefits, though supported by research, are indirect and rely on assumptions about the changes in soil processes and characteristics, fate of waste biomass, and market substitutions for fertilizer, fossil fuels and other carbon-intensive inputs.

Last, in conjunction with end uses and what was stated above, most biochar research has focused on short-term impacts of biochar applications on soil chemical, physical, and biological properties. However, future work on biochar additions to forest sites should focus on long-term field research that determines changes to nutrient availability, microbial community changes, net GHG emissions, and net C sequestration. Furthermore, to produce a sustainable supply of biochar derived from wood, sustainable production of the feedstock (i.e. raw material) itself must be considered. In the next section of this book, the authors will discuss potential sustainable feedstocks for pyrolysis.

Acknowledgements

Funding for much of the research and analysis described in this chapter was provided to the authors by the US Department of Agriculture (USDA) National Institute of Food and Agriculture Biomass Research and Development Initiative (BRDI), award no. 2011-10006-30357, and is gratefully acknowledged. BRDI is a joint effort between the USDA and the US Department of Energy.

References

ASTM (2011). ASTM D3588-98, Standard Practice for Calculating Heat Value, Compressibility Factor, and Relative Density of Gaseous Fuels, ASTM International, West Conshohocken, PA.

ASTM (2014). ASTM D1945-14, Standard Test Method for Analysis of Natural Gas by Gas Chromatography, ASTM International, West Conshohocken, PA.

Bare, J. C. (2011). TRACI 2.0: the tool for the reduction and assessment of chemical and other environmental impacts 2.0. *Clean Technologies and Environmental Policy*, 13, pp. 687–696.

Bergman, R. D. and Bowe, S. A. (2011). Life-cycle inventory of manufacturing prefinished engineered wood flooring in the eastern United States with a comparison to solid strip wood flooring. *Wood and Fiber Science*, 43, pp. 421–441.

Bergman, R. D. and Bowe, S. A. (2012). Life-cycle inventory of hardwood lumber manufacturing in the southeastern United States. *Wood and Fiber Science*, 44, pp. 71–84.

Bergman, R. D., Han, H-S., Oneil, E. and Eastin, I. (2013). Life-cycle assessment of redwood decking in the United States with a comparison to three other decking materials. University of Washington. Seattle, WA. Consortium for Research on Renewable Industrial Materials.

Bergman, R. D. and Gu, H. (2014). Life-cycle inventory analysis of bio-products from a modular advanced biomass pyrolysis system. In: *Proceedings, Society of Wood Science and Technology 57th International Convention*, 23–27 June 2014. Zvolen, Slovakia, pp. 405–415.

Brewer, C. E., Hu, Y. Y., Schmidt-Rohr, K., Loynachan, T. E., Laird, D. A. and Brown, R. C. (2012). Extent of pyrolysis impacts on fast pyrolysis biochar properties. *Journal of Environmental Quality*, 41, pp. 1115–1122.

DeLuca, T. H. and Aplet, G. H. (2008). Charcoal and carbon storage in forest soils of the Rocky Mountain West. *Frontiers in Ecology and Environment*, 6, pp. 18–24.

Dutta, B. and Raghavan, V. (2014). A life cycle assessment of environmental and economic balance of biochar systems in Quebec. *International Journal of Energy and Environmental Engineering*, 5, pp. 1–11.

Ekvall, T. and Andrae, A. S. G. (2006). Attributional and consequential environmental assessment of the shift to lead-free solders. *International Journal of Life Cycle Assessment*, 11, pp. 344–353.

Ekvall, T. and Weidema, B. P. (2004). System boundaries and input data in consequential life-cycle analysis. *International Journal of Life Cycle Assessment*, 9, pp. 161–171.

FPL (2004). Fuel value calculator. Techline WOE-3. Madison, WI: United States Department of Agriculture, Forest Service, Forest Products Laboratory. [online] Available at: www.fpl.fs.fed.us/documnts/techline/fuel-value-calculator.pdf. [Accessed 12 March 2015].

Gaudreault, C., Samson, R. and Stuart, P. R. (2010). Energy decision making in a pulp and paper mill: selection of LCA system boundary. *International Journal of Life Cycle Assessment*, 15, pp. 198–211.

Gaunt, J. L. and Lehmann, J. (2008). Energy balance and emissions associated with biochar sequestration and pyrolysis bioenergy production. *Environmental Science & Technology*, 42, pp. 4152–4158.

Gaunt, J. L. and Cowie, A. (2009). *Biochar for Environmental Management: Science and Technology*. Chapter 18. Biochar, greenhouse gas accounting, and emissions trading. London: Earthscan.

Gu, H. and Bergman R. (2015). Life-cycle GHG emissions of electricity from syngas by pyrolyzing woody biomass. In: *Proceedings, Society of Wood Science and Technology 58th International Convention*, 7–12 June 2015. Jackson Hole, WY, pp. 376–389.

Guo, Z., Sun, C. and Grebner, D. L. (2007). Utilization of forest derived biomass for energy production in the USA: status, challenges, and public policies. *International Forestry Review*, 9, pp. 748–758.

Hammond, J., Shackley, S., Sohi, S. and Brownsort, P. (2011). Prospective life cycle carbon abatement for pyrolysis biochar system in the UK. *Energy Policy*, 39, pp. 646–655.

Homagain, K., Shahi, C., Luckai, N. and Sharma, M. (2014). Biochar-based bioenergy and its environmental impact in Northwestern Ontario Canada: a review. *Journal of Forestry Research*, 25, pp. 737–748.

ILCD (2010). *ILCD System Handbook – General Guide for Life Cycle Assessment – Detailed Guidance. EUR 24708 EN*. European Commission – Joint Research Centre – Institute for Environment and Sustainability, Luxembourg. *International Reference Life Cycle Data System*. Publications Office of the European Union.

Ince, P. J. (1979). How to estimate recoverable heat energy in wood or bark fuels. *General Technical Report FPL–GTR–29*. Madison, WI: United States Department of Agriculture, Forest Service, Forest Products Laboratory.

Ippolito, J. A., Laird, D. A. and Busscher, W. J. (2012). Environmental benefits of biochar. *Journal of Environmental Quality*, 41, pp. 967–972.

ISO (2006a). *Environmental Management – Life-cycle Assessment – Principles and Framework*. ISO 14040. International Organization for Standardization, Geneva, Switzerland.

ISO (2006b). *Environmental Management – Life-cycle Assessment – Requirements and Guidelines*. ISO 14044. International Organization for Standardization, Geneva, Switzerland.

Kammann, C., Ratering, S., Eckhard, C. and Müller, C. (2012). Biochar and hydrochar effects on greenhouse gas (carbon dioxide, nitrous oxide, methane) fluxes from soils. *Journal of Environmental Quality*, 41, pp. 1052–1066.

Lal, R. (2004). Soil carbon sequestration impacts on global climate change and food security. *Science*, 304, pp. 1623–1627.

Lehmann, J., Gaunt, J. and Rondon, M. (2006). Bio-char sequestration in terrestrial ecosystems – a review. *Mitigation and Adaptation Strategies for Global Change*, 11, pp. 403–427.

Lentz, R. D. and Ippolito, J. A. (2012). Biochar and manure affect calcareous soil and corn silage nutrient concentrations and uptake. *Journal of Environmental Quality*, 41, pp. 1033–1043.

Loeffler, D. R. and Anderson N. M. (2014). Emissions tradeoffs associated with cofiring forest biomass with coal: a case study in Colorado, USA. *Applied Energy*, 113, 67–77.

Lorenz, K. and Lal, R. (2014). Biochar application to soil for climate change mitigation by soil organic carbon sequestration. *Journal of Plant Nutrition and Soil Science*, 177, pp. 651–670.

McKechnie, J., Colombo, S., Chen, J., Mabee, W. and Maclean, H. L. (2011). Forest bioenergy or forest carbon? Assessing trade-offs in greenhouse gas mitigation with wood-based fuel. *Environmental Science & Technology*, 45, pp. 789–795.

NETL (2013). *NETL Life Cycle Inventory Data – Unit Process: Burning Crowns in Slash Piles*. US Department of Energy, *National Energy Technology Laboratory*. Last Updated: March 2013 (version 01). [online] Available at: www.netl.doe.gov/LCA [Accessed 23 July 2015].

NREL (2012). Life-cycle inventory database project. *National Renewable Energy Laboratory*. [online] Available at: www.lcacommons.gov/nrel/search [Accessed 23 July 2015].

Pennington, D. W., Potting, J., Finnveden, G., et al. (2004). Life-cycle assessment part 2: current impact assessment practice. *Environment International*, 30, pp. 721–739.

Pierobon, F., Ganguly, G., Anfodillo, T. and Eastin, I. L. (2014). Evaluation of environmental impacts of harvest residue-based bioenergy using radiative forcing analysis. *The Forestry Chronicle*, 90, pp. 577–585.

Roberts, K. G., Gloy, B. A., Joseph, S., Scott, N. R. and Lehmann, J. (2010). Life cycle assessment of biochar systems: estimating the energetic, economic, and climate change potential. *Environmental Science & Technology*, 44, pp. 827–833.

Sackett, T. E., Basiliko, N., Noyce, G. L., et al. (2014). Soil and greenhouse gas responses to biochar additions in a temperate hardwood forest. *Bioenergy*, 7, 1062–1074.

Spokas, K. A. and Reicosky, D. C. (2009). Impacts of sixteen different biochars on soil greenhouse gas production. *Annals of Environmental Science*, 3, pp. 179–93.

Spokas, K. A., Cantrell, K. B., Novak, D. A., et al. (2012). Biochar: a synthesis of its agronomic impact beyond carbon sequestration. *Journal of Environmental Quality*, 41, pp. 973–989.

Thomassen, M. A., Dalgaard, R., Heijungs, R. and de Boer, I. (2008). Attributional and consequential LCA of milk production. *International Journal of Life Cycle Assessment*, 13, pp. 339–349.

US EPA (1995). *Compilation of Air Pollutants Emission Factors. Vol. I. Stationary Point and Area Sources*. Washington, DC: US Environmental Protection Agency.

US EPA (2015). eGRID 2012 Summary Tables. Available at: www.epa.gov/sites/production/files/2015-10/documents/egrid2012_summarytables_0.pdf [Accessed 11 August 2016].

Wang, Z., Dunn, J. B., Han, J. and Wang, M. Q. (2014). Effects of co-produced biochar on life cycle greenhouse gas emissions of pyrolysis-derived renewable fuels. *Biofuels, Bioproducts and Biorefining*, 8, pp.189–204.

Woolf, D., Amonette, J. E., Street-Perrott, F. A., Lehmann, J. and Joseph, S. (2010). Sustainable biochar to mitigate global climate change. *Nature Communications*, 1, Article No. 56.

Yoo, G. and Kang, H. (2012). Effects of biochar addition on greenhouse gas emissions and microbial responses in a short-term laboratory experiment. *Journal of Environmental Quality*, 41, pp. 1193–1202.

4

Systems Integration for Biochar in European Forestry: Drivers and Strategies

SARAN P. SOHI AND TOM KUPPENS

Abstract

Biochar is a product with multiple functions and a range of uses, and which could be manufactured from a range of biomass types, including wood and forest residues. Economic production and use of biochar may take place within spatial boundaries that contrast greatly in total area and patchiness. At the moment, however, biochar is generally considered at a project level and assessed quite narrowly from the perspective of soil or crop effects or energy yield. This chapter presents a set of scenarios for how biomass from trees grown in different contexts might lead to production of biochar used and deployed for contrasting purposes in multiple markets. The integration required across markets and sectors is considered, distinguishing options that are more or less spatially contained (circular versus directional), with implications for environmental sustainability. Four types of forestry are considered as a source of biomass: brownfield sites, short rotation forestry, short rotation coppice, and trees on amenity land. As well as forestry, horticulture, agriculture and the urban landscape are considered as consumers of biochar, often as a formulated product. The scenarios emphasize the range of opportunities that may be available, but also the complexity of the systems fit, which includes aspects of spatial logistics and questions of scale.

4.1 Introduction

Biochar is often presented in a project context. A project tends to focus on a homogenous, contiguous area of land with defined spatial boundaries and a specific geographic location. Projects may draw on resources outside these boundaries, but these are seen simply as inputs. The outputs of the project are the products of the land, focused on the harvested crop. In contrast, systems have diffuse conceptual boundaries that span the supply chain. In a systems context, biochar use is viewed with consideration to biomass production and biomass conversion, sometimes extending to the markets for products from the systems where biochar might be used.

Pyrolysis equipment technology will be expensive and biochar production also incurs operating costs. Existing analysis suggests that the profitability of investment will be highly scale-dependent (Kuppens et al., 2015). Moreover, adding biochar to soil is a relatively radical intervention that requires care and precision. For these reasons, the technology will best fit use on actively managed land. In agriculture, food dominates. The value chain in forestry differs markedly, but has been less intensively studied in the context of biochar. This chapter focuses on systems that involve the potential integration of biochar into a range of forestry contexts, from conventional plantations and short-rotation forestry to coppice as a woody agricultural crop. The scenarios put forward are necessarily illustrative: a set of theoretical situations where biochar is potentially viable.

The full range of biophysical and socioeconomic contexts that encompass these scenarios will lead to patchiness in adoption of the technology and its proliferation. Extrapolating such patterns across sectors should allow its realisable potential of the technology to be distinguished from the technical and theoretically sustainable potential (Woolf et al., 2010). A full systems analysis should draw on social science to capture the true nature of decision making among individuals and institutions. It will then be possible to re-assess technologies in the light of proposed new economic or social policies. Systems analysis should not consider a single technology innovation in isolation: as well as the dynamic interaction between multiple components of the wider system and their co-dependence, parallel innovations and interventions should be considered. These will offer both competition and potential synergies.

4.2 Contexts for Biochar in Forestry

4.2.1 Bioeconomy

The benefits of directly adding plant biomass into soil to build soil quality is now well recognised. Agricultural soils have been improved as a result of policies to, for example, ban open burning of crop straw in the 1980s (Nicholson et al., 2014). But such policies also initiate investment into alternative uses for a resource previously viewed as waste, such as facilities for energy recovery – notably straw-burning power stations (Edwards et al., 2005). Growing agricultural crops specifically as fuel or for other non-food biomass yield has become a recognised modern-day option for land use (Sheppard et al., 2011). The adoption of perennial non-tree crops such as *Miscanthus* has blurred the boundaries between arable- and pasture-orientated agriculture, as well as the focus on food versus non-food products. At the same time the development of short-rotation cropping practices for trees has blurred the boundary between agriculture and forestry, with commercial planting of *Salix* (willow) and *Populus* (poplar) on agricultural land (Defra, 2007).

A range of new processes have been investigated to gain non-energy product value from biomass, whether harvest has an energy focus or not (de Jong et al., 2012).

This leads to the biorefinery concept, where synergies are sought from the co-location of different processing options applied to multiple feedstock (Cherubini, 2010; Van Dael et al., 2014). A specific end-use no longer defines a crop: each biomass will be used for a range of purposes and products, some stimulated by environmental drivers such as the principle of cascading (Keegan et al., 2013). The optimal crop mix at a particular location will be driven by a combination of markets and biophysical conditions. In some situations farming could be economic even where the main crop is non-food.

In contrast to agriculture, forestry has traditionally been orientated towards markets that are non-food, that is the supply of constructional timber, wood-based building materials, paper pulp and also fuel. The current EU Forest Strategy (European Commission, 1999) emphasises the synergies that exist between forestry and other sectors, highlighting their co-dependence. The emerging new strategy has also proposed that current and emerging policies on issues that affect forests should be made more coherent (European Commission, 2013a). Commonalities that are emerging between forestry and agriculture should foster consolidation of research and development. Technologies that facilitate the conversion of woody biomass into liquid fuels is an obvious example (Margeot et al., 2009); use of wood fibre in growing media is one that specifically links back to crop production (Carlile et al., 2015).

There are major logistical, practical and economic challenges associated with bio-refining that challenge more elaborate visions (Kokossis and Yang, 2010). Pyrolysis, including biochar production and use, has potential as an entry point for bio-refining and for catalysing a bioeconomy (Laird et al., 2009).

4.2.2 Definition, Management and Policy

Forests are generally understood as large areas of land where the vegetation is dominated by trees, but official definitions vary (Sasaki and Putz, 2009). In the Global Forest Resources Assessment of the UN Food and Agriculture Organization (FAO), units of land as small as 0.50 ha may be declared forest; areas of sparse cover (as low as 10% canopy density) are also categorised this way (FAO, 2006). The FAO definition of forests does not align precisely with that used under the UN Framework Convention on Climate Change (UNFCCC). This is important if valorisation of carbon sequestration and storage is sought, since it affects methodologies that support carbon storage schemes (which are land use specific). Under the UNFCCC, national reporting can include stands of trees as low as 2 m height and areas of trees as small as 10 m × 10 m (0.01 ha). The UNFCCC distinguishes areas of tree canopy density (>40%) as closed forest from 'open' forests, which are commonly also referred to as woodland (UNFCCC, 2002).

Some definitions require or imply that trees in forests are managed for their productivity in some way. Differentiating plantation forest can be useful in this context as plantations involve forests where trees are not only harvested, but also cultivated. This implies sustainability of forest cover through replanting, but also active intervention in the natural function of the land and soil. Indeed, plantations can be established on land that would historically have not supported tree growth or on degraded land, for instance for immobilization or extraction of heavy metals (Vangronsveld et al., 2009). However, the FAO definition of forestry does not depend on tree harvesting. It does, however, exclude land where trees are grown specifically for their non-wood products – orchards and areas of 'agricultural trees' managed for fruit, nuts or seeds (FAO, 2006). Areas where trees are mixed with traditional agriculture (agroforestry) are also excluded regardless of their use and even if average canopy cover exceeds 10%. The FAO definition similarly excludes trees in urban parks; owing to their amenity focus, the presence of trees is in some way incidental. However, some reporting to UNFCCC is based on national definitions that can include parkland trees as well as other urban forest, plantation and woodland.

4.2.2.1 Management

In Europe, forest and woodland are considered to be the natural climax vegetation: predominantly mixed broadleaves, but coniferous in the north and at higher altitude (e.g. in the Alps), and transitioning into scrubland in the south. Human intervention in forest function and growth has a long history, leaving three-quarters of forests in a semi-natural rather than pristine state. In the past forest function has been affected by interventions such as coppicing (see Chapter 7) or the incidental effects of atmospheric pollution. Intensity of management is typically low, since physical access can be limited by geographic and topographic factors. Only 3% of all forest in Europe is categorised as plantation, though the proportion is much higher in specific countries (44% in the UK for example). Plantations established on non-forest land are targeted towards land that is non- or less productive in agricultural use. On individual farms, tree planting tends to focus on parts of a farm that perform less well in grain or grass production.

Plantation management is less intensive than that for arable cropping. The physical size of tree crops, the length of the cropping cycle and factors linked to access lead to high actual or opportunity cost to additional operations. This means that a key objective of innovations such as the use of biochar should be improvement of economic performance. Increasingly, however, forestry has objectives beyond wood production. The design and management of plantations is increasingly driven or incentivised by the recognition of ecosystem services provided by trees, such as amenity use, watershed management and increasing the biodiversity of predominantly agricultural landscapes. Carbon storage and improved flow and quality of water are examples of ecosystem function that could be enhanced through the integration of biochar into plantation management. Monetisation of these services should therefore accelerate adoption within the sector.

4.2.2.2 Policy

Biochar connects several key areas of policymaking: land, energy and climate change. In industrialised countries, policy on climate change has addressed power generation for carbon emissions and agriculture for other greenhouse gases and biofuel provision, while forestry has been the focus for carbon storage. This is consistent with commitments and reporting arrangements under the Kyoto Protocol.

Taxation of carbon emissions has been implemented (Sumner et al., 2009), but subsidies have been focused on renewable energy technologies (European Commission, 2009b; Adeyeye et al., 2009). This discriminates against the use of biomass for biochar production, even though carbon abatement is potentially higher than for bioenergy. Investment in biomass carbon removal or negative emissions technology (McGlashan et al., 2012) has so far focused on power station carbon capture (Gibbins and Chalmers, 2008), where the eventual cost would be borne by electricity consumers. Although certified emission reductions have been available for afforestation and reforestation, carbon sequestration in soil or related to agricultural crops has been excluded (Galinato et al., 2011).

4.2.3 Biomass and Resource Sustainability

If it is near-term benefits that will drive biochar deployment in the first instance, it should be possible to engineer biochar to provide for these. Longer-term negative effects cannot be automatically precluded, however. Despite the basic material permanence of biochar, physical and chemical changes will result from delivery of the intended function, or independently (supply of mineral nutrients, through weathering, etc.). For example, large particles of biochar added to coarse soil to reduce bulk density over time weather into fine particles with the opposite outcome and potentially negative effects on drainage. Possible future effects have to be considered alongside delivery of the target function, possibly as part of a regulatory framework.

In terms of biomass resources, sustainability can be assessed in two ways. It can be assessed more narrowly, focusing on the resources used to make specific biochar products, that is if these were wastes, if they were grown specifically for biochar production, or if they would be regenerated. More completely the assessment can be made within the spatial boundaries that encompass biomass acquisition, biomass conversion and the use of biochar.

Assessed spatially, sustainability is a function of: (1) the total area of land encompassed by the system boundaries; (2) the proportion of the land on which biochar is used; (3) the dose of biochar that is applied in order to achieve its effect and (4) the frequency of treatment necessary to sustain results. Area-based yields assumed for biomass have to be representative of current practice and the efficiency for conversion of biomass to biochar must be realistic. As a minimal condition, biochar use is sustainable if: use frequency (yr^{-1}) × dose (t ha^{-1}) × area treated (ha) < average feedstock yield (t ha^{-1} yr^{-1}) × feedstock source area (ha) × pyrolysis efficiency (fraction). Also, soils should not receive organic matter only from biochar and roots: a fraction of available biomass should be reserved for direct addition to soil. Even if increasing biomass productivity is a key objective in biochar deployment, sustainability of the system should still also be assessed assuming no change.

If biochar were to be used in the remediation of degraded land or to restore normal function to soils that had been contaminated, there would be limited or no prior biomass productivity. In this case biochar would be used as part of a broader set of measures that enables transition of land back to a more productive state. The alternative mode of assessing sustainability should be more appropriate, though once land has been successfully restored a different scenario could emerge. Gaining assurance that the biomass used to create imported biochar has been replaced would be a key step in assessing sustainability, but plants require nutrient resources to build biomass.

Biomass harvested and transported between spatial locations transfers the nutrients previously acquired in one location to a new system. In assessing the environmental and economic benefits of biochar use at the point of use, the wider impacts of such transfers must be considered. Although biochar could provide an efficient means to recycle and supplement plant nutrients that are present in a particular soil, the potential economic benefit has yet to be investigated.

Carbon stabilised in biochar represents energy captured via photosynthesis. Removing biomass to make biochar therefore denies ecosystems part of a resource otherwise available in a food web. This should not be viewed as a barrier to deployment, not least since net primary productivity has historically been maximised for harvest and removal. However, any apparent 'gain' in a system where biochar is used has to be weighed against any loss over a potentially larger spatial area from which feedstock has been acquired (Monforti et al., 2015; Sohi et al., 2015).

4.2.4 Regulation and Legislation

As well as potential value in supporting ecosystem service provision, biochar could also pose certain risks to humans and the environment (Hilber et al., 2012; Buss et al., 2015). Legislation may be necessary to regulate the use of biochar, in the same way as for other agronomic inputs and soil amendments. Other controls should be necessary to ensure its clean and safe manufacture (Basu, 2010). Legislation displays a dichotomy according to the designation of biomass feedstock as waste or non-waste. A further distinction is made between waste arising from the isolation of usable crop and that from product consumption. In the context of forestry, the comparison could be between crop residues arising in the field and sawdust from mills, versus wood recovered from construction or used consumer products. In terms of waste management and energy recovery, biomass pyrolysis and gasification are covered by the same EU directives that govern combustion. As a product destined for soil, however, biochar is not well integrated into national or regional law (Montanarella and Lugato, 2013; Van Laer et al., 2015).

Combusting the volatile product of pyrolysis creates a new set of gases. Although many of the non-hydrocarbon (as well as hydrocarbon) components of biomass are retained in biochar, the balance is liable to be present in the final flue gas. Nitrogen presents a general potential pollution risk, but toxic trace elements and precursors such as mercury and chlorine can also be present. Isolating condensable products between pyrolysis and combustion leaves only permanent gases, which may be easier to combust and limit the contaminants in flue gas. A condensed liquid fraction contains an array of chemical constituents, some of which may be valuable if they can be isolated (Christis, 2012). However, condensates that cannot be viably refined may also involve a high cost of disposal. Pyrolysis of biomass is regulated as thermal processing, together with gasification and combustion, that is under the EU Directive on Integrated Pollution Prevention and Control (IPPC) (European Commission, 2009a). In the UK, virgin wood has recently been exempt from waste regulation, even when comprising residues from forests and primary processing (Environment Agency, 2014). Regulation under IPPC therefore applies only to facilities with a capacity exceeding 100 t d^{-1}, applied by a national inspectorate. Importantly, if processing is construed as a disposal operation then a different regime applies, so demonstrable energy recovery is relevant to the regulatory context. If the purpose is found to be disposal, controls will apply to units of capacity exceeding only 1 t d^{-1}, regardless of wood waste being virgin or non-virgin. The Waste Incineration Directive (WID) is relevant to non-wood wastes. The high cost of compliance favours larger scales of material processing under WID (European Commission, 2000).

Since the purpose of biomass thermal processing has been waste disposal and energy capture in the past, the status of biochar is not surprisingly subject to scrutiny. If biochar has been created from virgin biomass without capture of energy or other major co-product it is liable to be viewed as non-waste. If there is an important energy or other co-product, biochar may be perceived as a waste. This is problematic for waste virgin biomass as the same criterion that favours its regulation as non-waste in manufacture, favours a designation as a waste in its end use (and *vice versa*). The surest option to avoid waste designation is therefore to process the main crop (wood) with a limited focus on energy recovery. An alternative is to modify biochar after manufacture, such that it is viewed as a product and satisfies end of waste criteria (achieves non-waste status).

Economics may dictate that bulk materials used at field scale are wastes or co-products: these carry a low price, are poorly specified and have low effectiveness or precision. In contrast, manufactured inputs have a high price and are designed for use in small doses or on small areas (e.g. horticulture) regulated as fertilisers, growing media or other specific product classes. The current revision of EU fertiliser regulations is pertinent in this regard (European Commission, 2013b) since soil conditioners (bulk amendments targeting soil physical properties, e.g. paper wastes) are likely to enter a common regulatory framework with chemical nutrients (conventional fertilisers), materials that release nutrients mainly through biological processes (composts, etc.), and potentially materials that are used as soil substitutes (i.e. growing media). In the meantime member states retain the authority to declare non-EC fertilisers as national fertilisers. It is by this route that biochar has become tradeable as a permitted input to soil in Italy.

Risks presented by biochar depend on control of the manufacture process, independently of emissions to air or the use of waste versus non-waste feedstock. The presence of toxic organic constituents may also be affected by control of the cooling phase, when condensation of tars can occur. The mobility of metals is affected by the carbon retained for its effects on physical occlusion, abundance and accessibility of binding surfaces and surface properties. The greatest direct exposure to biochar would be to the user, through direct inhalation and ingestion of dust, or by skin contact. These exposures should be simple to assess and manage. Risks to the general public, however, depend on how and where biochar is used, that is for the management of water quality versus the quality of amenity land, or farmland used to grow food crops, and so on. Biochar created from wood but deployed on farms would connect otherwise relatively isolated systems in forestry and agriculture.

As a substance created by industry, biochar may be affected by increasing levels of control under the system for registration, evaluation, authorisation and restriction of chemicals (REACH; European Commission, 2006). This requires companies to identify and manage risks associated with all of the substances manufactured and marketed in quantities that exceed 1 t yr^{-1}. The onus is on producers to demonstrate safe use of their products to the European Chemicals Agency (ECHA) in Helsinki, Finland, together with risk management measures. The definition of a substance in this context is not restricted to homogenous or chemically pure products; it includes heterogeneous structured solids such as biochar. Trade associations can act to register materials on behalf of their members. However, no such association currently represents biochar producers and biochar has yet to be registered. There are, however, over 100 entries for charcoal as a material analogous to some types of biochar (EC list no. 240-383-3, CAS no. 16291-96-6).

4.2.5 Life Cycle Analysis

The environmental impact of a product can be assessed through all stages of its production, use and disposal, beginning with the source of its raw materials (see Chapter 3 for a life cycle analysis of biochar). Carbon provides a common currency in LCA. It can be used to compare products, policies and strategies even without a specific climate change focus. In the case of biochar, most definitions emphasise a high carbon content and chemical aromaticity that is pertinent to biological recalcitrance and carbon storage. Stabilising plant-derived carbon by pyrolysis prevents its return to the atmosphere, with a demonstrable sequestration into the biosphere. However, in proposing to use biochar as a global strategy to mitigate climate change, the merits of alternative innovations have to be considered. A counterfactual should be invoked, where alternative scenarios can be assessed for their carbon abatement value and costs.

The alternatives may use the biomass that is potentially available for manufacturing biochar, but seeking to maximise energy recovery. In this alternate scenario there can be larger avoided emissions of fossil carbon, as energy capture from pyrolysis must be less than from complete combustion. This affects the relative merits of pyrolysis over combustion (or gasification), along with four other factors:

(i) the CO_2 emissions that can be avoided by fossil fuel substitution, per unit of useful energy recovered;
(ii) the efficiency of conversion of dry biomass carbon into useful energy, considering the moisture content of available biomass as well as the equipment used;
(iii) the efficiency of conversion of dry biomass carbon into a stabilised form in pyrolysis (defining 'stabilised' with reference to a specific time horizon);
(iv) any quantifiable, certain effects on the emission of other greenhouse gases, effects on indirect emission CO_2, or avoidance of direct CO_2 emission.

If a biomass resource can be used to recover much more useful energy per unit of CO_2 emitted than the prevailing energy source, probably dominated by fossil fuels, then it may be preferable to burn biomass than to stabilise it. By convention, LCA uses the average fuel mix of an electricity grid to assess the emissions avoided when additional electricity can be supplied to the same grid. In systems where renewable energy is a major contributor to a grid, the life cycle benefit of generating electricity from bioenergy is therefore decreased. Combustible volatiles and gases created in gasification and pyrolysis represent higher grade fuel for combustion than biomass alone, giving the potential for higher conversion efficiencies.

However, pyrolysis can be more efficient in stabilising carbon than the typical conversion of biomass carbon to electricity. So if there is low potential to avoid CO_2 emissions from fossil fuel, then pyrolysis is preferable provided biochar carbon is stable. In off-grid situations where biomass is available but not used effectively (if at all), or where renewable energy is abundant, pyrolysis should be preferred. Biochar then becomes an open-path carbon capture and storage (CCS) option, where fossil carbon emitted at one location is stored into biochar at another. Biochar production may have particular attraction in situations where there are limited options for geological storage required for CCS.

The extensive and often remote nature of most forests is an obstacle to efficient energy capture from biomass residues, especially in climates that favour slower tree growth. Efficient generation of electricity from biomass requires a certain unit scale that may not be supported by biomass productivity (given a certain transportation cost). This is essentially an off-grid situation, which favours biochar production. Biochar created in the vicinity of forests is relatively transportable; costs of transportation are relatively scalable. However, the proximity of end users has to be considered.

Biomass moisture content has a considerable effect on the outcome of comparisons of pyrolysis and a counterfactual. The evaporation of water before pyrolysis or combustion temperatures are reached consumes energy. Although this energy is potentially recoverable from the steam that is generated, heat is a low grade (diffuse) form of energy. Wet biomass is therefore associated with lower efficiency of energy capture, benefitting the case for stabilisation. Pyrolysis of wet biomass may offer no exportable energy but half of the carbon should be stabilised. This can present a marketable output with limited infrastructure, where energy yield from combustion would not warrant electricity generation.

Importantly, the clearest case for biomass pyrolysis is where biomass carbon is not stabilised only to create a new store or as an alternative to a fossil fuel emission, but where biochar simultaneously avoids the use of another degradable carbon from a second source. Using biochar to part-substitute peat in the formulation of growing media has a unique advantage in terms of carbon balance. The decomposition of already-sequestered carbon is avoided, at the same time as the biochar source material has been stabilised. Indirect gains that arise from suppression of soil-derived greenhouse gases and improved use-efficiency of fertilisers (with consequent decrease in doses) should become an important part of the equation.

It is important to integrate life cycle analysis with economic analysis, for which the scope and functional unit of assessment must also be harmonized (Thomassen et al. 2016). To monetise the most basic component of carbon abatement (carbon storage) requires an accounting methodology to be approved in a major carbon market (Whitman et al., 2010). Achieving this depends on assurance of time horizons for biochar carbon storage. It may be relevant that biomass in forests is conventionally part of accounting methodologies, whereas in arable crops it is ignored, assumed neutral. It means that biochar storage arising out of forest harvesting can already be presented as an avoided carbon emission; to do so in agriculture requires some basic principles of carbon accounting to be changed.

4.3 System Scenarios for Biochar in European Forestry

For the purposes of the present chapter, four types of forestry are distinguished: (1) trees planted to assist in the remediation of contaminated or otherwise degraded land; (2) trees cropped on a rotation cycle oriented towards energy use rather than wood production (shortened, see Chapter 7); (3) growing of trees on farmland for biomass production as coppice; and (4) amenity woodland planted in an urban context. These capture a range of contexts in terms of land use policy and legislation. In each case biochar could be targeted at economic gain in one of the following phases of forest management: tree propagation, planting operations or crop establishment.

Since the evidence for relevant functions in biochar is not always field based, may be location- or context-specific, and is not oriented specifically to forest soils or tree crops, targeted research is required for validation. It is an appropriate time to make a first iteration between the properties and functions of biochar and their systems fit. Over time pilot projects should emerge to test their value and viability.

4.3.1 Land Remediation

The general affinity of biochar for organic contaminants and metal cations has been reported extensively in the literature (Beesley et al., 2015). Uncertainties remain concerning sources for feedstock to make biochar that could be used to remediate unproductive land, the permanence of binding metal contaminants to biochar, the permanency of the biochar itself and adsorbed organic compounds, and the physical mobility of biochar particles through the soil profile. There is a potential role for biochar from forest resources to create biochar that assists in remediation (Figure 4.1a), for biochar to support growth of tree species planted on contaminated land (Figure 4.1a), or for biochar to be made from trees grown on contaminated land and recycled to land or cleaned and used elsewhere (Figure 4.1b). Urban land has a high value, giving the possibility of large investment in remediation projects, but this is sensitive to the speed and certainty of the approach adopted.

4.3.1.1 Scenarios and Systems

Plants and trees with tolerance for toxic metals transfer toxic metals from soils into biomass (Pulford and Watson, 2003; Thewys and Kuppens, 2008; Vangronsveld et al, 2009; Evangelou et al., 2012). Species that hyper-accumulate have shown potential to mitigate soil contamination by phyto-extraction, especially where contamination concerns mainly one element (Ghosh and Singh, 2005). Biomass of hyper-accumulators harvested from

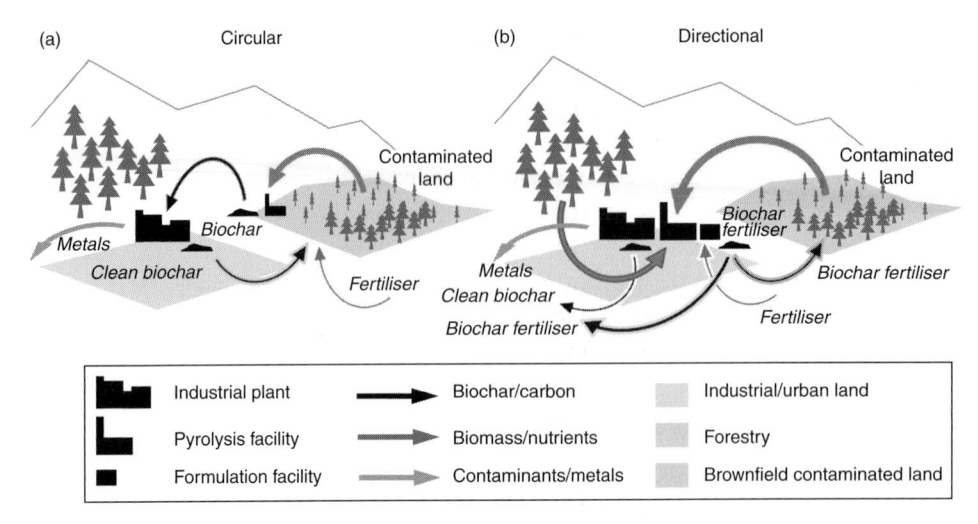

Figure 4.1. Biochar in scenarios for the remediation of contaminated brownfield land: (a) biomass of wood hyper-accumulator species grown on contaminated sites is pyrolysed with the biochar extracted for metals recovery, amended with nutrients or microbial inoculums and used to support further phyto-extraction-/localisation (or facilitated degradation) of organic contaminants where present (circular system); (b) remediation of land accelerated by optimising biochar using other forest sources of biomass, with cleaned biochar from hyper-accumulators used in other fertiliser products (directional system). (A black and white version of this figure will appear in some formats. For the colour version, please refer to the plate section.)

polluted sites can also be viewed as 'bio-ore', where combustion enables recovery of metals accumulated in plant biomass along with energy; selling the recovered elements into the global market can be described as phyto-mining (Jiang et al., 2015). Pyrolysis rather than combustion of bio-ore should yield a metal-enriched matrix from which metals can be chemically extracted. The peak temperature in pyrolysis is lower and more controlled than in combustion so higher proportions of metals should remain in the solid phase, simplifying emissions control (Stals et al., 2010). Melting of metals during pyrolysis might be catalysed by other mineral constituents of ash, reaching equilibrium concentration and distribution within the porous structural matrix dominated by carbon.

Creating clean biochar from hyper-accumulators depends on cultivation of suitable woody species and establishing effective low-cost methods to solubilise for metal extraction and recovery (Chaney et al., 2007). Clean, homogenous carbon structures have a high value if they support modification, functionalisation and safe use as activated carbon or bespoke products for use in soil. In a circular model for biochar use (Figure 4.1a) the conversion of hyper-accumulator biomass would create biochar for deployment back onto the same site. It would be counter-productive for biochar to diminish the efficiency of hyper-accumulator uptake of metals, but biochar could also localise persistent organic pollutants (POPs) where both metal and organic contaminants present risks (Zhang et al., 2013).

Co-applying biochar with chemical fertiliser nutrients should optimise the performance of hyper-accumulators and the rate at which remediation occurs.

Localisation of POPs as an acceptable outcome has to be considered. The bio-accessibility of pollutants to plants might be diminished, along with risk of human exposure (Beesley et al., 2010). The contaminant load would not be changed and degradation would be slowed down, at least in the short term, however. In the longer term, high local concentrations of organic contaminants should support their accelerated degradation, by increasing the access of specialist microorganisms and enabling higher plant productivity. The presence of productive plants supports higher background biological activity that is considered to prime for the degradation of POPs.

Using biochar could be an important aspect of a wider strategy to support hyper-accumulators and other plants, thus to increase the productivity of the land and the speed at which phyto-extraction can proceed, such as the delivery of fertiliser nutrients mentioned above. It may also be possible to optimise the macro- and micro-structure of biochar such that the adsorption of degradable contaminants could be synchronised in time and space with the delivery of microbial inoculants. Species and consortia of microorganisms that have a higher capacity to degrade POPs have been identified, but their longevity in bulk soil has been generally too short to have meaningful impacts. The use of biochar as a carrier for microorganisms has been reported but is not yet predictable or validated in the field (Chen et al., 2012).

In a directional model for biochar use in remediation of land (Figure 4.1b), biomass resources from another location would be used. Deployment of biochar on contaminated sites is less contentious from a regulatory perspective, although the concentration of any contaminants created in pyrolysis (such as polycyclic aromatic hydrocarbons, PAHs) should be below soil guidance values rather than initial site concentrations (Freddo et al., 2012).

Biochar made from intact biomass inherits the micro-porosity of the plant material from which it is derived, but biochar made from wood has a higher physical integrity and lower native mineral content than that of other virgin biomass. It may be selected to achieve the desired interaction of contaminants in soil and to restrict the mobility of biochar within the soil profile. The composition, concentration and distribution of biochar minerals can be more simply and reliably manipulated. The native composition should be predictable and homogeneous and a prior treatment to remove ash should be unnecessary. Biochar provided from woody feedstock is also likely to have a capacity and affinity for adsorption of POPs owing to a predominance of hydrophobic surfaces and a suitable micro-structure.

Localisation is a fundamentally different outcome to facilitated degradation, although both demonstrably mitigate risk in the short term. Biochar could directly assist in managing contaminated land by decreasing bioavailability; it may also indirectly assist through support of plants in phyto-stabilisation. The main uncertainty is the permanence of the association between metals and biochar, especially where the prime mechanism does not include an association of stable aromatic carbon.

4.3.1.2 Value Chain Logistics

The siting and ownership of a pyrolysis facility for the primary conversion of biomass to biochar would have to be carefully considered (see Chapter 2). Multiple criteria can be used in macro-screening. Distance costs must then be considered in local site screening (Van Dael et al., 2012), drawing on geographical information systems (Voets et al. 2013). Based on rates of biomass production and logistics of transporting contaminated feedstock, it should be preferable to site the pyrolysis facility near the biomass production area. The operational life of pyrolysis equipment would be longer than the duration of the remediation process; a portability facility would be appropriate. Owing to this requirement and the specialised nature of the process, pyrolysis might be undertaken by a third party rather than the site owner or project managing company. Investment in this equipment with the necessary emissions control should act as a hub for a suite of similar projects, possibly also for other processing. Access to a chemical plant to provide secondary processing might be a relevant consideration.

If biomass productivity for hyper-accumulators is assumed to sustain 10 t ha^{-1} yr^{-1} and the boundaries of the project reflect only contaminated land, four years might elapse before all land has been first treated with biochar (ratio in area treated to area harvested annually of 1:4). This calculation is based on biochar mass yield in pyrolysis of 25% and a viable biochar dose of 10 t ha^{-1}. A period of equilibration during which sorbed POPs may be degraded would follow. In the longer term, on completion of remediation, a different permanent crop might be established.

In the circular model, remediation depends on establishing a hyper-accumulator crop. The viability will be improved if the crop reaches maturity more quickly and continues to extract metals in subsequent harvest cycles. Valorisation of the investment in land is only possible once permissible levels of contaminants have been met in the relevant soil profile. Biochar production can only begin once the crop has first been harvested and biomass is available somewhere on the site. It can also be incorporated back into the soils only if it will not interfere with uptake of the relevant metals by the hyper-accumulator plants, or once at the permissible level. Increased productivity of the land should benefit the removal of metals from soil and potentially support degradation of POPs if they are present. Depending on the discount rate used, the uncertainty around the efficacy and duration of each stage can compromise the return available and increase the risk of investment.

As is typical, the physical scale of the project would be important. The biochar that remains after extraction of metals would be used in the manufacture of distinct non-waste products. The amount of biochar required to profitably sustain such post-processing activities would determine the minimum size of project viable within the whole system. This affects the area over which biomass must be collected, depending in turn on assumptions of biomass yield. Owing to the limited duration of remediation projects and the areas of plots to be remediated, recovery of metals and functionalisation of biochar should be undertaken off-site, by an independent contractor. Biochar may, like the harvested unpyrolysed biomass, constitute waste when moved off-site. This means that end-of-waste criteria should be addressed in product specification and manufacture.

In the directional model, options for manufacture and sale of higher-value products off-site depend on demonstrable efficacy in recovering metals from biochar. The economic viability of the proposition depends on the cost of this recovery. Post-pyrolysis processing, which includes any modifications to optimise for metal adsorption, requires additional investment and has its own sensitivity to unit scale of operation.

The economic case for including biochar in a land remediation strategy depends on: (a) any higher value use options for the site that are not compatible with remediation using biochar, and (b) the cost of alternative remediation options. In general, the most complete means of land remediation is the removal and replacement of topsoil. This is an extremely expensive option that depends on the highest value end-use options for the land, that is housing or retail. The remediation approaches outlined may be more compatible with amenity use and delivery of ecosystem services than housing, where soil removal and replacement may be required. Opting for moderate recovery and stabilisation is a more sustainable option for such uses, avoiding landfill and creating a new value chain and associated economic activities.

Alternative strategies for remediation using hyper-accumulators could generate bioenergy through combustion of harvested biomass, with collection of ash for metals recovery. Energy is a relatively low-value product, especially for small-scale electricity generation. Under combustion conditions some metal contaminants may be volatilised and there would also be no creation of stored carbon to build soil function. The extended value chain available from pyrolysis co-products would not be available.

4.3.2 Short-Rotation Forestry Bioenergy

Short-rotation forestry (SRF) has been explored as an option for supplying bioenergy (Drake-Brockman, 1996). It is an adaptation of conventional plantation forestry, where biomass productivity is optimised for fuel rather than for timber, other construction materials or paper. Trees would be harvested in their most rapid growth stage, as they reach stem diameters of 10–20 cm (about 8–20 yr under European conditions). The stems of trees harvested at this growth stage are not marketable for other uses. As the bark is thin at this stage, it might be removed in the field and left on the soil surface. A specific kind of SRF, that is traditional coppice forestry, is discussed as a potential feedstock resource in Chapter 7.

Planting for SRF is likely to be high density and focused on low-grade agricultural land or current lowland forest. The viability of SRF will be affected by achieving the target rotation length and is therefore sensitive to rapid establishment and good early-stage growth. It depends on effective planting that also minimises the number of subsequent operations. Land preparation is important in this context. Post-planting control of pests and management of soil nutrient status are also major considerations.

4.3.2.1 Scenarios and Systems

The likely high price of biochar may preclude targeting a potential change in bulk soil properties of marginal land. Bulk application of biochar from point-source feedstock such as sewage sludge could be considered in arable or horticultural crops, where it might have a particular value in the direct supply of P and K, possibly on an annual basis (Méndeza et al., 2013). In forestry, biochar is likely to be most effective as a bespoke medium for the direct controlled delivery of fertiliser nutrients to tree seedlings with any simultaneous benefit to root health.

Adding nutrient-enriched biochar as part of the planting operation could provide for replenishment of soil nutrients from the previous rotation, as a single controlled-release dose. This mitigates uncertainty of adequate nutrient supply or the cost of multiple applications. Matching the supply of nutrients to the demand of the growing tree also mitigates the risk of excess P impacting negatively on sensitive freshwater ecosystems.

Types of biochar exhibiting a micro-structure that supports delivery of fertiliser nutrients might also later support the retention of water in close proximity to tree roots. The mitigation of disease during and after crop establishment is valuable on a much longer timeframe than is the case for arable crops. Reports of positive benefits for certain types of biochar on tree root health could be important to tree establishment where planting uses bare root stock. For seedlings that are container grown, enriched biochar should replace existing slow-release or regular fertiliser products in the nursery growing medium. Using biochar to optimise nursery operations could partially or entirely support the cost of delivering the benefits arising in-field after planting.

In the circular model (Figure 4.2a) a portion of biomass from SRF would be diverted into biochar production, and, after modification, used in new planting operations. Since biomass combustion would be the main basis for forest investment, a fraction not compatible with combustion would have to be targeted for biochar manufacture. Bark has a low value in bioenergy. It has high mineral ash content, about five times higher than that of wood. Alkaline elements promote corrosion in combustion systems and present a major constraint to economic generation of electricity from biomass. Ashy feedstocks require burners to be manufactured from expensive alloys and can result in greater down-time and operating costs. Pyrolysis is not susceptible to the same constraints, since process temperatures do not reach the point at which ash melts and becomes volatile. Ash instead remains embedded within the physical carbon-based structure provided by the char. The rate at which plants can recycle nutrients in the biochar is then controlled by the complexity of the physical structure and its connections to the soil.

Bark left in the field returns nutrients to soil through microbial decomposition rather than a purely physical process. In sensitive environments an uncontrolled flush of N and P from bark decomposition can be detrimental to the ecosystem. Converting bark to biochar offers the potential for P and K to be recycled directly to growing trees, rather than the bulk soil. It may also be possible to manipulate the biochar such that nutrients are supplied at a rate matched to root demand, rather than relying on an asynchronous biological process. The mineral content of biochar may be manipulated by adjusting the yield of biochar from the process: a lower yield of biochar provides a higher mineral ash content and also greater potential for recovery of energy products. Carbon stabilised by pyrolysis should be eligible for payments under carbon accounting schemes, once methodologies have been agreed.

Nitrogen is volatile at pyrolysis temperatures and is also present at higher concentration in bark than in wood. Nitrogen is not a major consideration in nutrient management in the

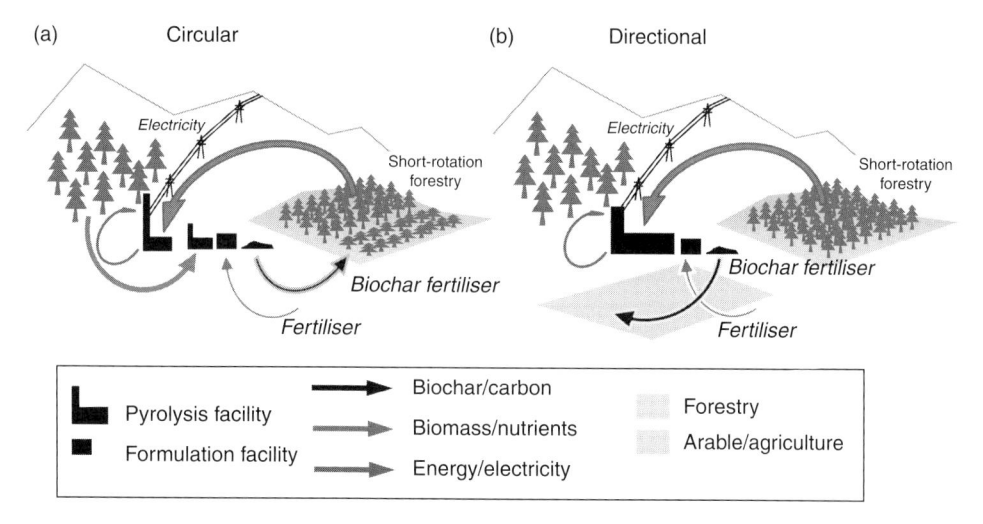

Figure 4.2. Biochar in scenarios involving short-rotation forestry: (a) processing of minor ash-rich fractions of trees grown mainly for bioenergy using pyrolysis (possibly augmented by the same from other forestry), with biochar used to convey and slowly recycle minerals to newly established stands (circular system); (b) ash-rich biochar used in horticulture as a slow-release, nutrient-enhanced ingredient in growing media, substituting peat and chemical nutrients (directional system). (A black and white version of this figure will appear in some formats. For the colour version, please refer to the plate section.)

field or nursery, so its elimination by pyrolysis is unlikely to limit the use of biochar manufactured from it. However, the potential emission of nitrogen-rich gases to the atmosphere has to be avoided.

In the directional model (Figure 4.2b) biochar from SRF would be used in the manufacture of soil conditioning products destined for forestry, horticulture or agriculture more generally. Biochar from wood offers the defined, regular structure required for the precision products demanded by these markets. The low-ash, low-contaminant load of wood from SRF offers a readily manipulated base for the production of bespoke biochars. The biochar should remain amenable to manipulation through crushing, grinding and reforming into granules, prills or pellets. It is the form that should be most readily integrated into bulk nitrogen fertilisers, to mediate the supply of nitrate to the plant root.

4.3.2.2 Value Chain Logistics

In the circular model, a byproduct of wood harvesting is invested into specialist side-product, as a nutrient-rich input that supports the propagation or transplanting of trees, that is, essentially spot application. Biochar created from bark of P content approximately 1% by mass should contain total P at a concentration comparable to a conventional fertiliser. This uses 1–3% of the tree biomass that would otherwise be directly returned to the soil. Using biochar to add both function and volume to nursery growing media is consistent with policy themes of circularity in the economy and of peat substitution.

The high lignin content of bark may also offer good potential for a value stream to be developed around volatiles condensed into liquid phase. Pyrolytic recovery of phenolic resins is known to be highest when applied to feedstock of highest lignin content. Co-locating pyrolysis with chipping, pulping or pelleting of the wood fraction would assist in the logistics of transportation, while also allowing waste heat to be recovered and used.

Elimination of nitrogen by pyrolysis does not constrain forestry use of biochar, owing to the low nitrogen requirement of trees and the problem of nitrogen leaching when fresh residues are left to decay naturally. It does, however, affect the design for treatment of flue gas. This could be integrated with systems for wood combustion if co-location is appropriate. A combustion plant could also utilise biochar fines resulting from size-grading of biochar that might otherwise be used in the manufacture of charcoal briquettes.

Since horticulture is gradually adapting to use of less peat, the growing media industry may cease to locate around peatlands. The location and scale of investment is changing in response to policy drivers. If wood fibre and biochar from bark emerge as key ingredients of future mixes, a rationale emerges for the location of manufacturing plants in proximity to forests.

The directional model is a high-risk proposition, putting biochar close to the centre of the value chain and committing large investment in pyrolysis technologies, for which technical maturity is required. So far, the maturity of pyrolysis at the relevant unit scales is rather tenuous. In a scenario where biochar would be supplied as an additive to an N-fertiliser manufactured in Europe at a rate of 1 Mt yr^{-1}, approx. 1000 t d^{-1} biomass would need to be converted to biochar. At an SRF productivity of 7 t ha yr^{-1} biomass, an estate of 50,000 ha would be required to satisfy this demand – the approximate size of the extensive Kielder Forest in northern England. Assuming facilities in which throughput could be 10 times higher than current examples, 50 pyrolysis units would need to be built and maintained for this purpose.

Although commercially SRF might be established primarily to supply wood for bioenergy, it should also contribute to other ecosystem services through careful management. In addition to facilitating the effective use of external inputs, the gradual accumulation of biochar in the land through successive planting operations could build function in the forested landscape over the longer term, not least in the management of water quality. In a policy context, multi-functionality is relevant.

4.3.3 Bioenergy from Perennial Crops on Farmland

Investment in biomass crops on-farm is on a shorter time horizon than typical forestry, so harvests should begin sooner and be made more frequently. *Salix* spp. (willow) are seen as a key option for bioenergy production, harvested on a 2–3 year cycle by coppicing and chipping. Willow have been planted commercially and used in bioenergy, with optimisation of cultivars and planting practices.

4.3.3.1 Scenarios and Systems

Biochar could have two contexts in the production of biomass for bioenergy. The first addresses an agronomic objective and is integrated into crop establishment. The second

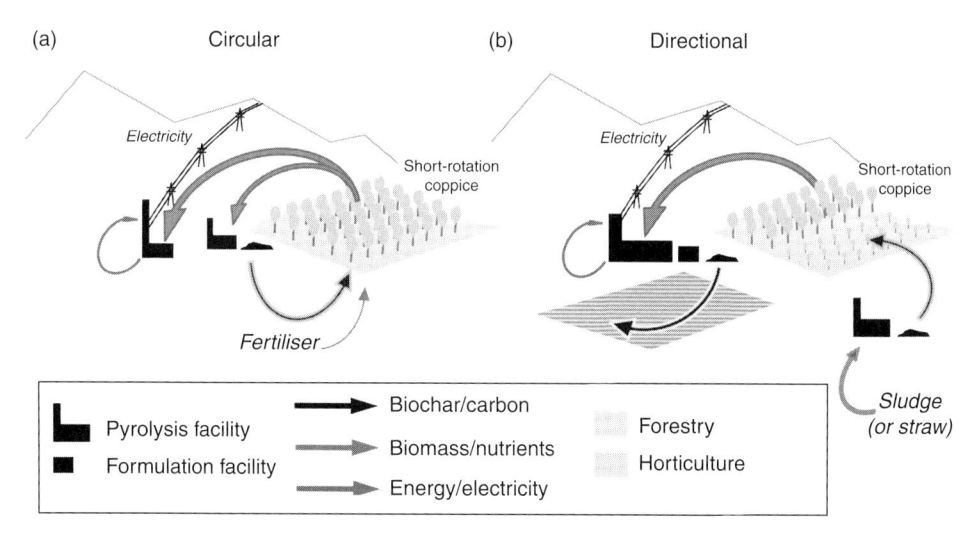

Figure 4.3. Biochar in scenarios involving short-rotation coppice: (a) part of the harvested biomass is pyrolysed rather than combusted, the biochar used to compensate for any carbon loss in land conversion, return mineral nutrients and improve nitrogen cycling/use efficiency (co-applied with chemical fertiliser at each harvest) (circular system); (b) harvest biomass pyrolysed in large facilities with lower electricity yield than the combustion alternative, but biochar used as a nitrogen carrier in distributed mainstream fertiliser; and coppice establishment supported by biochar from alternative feedstock (directional system). (A black and white version of this figure will appear in some formats. For the colour version, please refer to the plate section.)

insures against potential short-term decrease in soil carbon content that can occur during change in land use, especially from grassland. Effects on soil carbon are a key uncertainty in life cycle analysis and present a barrier in planning or zoning processes.

In the circular model (Figure 4.3a) a small portion of the biomass crop would be pyrolysed in order to stabilise carbon, prior to its return to the soil. The increase in soil carbon should be demonstrable and permanent. A fraction of the nutrients used by the crop would also be recycled to the soil. At the interface between pyrolysis and gasification, biomass conversion can yield a small amount of biochar with a useful content of entrained nutrients, even when derived from wood. Precise delivery of nutrient elements from the rigid structure of wood-derived biochar has the potential to support the establishment of a biomass crop, while avoiding eutrophication of the environment.

The directional model (Figure 4.3b) would invest biochar from a source and process that is optimised for agronomic purposes. Used in the establishment of a tree crop, the carbon content of the land in conversion would be permanently increased at the same time. Successful establishment of the crop should maximise the density and yield of the biomass crop as well as minimising the time to maturation. Ash-rich biochar from pyrolysis or gasification, particularly of straw, might be beneficial in planting operations, or in the production of willow stock. Biochar from gasification of non-wood substrate such as sewage sludge has higher concentrations of nutrients and a lower anticipated price, which may afford higher blanket applications in land preparation.

4.3.3.2 Value Chain Logistics

Biomass crop planting has been driven by subsidies for energy (heat or electricity) captured at the point of conversion, as well as planting grants. Subsidies for bioenergy increase the opportunity cost associated with diversion of biomass crop to non-energy use, including biochar production. Subsidies that are more explicitly oriented to carbon abatement would be required to support the general use of biochar in the circular mode outlined above. A directional model based on an extended value chain looks more feasible. It should be noted that there are current precedents for non-energy use of biomass crops under current regimes – sale of *Miscanthus* straw for animal bedding is already a preferred option for growers in some parts of the UK.

Recent biomass crop planting has generally involved a minor proportion of land within the farm enterprise and therefore been fragmented within a landscape. This confers a high cost to the aggregation of biomass necessary for power generation, particularly since electricity generation is efficient only at large unit scale. Co-combustion of available biomass and coal addresses this constraint. The use of existing coal-fired plants also defrays the risk associated with investing capital against the uncertainty of future support for biomass from grants and subsidies. In principle, planting rates should increase once co-combustion has been initiated.

The optimal siting for facilities dedicated to biomass pyrolysis is less obvious in this context, as they are processing only a small portion of the crop. Options for integration or aggregation of a heat product may be more effective at a site that is ostensibly used for biomass combustion. Reconciling biomass sources and destinations for biochar could be simpler as well, when conducted alongside a combustion stream. Conversely, local processing would minimise transport emissions associated with the secondary processing option. Local investment might provide a facility that could be used in processing other biomass, where innovative niche options for biochar utilisation could increase activity in the local economy. Small uses for heat may be identified locally and used in the siting of plants, including biomass pre-drying.

In the circular scenario of biochar used in crop establishment, a prior harvest is required within the physical boundaries of the production system, so returned to land within the same project catchment rather than the exact same parcel of land. This would ideally be implemented as part of a rolling programme of planting during biomass crop expansion, sustained in the longer term by re-planting after productivity has started to decline. It implies a minimum size for the biomass system in terms of the spatial area involved. Deployment of biochar during establishment provides a buffer against loss of carbon below ground during the project lifetime. Smaller targeted doses applied at the outset could be supplemented by further applications in synchrony with the harvest cycle. This would provide for part replacement of carbon removed in above-ground biomass.

4.3.4 Amenity Forestry in an Urban Context

Urban society is affected by commercial and residential development of land and its availability for amenity use. Since half of the global population is urban and urbanisation is a continuing trend, there is increasing focus on the ecological value of land in towns and cities and the services that urban land provides. Land development has implications for carbon balance within the urban boundary, both in construction and during use.

Legal precedent in the UK requires developers to define strategies to offset lifetime emissions from land development. This in turn creates incentives for adoption and integration of technologies such as pyrolysis that benefit carbon balance within the tight spatial boundaries of a single development project. In this context carbon can acquire a financial value that greatly exceeds its market value, mainly for providing leverage in land use planning and permitting.

Carbon storage can be built into the landscaping of individual development sites using biochar (Chapter 8 presents an example of biochar deployment in landscaping associated with a shopping centre). It can also be built into margins around linear infrastructure such as roads and waterways. In these contexts the desired functions of amended soil may be quite different from those required in an agricultural context. Biochar has potential to support many of these if specifically designed and manufactured to do so, potentially avoiding costs associated with alternative materials.

Biochar also has potential uses in public space, that is amenity or recreational space and architecture, for example as a planting medium or mulch ingredient in public parks. Intensive management of vegetation is typical in these areas for the purpose of aesthetics and safety. Biomass harvested in this way presents a labour, space or direct monetary cost that could be addressed locally using pyrolysis.

Biochar production by pyrolysis could be favoured over combustion partly on grounds of life cycle carbon balance and partly on a long-term improvement to ecosystem services provided by urban land after biochar addition. Passive treatment of urban water drainage around road margins is one such service, where road dust, organic contaminants and road salt can present a threat to ecosystem health. Surfaces in the built environment are predominantly engineered, sealing the soil surface. Increased water infiltration in open surfaces could be extremely valuable to manage flood risk, if storm flows can be attenuated. Biochar could be used in the management of amenity landscapes, as a vehicle for the slow release of fertiliser nutrients. The objective would be to reduce the costs of application while avoiding eutrophication of urban water bodies.

4.3.4.1 Scenarios and Systems

Boundaries in a circular deployment scenario (Figure 4.4a) might be drawn at the level of an entire town, one sector of a larger city or a single area of parkland (in Chapter 5 of this book the use of biochar within the botanical garden of Berlin is described). Within these boundaries biomass wastes arising from the management of amenity land might be aggregated and pyrolysed. This resource will be a heterogeneous mix of all types of plant material, not only tree biomass, but should be relatively free of contaminants. It will arise from defined locations and can be delivered on a defined schedule at a predictable rate. Composting is the increasingly established route for the disposal of this material, but on-site options are aesthetically problematic.

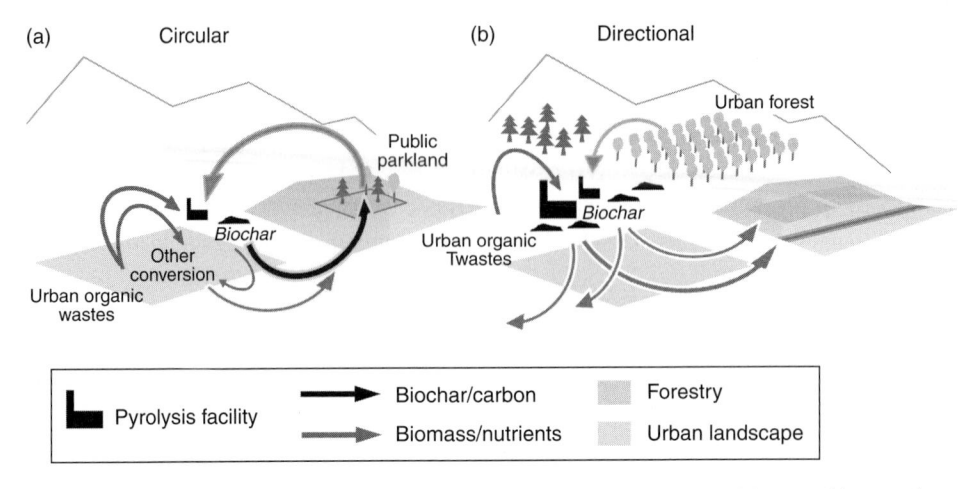

Figure 4.4. Biochar in scenarios involving urban forestry and green waste: (a) waste biomass from parkland management pyrolysed to create biochar for use in park landscaping in synergy with compost etc., produced also using biochar as an ingredient (circular system); (b) community pyrolysis facility used to convert urban forest and green waste, biochar used in enhancing services provided by urban environment (directional system). (A black and white version of this figure will appear in some formats. For the colour version, please refer to the plate section.)

Pyrolysis from point-source biomass resources in cities could provide an energy product alongside biochar. Biochar from pyrolysis of urban food waste within cities is not strictly circular, since foodstuffs and ingredients will originate from distant and diverse locations. Systems for the collection of these wastes are increasingly organised and the technology for automated separation increasingly efficient. Biochar from food waste is heterogeneous in physical properties and chemical composition compared to biomass from arboriculture. Wood waste arising in cities is similarly non-circular, but should create biochar with stronger resemblance to that of fresh woody biomass. The key limitation is that engineered timber and sheet materials (particle or fibre board) contain non-biomass ingredients that can be a source of contaminants in biochar.

A model for directional use of resources to produce biochar in a local context (Figure 4.4b) would include the use of biomass from management of urban woodland on other amenity land. Soft landscaping in public parks could use biochar in planting operations, blended with fertiliser or compost to mediate nutrient delivery, or as an alternative mulch to assist in capture and storage of irrigation water. Other public spaces managed by public administration could be suitable: road verges are costly to maintain but are the obvious place to intercept contaminated water draining from sealed surfaces; managing the health and performance of grass on sports fields is expensive and should benefit from soil structural properties using biochar.

The high value of space in commercial development projects places a high price on landscaped areas. The requirements of materials used to deliver specific and critical functions over the short and long term may be demanding. Carefully designed and specified biochar may provide carbon storage while supporting drainage, aesthetics and other amenity functions. Such biochar is likely to be based on particular homogeneous wood resources with specific physical modification to enhance macro- and microphysical properties.

In an urban context there may be many alternative interventions that present counterfactuals. Biomass arising from amenity land can be used in centralised composting operations, while anaerobic digestion is increasingly used to convert food waste. Hydrothermal carbonisation is another option for compostable and digestible materials and could be considered as another counterfactual. Although often discussed in the context of biochar, the residues of hydrothermal carbonisation are not sufficiently similar in chemical or physical terms to provide a direct analogue to biochar and the inputs and outputs of each process are also distinct.

Candidate technologies for organic waste management are not mutually exclusive: the products from one process may also provide an input to another. The potential to create a value chain around biochar that integrates pyrolysis with other conversion technologies must be considered.

4.3.4.2 Value Chain Logistics

The urban resource stream is variable in space and time, which precludes delivery of highly homogenous biochar. Feedstock may also be contaminated in various ways, with non-biomass components or polluting compounds from domestic and industrial sources. Although problematic these are also predictable and the resource streams subject to intensive management and control. There is a corresponding high monetary value on efficient use and management of these resources, however, supporting technological innovation and process optimisation. It should be conceived that, at least, efficient conversion technologies can be matched to well categorised resources and defined biochar products.

Gas-phase products of pyrolysis provide for recovery of energy that is not possible through composting. The urban situation provides diverse opportunities for waste heat in close proximity, avoiding the problem of minimum unit scale where electricity generation is required. Pyrolysis stabilises metals and degrades organic contaminants, while minimising volatilisation that is a risk with respect to air, a particularly important issue in densely populated areas. Process control and monitoring would be essential and gas would have to be treated to remove the polluting products of thermal oxidation (e.g. oxides of nitrogen), in order to meet regulatory requirements.

Biochar requires distribution, but this potentially substitutes for the import or movement of other soil inputs such as fertiliser or compost. In principle the same vehicles that are used to deliver feedstock could be used to distribute biochar, but there are practical and logistical limitations to this concept.

In urban areas there may be many public and private actors involved in the implementation of a pyrolysis and biochar project, as well as other stakeholders. Cross-sector investment and cooperation could be difficult to achieve. Institutional infrastructure designed to work cross sector in addressing climate change could be harnessed to address potential barriers related to capital cost, technology maturity, investment risk and public liability.

4.4 Summary

As a soil and land management technology, biochar materials could fit within a diverse range of complex systems that unite the biomass production phase with a use context, via conversion technologies and a range of socioeconomic and logistical considerations. In a systems context, the part of the value chain for which biochar accounts depends on uses of the product that are compatible with the source material and the value that can (or cannot) be added. The non-biochar components of the value chain may be correspondingly major or minor, and potentially essential to the economic viability of the system. Links to the wider bioeconomy provide further opportunity, but also add contingency and complexity. Other technologies also provide potential counterfactuals that cannot be neglected in consideration of long-term viability. Trees will provide some of the many opportunities that exist around biochar, in provision of feedstock, use of biochar or both. This chapter has sought to place pyrolysis and biochar in the context of forestry and trees in all their diversity, focusing on four quite specific scenarios. Extending the process of identifying and categorising potential systems may be useful to identify and rank a more complete range of options. It should engage a cross-sector audience to explore, define and realise the full potential of biochar technologies.

References

Adeyeye, A., Barrett, J., Diamond, J., Goldman, L., Pendergrass, J. and Schramm, D. (2009). *Estimating US Government Subsidies to Energy Sources 2002–2008*. Washington, DC: Environmental Law Institute.

Basu, P. (2010). *Biomass Gasification and Pyrolysis – Practical Design and Theory*. Amsterdam: Elsevier.

Beesley, L., Moreno, E., Fellet, G., Carrijo, L. and Sizmur, T. (2015). Biochar and heavy metals. In: *Biochar for Environmental Management: Science, Technology and Implementation*, Lehmann, J. and Joseph, S. (eds.). 2nd Edition. Abingdon: Routledge.

Beesley, L., Moreno-Jiménez, E. and Gomez-Eyles, J. L. (2010). Effects of biochar and green waste compost amendments on mobility, bioavailability and toxicity of inorganic and organic contaminants in a multi-element polluted soil. *Environmental Pollution*, 158, 2282–2287.

Buss, W., Masek, O., Graham, M. and Wüst, D. (2015). Inherent organic compounds in biochar – their content, composition and potential toxic effects. *Journal of Environmental Management*, 156, 150–157.

Carlile, W. R., Cattivello, C. and Zaccheo, P. (2015). Organic growing media: constituents and properties. *Vadose Zone Journal*, 14, doi:10.2136/vzj2014.09.0125

Chaney, R. L., Angle, J. S., Broadhurst, C. L., Peters, C. A., Tappero, R. V. and Sparks, D. L. (2007). Improved understanding of hyperaccumulation yields commercial phytoextraction and phytomining technologies. *Journal of Environmental Quality*, 36, 1429–1443.

Chen, B., Yuan, M. and Qian, L. (2012). Enhanced bioremediation of PAH-contaminated soil by immobilized bacteria with plant residue and biochar as carriers. *Journal of Soils and Sediments*, 12, 1350–1359.

Cherubini, F. (2010). The biorefinery concept: using biomass instead of oil for producing energy and chemicals. *Energy Conversion and Management*, 51, 1412–1421.

Christis, M. (2012). An exploration of the economic value of pyrolysis oil from short rotation wood. MSc thesis, Hasselt University, the Netherlands (in Dutch).

de Jong, E., Higson, A., Walsh, P. and Wellisch, M. (2012). *Biobased Chemicals – Value Added Products from Biorefineries*. IEA Bioenergy – Task 42 Biorefinery. Vienna: International Energy Agency.

Defra (2007). *Growing Short Rotation Coppice – Best Practice Guidelines for Applicants to Defra's Energy Crops Scheme*. London: Department for Environment, Food and Rural Affairs (Defra).

Drake-Brockman, G. R. (1996). *Establishment and Maintenance of a Woodfuel Resource*. Forest Research Technical Note 17/96. Rugeley: Forestry Commission.

Edwards, R. A., Šúri, M., Huld, T. A. and Dallemand J. F. (2005). GIS-based assessment of cereal straw energy resource in the European Union. In Proceedings of the 14th European Biomass Conference and Exhibition, Paris, France, 17–21 October.

Environment Agency (2014). *Regulatory Position Statement on the Environmental Regulation of Wood*. Bristol: Environment Agency.

European Commission (1999). Council Resolution of 15 December 1998 on a forestry strategy for the European Union. *Official Journal of the European Communities*, 26.2.1999, C 56.

European Commission (2000). Directive 2000/76/EC of the European Parliament and of the Council of 4 December 2000 on the incineration of waste. *Official Journal of the European Communities*, 28.12.2000, L 332/ 91–111.

European Commission (2006). Regulation (EC) No. 1907/2006 of the European Parliament and of the Council of 18 December 2006 concerning the Registration, Evaluation, Authorisation and Restriction of Chemicals (REACH), establishing a European Chemicals Agency, amending Directive 1999/45/EC and repealing Council Regulation (EEC) No 793/93 and Commission Regulation (EC) No 1488/94 as well as Council Directive 76/769/EEC and Commission Directives 91/155/EEC, 93/67/EEC, 93/105/EC and 2000/21/EC. *Official Journal of the European Union*, 30.12.2006, L396/1–849.

European Commission (2009a). Directive 2008/1/EC of the European Parliament and of the Council of 15 January 2008 concerning integrated pollution prevention and control. *Official Journal of the European Communities*, 29.1.2008, L24/8–29.

European Commission (2009b). Directive 2009/28/EC of the European Parliament and of the Council of 23 April 20009 on the promotion of the use of energy from renewable sources and amending and subsequently repealing Directives 2001/77/EC and 2003/30/EC. *Official Journal of the European Communities*, 5.6.2009, I140.

European Commission (2013a). *A New EU Forest Strategy: for Forests and the Forest-Based Sector. Communication from the Commission to the European Parliament, the Council, the European Economic and Social Committee and the Committee of the Regions*. Brussels: European Commission.

European Commission (2013b). *Proposal for a Regulation of the European Parliament and of the Council Relating to Fertilisers, Liming Materials, Soil Improvers, Growing Media and Plant Biostimulants and Repealing Regulation (EC) No 2003/2003*. Brussels: European Commission. [online] Available at: http://ec.europa.eu/smart-regulation/impact/planned_ia/docs/183_entr_fertilisers_en.pdf.

Evangelou, M. W. H., Conesa, H. M., Robinson, B. H. and Schulin, R. (2012). Biomass production on trace element-contaminated land: a review. *Environmental Engineering Science*, 29, 823–39.

FAO (2006). *Global Forest Resources Assessment 2006: Progress Towards Sustainable Forest Management*. Forestry Paper, 147. Rome: Food and Agriculture Organization of the United Nations (FAO).

Freddo, A., Cai, C. and Reid, B. (2012). Environmental contextualisation of potential toxic elements and polycyclic aromatic hydrocarbons in biochar. *Environmental Pollution*, 171, 18–24.

Galinato, S.P., Yoder, J. K. and Granatstein, D. (2011). The economic value of biochar in crop production and carbon sequestration. *Energy Policy*, 39, 6344–6350.

Ghosh, M. and Singh, S. P. (2005). A review on phytoremediation of heavy metals and utilization of its byproducts. *Applied Ecology and Environmental Research*, 3, 1–18.

Gibbins, J. and Chalmers, H. (2008). Preparing for global rollout: a 'developed country first' demonstration programme for rapid CCS deployment. *Energy Policy*, 36, 501–507.

Hilber, I., Blum, F., Leifeld, J., Schmidt, H.-P. and Bucheli, T. D. (2012). Quantitative determination of PAHs in biochar: a prerequisite to ensure its quality and safe application. *Journal of Agricultural and Food Chemistry*, 60, 3042–3050.

Jiang, Y., Lei, M., Duan, L. and Longhurst, P. (2015). Integrating phytoremediation with biomass valorisation and critical element recovery: a UK contaminated land perspective. *Biomass and Bioenergy*, 83, 328–339.

Keegan, D., Kretschmer, B., Elbersen, B. and Panoutsou, C. (2013). Cascading use: a systematic approach to biomass beyond the energy sector. *Biofuels, Bioproducts and Biorefining*, 7, 193–206.

Kokossis, A. C. and Yang, A. (2010). On the use of systems technologies and a systematic approach for the synthesis and the design of future biorefineries. *Computers and Chemical Engineering*, 34, 1397–1405.

Kuppens, T., van Dael, M., Vanreppelen, K., Thewys, T., Yperman, J., Carleer, R., Schreurs, S., van Passel, S. (2015). Techno-economic assessment of fast pyrolysis for the valorization of short rotation coppice cultivated for phytoextraction. *Journal of Cleaner Production*, 88, 336–344.

Laird, D. A., Brown, R. C., Amonette, J. E. and Lehmann, L. (2009). Review of the pyrolysis platform for coproducing bio-oil and biochar. *Biofuels, Bioproducts and Biorefining*, 3, 547–562.

Margeot, A., Hahn-Hagerdal, B., Edlund, M., Slade, R. and Monot, F. (2009). New improvements for lignocellulosic ethanol. *Current Opinion in Biotechnology*, 20, 372–380.

McGlashan, N. R., Workman, M. H. W., Caldecott, B. and Shah, N. (2012). *Negative Emissions Technologies*. Grantham Institute for Climate Change, Briefing paper No. 8. London: Imperial College.

Méndeza, A., Terradillos, M. and Gascó, G. (2013). Physicochemical and agronomic properties of biochar from sewage sludge pyrolysed at different temperatures. *Journal of Analytical and Applied Pyrolysis*, 102, 124–130.

Monforti, F., Lugato, E., Motola, V., Bodis, K., Scarlat, N. and Dallemand, J.-F. (2015). Optimal energy use of agricultural crop residues preserving soil organic carbon stocks in Europe. *Renewable and Sustainable Energy Reviews*, 44, 519–529.

Montanarella, L. and Lugato, E. (2013). The application of biochar in the EU: challenges and opportunities. *Agronomy*, 3, 462–473.

Nicholson, F., Kindred, D., Bhogal, A., Roques, S., Kerley, J., Twining, S., Brassington, T., Gladders, P., Balshaw, H., Cook, S. and Ellis, S. (2014). *Straw Incorporation Review*. Research Review No. 81, HGCA. Kenilworth: Agriculture and Horticulture Development Board.

Pulford, I. D. and Watson, C. (2003). Phytoremediation of heavy metal-contaminated land by trees – a review. *Environment International*, 29, 529–540.

Sasaki, N. and Putz, F. E. (2009). Critical need for new definitions of 'forest' and 'forest degradation' in global climate change agreement. *Conservation Letters*, 2, 226–232.

Sheppard, A. W., Gillespie, I., Hirsch, M. and Begley, C. (2011). Biosecurity and sustainability within the growing global bioeconomy. *Current Opinion in Environmental Sustainability*, 3, 4–10.

Sohi, S. P., McDonagh, J., Novak, J., Wu, W. and Miu, L.-M. (2015). Biochar systems and system fit. In: Lehmann, J. and Joseph, S. (eds.) *Biochar for Environmental Management: Science, Technology and Implementation*. 2nd Edition. Abingdon: Routledge, pp. 735–759.

Stals, M., Thijssen, E., Vangronsveld, J., Carleer, R., Schreurs, S. and Yperman, J. (2010). Flash pyrolysis of heavy metal contaminated biomass from phytoremediation: influence of temperature, entrained flow and wood/leaves blended pyrolysis on the behaviour of heavy metals. *Journal of Analytical and Applied Pyrolysis*, 87, 1–7.

Sumner, J., Bird, L. and Smith, H. (2009). *Carbon Taxes: A Review of Experience and Policy Design Considerations*. Technical Report NREL/TP-6A2-47312. Golden, CO: National Renewable Energy Laboratory (NREL).

Thewys, T. and Kuppens, T. (2008). Economics of willow pyrolysis after phytoextraction. *International Journal of Phytoremediation*, 10, 561–583.

Thewys, T., Vassilev, A., Meers, E., Nehnevajova, E., van der Lelie, D., and Mench, M. (2009). Phytoremediation of contaminated soils and groundwater: lessons from the field. *Environmental Science and Pollution Research*, 16, 765–794.

Thomassen, G., Van Dael, M., Lemmens, B. and Van Passel, S. (2016). A review of the sustainability of algal-based biorefineries: Towards an integrated assessment framework. Renewable and Sustainable Energy Reviews (in press) 10.1016/j.rser.2016.02.015.

UNFCCC (2002). *Report of the Conference of the Parties on its Seventh Session, Marrakesh 29 October to 10 November 2001*, FCCC/CP/2001/13/Add.1. Bonn: United Nations Framework Convention on Climate Change (UNFCCC).

Van Dael, M., Van Passel, L., Pelkmans, L., Guisson, R., Swinnen, G., and Schreurs, E. (2012). Determining potential locations for biomass valorization using a macro screening approach. *Biomass and Bioenergy*, 45, 175–186.

Van Dael, M., Márquez, N., Reumerman, P., Pelkmans, L., Kuppens, T. and Van Passel, S. (2014). Development and techno-economic evaluation of a biorefinery based on biomass (waste) streams - case study in the Netherlands. *Biofuels, Bioproducts & Biorefining*, 8, 635–644.

Van Laer, T., De Smedt, P., Ronsse, F., Ruysschaert, G., Boeckx, P., Verstraete, W., Buysse, J. and Lavrysen, L. J. (2015). Legal constraints and opportunities for biochar: a case analysis of EU law. *GCB Bioenergy*, 7, 14–24.

Voets, T., Neven, A., Thewys, T. and Kuppens, T. (2013). GIS-based location optimization of a biomass conversion plant on contaminated willow in the Campine region (Belgium). *Biomass and Bioenergy*, 55, 339–349.

Whitman, T., Scholz, S. M. and Lehmann, J. (2010). Biochar projects for mitigating climate change: an investigation of critical methodology issues for carbon accounting. *Carbon Management*, 1, 89–107.

Woolf, D., Amonette, J. E., Street-Perrott, F. A., Lehmann, J. and Joseph, S. (2010). Sustainable biochar to mitigate global climate change. *Nature Communications*, 1, 56. doi:10.1038/ncomms1053.

Zhang, X., Wang, H., He, L., Lu, K., Sarmah, A., Li, J., Bolan, N. S., Pei, J. and Huang, H. (2013). Using biochar for remediation of soils contaminated with heavy metals and organic pollutants. *Environmental Science and Pollution Research*, 20, 8472–8483.

5

Biochar as an Integrated and Decentralised Environmental Management Tool in the Botanic Garden Berlin-Dahlem

ROBERT WAGNER, RENÉ SCHATTEN, KATHRIN RÖSSLER, INES VOGEL AND KONSTANTIN TERYTZE

Abstract

Within the research project TerraBoGa, located at the Botanic Garden Berlin-Dahlem, biochar was explored as a means to achieve a closed-loop recycling system. The annual quantity of plant residues, as well as the potential amount of valuable nutrient resources like urine and faeces of the employees and visitors, was determined and an integrated sustainable sanitation system was developed. A carbonisation plant was installed to provide energy and to produce biochar from green waste. The addition of biochar to the composting process reduced the emission of greenhouse gases and showed substantial improvements in the moisture, odour and substrate structure parameters when compared with pure compost. In all plant trials undertaken, the amendment of biochar resulted in either better or similar plant growth when compared with the plant-specific standard substrates traditionally used. Biochar as an additive for horticultural substrates can reduce the use of peat by up to one-third without adversely affecting plant growth. The production and application of biochar as a nutrient carrier and nutrient storage medium has great potential to close the regional/small material cycles in conjunction with sustainable biomass and organic waste management.

5.1 Introduction

Utilisation of organic residuals and waste materials makes an important contribution to climate and environmental protection and the conservation of fossil resources such as peat. Partly due to ecological and economical relevance, organic waste is an important component in regional material flow management. The use of local, renewable potentials can unburden ecosystems while creating new and preserving old values (IfaS, 2008).

5.1.1 Terra Preta and Biochar

The rediscovery of Terra Preta do Indio in Central Amazonia points to a successful system of waste management. Compared with the natural soils of the tropical rainforest, the anthropogenic dark earth is a highly fertile soil with a remarkably high nutrient- and water-holding capacity. Terra Preta has a high charcoal content of approximately 50 Mg/ha in the first metre. Terra Preta research has been responsible for the increasing interest in charcoal (biochar) over the last 20 years (Jeffrey et al., 2011). It is scientifically proven that pre-Columbian inhabitants placed organic matter (e.g. excrements, kitchen waste and plant and animal remains) combined with charcoal into soil (Lehmann et al., 2003; Glaser and Woods, 2004; Glaser, 2007) to improve the soil fertility.

Pyrogenic carbon is a key component of Terra Preta (Glaser and Woods, 2004; Lehmann and Joseph, 2009). During pyrolysis, in contrast to combustion or natural rotting, only a portion (approximately one-third) of the CO_2 absorbed by the biomass is released back to the atmosphere. The final product, pyrogenic carbon (biochar), opens up a simple way to convert atmospheric CO_2 from biomass into a stable storage form (Lehmann and Joseph, 2009). Another important property of biochar is its porous structure with a large specific surface area (Lehmann and Joseph, 2009; Lehmann et al., 2009; Kammann et al., 2016). Therefore, biochar shows a high storage potential for nutrients and water. In this way, composts with biochar can positively influence the fertilising effect (Prost et al., 2013) and reduce nitrogen losses (Major et al., 2009). According to Blackwell et al. (2009), the application of biochar provides a possibility for reducing the amount of fertiliser used.

In recent years, the demand for, and thus the production of, biochar have significantly increased. The recycling of waste materials and the conversion of biomass to bind carbon to a more stable form has led to increasingly more sophisticated production techniques, although in this chapter, only the biochar produced by pyrolysis is considered. Different feedstock and production processes (production temperature, pyrolysis speed, handling of the synthesis gases, etc.) create different biochars with varying properties (carbon/nutrient/pollutant content, H/C_{org}- and O/C_{org}-ratio, surface area, water holding capacity, cation exchange capacity, etc.). For more details concerning biochar production please see Part 3.

Biochar is more than just a material for soil improvement. Lehmann and Joseph (2009) describe biochar as an environmental management tool, for utilisation of organic waste, for energy production and to mitigate climate change. The production of biochar must not, however, result in unregulated logging of precious woods. The biochar system must have an integrated, decentralised and sustainable approach.

This basic principle served as the foundation of a research project, called TerraBoGa (Complete material cycles by energy and material flow management by using the Terra-Preta-Technology in the Botanic Garden in terms of resource efficiency and climate protection, www.terraboga.de), in the Botanic Garden Berlin-Dahlem, Germany (BG). The objectives of the study were to close off the internal material cycle as well as to build up a local supply chain with respect to resource efficiency, climate change mitigation and environmental protection.

5.1.2 Green Waste Composting with Biochar

There is a growing need to maintain soil fertility in horticultural and agricultural areas as well as to reclaim new land that is suitable for food production by increasing soil fertility. This can potentially be achieved by a recycling strategy for organic matter that combines the benefits of compost and biochar application. It is well known that compost can be a considerable source of plant nutrients, act as a buffer for soil pH value and lead to a build-up of organic matter in the long term. Another advantage is a reduction of pathogens (Elad et al., 2011), in part due to the large number of beneficial microorganisms within it (Fuchs, 1996). Even though compost provides only a moderate potential for the reproduction of soil organic matter, strategies for further optimisation are required. Recent studies have shown that biochar addition to compost may promote its potential over the long term, leading to more sustainable management of residual biomass (Fischer and Glaser, 2012).

Biochar application can lower net CO_2 emissions, increase carbon sequestration, promote the soil buffering capacity for nutrients, water and air, and reduce leaching. Moreover, due to its high surface area and porosity, biochar is a favourable habitat for microorganisms, including those that increase the nutrient and water availability to plants or protect plants from soil-borne pathogens (e.g. rhizobia and arbuscular mycorrhiza fungi) (Lehmann et al., 2009). The potential of compost/biochar application to increase soil organic matter content is particularly high in coarse-textured soils with low fertility (Körschens et al., 2013). These soils are dominant in the Berlin-Brandenburg area and many other central European regions. In particular, the sensitivity of these soils to extreme weather events, such as drought (Amlinger et al., 2006), can be mitigated by the buffering effects of soil organic matter on water, air and nutrient availability.

Terra Preta research shows that biochar needs to be blended with organic residues to create a high-fertility and carbon-rich soil amendment. Composting accelerates the change of biochar properties during the course of bio-oxidative reaction processes such as mineralisation and humification (Bernal et al., 2009). Biochar is an excellent bulking and adsorbing agent of wet and nitrogen-rich materials (Dias et al., 2010; Steiner et al., 2011; Wang et al., 2013), which can reduce NH_3 losses and N_2O emissions during composting (Steiner et al., 2010; Hua et al., 2012; Wang et al., 2013). Ma et al. (2013) pointed out that biochar as a bulking agent reduces the odour during composting, which is an important property in urban areas. Additionally, Dias et al. (2010), Jindo et al. (2012) and Fischer and Glaser (2012) showed a reduction in C losses and better formation of stable humates and humic acids during composting. Thus, a biochar amendment at the beginning of a good composting process can yield a high-quality end product for horticulture and agriculture with environmental relief potential.

5.1.3 Material Flows in the Botanic Garden Berlin-Dahlem

The BG is one of the largest botanical gardens in the world. About 22,000 different species of plant are cultivated here on an area of 430,000 m². These plants produce a large amount of biomass that saves about 600 tonnes of CO_2 per year (BG, unpublished). In the

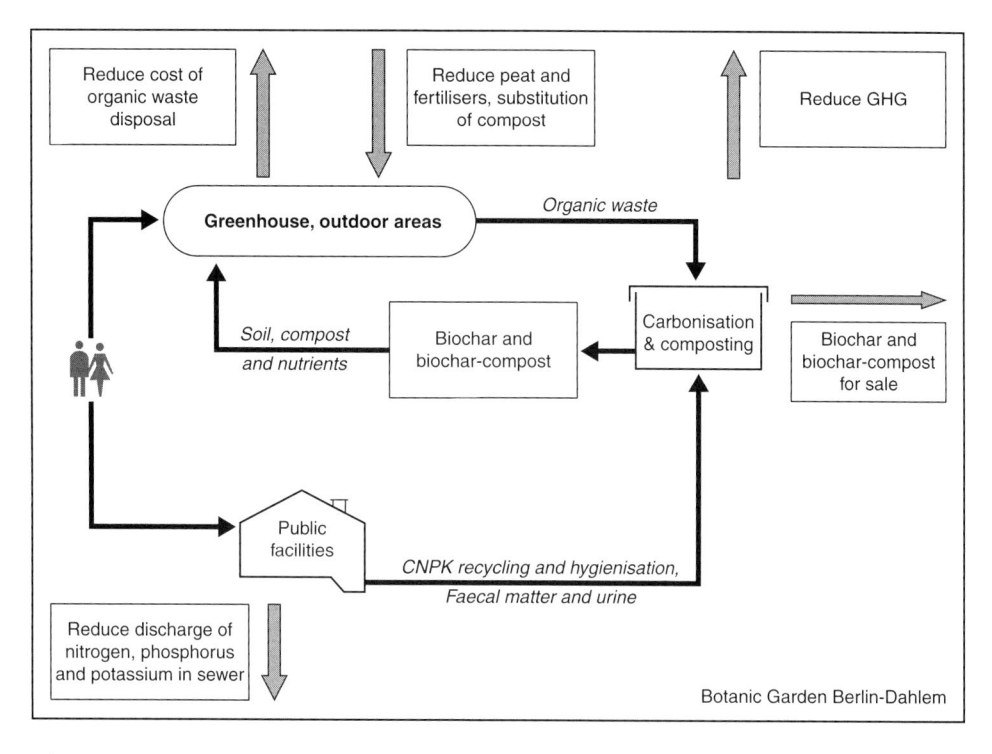

Figure 5.1. Material cycle in the Botanic Garden Berlin-Dahlem.

past, most of the biomass was unused and disposed of in a way that was both energy- and cost-intensive.

Figure 5.1 presents the material cycle in the BG. An investigation into the volumes of plant residues from maintenance operations (period considered: 2008–2010) showed that approx. 2,100 m³ or 840 Mg of fresh biomass was produced each year. This feedstock can be classified into two main categories: lignified residues (stem wood and prunings) and non-lignified residues (green waste, grass clippings and leaves).

In addition to the plant residues, the human faeces and urine that are left by approximately 260,000 visitors and 300 employees are also seen as important and valuable nutrient resources that have hitherto been disposed of as wastewater. To determine the potential amount of urine and faeces (including toilet paper), the number of visitors and staff as well as the consumption of water and toilet paper were analysed for the period from 2008 to 2010. The specific loads of N (nitrogen), P (phosphorus), K (potassium) and C (carbon) were estimated and analysed for every toilet facility, depending on the user frequency according to DWA (2008) and the flow rates of drinking water and black water. In the BG, there are six public toilet facilities that are differently frequented by visitors. The calculations indicate that the staff and visitors of BG produce an organic dry mass of about 2.8 Mg per year and nutrients such as nitrogen (1.6 Mg/a) and phosphate (0.2 Mg/a).

The plants in the BG require a large amount of compost, peat and fertilisers. In the past, these materials needed to be purchased. From 2008 to 2010, 180 m³/a of compost was obtained from nearby composting plants. In addition to compost, ca. 70 m³/a of peat from the Baltic States and Belarus were used. Furthermore, 2.8 Mg of fertiliser and 230 litres of liquid mineral fertiliser are purchased every year.

The main ways of utilisation of existing biomass are the carbonisation of lignified residues to biochar and the composting of non-lignified residues co-composting the biochar produced.

5.2 Biochar Production in the Botanic Garden Berlin-Dahlem

The aim must be to produce a biochar of high quality from locally available input materials and finally to re-use it on-site. Only in this way can biochar contribute to optimised material management. Fortunately, in the BG, a few of these conditions were already present, such as the required lignified biomass (stem wood and prunings), the demand for high-quality compost, nutrient carriers and peat substitution, a collecting/composting site and the possibility of heat exhaust. Using project funds, composting equipment, such as a shredder and turning machine, as well as a carbonisation plant, were purchased to enable the production of high-quality biochar and biochar-compost.

5.2.1 Carbonisation Plant and Process Description

For effective recycling of the dendromass, a carbonisation plant from the company BioMaCon (Rehburg, Germany, www.biomacon.com) was installed as part of the TerraBoGa project. In addition to high-quality biochar for composting, this carbonisation plant provides energy in the form of heat, which is used for heating a maintenance building.

The carbonisation plant consists of a wood chip container with a sliding floor, a biomass converter and a grinder-filling device that fills biochar in flexible big bags. The plant was situated on the courtyard of the BG and is connected to the local heating network.

A key component of the carbonisation plant is the biomass converter. Using a charging screw and a rotary feeder, the raw material reaches the converter screw, which then transports the wood chips through a combustion chamber. Within this chamber, the biomass is dried, heated and finally pyrolised at temperatures between 450 and 900°C. Similar to the production of active carbon, a very hot gas mixture (the synthesis gas) passes through the smouldering biomass, leading to constant carbonisation without cold spots. This is important because biochar without high levels of contaminants, such as dioxins, PAHs, or polychlorinated biphenyls (PCBs), is required. Finally, discharge screws carry the biochar into big bags for further use.

Technically, the plant is an autothermal fixed-bed gasifier with internal circulation of the pyrolysis gases. The combustible gas components (CO, H_2 and CH_4) are burned at the top and side-mounted in the large-scale gasifier chamber/combustion chamber. The air supply for combustion is provided via a servo motor and a fan retrofitted with a lambda-controlled flap. The combustion gases are passed through a heat exchanger under low pressure (constant depression) generated by a suction motor or exhaust gas fan and discharged via a double-walled chimney.

With the exception of the converter screw, which moves through the combustion chamber, the carbonisation plant has no moving metal parts. The converter and the gasifier consist of about 1.7 Mg of mortared firebrick.

The different feedstock, such as stem wood and prunings, are collected at the courtyard, chopped, stored and covered by a fleece. The wood chips are pre-dried (water content approx. 15–20%) through natural convection. The pre-drying process is followed by sieving the wood chips. Subsequently, the wood chips are ready for carbonisation.

5.2.2 Biochar Characterisation

The first test runs and production trials (carbonisation temperature approx. 550°C) with accruing dendromass from the BG delivered very good results in terms of biochar quality. The biochar produced in the BG had H/C_{org}-ratios of 0.16–0.27 and O/C_{org}-ratios of 0.03–0.12 (elemental analysis according to DIN 51 732), which are very comparable with other biochars and lie within the threshold values of the European Biochar Certificate (EBC, 2012) for the production of biochars (H/C_{org}: < 0.7; O/C_{org}: < 0.4). The H/C_{org}-ratio is one of the most important characterisation parameters of biochar because it is an indicator of the degree of carbonisation and of the biochar's stability. H/C_{org}-ratios > 0.7 indicate a low biochar quality and deficient pyrolysis processes (EBC, 2012; Schimmelpfennig and Glaser, 2012).

The biochars produced are fully carbonised and have high carbon contents (50–89%) along with low contents of heavy metals and organic pollutants (Table 5.1). The PAH contents of 0.05–0.053 mg/kg dry matter (DM) (according to DIN ISO 13877:2000–01, extraction solvent: n-hexane) are far below the precautionary values of the Federal Soil Protection Act and the EBC threshold values. PCBs are below the detection limit and dioxin and furan are below the threshold value of 20 ng/kg. In summary, the biochars produced in the BG are of premium quality according to the EBC.

In the test runs, the thermal efficiency of the carbonisation plant was between 52 and 61%, which is significantly higher than that required by the EBC. However, compared with woodchip boilers (minimum efficiency 80% (DIN EN 303–5, 2012)), which are designed for energy recovery, the determined efficiency is significantly lower.

Based on the hourly conversion and output, the carbonisation plant provided at least 5 kg of biochar and approximately 28 kW heat from 19.5 kg of woodchips at a carbonisation rate of almost 100%. Although higher carbonisation speeds are possible, this leads to lower biochar quality. An optimum combination of input, output, carbonisation speed, generated heat and biochar quality must be determined for every carbonisation plant.

Table 5.1. *Characterisation of biochars produced in the BG represented as ranges and compared with the requirements of the EBC and purchased biochar*

Parameter	Threshold value EBC*	Range of biochar BG	Purchased biochar
Carbon content [%]	>50	50–89	72
H/Corg-ratio	<0.7	0.16–0.27	0.18
O/Corg-ratio	<0.4	0.03–0.12	0.07
Lead [mg/kg DM]	basic: 150 mg/kg DM premium: 120 mg/kg DM	1–66	<1
Cadmium [mg/kg DM]	basic: 1.5 mg/kg DM premium: 1 mg/kg DM	<0.1–0.1	<0.1
Chromium [mg/kg DM]	basic: 90 mg/kg DM premium: 80 mg/kg DM	13–29	9.3
Copper [mg/kg DM]	100	16–36	17
Nickel [mg/kg DM]	basic: 50 mg/kg DM premium: 30 mg/kg DM	17–55	12
Zinc [mg/kg DM]	basic: 400 mg/kg DM premium: 300 mg/kg DM	33–154	52
pH-value [-]	–	8.55–8.79	8.50
16 EPA-PAH [mg/kg DM]	basic: 12 mg/kg DM premium: 4 mg/kg DM	0.050–0.056	5.0
Nitrogen [%]	only declaration needed	0.6–0.9	0.8
Phosphorus [mg/kg DM]	only declaration needed	1,567–2,164	1,909
Potassium [mg/kg DM]	only declaration needed	8,891–9,893	8,731
Magnesium [mg/kg DM]	only declaration needed	2,349–3,463	2,589
Calcium [mg/kg DM]	only declaration needed	23,565–34,210	27,020

* European Biochar Certificate 2013; Version 5 of 1 December 2014.

5.3 Biochar Application in the Botanic Garden Berlin-Dahlem

The design and implementation of the biochar system in the BG is motivated mainly by the need for high-quality compost and optimisation of in-house waste management with respect to the recycling and utilisation of organic waste. Besides the value gained from waste recycling, the focus includes the efficient use of waste, minimising CO_2 and other GHG emissions and preventing nutrient losses.

The biochar-composts are the basis for plant substrates required in the BG. Besides the possibility of co-composting, biochar can be activated with soluble nutrients from urine or used as a biochar-filter to absorb dissolved nutrients from the liquid phase of the separated blackwater (see Section 5.5). The production of a biochar-fertiliser with urine-activated biochar represents another opportunity for the utilisation of residual materials.

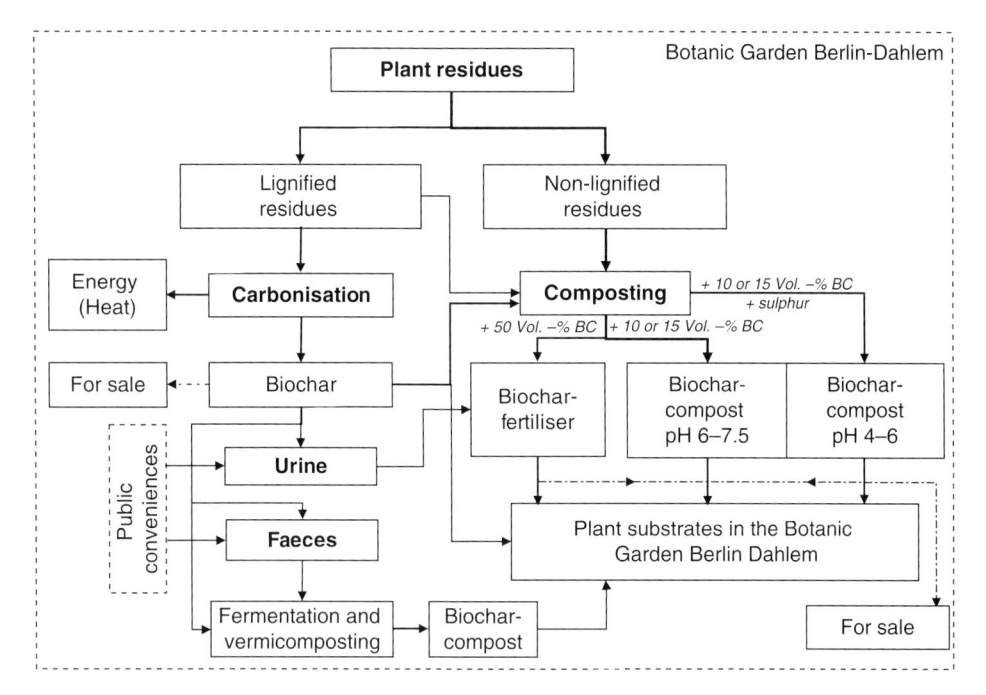

Figure 5.2. Production and use of biochar in the Botanic Garden Berlin-Dahlem.

The fermentation and vermicomposting of faeces by adding biochar completes the biochar system (Figure 5.2).

5.3.1 Compost Production

In the BG, four different biochar-compost trials were conducted between 2011 and 2014. The studies were performed at an industrial scale using piles with a volume of 10–20 m³. A feedstock blend of various plant residues, such as green waste, grass clippings, mown grass and leaves, was composted with rock powder (20 kg/m³) and clay powder (20 kg/m³) as well as with 10 or 15% biochar by volume. To investigate the influences of the biochar, every trial was run in parallel with and without biochar.

The resulting plant residues from the maintenance operations were collected for up to six weeks. Subsequently, the feedstock were shredded, blended with biochar and piled up in small triangular windrows (2.50 m wide and 1.50 m high; Figure 5.3a). Windrows were turned at least every three days during the first two weeks and later at least once a week (Figure 5.3b). The water content was regularly controlled gravimetrically and adjusted by watering the windrows if necessary.

Due to the small cross section of diameters and the turning of the windrows, an optimal oxygen supply was ensured. The composting process generally took six weeks. Subsequently, the windrows were combined into large heaps. In order to avoid contamination with seeds

Figure 5.3. (a) Piling up of shredded green waste materials and biochar, and (b) turning of compost windrows.

via air and uncontrolled waterlogging and eluviation by precipitation, the heaps were covered with a breathable compost-fleece. During composting, the temperature was regularly measured at six measuring points around the windrows. For determination of the physical and chemical parameters, the windrows were sampled and analysed according to the methods of the German Compost Quality Assurance Organisation.

5.3.2 Comparison of Compost and Biochar-Compost

The composting trials with biochar showed substantial improvements in the moisture, odour and substrate structure parameters compared with the trials with compost only. Excess water from the decomposition of the biomass was absorbed by the biochar, thus counteracting the waterlogging of the windrows. The odour from the biochar-compost was distinctly reduced in contrast to the compost (see Section 5.4.1) and an improvement in structure (crumbly) was noticeable.

In general, the biochar-compost (+BC) reached higher temperatures in the majority of cases during both the intensive heat-rotting phase and the post-rotting phase (Figure 5.4). In comparable trials, Kammann et al. (2016) detected significantly higher temperatures in the heat-rotting phase when biochar was used. Higher temperatures have an important impact with respect to the hygienisation of compost.

Changes in organic carbon (OC) in the biochar-compost trials show that biochar affects the release of non-biochar carbon. Composting without biochar showed on average an almost three times higher mineralisation rate compared with biochar-compost (40% reduction of OC in compost vs. 15% reduction of OC in biochar-compost, mean value of four compost trials). Respiration tests with the same compost and biochar-compost material indicate that biochar decreases microbial activity in the first weeks.

The changes in OC and the decrease of microbial activity can not only be explained by the dilution effect of 15% biochar addition. Erben (2011) showed likewise that an addition

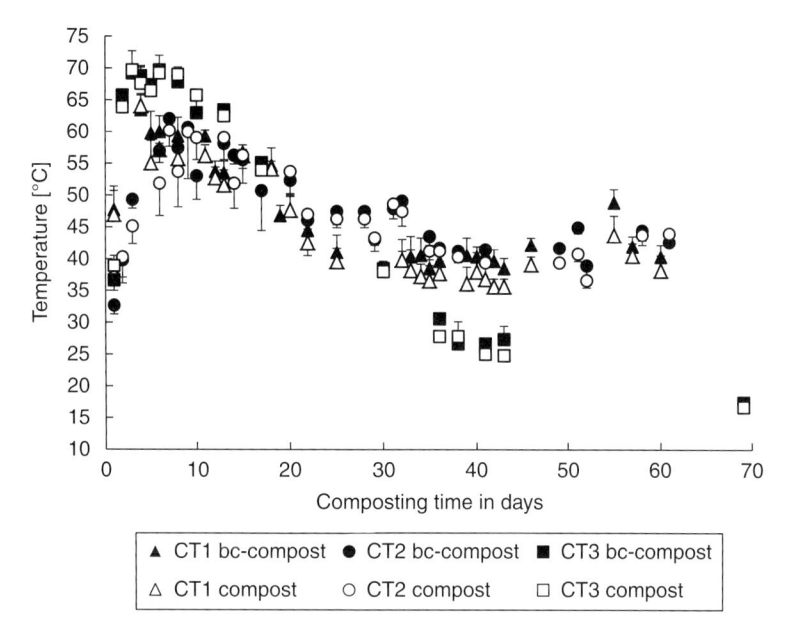

Figure 5.4. Changes of temperature in three different biochar-compost trials in the Botanic Garden Berlin-Dahlem. CT = compost trial. Values are means and standard deviation of n = 6 measuring points per windrow.

of 10% weight-for-weight biochar to the composting process reduces mineralisation significantly. It is not clear whether this represents a biochar effect on degradation and mineralisation, or on transformation of organic matter and its labile fractions. Fischer and Glaser (2012) pointed out that biochar could stabilise labile organic matter fractions during the composting process. Dias et al. (2010) and Jindo et al. (2012) observed a better formation of stable humates and humic acids during composting and a reduction of water-soluble carbon content. The apparently ability of biochar to stabilise organic carbon in compost opened an interesting opportunity to save CO_2 and reproduction of soil organic matter. Further investigations are needed.

After composting, the humified compost was analysed and evaluated according to the German Compost Quality Assurance Organisation guidelines. Table 5.2 shows the physical and chemical properties of the compost and biochar-compost from the four compost trials. Water holding capacity, pH, total nitrogen and total potassium generally increased, and bulk density and salinity generally decreased with the addition of biochar; a significant increase was only observed in the amount of organic matter and total organic carbon, caused by the higher stability of biochar relative to green waste material, the relative enrichment during the composting process and the inhibition of the degradation of non-biochar organic matter/carbon.

Table 5.2. *Characteristics of the compost and biochar-compost (n = 4)*

Parameter	Unit	Compost Mean	Compost SD	BC-Compost Mean	BC-Compost SD	Range according to BGK (Kehres et al., 2006)
pH$_{CaCl2}$	-	7.62	0.28	7.75	0.18	6.9–8.3
WHC	% FM	74.5	14.08	76.0	13.69	–
Density wet	g/L FM	875.25	50.02	813.25	87.28	500–820
Salt content	g/L FM	1.81	0.26	1.76	0.53	1.9–8.0
OM	% DM	21.20	4.60	32.28	8.95	24–51
TOC	% DM	10.83	2.21	21.25	6.98	16–37
Total N	% DM	0.71	0.13	0.86	0.16	0.5–1.5
Total P	mg/kg DM	1386.50	73.27	1381.75	128.46	–
Total K	mg/kg DM	6712.86	1180.25	7353.75	2043.78	–
Available N (NH$_4$-N + NO$_3$-N)	mg/L FM	36.20	13.36	36.95	11.87	0–740
Available P	mg/L FM	364.48	322.17	324.28	300.82	176–704
Available K	mg/L FM	2143.50	761.30	2231.68	853.27	1245–4565

Note: BGK = German Compost Quality Assurance Organisation, WHC = water holding capacity, OM = organic matter, TOC = total organic carbon, FM = fresh matter, DM = dry matter, SD = standard deviation.

5.3.3 Pot Trials with Ornamental Plants

Biochar is being increasingly studied for its use as an additive for plant substrates in addition to compost, peat, cocoa, wood fibres and rice husks. The effects of biochar-compost on plant growth were studied over a long period (up to 30 months) in pot trials with 11 different plant species from three different climate zones (tropical, subtropical and temperate) at the BG in greenhouses. The following hypotheses were investigated: (1) biochar-compost and compost produced at the BG can replace the previously purchased compost; (2) biochar-compost has positive effects on plant growth compared with the control; (3) using biochar-compost and compost as an additive in plant substrates can reduce the use of peat. The results of the pot trials are presented for two tropical plant species (*Carica papaya* and *Coffea arabica*) and two subtropical plant species (*Geranium maderense* and *Nerium oleander*). The pot trials were set up with two substrate treatments and a control. The two substrate treatments were biochar-compost and compost each mixed with different additives as required for the plants. Table 5.3 shows the composition of the mixed plant substrates. All parts of the different additives used are shown as well as the calculated parts of compost, biochar-compost, peat and biochar. Plant substrates with biochar-compost and substrates with compost had only 23.5% peat by volume instead of 30–35.3% in the controls. The

Table 5.3. Composition of the different mixed plant substrates for Carica papaya, Coffea arabica, Geranium maderense, Nerium oleander and the controls

Substrate composition for Carica papaya and Coffea arabica

Substrate	Co./bc+co.	Peat	Pine bark	Clay powder	Xylitol	Pumice	Parts in total	Co./bc+co. [%]	Peat [%]	Biochar [%]	Other additives [%]
Control	–	3	1.5	1.5	2	2	10	0	30	0	70
Compost	3	2	1.5	–	–	2	8.5	35.3	23.5	0	41.2
Biochar-compost	3	2	1.5	–	–	2	8.5	28.6	23.5	6.7	41.2

Substrate composition for Geranium maderense and Nerium oleander

Substrate	Co./bc+co.	Peat	Purch. comp.	Sand	Lava slack	Pumice	Parts in total	Co./bc+co. [%]	Peat [%]	Biochar [%]	Other additives [%]
Control	–	6	6	2	2	1	17	35.3	35.3	0	29
Compost	8	4	–	2	2	1	17	47.1	23.5	0	29.4
Biochar-compost	8	4	–	2	2	1	17	38.2	23.5	8.9	29.4

Note: Co./bc+co. = compost/biochar-compost; Purch. comp. = purchased compost.

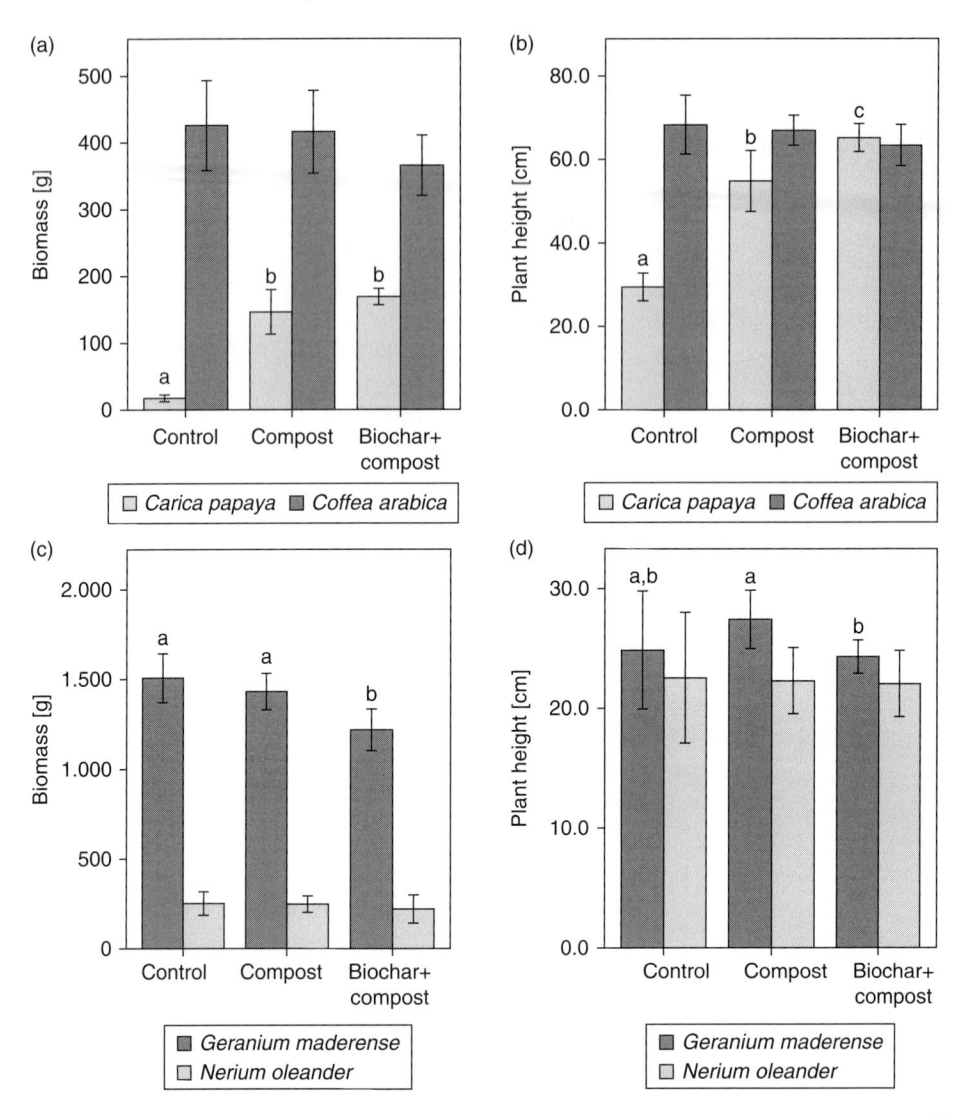

Figure 5.5. Effects of different substrate treatments on plant growth measured by plant height in cm and biomass in g for *Carica papaya* (n = 10) and *Coffea arabica* (n = 8) (a and b) and *Geranium maderense* (n = 8) and *Nerium oleander* (n = 8) (c and d). The bar above the block represents the standard deviation; different letters above the blocks indicate significant differences (Tukey's honest significant difference (HSD) $p < 0.05$) between the treatments. No letters indicate no significant differences for the plants under the different treatments. Symbols and bars represent the mean ± SD.

biochar-compost had 6.7–8.9% biochar. The biochar, biochar-compost and compost were produced from plant residues as described in Sections 5.2 and 5.3.

The plant growth was compared with a control (previously used substrate in the BG). During the growth period several plant parameters, such as plant height and width, number of leaves and biomass, were determined. Significant differences in plant height and biomass between treatments were found for *C. papaya* and *G. maderense* (Figure 5.5).

The plant height for *C. papaya* in the biochar-compost treatment was significantly higher (p = 0.000) compared with the control and also significantly higher than in the compost treatment (p = 0.014). Relative to the control, a significant increase (p = 0.000) in the biomass of *C. papaya* was found in the compost and biochar-compost treatment, although there was no significant difference between the compost and biochar-compost. No significant differences in plant height and biomass were observed for *C. arabica*. All treatments displayed similar plant growth, but plants in the compost treatment and in the control showed better plant growth than in the biochar-compost treatment.

The plant height for *G. maderense* in the compost treatment was not only significantly higher (p = 0.041) compared with the biochar-compost treatment but also higher compared with the control. A significant increase in the biomass of *G. maderense* was found in the compost treatment (p = 0.009) and in the control (p = 0.008) when compared with the biochar-compost treatment. For *N. oleander*, no significant differences in plant height and biomass were observed (Figure 5.5c, d). All treatments showed similar plant growth.

The results of the pot trials revealed different effects of biochar-compost and compost on plant growth when compared with the control. A significant positive effect of biochar-compost on plant growth was only found for *C. papaya* when compared with the control and compost. A significant negative effect of biochar-compost on plant growth was only observed for *G. maderense* when compared with the compost treatment, but compared with the control, there was similar plant growth. *C. arabica* and *N. oleander* showed similar plant growth in the biochar-compost and compost treatments when compared with the control. None of the four plants in the biochar-compost and compost treatments had a negative effect on plant growth when compared with the control, and therefore the initial hypotheses (1) and (3) were shown to be accurate. The previously purchased compost at the BG can be replaced by the newly produced biochar-compost and compost. Furthermore, the amount of peat can be reduced through mixing plant substrates with biochar-compost or compost, without having negative effects on plant growth. The different results of the plant growth in the biochar-compost treatment only partly support hypothesis (2).

5.3.4 New Ericaceous Plant Substrates with Less Peat

Within the TerraBoGa project, the substitution for peat was analysed through a dedicated study. The objective was the production of a special ericaceous plant substrate with only 25 Vol.-% peat using biochar-compost or compost acidified with bentonite sulphur. The effects on plant growth were studied using *Rhododendron simsii* in four different treatments and the results were compared with a control. The treatments were biochar-compost without bentonite sulphur, compost without bentonite sulphur, biochar-compost acidified with bentonite sulphur and compost acidified with bentonite sulphur. The control was a commercial

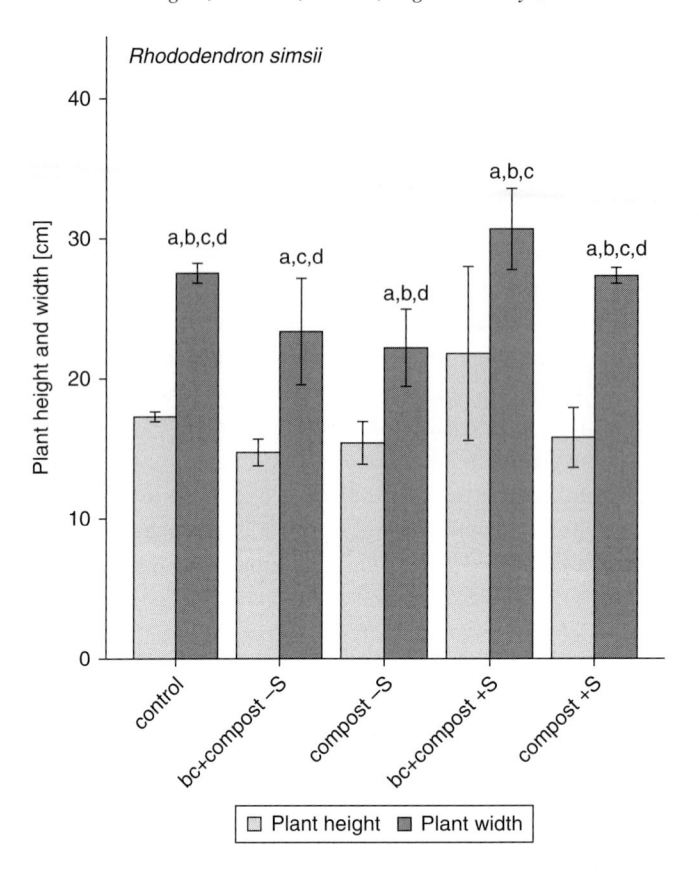

Figure 5.6. Effects of different substrate treatments on plant growth measured by plant height/width in cm for *Rhododendron simsii* (n = 3). The bar above the block represents the standard deviation; different letters above the blocks indicate significant differences (Tukey's HSD $p < 0.05$) between the treatments. No letters indicate no significant differences for the different treatments. Symbols and bars represent the mean ± SD. All substrates had 25% peat and 75% treatment. bc+compost -S = biochar-compost without bentonite sulphur; compost -S = compost without bentonite sulphur; bc+compost +S = biochar-compost with bentonite sulphur; compost +S = compost with bentonite sulphur.

ericaceous plant substrate. Figure 5.6 shows the plant growth measured in plant height and width for *Rhododendron simsii* in the different treatments. Plants treated with biochar-compost acidified with bentonite sulphur showed a significant difference ($p < 0.05$) in width growth when compared with those treated with biochar-compost without bentonite sulphur and compost without bentonite sulphur. Similar height and width growth was observed for plants treated with biochar-compost acidified with bentonite sulphur and compost acidified with bentonite sulphur when compared with the control.

Tian et al. (2012) demonstrated an 85% increase in leaf biomass for *Calathea rotundifola cv. Fasciata* when treated with a peat-biochar substrate mixture (1:1) as compared with a pure peat substrate. Zhang et al. (2014) stated that it is possible to optimise compost substrates in the cultivation of ornamentals by adding further organic materials with similar properties to peat. A test with *Calathea insignis* showed a 20% increase in growth when biochar was added to the compost. Fascella et al. (2013) observed the best growth for *Euphorbia x lomi Rauh cv. 'Chiara'* when it was treated with a peat-biochar substrate with 60% biochar.

Those results demonstrate that it is possible to use substrates with lower peat amounts without causing negative effects on plant growth.

Results from the TerraBoGa project show that the use of biochar-compost and compost acidified with bentonite sulphur ericaceous plant substrates can reduce the need for peat by up to 75% without negative effects on plant growth. These substrates can therefore contribute significantly to reducing the ecologically harmful extraction of peat without loss of productivity. It has to be mentioned that these results are only from a nine-month test duration. However long-term data have to our knowledge not been published and further trials of a longer duration focusing on the long-term effect on plant growth using biochar-compost acidified with bentonite sulphur are needed. In the BG such tests are ongoing with pending results.

5.4 Environmental Relief Potential

With respect to climate protection, Berlin is aiming to reduce CO_2 emissions by at least 40% by 2020 as compared with 1990. To achieve this, increases in energy efficiency and the use of renewable energy will be promoted. The utilisation of organic residuals and waste materials makes an important contribution to climate and environmental protection and the conservation of fossil resources such as peat. Biochar could additionally help to improve the environment through utilisation of organic residuals. Investigations in the BG have shown that not only does the conversion of lignified plant residues into biochar have environmental relief potential but also that the application of biochar in composting and in plant substrates can reduce the emission of greenhouse gases (GHGs), the need for chemical fertilisers and the release of nutrients. This section presents the results of leaching tests and the determination of GHG emissions during composting. The possibility of reducing peat and mineral fertilisers in plant substrates has already been demonstrated in Sections 5.3.3 and 5.3.4.

5.4.1 Impact of Biochar on Non-CO$_2$ GHG Emissions During Composting

In order to measure GHG emissions, a flexible flux chamber system was used based on Andersen et al. (2010). The flux chamber was placed at three positions on top of the covered windrows. To determine the large emissions during the turning of compost the gas inside the windrows was measured using lances. The difference in concentrations before and after the turning of the compost windrows implies the possible emission rate (evaluation is still in process). The quantification of the non-CO_2 GHGs (N_2O, NH_3 and CH_4) was performed using the Photoacoustic Gas Monitor INNOVA 1412i manufactured by Luma Sense Technologies. The measurements are based on photoacoustic infrared detection.

The compost trial was conducted as mentioned above between August and October 2014. The impact of biochar on non-CO_2 GHG emissions was investigated in two parallel trials (A and B) differing in turning frequency and measuring time. Each trial included one windrow with biochar and one without biochar. Additional GHG emissions after the application of urine during the post-rotting phase were measured in one biochar-compost windrow and one compost windrow.

The biochar-compost (bc-compost) showed, in most cases, lower CH_4, N_2O and NH_3 emissions during the heat-rotting and post-rotting phases in comparison with the compost. The CH_4 emission in trial A was lower from the biochar-compost during the entire composting period except for the first day. In trial B, the CH_4 emission from the biochar-compost was lower at the beginning and at the end of composting (Figure 5.7a). The N_2O emission from the biochar-compost was higher in the heat-rotting phase and lower in the post-rotting phase in trial A; in trial B it was lower in most cases during the entire composting period (Figure 5.7b). The measured NH_3 concentrations from the biochar-compost were lower at all measuring times (Figure 5.7c). The addition of urine to the biochar-compost after the heat-rotting phase resulted in lower volatilisation of NH_3 compared with its addition to the compost (Figure 5.7d).

The reduced emissions of GHGs in the biochar-composts could have been caused by the influence of parameters such as moisture and density. Reinhold (2013) showed in a wide-ranging study (n = 28) to optimise open windrow composting that the density of the compost significantly affects the emissions of GHGs and odours. Emissions reduced with reducing the density of the compost. Thus, biochar has the potential to reduce GHGs because the addition of biochar reduces the density of compost. Dias et al. (2010), Steiner et al. (2010, 2011), Wang et al. (2013) and Hua et al. (2012) also showed that biochar as a bulking and adsorbing agent of wet and nitrogen-rich materials reduces NH_3 losses and N_2O emissions as well as the odour during composting.

5.4.2 Release Behaviour of Nutrients (Leaching Tests)

From the produced biochar and biochar-composts, nutrient contents may be mobilised and washed out with the leachate. These nutrient contents are no longer available for plants and are a potential hazard to the environment (e.g. groundwater pollution with nitrate), similar to the release of pollutants. The objective of leaching tests is to determine the release behaviour and evaluate the materials above in terms of their environmental and fertiliser effects. The charging/loading of biochar with nutrients and the sorption of nitrogen (NH_4^+/NH_3, NO_3^-) in biochar has become of significant interest to the scientific community (Kammann et al., 2012). The 'ageing' of biochar apparently has a significant impact on the loading/charging of biochar with nutrients and, therefore, also on the nitrogen retention. Reduced leaching of nitrate, ammonium and phosphorus are reported in Yao et al. (2012) and in Major et al. (2009) for calcium, magnesium and phosphorus.

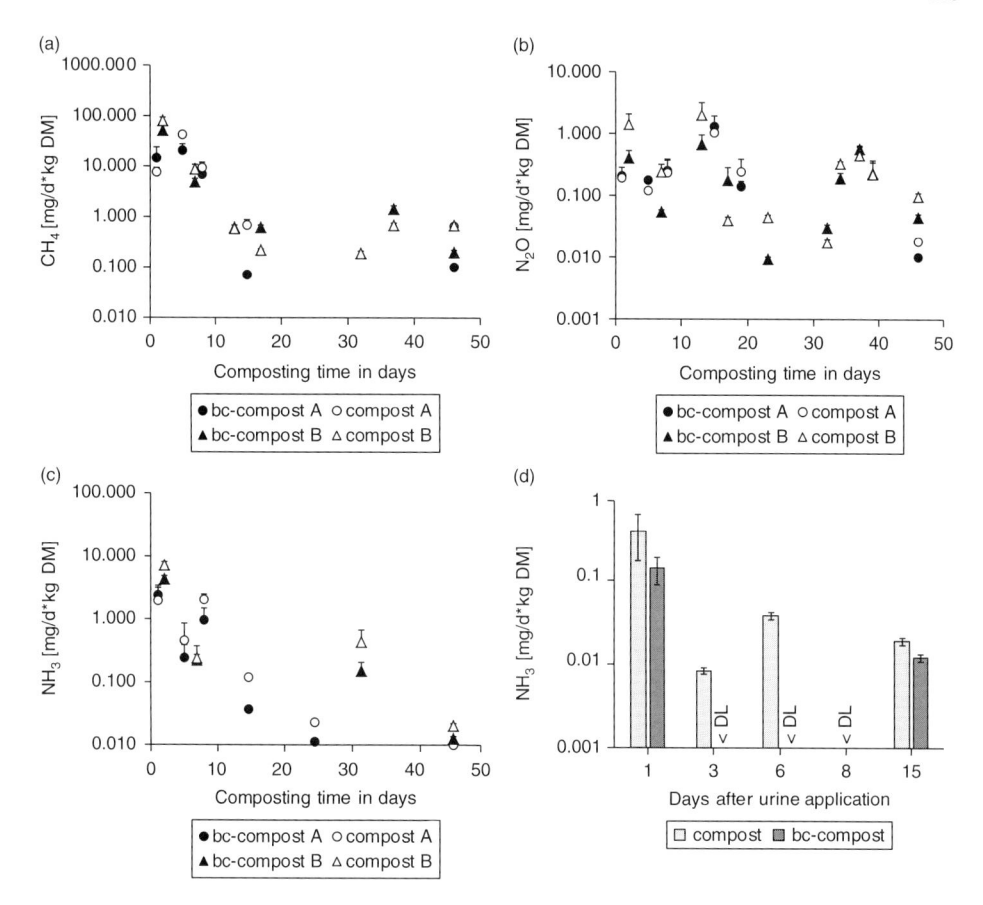

Figure 5.7. Emissions of CH_4 (a), N_2O (b) and NH_3 (c) during composting as well as NH_3 (d) after application of urine. A (circle) and B (triangle) differ in turning frequency and measuring time (data from emissions study from Horneber, bachelor thesis, unpublished, 2015), values are means and standard deviation of n = 3 measuring points per windrow.

In order to assess the release behaviour and finally the availability of nutrients and pollutants, static and dynamic leaching tests, such as batch tests based on DIN 19527 (2012) and DIN 19529 (2009) and column experiments (percolation tests) based on DIN 19528 (2009) were performed in the TerraBoGa project. Column tests enable the determination of release over time, measured as liquid/solid ratio (L/S).

The application of biochar usually leads to a reduced nutrient release in the form of 10% nitrate, 25% phosphorus, 30% potassium, 16% magnesium and 17% sulphur from freshly produced biochar-composts compared to composts without biochar (Figure 5.8; magnesium and sulphur not shown). In the case of potassium, leaching is partially and temporarily increased because the biochar itself has a high potassium content.

The 'ageing' of the investigated biochar-composts shows positive effects of biochar on the nutrient release in the form of a significant increase in nutrient retention, indicating a

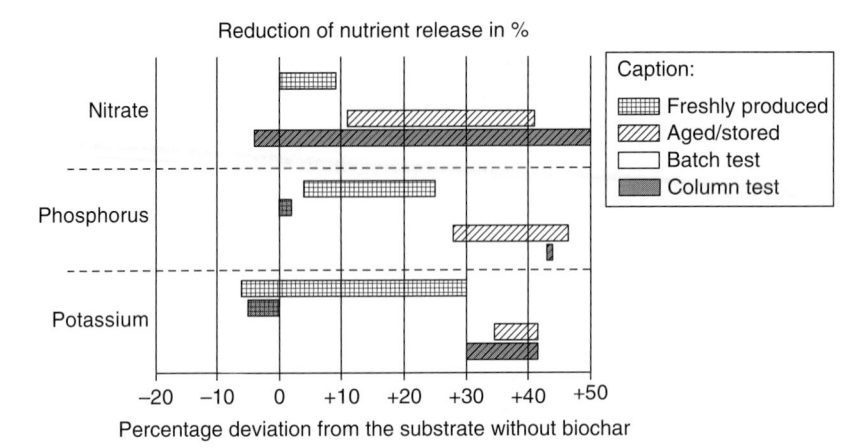

Figure 5.8. Reduction of nutrient release (percentage deviation from the substrate without biochar indicated as ranges (n = 4–6); pattern and filling of the bars represent age and leaching method (batch test; column test)).

delayed effect of biochar and biochar-composts concerning nutrient retention potential. The aged co-composted biochar leads to a reduction in nutrient leaching of up to 50% of the total content of nitrogen, 45% of phosphorus and 40% of potassium compared to compost without biochar (Figure 5.8). In particular, the considerable reduction of nitrate release with the application of biochar indicates an increased sorption on biochar. The release (leaching) of the macronutrients mentioned above and additionally selected micronutrients (e.g. manganese, copper, molybdenum) usually compares well with the few studies on composts.

In summary, biochar reduces nutrient leaching and is, therefore, suitable as a nutrient carrier or nutrient storage medium. Thus, biochar has a positive effect on fertiliser efficiency and environmental relief.

5.5 Biochar and Sanitation

Conventional sanitation systems, which are commonly based on centralised sewerage systems, do not provide a sustainable solution. Different sewage streams with different properties and pollution levels are mixed and treated together in sewage treatment plants. Conventional centralised systems do not facilitate the reuse of contained resources. Nutrients such as nitrogen and phosphorus are eliminated to protect water bodies (Schuetze and Thomas, 2013). In contrast, sustainable sanitation systems, such as Terra Preta Sanitation (TPS) systems, aim to reuse resources such as urine and faeces. The TPS enables sustainable toilet and wastewater management. This approach aims to generate a high-fertility soil similar to the Terra Preta do Indio (Schuetze and Thomas, 2013). An advantage of such a local system is that no industrial water with a high level of toxic chemicals and pollutants can contaminate the relatively clean 'domestic' sewage stream (toilet wastewater). Other critical substances, for example pharmaceutical residuals and pathogens, which may be present in urine and faeces, will be taken into account in the further remarks.

To evaluate the potential of TPS in urban infrastructure systems, an integrated sustainable sanitation system was developed and operated in the BG. The sustainable sanitary system is based on the separation of faeces from blackwater and the collection of urine from waterless urinals. Contained solids and valuable nutrients (P, N and K) are recycled as much as possible and can be used for the production of biochar substrates. The blackwater is separated into a liquid and solid phase by means of a wedge wire filter. The liquid phase can be led through a container filled with biochar (biochar-filter) to absorb the dissolved nutrients as much as possible. The potential of biochar as a wastewater filter is currently being investigated. Faeces and toilet paper are transported out of the separator unit in a plastic barrel with a layer of biochar inside to absorb the remaining liquids at the bottom of the barrel. Before exchanging the barrel, a further layer of biochar is added on the top. The collected solids are fermented, blended with green waste and composted (vermicomposting) to obtain a humified and fertile biochar-compost. Böttger et al. (2013) have demonstrated the disinfection potential of both processes, among others, on *Enterococci* and *Salmonella* Senftenberg. Additionally, the TPS approach could minimise the health risks, which originated from contamination with residues and derivatives from pharmaceuticals, by both binding of harmful substances to the biochar-humus matrix and degradation by high microbial activity, as reported by Bettendorf et al. (2015).The urine from waterless urinals is collected in a flexible and non-ventilated bag-in-box container. This container type unfolds automatically when it is filled and folds automatically when it is emptied. The advantage of this container type is that nitrogen losses from the urine by evaporation can be effectively prevented (Schuetze and Thomas, 2013). The collected urine was used in investigations to activate biochar and to balance the C/N ratio in composting (Section 5.4.1). Urine is considered nearly sterile when faecal contamination can be ruled out (Heinonen-Tański and Wilk-Sijbesma, 2004).

The calculations of specific loads of nutrients in the wastewater indicated 1.6 Mg of nitrogen and 0.2 Mg of phosphorus. These amounts can completely cover the annual needs of these nutrient elements in the BG. However, possible complications can arise in salt-sensitive plants due to the high salt content of urine (Boh et al., 2013; Mnkeni et al., 2008). Biochar can help to reduce the salt content. A closed study in the BG investigated the effect of urine on salt content under various concentrations of biochar in a horticulture substrate. Addition of fresh biochar to a green waste compost led to a decrease in salt content. An addition of 30% biochar showed a reduction in salt content of up to 55% (S. Tietjen, bachelor thesis, unpublished, 2014). Investigations in respect to restoration of saline soils show comparable results. Wu et al. (2014) observed a reduction in soil exchangeable sodium after biochar application in saline soils of up to 51% at the end of a 56-day incubation experiment. Lashari et al. (2013) indicates that biochar poultry manure compost ameliorates the soil salinity significantly in a two-year field experiment. Results of Thomas et al. (2013) show that biochar can reduce salt stress effects on plants through salt sorption.

Ammonium and ammonia exist in equilibrium; however, the formation of ammonia is preferred even under slightly alkaline conditions. This leads to nitrogen losses due to the release of ammonia. Germer et al. (2011) pointed out that 6–49% of the applied nitrogen is lost through volatilisation. Due to the high pH of biochar, an intensified volatilisation of ammonia was sensorially observed. Therefore, pre-treatment of biochar (e.g. with acid) for alkalinity reduction is essential.

5.6 Legal Framework for Biochar Application in Horticulture in Germany

The application of biochar in horticulture or for agriculture purposes is regulated by the German Fertiliser Ordinance. However, only charcoal made of untreated trunk wood is allowed as a component of fertilisers or plant cultivation substrates. There are some restrictions concerning maximum contents of heavy metals and certain organic pollutants such as PCB and dioxins, but there is no regulation concerning PAH content (DüMV, 2012).

However, it would be useful to have a more precise definition of input material. Roots, prunings from trees, vines and shrubs, wood from biomass plantations, bark, wood chippings and shavings, sawn wood, wood waste, sawdust, wood chips, wood wool and husks, such as those used as certified biochar in Swiss agriculture, as well as other uncontaminated, dry substance-rich materials from landscape work, such as common reeds, could be valuable input material for biochar production.

In particular, according to the temperature and type of production process as well as the quality of input material, the PAH content of biochar has to be regulated. The German Federal Soil Protection and Contamination Ordinance defines precautionary soil values for PAH that guarantee good soil quality; these values can be a quality goal for biochar, which needs to be present in soil for decades (BBodSchV, 1999).

Furthermore, the EBC has established diverse quality rules for the application of biochar in agri- and horticulture, including threshold values for pollutants, quality criteria for high value biochar, requirements for the production process and the sustainable production of input materials (EBC, 2012). Switzerland was the first country in Europe to officially approve the use of certified biochar in agriculture, in 2013, with the Ithaka Institute given responsibility for controlling biochar quality and the sustainability of its production (Schmidt, 2013). In Germany, there is still discussion concerning these aspects.

5.7 Conclusions

The TerraBoGa project is based on an integrated zero-emission approach, aiming for the realisation of a closed-loop recycling system for organic materials within the BG. Hence, in the BG, the implementation of a biochar system is motivated mainly by the need for high-quality compost and the optimisation of the in-house waste with respect to the recycling and utilisation of organic waste. In addition to the value gained from waste recycling, the focus includes the efficient use of waste, minimising CO_2 and other GHG emissions and preventing nutrient losses.

The main ways of utilising the existing biomass are the carbonisation of lignified residues to biochar and the combined composting of non-lignified residues to biochar-compost. Biochar-composts are the basis for plant substrates required in the BG.

The addition of biochar to the composting process helps to improve the hygienisation of green waste, reduce bad odours and GHG emissions from composting, and reduce nutrient losses. Biochar-compost optimises the soil structure for plant substrates and can help reduce the amount of peat and other additives used. The use of biochar-compost did not show any negative effects on plant growth. Therefore, biochar-compost is a promising horticulture substrate.

In summary, in the BG, the production and application of biochar to produce biochar-composts and -fertilisers, or the use of biochar as a nutrient carrier and nutrient storage medium, shows promise to close off the regional/small material cycles in conjunction with sustainable biomass and organic waste management.

Acknowledgement

We would like to thank Berlin's Senate Department for Urban Development and the Environment and the European Union for their financial support within the framework of the Berlin Environmental Relief Programme II (ERP II).

References

Amlinger, F., Peyr, S., Geszti, J., Dreher, P., Weinfurther, H. and Nortcliff, S. (2006). *Evaluierung der nachhaltig positiven Wirkung von Kompost auf die Fruchtbarkeit und Produktivität von Böden*. Bundesministerium für Land- und Forstwirtschaft, Umwelt und Wasserwirtschaft, Wien.

Andersen, J. K., Boldrin, A., Christensen, T. H. and Scheutz, C. (2010). Greenhouse gas emissions from home composting of organic household waste. *Waste Management*, 30, pp. 2475–2482.

BBodSchV (1999). Federal Soil Protection and Contaminated Sites Ordinance (BBodSchV).

Bernal, M. P., Alburquerque, J. A. and Moral, R. (2009). Composting of animal manures and chemical criteria for compost maturity assessment. A review. *Bioresource Technology*, 100, pp. 5444–5453.

Bettendorf, T., Wendland, C. and Schuetze, T. (2015). Terra Preta sanitation systems and technologies. *Terra Preta Sanitation*, 1, pp. 62–85.

Blackwell, P., Riethmuller, G. and Collins, M. (2009). Biochar application to soil. In: Lehmann, J. and Joseph, S. (eds.) *Biochar for Environmental Management: Science and Technology*. London: Earthscan, pp. 207–226.

Böttger, S., Töws, I., Bleicher, J., Krüger, M., Scheinemann, H., Dorgeloh, E., Khan, P. and Philipp, O. (2013). Applicability of Terra Preta produced from sewage sludge of decentralized wastewater systems in Germany. In: Bettendorf, T., Wendland, C. and Otterpohl, R. (eds.) *Terra Preta Sanitation*. Deutsche Bundesstiftung Umwelt, ISBN 978-3-00-046586-4.

Boh, M. Y., Germer, J., Müller, T. and Sauerborn, J. (2013). Comparative effect of human urine and ammonium nitrate application on maize (*Zea mays* L.) grown under various salt (NaCl) concentrations. *Soil Science*, 176, pp. 703–711.

Dias, B. O., Silva, C. A., Higashikawa, F. S., Roig, A. and Sanchez-Monedero, M. A. (2010). Use of biochar as bulking agent for the composting of poultry manure: effect on organic matter degradation and humification. *Bioresource Technology*, 101, pp. 1239–1246.

DIN 19527 (2012). *Leaching of solid materials – Batch test for the examination of the leaching behaviour of organic substances at a liquid to solid ratio of 2 l/kg.*

DIN 19528 (2009). *Leaching of solid materials – Percolation method for the joint examination of the leaching behaviour of inorganic and organic substances.*

DIN 19529 (2009). *Leaching of solid materials – Batch test for the examination of the leaching behaviour of inorganic substances at a liquid to solid ratio of 2 l/kg.*

DIN EN 303–5 (2012). *Heating boilers – Part 5: Heating boilers for solid fuels, manually and automatically stoked, nominal heat output of up to 500 kW – Terminology, requirements, testing and marking*; German version EN 303–5:2012.

DIN EN ISO 14240–1 (2010). *Soil quality – Determination of soil microbial biomass – Part 1: Substrate-induced respiration method (ISO 14240–1:1997).*

DüMV (2012). *Verordnung über das Inverkehrbringen von Düngemitteln, Bodenhilfsstoffen, Kultursubstraten und Pflanzenhilfsmitteln.* (Düngemittelverordnung) (BGBl. I S. 2482).

DWA (2008). *Neuartige Sanitärsysteme*. Deutsche Vereinigung für Wasserwirtschaft, Abwasser und Abfall e. V., Hennef.

EBC (2012). European Biochar Certificate – Guidelines for a Sustainable Production of Biochar. *European Biochar Foundation* (EBC), Arbaz, Switzerland. [online] Available at: www.european-biochar.org/en/download. Version 4.8 of 13 December 2013 [Accessed 3 February 2015].

Elad, Y., Cytryn, E., Meller Harel, Y., Lew, B. and Graber, E. R. (2011). The biochar effect: plant resistance to biotic stresses. *Phytopathologia Mediterranea*, 50, pp. 335–349.

Erben, G. A. (2011). *Carbon dynamics and stability of biochar compost – an evaluation of three successive composting experiments*. Bachelor Thesis, University of Bayreuth.

Fascella, G., Dispensa, V., De Pasquale, C., Fontana, G. and Zizzo, G. (2013). *Evaluation of biochar as growing substrate for ornamental plants*. 1st Mediterranen Biochar Symposium. [online] Available at: http://meditbiochar2.weebly.com/uploads/1/1/0/8/1108765/sp1_09_poster_fascella_et_al.pdf [Accessed 15 June 2015].

Fischer, D. and Glaser, B. (2012). Synergisms between compost and biochar for sustainable soil amelioration. In: Sunil, K. and Bharti, A. (eds.) *Management of Organic Waste*. Rijeka, Croatia: In Tech, pp. 167–198.

Fuchs, J. (1996). Komposteinsatz im Gartenbau: Möglichkeiten und Limiten aus der Sicht der biologischen Komposteigenschaften. *Branchenmagazin G'plus*, 12, p. 19.

Germer, J., Addai, S. and Sauerborn, J. (2011). Response of grain sorghum to fertilisation with human urine. *Field Corps Research*, 122, pp. 234–241.

Glaser, B., and Woods, W. I. (eds.) (2004). *Amazonian Dark Earths: Explorations in Space and Time*. Berlin: Springer-Verlag.

Glaser, B. (2007). Prehistorically modified soils of central Amazonia: a model for sustainable agriculture in the twenty-first century. *Philosophical Transactions of the Royal Society, B: Biological Sciences*, 362, pp. 187–196.

Heinonen-Tanski, H. and Van Wijk-Sijbesma, C. (2004). Human excreta for plant production. *Bioresource Technology*, 96, pp. 403–411.

Hua, L., Chen, Y. and Wu, W. (2012). Impacts upon soil quality and plant growth of bamboo charcoal addition to composted sludge. *Environmental Technology*, 33, pp. 61–68.

IfaS (2008). [online] Available at: www.stoffstrom.org/fileadmin/userdaten/dokumente/Veroeffentlichungen/2008-09-Broschuere_Info-Plattform_Reg_SSM_IfaS.pdf
Jeffreys, S., Verheijen, F. G. A., van der Velde, M. and Bastos, A. C. (2011). A quantitative review of the effects of biochar application to soils on crop productivity using meta-analysis. *Agricultures, Ecosystem & Environment*, 144, pp. 175–187.
Jindo, K., Suto, K., Matsumoto, K., Garcia, C., Sonoki, T. and Sanchez-Monedero, M. A. (2012). Chemical and biochemical characterisation of biochar-blended composts prepared from poultry manure. *Bioresource Technology*, 110, pp. 396–404.
Kammann, C., Ratering, S., Eckhard, C. and Muller, C. (2012). Biochar and hydrochar effects on greenhouse gas (carbon dioxide, nitrous oxide, and methane) fluxes from soils. *Journal of Environmental Quality*, 41, pp. 1052–1066.
Kammann, C., Glaser, B. and Schmidt, H. P. (2016). Combining biochar and organic amendments. In: Shackley, S. Ruysschaert, G. and Zwart, K. (eds.) *Biochar in European Soils and Agriculture: Science and Practice*. London: Earthscan, Routledge, pp. 136–164.
Kehres, B., Kirsch, A., Luyten-Naujoks, K. et al. (2006). *Methodenhandbuch zur Analyse organischer Düngemittel, Bodenverbesserungsmittel und Substrate*. Bundesgütegemeinschaft Kompost e.V.
Körschens, M., Albert, E., Armbruster, M. et al. (2013). Effect of mineral and organic fertilization on crop yield, nitrogen uptake, carbon and nitrogen balances, as well as soil organic carbon content and dynamics: results from 20 European long-term field experiments of the twenty-first century. *Archives of Agronomy and Soil Science*, 59, pp. 1017–1040.
Lashari, M. S., Liu, Y., Li, L., Pan, W., Fu, J., Pan, G., Zheng, Ju, Zheng, Ji, Zhang, X. and Yu, A. (2013). Effects of amendment of biochar-manure compost in conjunction with pyroligneous solution on soil quality and wheat yield of a salt-stressed cropland from Central China Great Plain. *Field Crops Research*, 144, pp. 113–118.
Lehmann, J., Czimczik, C., Laird, D. and Sohi, S. (2009). Stability of biochar in soil. In: Lehmann, J. and Joseph, S. (eds) *Biochar for Environmental Management: Science and Technology*. London: Earthscan, pp. 183–205.
Lehmann J., da Silva Jr., J. P., Steiner, C., Nehls, T., Zech, W. and Glaser, B. (2003). Nutrient availability and leaching in an archaeological Anthrosol and a Ferralsol of the Central Amazon basin: fertilizer, manure and charcoal amendments. *Plant and Soil*, 249, pp. 343–357.
Lehmann, J. and Joseph, S. (2009). Biochar for environmental management: an introduction. In: Lehmann, J. and Joseph, S. (eds.) *Biochar for Environmental Management: Science and Technology*. London: Earthscan, pp. 1–12.
Ma, J., Wilson, K., Zhao, Q., Yorgey, G. and Frear, C. (2013). Odor in commercial scale compost: literature review and critical analysis. *Washington State Department of Ecology*, p. 74.
Major, J., Steiner, C., Downie, A. and Lehmann, J. (2009). Biochar effects on nutrient leaching. In: Lehmann, J. and Joseph, S. (eds.) *Biochar for Environmental Management: Science and Technology*. London: Earthscan, pp. 271–287.
Mnkeni, P. N. S., Kutu, F. R., Muchaonyerwa. P. and Austin, L. M. (2008). Evaluation of human urine as a source of nutrients for selected vegetables and maize under tunnel house conditions in the Eastern Cape, South Africa. *Waste Management & Research*, 26, pp. 132–139.
Prost, K., Borchard, N., Siemens, J., Kautz, T., Möller, A. and Amelung, A. (2013). Biochar affected by composting with farmyard manure. *Journal of Environmental Quality*, 42, 164–172.

Reinhold, J. (2013). *Betrachtung zu Möglichkeiten der Optimierung der offenen Mietenkompostierung.* [online] Available at: www.guetegemeinschaftkompostbbs.de/wordpress/wp-content/uploads/2013/03/02-Mietenkompostierung-Dr.Reinhold.pdf [Accessed 03 February 2015].

Schmidt, Hans-Peter (2013): *Schweiz bewilligt Pflanzenkohle zur Bodenverbesserung. Journal für Terroirwein und Biodiversität, 2010,* ISSN 1663-0521. Available at: www.ithaka-journal.net/schweiz-bewilligt-pflanzenkohle-zur-bodenverbesserung

Schuetze, M. and Thomas, P. (2013). Terra Preta sanitation – a key component in sustainable urban resource management systems. In: Bettendorf, T., Wendland, C. and Otterpohl, R. (eds.) *Terra Preta Sanitation.* Deutsche Bundesstiftung Umwelt, ISBN 978-3-00-046586-4.

Schimmelpfennig, S. and Glaser, B. (2012). One step forward toward characterization: some important material properties to distinguish biochar. *Journal of Environmental Quality*, 41, pp. 1001–1013.

Steiner, C., Das, K. C., Melear, N. and Lakly, D. (2010). Reducing nitrogen loss during poultry litter composting using biochar. *Journal of Environmental Quality*, 39, pp. 1236–1242.

Steiner, C., Melear, N., Harris, K. and Das, K. C. (2011). Biochar as bulking agent for poultry litter composting. *Carbon Management*, 2, pp. 227–230.

Thomas, S. C., Frye, S., Gale, N., Garmon, M., Launchbury, R., Machado, N., Melamed, S., Murray, J., Petroff, A. and Winsborough, C. (2013). Biochar mitigates negative effects of salt additions on two herbaceous plant species. *Journal of Environmental Management*, 129, pp. 62–68.

Tian, Y., Sun, X., Li, S. et al. (2012). Biochar made from green waste as peat substitute in growth media for *Calathea rotundifola* cv. *Fasciata. Scientia Horticulturae*, 143, pp. 15–18.

Wang, C. Lu, H., Dong, D., Deng, H., Strong, P. J., Wang, H. and Wu, W. (2013). Insight into the effects of biochar on manure composting: evidence supporting the relationship between N_2O emissions and denitrifying community. *Environmental Science & Technology*, 47, pp. 7341–7349.

Wu, Y., Xu, G. and Shao, H. B. (2014). Furfural and its biochar improve the general properties of a saline soil. *Solid Earth*, 5, pp. 665–671.

Yao, Y., Gao, B., Zhang, M., Inyang, M. and Zimmerman, A. R. (2012). Effect of biochar amendment on sorption and leaching of nitrate, ammonium and phosphate in a sandy soil. *Chemosphere*, 89, pp. 1467–1471.

Zhang, L., Sun, X.-Y., Tian, Y. et al. (2014). Biochar and humic acid amendments improve the quality of composted green waste as a growth medium for the ornamental plant *Calathea insignis. Scientia Horticulturae*, 176, pp. 70–78.

Part II

Sustainable Biomass Resources

6

An Integrated Approach to Assess Sustainable Forest Biomass Potentials at Country Level

MICHAEL ENGLISCH, THOMAS GSCHWANTNER,
THOMAS LEDERMANN AND KLAUS KATZENSTEINER

Abstract

Forests are important for providing wood for products and energy, and the demand for wood is expected to increase over the next decades. The potential woody biomass supply was estimated for the period 2000–2020 for stem wood as well as residues, taking into account economic, environmental and technical restrictions. Constraints reducing the availability of forest biomass were defined and quantified for three mobilisation scenarios and five wood price scenarios in order to estimate the realisable potentials. The theoretical biomass potential was estimated from Austrian forest inventory data and applying the PROGNAUS forest growth simulator. It lies between 32.7 and 38.4 million m³ equivalents yr^{-1} over bark for the period 2000–2020. The realisable potential in Austria was estimated in a range between 23.9 and 31.1 million m³ equivalents yr^{-1} over bark for the period 2000–2020. These potentials represent 73–84% of the theoretical potential. Nutrient sustainability in the context of whole-tree harvesting appeared to be an important constraint when considering how much biomass is realisable from forests. The attitude of private forest owners towards increased harvest of forest biomass is also of major importance for the realisable potential, given the small-scale structure of forest ownership in Austria.

6.1 Introduction

The European Community aims at a share of 20% of renewable energy in EU gross final energy consumption in 2020 (CEC, 2010), assigning a specific share to each member state. The national 2020 goals range between 10% for Malta and 49% for Sweden and amounts to 34% in Austria of Member States' individual gross final energy consumption. This goal is mirrored in Austria's National Renewable Energy Action Plan (Indinger et al., 2006).

In Austria forests are considered an important resource for material supply and feedstock for meeting the binding renewable energy targets. Wood-based biomass contributed

45% (Nemestothy, 2012) of the 390 petajoule (PJ) gross inland consumption of renewable energy in 2010, the gross inland energy consumption amounting to 1,448 PJ (European Union, 2015). Nemestothy (2012) estimated an increase in the demand for woody biomass of three million m³ equivalents yr⁻¹ from the year 2010 until 2020, mainly in the chips and pellets segments.

Since annual increment regularly exceeds annual forest harvest in Austria (BFW (Austrian Research Centre for Forests) 2015) it is often suggested that wood harvests could be increased to meet renewable energy targets. Greenhouse gas mitigation effects from forestry, however, are considered to be highest if harvest does not exceed increase and if wood is brought to a material use, conveying mainly byproducts and small-dimensioned wood for energetic use (Eitzinger et al., 2014).

The aim of the Austrian HOBI study (Holz- und Biomassenaufkommensstudie = The Potential of Austrian Forests for Wood and Biomass Supply) was to estimate the potential supply of woody biomass for all uses from the forests in Austria considering constraints with respect to economic and technical, as well as environmental, issues up to the year 2020. Different scenarios were defined taking into account various mobilisation rates and different prices for timber grades, industrial wood and energy wood biomass.

6.2 Material and Methods

6.2.1 Calculation of Theoretical Potentials

In order to estimate the theoretical potential of forest biomass supply in Austria for the period from 2000 to 2020, we used data from the Austrian National Forest Inventory (ANFI) from the assessment period 2000/02. In the first step the theoretical potential was defined as the maximum amount of forest biomass that could be harvested annually under the given bio-physical limits taking into account growth, age structure, stocking level and tree species mixture. It is a fundamental condition that harvest does not exceed increment in the long term. Forest biomass was defined as including stem wood and bark, branches, twigs, and needles. Stumps and roots are not likely to be harvested in Austria and were not included. In the HOBI study forest biomass from all harvesting operations, from thinnings to final felling, was considered.

The measurements of the ANFI are conducted in a systematic sampling grid throughout Austria. In total, we used measurements from 11,518 sample plots situated in forests available for wood production, including 'protective forests with yield' (that is protective forests still managed for timber production) according to the national definition of these areas (Gschwantner et al., 2010). 'Protective forests without yield', which are situated in barely accessible or inaccessible locations, or on very poor sites (national definition of these areas, see Gschwantner et al., 2010), were excluded (Table 6.1).

In order to illustrate trends in the development of key indicators of Austrian forests we used data from ANFI inventory periods, 92/96, 00/02 and 07/09 (BFW, 2015).

The estimates from the ANFI (Table 6.1) show a steady increase of forest land and growing stock over the last three inventory periods, 92/96, 00/02 and 07/09. This continues

Table 6.1. *Forest area, total growing stock over bark, annual increment and harvest in Austria*

	ANFI 92/96	ANFI 00/02	ANFI 07/09
Forest area (million ha)	3.92	3.96	3.99
Productive forest land (million ha)	3.35	3.37	3.37
Growing stock over bark (million m^3)	988	1095	1135
Annual increment (million m^3 yr^{-1})	27.34	31.26	30.37
Annual harvest (million m^3 yr^{-1})	19.52	18.80	25.89

Note: Figures according to estimates of the Austrian National Forest Inventory (ANFI) (BFW, 2012), inventory periods 1992/96, 2000/02 and 2007/09. All estimates exclude 'protective forests without yield'. See national definition in Gschwantner et al. (2010).

a trend from the last decades (Gschwantner et al., 2010). Annual increment was highest in the ANFI period of 00/02, followed by a slight decrease for the following inventory period, mainly due to increased harvest following large-scale windthrows. Annual harvest decreased in the ANFI period 00/02 compared to 92/96 due to low prices for stem wood, resulting in a rather steep increase of growing stock.

The volume of ANFI sample trees with diameters at breast height (DBH) >10.5 cm is estimated using the angle count method (Bitterlich, 1948, 1952; basal area factor k = 4 m² ha^{-1}), while trees 5.0 < DBH < 10.4 cm are assessed within a circular plot of radius = 2.6 m. Trees with a DBH < 5.0 cm were not assessed in the period 2000/02. The sample tree measurements include the DBH, tree height and the height to the living crown base. Tree species, quality class and damage are among the recorded tree characteristics. Additionally, site-relevant parameters such as growth region, altitude, slope, aspect, slope position and curvature, soil type, thickness of humus layers and water regime are assessed.

For calculating the potential of forest biomass supply a modified version of the forest growth simulator PROGNAUS for Windows 2.2 (Ledermann, 2006) was used, which is based upon the implementation of the distance-independent individual-tree growth model PROGNAUS (PROGNosis for AUStria). The modified version includes a basal area increment model (Monserud and Sterba, 1996), a height increment model (Nachtmann, 2006), a mortality model (Monserud and Sterba, 1999) and an ingrowth model for trees reaching a DBH threshold of 5 cm (Ledermann, 2002). Site-relevant parameters are used as input parameters for increment predictions at plot level. Allometric functions for calculating dry mass of branches and needles are also implemented in PROGNAUS. Details about these functions can be obtained from a special issue of the *Austrian Journal of Forest Science* (Weiss (ed.), 2006). A timber assortment model was applied for the prognosis of wood quality and diameter classes as well as for harvest loss (Eckmüllner et al., 2007; Eckmüllner and Schedl, 2009).

6.2.2 Harvest Constraints

The theoretical forest biomass potential is higher than the amounts that can be harvested due to various constraints. In the following section the selected constraints are described. However there are relevant restrictions which were not implemented in this study due to a lack of information, including, for example, soil compaction and soil erosion caused by heavy harvesting machinery (e.g. Horn et al., 2004; Cambia et al., 2015; Fernholz et al., 2009), possible issues with the deterioration of water quality (e.g. Siemion et al., 2011) after intensive logging operations or socio-economic constraints due to small-scaled distributed forest ownership (Amacher et al., 2003).

6.2.2.1 Technical and Economic Constraints

We used harvesting priorities predicted by the silvicultural management model and the ranking according to the economic value of increments as implemented in the forest growth simulator to estimate the amount of harvested volume. The application of the assortment model returned the classification of harvest amounts into wood quality and diameter classes which were used for calculating harvesting costs and profits at plot level. It was assumed that even if there are high silvicultural priorities for harvesting, this was done only if at least the costs of harvesting operations are covered. The amount of forest biomass potentially removed therefore depended on wood price on the one hand and, on the other hand, the costs of the harvesting systems that could be applied in an economically justifiable way. When there was the option of choosing between two different harvesting systems the less expensive was selected.

The applicability of different harvesting systems is dependent on the terrain as well as on the existing logging infrastructure (i.e. roads, skidding trails). Slope and skidding distance to the next forest road are the main criteria and were assessed for each plot by aerial photograph interpretation and ANFI field assessments. Based on this information the applicable skidding technologies and harvesting techniques were determined for each plot. The productivity of the harvesting systems, including sub-systems (for cutting, chopping off branches, dressing, hauling, banking, machine transfer and setting up), was calculated following Erni et al. (2003), and models from Jacke (2004), Stampfer et al. (2003) and FHP (1996). For highly mechanised systems upper and/or lower single-stem volume limits (harvester: upper limit 45 cm dbh, lower limit 0.05 m^3; mountain harvester: lower limit 0.12–0.2 m^3) were defined reflecting technical and economic system constraints. Machine costs were calculated after Bauer (2001) and labour costs according to the general labour agreement of 2007 for skilled forest workers and piece work tariffs for other workers. Total harvesting costs were thus calculated for each plot considering the harvesting system and stand-related criteria of mean tree volume, terrain and skidding distance as total harvesting costs per cubic metre of harvested mass.

6.2.2.2 Ecological Constraints

The nutritional impact of forest biomass harvesting increases with the degree of removal of non-stem wood biomass and increasing technical recovery rate. Increased removal of forest residues may cause changes in nutrient availability, nutrient leaching to watercourses and groundwater, soil carbon sequestration potential and soil biodiversity, with possible long-term consequences for soil fertility and thus the soil's potential to produce forest biomass. However, the effects are site-, soil- and practice-specific. There is a vast literature on the effects of whole-tree harvesting (WTH) on productivity (e.g. Olsson et al., 1993; Jacobson et al., 1996; Nord-Larsen, 2002; Egnell and Valinger, 2003; Helmisaari et al., 2011; Kaarakka et al., 2014) and parallels to negative effects of historic land use such as litter raking and lopping of trees on soil status and productivity of forests being drawn (Kreutzer, 1979; Krapfenbauer, 1983; Glatzel, 1991). An Austrian experiment conducted at several different sites shows increment losses after WTH in early thinnings of more than 20% compared to stem only harvesting (Sterba et al., 2003).

We estimated nutritional sustainability by comparing nutrient extractions to short-to-medium term available stocks of Ca, Mg and K (exchangeable soil stocks + atmospheric input + weathering input – leaching) and to the long-term available N and P stocks using a nutrient balance approach. Exchangeable soil nutrient stocks were calculated using the dataset of the Austrian Forest Soil Inventories of 1988/89 FBVA (ed., 1992) and 2007/09 (Mutsch et al., 2013) and extrapolated to the ANFI sample plots on the productive forest area using transfer functions (e.g. Lexer et al., 1999).

In order to calculate a plot-specific mass of atmospheric deposition of N, P, K, Ca and Mg we used a set of equations by Schneider (1998). These equations correlate element concentrations in wet precipitation to altitude. Data for atmospheric element deposition came from 20 Austrian monitoring sites of ICP Forests Level II (Fürst and Kristöfel, 2012), and precipitation data in high resolution were supplied by a climate model (Kindermann, 2010). The weathering input was estimated with the soil type–texture approximation (De Vries et al., 1993) using the site and soil information from the ANFI and the soil inventories, and the geological map by Schubert (2003) for deriving parent material classes for the ANFI plots and a soil temperature–altitude regression from long-term soil temperature measurements (Englisch, unpublished). The actual weathering rates were calculated by an equation from Sverdrup (1990) and regression equations for individual weathering rates of Ca, Mg, K and Na from van der Salm et al. (1998) and De Vries et al. (1994). Leaching was estimated using literature values from Krapfenbauer and Buchleitner (1981), Rehfuess (1990) and von Wilpert (2006). For nutrient extraction estimates, data from the Austrian Bio-Indicator grid (Fürst, 2009) as well as literature values were applied for nutrient concentrations of different tree compartments.

Nutrient pools of forest biomass on plot-level were calculated by combining region-specific nutrient concentration values with extracted biomasses for the different tree compartments in whole-tree harvest and timber over bark scenarios from PROGNAUS over one stand rotation period (extraction of timber with bark minus harvest losses and where applicable, extraction of 70% of branches, twigs and needles both in thinning and final harvest operations). Finally, a sustainability classification was done for each nutrient separately at plot level, based on the ratio of nutrient amounts extracted via WTH to available soil nutrient stocks over the rotation time (Table 6.2). The

Table 6.2. *Sustainability classification of forest harvesting based on the percentage of nutrients extracted via harvesting relative to the available soil nutrient pool*

	N	P	Ca	Mg	K
Not sustainable	> 60	> 40	> 100	> 100	> 100
Sustainable with restrictions	30–60	25–40	50–100	50–100	50–100
Sustainable	< 30	< 25	< 50	< 50	< 50

Note: N measured by dry combustion (ISO, 1998), P in *Aqua regia* by reflux digestion (ISO, 1995), Ca, Mg, K in 0.1 mol L^{-1}BaCl$_2$ extract (ISO, 1994).

classification system is according to Englisch and Reiter (2009), based on concepts from Meiwes et al. (2008).

If no nutrient fell into one of the classes 'not sustainable' or 'sustainable with restrictions', harvesting was considered to be sustainable with respect to soil nutrition. If at least one nutrient was classified as 'sustainable with restrictions', harvesting was seen as problematic. If even one nutrient met the class 'not sustainable', harvest was considered to be not (nutrient) sustainable.

Due to legal restrictions productive forest land is not equal to the forest area available for wood supply. Harvests on forest areas under legal restriction may be either entirely or partially prohibited. Our study separated forest land exempt from timber harvesting, like national parks and wilderness areas, and forest land where constraints due to conservation have to be applied. While forest biomass potentials from the first category were excluded from the realisable potential, a set of rules was designed in order to meet the targets of the Habitats and Birds Directives of the EU (CEC, 1992; CEC, 2009) for the second category. The rules were based on the indicators 'mass of dead wood per hectare', 'stand structure' and 'number of veteran trees' (Schadauer, 2009).

6.2.3 Scenarios

The study defined three mobilisation scenarios and five price scenarios. Each mobilisation scenario was calculated with five different assumptions for wood price levels. Prices stated here refer only to the leading assortment of the Austrian standard rules for timber classification (Norway spruce log-wood, A/B quality, diameter size class 2b; Kooperationsplattform Forst Holz Papier (2006)). The prices of all other timber grades are heavily dependent on this leading assortment.

Scenario **Constant Standing Volume** (CSV): In this scenario all thinnings were done as prescribed by the PROGNAUS simulations at the beginning of the simulation period. The time of the final harvest was determined targeting a constant mean growing stock of 325 m^3 over bark per hectare within the simulation period (2000–2020). This corresponds to the mean growing stock as determined by ANFI for the period 2000/02.

Scenario **Silviculture (SC)**: Within this scenario final harvesting was carried out for plots where the economic value of increment within the simulation period was negative and thinning was intensified. Not only prescribed thinnings at the beginning of the simulation period were done but thinning necessities arising within the simulation period (as indicated by PROGNAUS) were considered. Additionally, stand density reductions within the 5–10 cm DBH range were to be carried out. A possible reduction of mean growing stock was condoned.

Scenario **Adaptation of Standing Volume (ASV)**: Here, the aim was to lower the mean growing stock to 280 m^3 over bark per hectare, which corresponds to the mean growing stock of the early 1980s according to ANFI. This was implemented by increasing the number of plots where final harvest was prescribed in the simulation.

The price scenarios were defined as follows:

Scenario 1: price for the leading assortment Norway spruce log-wood, A/B quality, diameter size class 2b corresponding to mean 2004–2006, 71 € m^{-3} over bark.

Scenario 2: price for the leading assortment Norway spruce log-wood, A/B quality, diameter size class 2b corresponding to price end of the year 2006, 81 € m^{-3} over bark.

Scenario 3: price for the leading assortment Norway spruce log-wood, A/B quality, diameter size class 2b considerably increased above long-term price level, 100 € m^{-3} over bark.

Scenario 4: price for the leading assortment Norway spruce log-wood, A/B quality, diameter size class 2b based on oil price between 1985 and 2005, lower range, 162 € m^{-3} over bark.

Scenario 5: price for the leading assortment Norway spruce log-wood, A/B quality, diameter size class 2b based on oil price between 1985 and 2005, higher range 243 € m^{-3} over bark.

6.2.4 Calculating Realisable Potentials

A key element of a simulation run within PROGNAUS at plot level is the model-based decision at which point in time regular harvest is done. For all mobilisation and wood price scenarios harvesting actions were set in a two-step procedure, firstly defining the necessity of harvesting based on a silvicultural management model at plot level as well as defining different treatments (e.g. selective thinning, regeneration cut, clear cut). Secondly, the model prioritised the biomass having the lowest accretion of economic value in the final harvest and biomass from stands having the highest density in thinnings.

For simulation of a mobilisation scenario a preliminary number of prioritised plots for final harvest and thinning were stipulated in order to meet the respective scenario specifications (6.2.3). This preliminary number of plots was selected within a simulation run of the 20-year period from 2000 to 2020 in five-year time-steps. At the end of each time-step the ingrowth of new trees was considered and competition factors were updated. At the end of the simulation period it was also checked whether the specifications of the respective

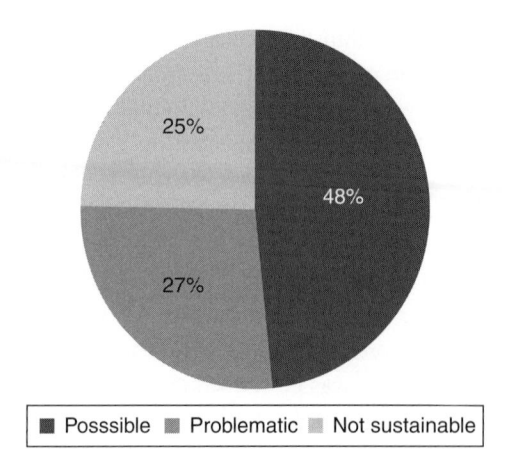

Figure 6.1. Percentage of ANFI plots where whole-tree harvest (WTH) was classified as 'possible', 'problematic' or 'not sustainable' (cf. Table 6.2).

mobilisation scenario were met, possibly running further iterations if this was not the case. The results provided direct estimates of the theoretical harvest potentials for stemwood as well as for branches, twigs and needles from final fellings and thinnings separately.

Combining the theoretical harvest potential with the pre-conditions from the five different price and three mobilisation scenarios (6.2.3) and the defined environmental, technical and economic constraints (6.2.2) a reduced amount of forest biomass was calculated, which was termed the realisable potential.

6.3 Results

Within the SC scenario and the ASV scenario the growing stock decreased from 325 m^3 ha^{-1} to 297 m^3 ha^{-1} and 280 m^3 ha^{-1}, respectively. Depending on the scenario between 23% (CSV) and 33% (SC) of the harvesting potential originated from thinnings.

Across the three mobilising scenarios the theoretical potentials of forest biomass range from 32.7 to 38.4 million m^3 stem wood equivalents yr^{-1} (Table 6.4).

The nutrient sustainability constraints resulted in 48% of ANFI plots being suitable for sustainable WTH. This corresponds to 1.65 million ha of the forest land excluding 'protected forests without yield'. On 25% of the ANFI plots (0.90 mill ha) WTH is considered 'problematic' and on 27% of the plots (0.79 mill ha) WTH is to be ruled out as 'not sustainable'. On 1.8% of all plots (0.03 mill ha) harvest of stem wood only was classified as 'not sustainable'. Mostly, the element pools of phosphorus (38% of plots) and potassium (23% of plots) were decisive for 'problematic' and 'not sustainable' classifications, while nitrogen was of minor importance (11%) in this respect. Soil type alone proved to be no suitable indicator for nutrient sustainability although there are considerable differences between more robust and more vulnerable soil types. The percentage of plots classified as 'not sustainable' in respect of WTH ranged between 11% (Gleysols

Table 6.3. *Theoretical and realisable potentials of branches, twigs and needles for the period 2000–2020, when economic, technical and ecological constraints and harvest losses are considered for the silviculture (SC) scenario (mill m^3 stem wood equivalents yr^{-1})*

Mobilising scenario	Theoretical potential	Realisable potential		
		Price scen. 1	Price scen. 2	Price scen. 3
Silviculture (SC)	6.7	1.0	1.1	1.1

Table 6.4. *Theoretical and realisable potentials of forest biomass, including bark, branches, twigs and needles for the period 2000–2020, when economic, technical and ecological constraints and harvest losses are considered (mill m^3 stem wood equivalents yr^{-1}).*

Mobilising scenario	Theoretical potential	Realisable potential				
		Price scen. 1	Price scen. 2	Price scen. 3	Price scen. 4	Price scen. 5
Constant Standing Volume (CSV)	32.7	23.9	24.9	25.7	26.7	27.0
Silviculture (SC)	35.7	25.0	26.2	27.2	28.4	28.8
Adaptation of Standing Volume (ASV)	38.4	27.4	28.5	29.5	30.7	31.1

and Stagnic Cambisols) and 58% (Leptosols) of the plots for all the major forest soil types of Austria.

Depending on mobilisation and price scenarios 0.5 to 0.8 mill m^3 equivalents yr^{-1} are not harvested as a consequence of constraints due to nature conservation.

In conjunction with economic constraints exclusion of WTH from 52% of forest land available for wood production resulted in a drastic reduction of the theoretical potential of branch, twig and needle biomass from 6.7 million m^3 equivalents yr^{-1} to a range of 1.04 to 1.13 million m^3 equivalents yr^{-1} for price scenarios 1 to 3 within the SC scenario (Table 6.3).

Realisable potentials considering all constraints defined in Section 6.2.2 range between 23.9 and 31.1 mill m³ stem wood equivalents yr⁻¹ including bark, branches, twigs and needles considering harvest losses for the period 2000–2020 (Table 6.4).

6.4 Discussion and Conclusions

The annual increment of the Austrian forest reaches 30.4 million m³ equivalents and exceeds significantly the total annual forest harvest according to ANFI (Table 6.1, period 07/09). Verkerk et al. (2011) calculated a slightly higher theoretical biomass potential from forest inventory data using the EFISCEN model for Austria, of 31.9 million m³ yr⁻¹ over bark for 2010 within the framework of a study comprising 27 EU member states.

For 2020 Verkerk et al. (2011) estimated 27.2 mill m³ yr⁻¹ over bark for a low, 31.2 for a medium and 37.8 for a high mobilisation scenario for Austria. While the results for the low and the medium mobilisation scenarios correspond to the results from our study, the results from the high mobilisation scenario come close to the theoretical potential of the ASV scenario of the HOBI study.

The defined technical, economic and environmental constraints within the HOBI study reduced the realisable potential to 73–75% of the theoretical potential for price scenario 1, and to 83–84% for price scenario 5 in the 2000 to 2020 period.

Actually realised potentials in Austria are significantly higher (60–85%) than the EU-27 mean of 58% according to Verkerk et al. (2011) for 2010. This may be due to high mechanisation of Austrian forestry, disturbances by windthrow and a high demand for forest biomass for material and energy use, amounting to 41.9 million m³ equivalents yr⁻¹.

Environmental considerations related to nutrient sustainability and technical-economic issues had the largest impacts among the constraints. Overall, the differences between the mobilisation scenarios were about the same as the differences between the price scenarios within a given mobilisation scenario itself. In particular, the resulting realisable forest biomass potential varies by about 3 to 4 million m³ equivalents (Table 6.4).

Uncertainties in respect of realisable potentials mainly concern ecological constraints. Besides the Austrian Forest Act § 16.1 (ForstG 1975, BGBl. Nr. 440/1975), which prohibits actions damaging the productive capacity of forest soils and sets a minimum rotation time for high forests as well as a minimum limit for canopy cover after thinnings, there is currently no binding regulation for extraction intensity in practical forest operations. Though rarely applied in practical forestry, guidelines for forest fertilisation and for the application of biomass combustion ash have been elaborated/published by the Austrian Advisory Council for Soil Protection and Soil Fertility (Baumgarten, 2012). Guidelines for PEFC certification (PEFC, 2014) as well as guidelines for silviculture with ecological considerations (Weinfurter, 2013), both advertising restrictions on whole-tree harvest, are voluntary. Due to steep terrain, timber is forwarded with cable cranes in more than 26% of the Austrian forests and often processed on the forest road. Whole-tree harvesting is therefore applied frequently on poor sites independently of the later utilisation of marginal resources. In consequence of these facts the extent of WTH in Austria can be assessed ex post only. Nutritional sustainability estimates for the decision support at the forest management unit level (e.g. Katzensteiner, 2011) are possible only if sufficient site information is made available. However, for Austria there exists no institutionalised forest site mapping system. Except for the Austrian Federal Forests and a few private enterprises no forest site maps with proper resolution and background data are available.

Further uncertainty arises from the fact that mass-balances may lead to incorrect conclusions as there are considerable uncertainties especially in weathering estimates (Stutter et al., 2003) as well as in deposition and leaching rates over time.

The general relationship between size of forest holdings and wood harvesting is known (e.g. Amacher et al., 2003), but there is great uncertainty for small-scale forest owners due to forests being used as 'savings', the existence of price elasticity, and infrastructural problems. Fifty-seven per cent of the forest land in Austria, excluding 'protected forest without yield', is located on properties < 200 ha.

Although multiple constraints on wood supply and mobilisation have been identified (e.g. MCPFE et al., 2010), it is not always possible to quantify these constraints, partly due to lack of data and partly due to the high complexity of interaction between different groups of constraints.

A severe recent threat is soil compaction due to a sometimes uncontrolled and unrestricted use of heavy machinery in forest operations in all forest property types. Another factor that is currently rarely considered in practical guidelines and models is soil organic matter management. In some eco-regions of Austria, soils with high organic matter content or thick forest floor layers probably rely on a certain input of woody debris to retain their function.

As of 2010, 24.3 million m^3 equivalents of woody biomass were conveyed to energy use, thus contributing about 12% to gross inland energy consumption (Nemestothy, 2012), while the net material use amounts to 17.6 million m^3 equivalents (Katzensteiner et al., 2013). Forest biomass contributed the largest share to renewable energy consumption in Austria (45%), followed by hydropower (34%) and other sources (16%) (Katzensteiner et al., 2013). The demand was met by Austrian forestry (23.6 million m^3 equivalents), imports of 10.1 million m^3 equivalents and other wood sources of 6 million m^3 equivalents (recycled wood, etc.) (Nemestothy, 2012).

The highest realisable potential from the scenarios calculated within the HOBI study indicate a possible increase of harvest of about 7 million m^3 stem wood equivalents yr^{-1} or about 30% (ASV, price scenarios 4 and 5) compared to Austrian forest production in 2010. The prices within these scenarios, however, lie beyond the long-time price level or the current price level. Most of the other scenarios indicate an increase of less than 20%.

Acknowledgements

The HOBI study was funded by the Austrian Federal Ministry of Agriculture, Forestry, Environment and Water Management. Data used were partly co-financed within the Life+ Programme of the EU (LIFE 07 ENV/DE/000218; project FUTMON) and the *BioSoil* project, a demonstration project co-financed under Regulation (EC) No 2152/2003 concerning the monitoring of forest and environmental interactions in the Community (Forest Focus). We thank two anonymous reviewers for helpful comments on the manuscript.

References

Amacher, G. S., Conway, M. C. and Sullivan, J. (2003). Econometric analyses of nonindustrial forest landowners: is there anything left to study? *Journal of Forest Economics*, 9, pp. 137–164.

Bauer, R. (2001). Maschinenkostenrechnung nach dem FBVA-Schema. *FBVA-Berichte*, 124, pp. 35–38.

Baumgarten, A. (2012). *Richtlinie für den sachgerechten Einsatz von Pflanzenaschen zur Verwertung auf land- und forstwirtschaftlich genutzten Flächen*. Fachbeirat für Bodenfruchtbarkeit und Bodenschutz. [online] Available at: https://dx.doi.org/10.2785/388553 [Accessed 15 February 2015]

BFW (2012). *The Austrian Forest Inventory 2007/09*. [online] Available at: http://bfw.ac.at/rz/wi.home. [Accessed 15 February 2015]

BFW (2015). *Ergebnisse der österreichischen Waldinventur (= Results of the Austrian Forest Inventory)*. [online] Available at: http://bfw.ac.at/rz/wi.home [Accessed 15 February 2015]

Bitterlich, W. (1948). Die Winkelzählprobe. *Allgemeine Forst- und Holzwirtschaftliche Zeitung*, 59, pp. 4–5.

Bitterlich, W. (1952). Die Winkelzählprobe. *Forstwissenschaftliches Centralblatt*, 71, pp. 215–225.

Cambia, M., Certini, G., Neri, F. and Marchi, E. (2015). The impact of heavy traffic on forest soils: a review. *Forest Ecology and Management*, 338, pp. 124–138.

CEC (1992). *Council Directive 92/43/EEC of 21 May 1992 on the conservation of natural habitats and of wild fauna and flora*. [online] Available at: http://eur-lex.europa.eu/LexUriServ/LexUriServ.do?uri=CELEX:31992L0043:EN:HTML [Accessed 15 February 2015]

CEC (2009). *Directive 2009/147/EC of the European Parliament and of the Council of 30 November 2009 on the conservation of wild birds*. [online] Available at: http://eur-lex.europa.eu/legal-content/EN/ALL/?uri=CELEX:32009L0147 [Accessed 15 February 2015]

CEC (2010). *Directive 2009/28/EC of the European Parliament and of the Council of 23 April 2009 on the promotion of the use of energy from renewable sources and amending and subsequently repealing Directives 2001/77/EC and 2003/30/EC*. [online] Available at: http://ec.europa.eu/energy/renewables/action_plan_en.htm [Accessed 15 February 2015]

De Vries, W., Posch, M., Reinds, G. J. and Kämäri, J. (1993). *Critical Loads and Their Exceedance on Forest Soils in Europe. Report 58 (revised version)*. Wageningen, NL: DLO Winand Staring Centre.

De Vries W., Reinds, G. J. and Posch, M. (1994). Assessment of critical loads and their exceedances on European forests using a one-layer steady-state model. *Water, Air and Soil Pollution*, 72, pp. 357–394.

Eckmüllner, O. and Schedl, P. (2009): Neue Ausformung in marktkonforme Sortimente. *BFW-Praxisinformation*, 18, pp. 8–9.

Eckmüllner, O., Schedl, P. and Sterba, H. (2007). Neue Schaftkurven für die Hauptbaumarten Österreichs und deren Ausformung in marktkonforme Sortimente. *Austrian Journal of Forest Science*, 124, pp. 215–236.

Egnell, G. and Valinger, E. (2003). Survival, growth, and growth allocation of planted Scots pine trees after different levels of biomass removal in clear felling. *Forest Ecology and Management*, 177, pp. 65–74.

Eitzinger, J., Haberl, H., Amon, B. et al. (2014). *Land- und Forstwirtschaft, Wasser, Ökosysteme und Biodiversität. Österreichischer Sachstandsbericht Klimawandel 2014 (AAR14)*. Austrian Panel on Climate Change (APCC). Vienna: Austrian Academy of Sciences, pp. 771–856.

Englisch, M. and Reiter, R. (2009). Standörtliche Nährstoff-Nachhaltigkeit der Biomassenutzung. *BFW Praxisinformation*, 18, pp. 13–15.

Erni, V., Lemm, R., Frutig, F., Breitenstein, M., Riechsteiner, D., Oswald, K. and Thees, O. (2003). *HeProMo – Produktivitätsmodelle für Holzerntearbeiten*. Windows-Software. Version 1.xx. Birmensdorf, Eidg. Forschungsanstalt WSL.

European Union (2015). *Eurostat statistical books. Energy balance sheets edition 2015*. doi: 10.2785/388553. [Accessed 31 August 2015]

FBVA (ed.) (1992). Österreichische Waldboden-Zustandsinventur, Ergebnisse. *Mitt. FBVA*, 168.

Fernholz, K., Bratkovich, S., Bowyer, J. and Lindburg, A. (2009). *Energy from Woody Biomass: A Review of Harvesting Guidelines and a Discussion of Related Challenges*. Minneapolis: Dovetail Partners.

FHP (Forst Holz Papier) (1996). *Holzernte im Schleppergelände 1.3: ÖBf-Schleppertabelle Nadelholz*. Vienna: FHP.

ForstG (1975). (= Forstgesetz 1975) BGBl. Nr. 440/1975 as amended on 9.9.2015. [online] Available at: www.ris.bka.gv.at/GeltendeFassung.wxe?Abfrage=Bundesnormen&Gesetzesnummer=10010371 [Accessed 9 September 2015]

Fürst, A. (2009). *Austrian Bioindicator Grid*. [online] Available at: http://bfw.ac.at/rz/bfwcms.web?dok=2824 [Accessed 15 February 2015]

Fürst, A. and Kristöfel, F. (2012). *Level II online*. [online] Available at: http://bfw.ac.at/rz/bfwcms.web?dok=2824 [Accessed 15 February 2015]

Glatzel, G. (1991). The impact of historic land use and modern forestry on nutrient relations of Central European forest ecosystems. *Fertilizer Research*, 27, pp. 1–8.

Gschwantner, T., Gabler, K. and Schadauer, K. (2010). Austria. In: Tomppo, E. et al. (eds) *National Forest Inventories*. Dordrecht: Springer Science Business Media B.V., pp. 57–71.

Helmisaari, H.-S., Hanssen, K. H., Jacobson, S., Kukkola M., Luiro, J., Saarsalmi, A., Tamminen, P. and Tveite, B. (2011). Logging residue removal after thinning in Nordic boreal forests: long-term impact on tree growth. *Forest Ecology and Management*, 261, pp. 1919–1927.

Horn, R., Vossbrink, J. and Becker, S. (2004). Modern forestry vehicles and their impacts on soil physical properties. *Soil and Tillage Research*, 79, pp. 207–219.

Indinger, A., Leutgöb, K., Lutter, E. et al. (2006). *Vorstudie für einen nationalen Biomasseaktionsplan für Österreich*. Vienna: Austrian Energy Agency.

ISO, 1994. ISO 11260. *Soil Quality – Determination of Effective Cation Exchange Capacity and Base Saturation Level Using Barium Chloride Solution*. Geneva: International Organization for Standardization.

ISO, 1995. ISO 11466. *Soil Quality – Extraction of Trace Elements Soluble in Aqua Regia*. Geneva: International Organization for Standardization.

ISO, 1998. ISO 13878. *Soil Quality – Determination of Total Nitrogen Content by Dry Combustion ('Elemental Analysis')*. Geneva: International Organization for Standardization.

Jacke, H. (2004). Holzernte im Hochlohnland. *Forst und Holz*, 59, pp. 69–74.

Jacobson, S., Kukkola, M., Mälkönen, E., Tveite, B. and Möller, G. (1996). Growth response of coniferous stands to whole-tree harvesting in early thinning. *Scandinavian Journal of Forest Research*, 11, pp. 50–59.

Kaarakka, L., Tamminen, P., Saarsalmi, A., Kukkola, M., Helmisaari, H.-S. and Burton, A. J. (2014). Effects of repeated whole-tree harvesting on soil properties and tree growth in a Norway spruce (*Picea abies* (L.) Karst.) stand. *Forest Ecology and Management*, 313, pp. 180–187.

Katzensteiner, K. (2011). Nährstoffbilanzmodelle und forstliche Standortskarten als Basis für eine nachhaltige Biomasseproduktion (Combining nutrient balance models and site maps as a basis for sustaining forest biomass production). In: Gerzabek, M. I. and Glößl, J. (eds) *Präsentation von geförderten Projekten der Stiftung '120 Jahre Universität für Bodenkultur'*. Elektronische Publikation zur Vortragsveranstaltung der Stiftung am 25.10.2011. [online] Available at: www.boku.ac.at/fileadmin/_/H13/Stiftungen_und_Preise/Elektronische_Publikation_120_Jahre_BOKU.pdf [Accessed 15 February 2015]

Katzensteiner, K., Englisch, M. and Nemestothy, K. P. (2013). Impacts of increased biomass utilization on soil sustainability in Austria. In: Helmisaari, H-S. and Vanguelova, E. (eds) *Proceedings of the Workshop W6.1 Forestbioenergy and Soil Sustainability at EUROSOIL Congress 2nd July to 6th July 2012, Bari, Italy*. [online] Available at: www.helsinki.fi/forestsciences/eurosoil/index.html and www.oecd.org/tad/crp/ [Accessed 15 February 2015]

Kindermann, G. (2010). A climate sensitive refining on the basal area increment model in PROGNAUS. *Austrian Journal of Forest Science*, 127, pp. 147–178.

Kooperationsplattform Forst Holz Papier (2006). *Österreichische Holzhandelsusancen 2006 (= Austrian Standard Rules for Timber Classification)*. Vienna: Service-GmbH der Wirtschaftskammer Österreich.

Krapfenbauer, A. (1983). Von der Streunutzung zur Ganzbaumnutzung. *Austrian Journal of Forest Science*, 100, pp. 143–174.

Krapfenbauer, A. and Buchleitner, E. (1981). Holzernte, Biomassen- und Nährstoffaustrag, Nährstoffbilanz eines Fichtenbestandes. *Austrian Journal of Forest Science*, 98, pp. 193–223.

Kreutzer, K. (1979). Ökologische Fragen zur Vollbaumernte. *Forstwissenschaftliches Centralblatt*, 98, pp. 298–308.

Ledermann, T. (2002). Ein Einwuchsmodell aus den Daten der Österreichischen Waldinventur 1981–1996. *Austrian Journal of Forest Science*, 119, pp. 40–77.

Ledermann, T. (2006). Description of PROGNAUS for Windows 2.2. In: Hasenauer, H. (ed.) *Sustainable Forest Management – Growth Models for Europe*. Berlin, Heidelberg, New York: Springer-Verlag, pp. 71–78.

Ledermann, T. and Neumann, M. (2009). Prognose des Waldwachstums und des Nutzungspotentials. *BFW Praxisinformation*, 18, pp. 5–7.

Lexer, M. J., Honninger, K. and Englisch, M. (1999): Estimating chemical soil parameters for sample plots of the Austrian Forest Inventory. *Forstwissenschaftliches Centralblatt*, 118, pp. 212–227.

MCPFE, EC DG Agriculture and Rural Development and UNECE/FAO (2010). *Good Practice Guidance on the Sustainable Mobilisation of Wood in Europe*. Brussels: EC DG Agriculture and Rural Development.

Meiwes, K. J., Asche, N., Block, J., Kallweit, R., Raben, G. and von Wilpert, K. (2008). Potenziale und Restriktionen der Biomassenutzung im Wald. *Allgemeine Forstzeitung*, 63, pp. 598–604.

Monserud, R. A. and Sterba, H. (1996). A basal area increment model for individual trees growing in even- and uneven-aged forest stands in Austria. *Forest Ecology and Management*, 80, pp. 57–80.

Monserud, R. A. and Sterba, H. (1999). Modeling individual tree mortality for Austrian forest species. *Forest Ecology and Management*, 113, pp. 109–123.

Mutsch, F., Leitgeb, E., Hacker, R., Amann, C., Aust, G., Herzberger, E., Pock, H. and Reiter, R. (2013). Projekt BioSoil – Europäisches Waldboden-Monitoring (2006/07) Datenband Österreich Band I: Methodik, Standort- und Bodenbeschreibung, Bodendaten aus Burgenland, Kärnten, Niederösterreich und Oberösterreich Band II: Bodendaten aus Salzburg, Steiermark, Tirol und Vorarlberg, Deskriptive Statistik. *BFW-Berichte*, 145.

Nachtmann, G. (2006). Height increment models for individual trees in Austria depending on site and competition. *Austrian Journal of Forest Science*, 123, pp. 199–222.

Nemestothy, K. P. (2012). Die Bedeutung von Holz als erneuerbarer Energieträger. *BFW Praxisinformation*, 28, pp. 5–8.

Nord-Larsen, T. (2002). Stand and site productivity response following whole tree harvesting in early thinnings of Norway spruce (*Picea abies* (L.) Karst.). *Biomass and Bioenergy*, 23, pp. 1–12.

Olsson, M., Rosén, K. and Melkerud, P.-A. (1993). Regional modelling of base cation losses from Swedish forest soils due to whole-tree harvesting. *Applied Geochemistry*, 8, pp. 189–194.

PEFC (Programme for the Endorsement of Forest Certification) (2014). *PEFC Leitlinien für die nachhaltige Waldbewirtschaftung in Österreich (Appendix 2)*. Version 2014. [online] Available at: www.pefc.at/content/downloadcenter/technische_dokumente.php [Accessed 31 August 2015]

Rehfuess, K. E. (1990). Waldböden. Entwicklung, Eigenschaften und Nutzung. *Pareys Studientexte* 29. Hamburg and Berlin: Verlag Paul Parey.

Schadauer, K. (2009). Naturschutz – Wie können Nutzungseinschränkungen ermittelt werden? *BFW Praxisinformation*, 18, pp. 18–19.

Schneider, J. (1998). *Kartierung der nassen Deposition in Österreich*. Wien: Umweltbundesamt.

Schubert, G. (ed.) (2003). *Hydrogeologische Karte der Republik Österreich im Maßstab 1:500.000*. Wien: Geologische Bundesanstalt.

Siemion, J., Burns, D. A., Murdoch, P. S. and Germain, R. H. (2011). The relation of harvesting intensity to changes in soil, soil water, and stream chemistry in a northern hardwood forest, Catskill Mountains, USA. *Forest Ecology and Management*, 261, pp. 1510–1519.

Stampfer, K., Limbeck-Lilienau, B., Kanzian, C. and Viertler, K. (2003). *Baumverfahren im Seilgelände - Verfahrensbeispiele: Wanderfalke mit Prozessor Woody 50, Syncrofalke mit Prozessor Wolf 50 B*. Wien: Eigenverlag, Kooperationsabkommen Forst-Platte-Papier.

Sterba, H., Brunner, H., Gugganig, H. and Hauser, B. (2003). Stammzahlreduktion ja, aber nicht als Ganzbaumnutzung. *Österreichische Forstzeitung*, 10, pp. 18–19.

Stutter, M., Langan, S. and Cresser, M. (2003). Weathering and atmospheric deposition signatures of base cations in upland soils of NE Scotland: their application to critical load assessment. *Geoderma*, 116, pp. 301–324.

Sverdrup, H. (1990). *The Kinetics of Base Cation Release due to Chemical Weathering*. Lund: Lund University Press.

Van der Salm, C., Köhlenberg, L. and De Vries, W. (1998). Assessment of weathering rates in Dutch loess and river-clay soils at *pH* 3.5, using laboratory experiments. *Geoderma*, 85, pp. 41–62.

Verkerk, P. J., Anttila, P., Eggers, J., Lindner, M. and Asikainen, A. (2011). The realisable potential supply of woody biomass from forests in the European Union. *Forest Ecology and Management*, 261, pp. 2007–2015.

Von Wilpert, K. (2006). Waldbauliche Steuerung des Stoffhaushalts von Waldökosystemen. *FVA-Einblick*, 10, pp. 2–4.

Weinfurter, P. (2013). *Waldbau in Österreich auf ökologischer Grundlage. Eine Orientierungshilfe für die Praxis*. Wien: Landwirtschaftskammer Österreich, Ländliches Fortbildungsinstitut Österreich.

Weiss, P. (ed.) (2006). Austrian Biomass Functions. *Austrian Journal of Forest Science*, 123, pp. 1–102.

7

Sustainable Biomass Potentials from Coppice Forests for Pyrolysis: Chances and Limitations

VALERIU-NOROCEL NICOLESCU, EDUARD HOCHBICHLER
AND VIKTOR J. BRUCKMAN

Abstract

Coppice forests, originating from vegetative propagation (stump stools or root suckers), are an important component of forest ecosystems worldwide. Even though their economic importance has been reduced in Europe, especially since the Second World War, they still serve as important sources of raw materials (mostly firewood) for local communities. In addition, coppice forests could be considered as 'hotspots of biodiversity', having high habitat, historical and genetic resource values while being relatively resistant to environmental impacts such as droughts.

In this context, our chapter emphasizes the main characteristics of silvicultural coppice systems (e.g. simple coppice, short-rotation coppice, high coppice, coppice selection and coppice with standards), their ecology, history and current significance in Europe.

The two case studies on carbon stocks of coppice with reserves and coppice with standards in Austria are important arguments for considering coppice forests as a sustainable source of sawlogs for highly valuable wood products and of biomass (energy wood) that can be used for firewood as well as in pyrolysis processes.

7.1 Introduction

7.1.1 Characteristics of Coppice Forests

Coppice is a forest stand composed of stools (stumps) repeatedly cut that produce coppice shoots which form the major part of the crop (Hartig, 1877; Harmer, 1995). Under special conditions, such stands can also originate from root suckers. For centuries, coppice forests served as a sustainable source of raw materials (Buckley, 1992), such as fuelwood, charcoal or small-diameter construction wood for local communities.

Coppice woodlands represent unique ecosystems as a consequence of stand structure and disturbance regimes. A common high forest management system is typically characterized by a small number of stochastic events, such as thinning operations, and occasional windthrows, insect attacks and so on which create disturbance and thus contribute to diversity. The relatively short rotation periods in coppice-managed forests lead to a development pattern that faces a major disturbance at the time of harvesting followed by a period of regeneration. Light-demanding species are therefore able to regenerate after every coppice harvesting. Although the stand structure is relatively even in terms of age (with some exceptions as mentioned below), coppice forests usually show a high diversity of plant and animal species. The fully functional root system of a coppiced stem, which enables rapid resprouting and growth starting in the first vegetation period, can be a major advantage. In areas suffering from seasonal drought, it enables a high success rate in regeneration after coppicing as the root system may access deeper soil water tables. As a consequence, coppice stands do not need any weed control as the coppice species typically have this advantage over relatively shallow rooting herb species and grasses. In circumstances where fuelwood was the only source for (thermal) energy, as was the case in Vienna until about 1850 (Krausmann, 2013), a constant supply was essential for the development and welfare of local inhabitants. As coppiced woodlands show stable patterns of regeneration, nearly regardless of environmental conditions, land management was more foreseeable.

Species composition of coppice forests may change only gradually over a number of rotation periods, since coppicing removes only parts of the aboveground shoot while intact stools remain at the same location. For this reason, the age of coppice stools may be equivalent to a number of rotation cycles.

Coppice forests are often described as 'hotspots of biodiversity' (Terada et al., 2010; Nyland, 2002), although the habitat quality is determined by the current management practices. For instance, coppice with standards or over-matured (outgrown) coppice woodlands may offer a larger number of ecological niches as the stand structure tends to be more heterogeneous (Bruckman et al., 2011). Planted short-rotation coppices (SRC) represent the other extreme case, where a homogeneous stand structure is targeted in order to enable a high degree of mechanization, from site preparation to harvesting.

There is a significant interaction between coppice woodlands and the surrounding landscape in terms of habitat quality as shown in the case of bird communities, for instance (Berg, 2002).

The relatively high number of stems and resulting dense stand, specifically in the early stages of development after resprouting, offer favourable conditions for birds and mammals as there are many possibilities to hide from natural predators. On the other hand, dense stands inhibit or limit the development of herbaceous ground vegetation and therefore decrease diversity of herb species after crown closure. This implies very strong temporal diversity dynamics as compared to relatively stable high forest ecosystems.

7.1.2 Rediscovering Coppice Forestry as a Reliable Source for Biomass

Especially in the last two decades there is a growing interest in coppicing in Europe due to the increasing demand for energy production from renewable resources, supported by EU and national policies. This development was triggered by climate change mitigation efforts in the wake of the Kyoto Protocol (Zlatanov and Lexer, 2009). In addition, the increase in firewood prices in the last few years makes coppicing an interesting forest management alternative in the close perspective (Kneifl et al., 2011). It has been shown that biomass can be economically harvested from traditional coppice forest systems, using modern machines (Spinelli et al., 2014). This makes coppice forests an interesting alternative source for obtaining woody biomass for biochar production.

The significant contributions of coppice forests (e.g. provision of biomass, biodiversity preservation while being relatively resistant against environmental impacts such as droughts) were recently explored scientifically by introducing a European Cooperation in Science and Technology (COST) Action. The Action titled 'Innovative management and multifunctional utilization of traditional coppice forests – an answer to future ecological, economic and social challenges in the European forestry sector' (EUROCOPPICE) emphasizes very well the interdisciplinary research context and the multi-functionality of such forest systems. Coppice management was stopped as a consequence of decreased demands for fuelwood in the nineteenth century, but especially after the Second World War, and large areas were converted into high forest. Remaining patches of coppice were often not coppiced at regular intervals leading to a large number of non-typical coppice stands and a shift towards shade-tolerant species, a process which was recently observed not only in Europe but also in Japan (Bruckman et al., 2016). Under these circumstances, and considering the current situation of coppice forests in Europe, one may emphasize the need for their continued management. By doing so, their long-term survival as well as good health state can be ensured (Sjölund and Jump, 2013).

7.2 Silvicultural Coppice Systems

7.2.1 Simple (Low) Coppice

Simple, or low, coppice (SC) consists of a clear felling and the reproduction is based on stool shoots or root suckers. It has *cultural tasks* (vegetative regeneration) as well as *economic tasks* (production of small- and medium-sized material, easily marketable at short intervals).

Simple coppice is used especially for broadleaved tree species such as oaks (*Quercus* spp.), sweet chestnut (*Castanea sativa*), hornbeam (*Carpinus betulus*), limes (*Tilia* spp.), eucalypts (*Eucalyptus* spp.), ash (*Fraxinus* spp.), alders (*Alnus* spp.), black locust (*Robinia pseudoacacia*), poplars (*Populus* spp.) and so on, which withstand repeated coppicing.

Figure 7.1. Wood products from sweet chestnut (a – Great Britain) and black locust coppices (b – France) (photos V.-N. Nicolescu).

Birches (*Betula* spp.) and European beech (*Fagus sylvatica*) are less responsive to coppice (Broilliard, 1881; Fankhauser, 1921) so that the use of these tree species in simple coppices is less recommended.

As the majority of broadleaved species coppice well for only about 40 years, the rotation of stands treated as simple coppice ranges generally between 15 and 25 years (Hartig, 1877; Hamm, 1900). Such stands produce small-diameter trees used for firewood, basket work, pea and bean sticks, hoops, hurdles, fascines, fencing, vine and hop poles, handles for tools and implements, pulpwood and, etc. (Matthews, 1991, see Figure 7.1).

The rotation can be longer – up to 30–35 years – in the case of oak, sweet chestnut and black locust when production of large timber, used for wood barrels, flooring, mining timber, solid furniture and so on, is targeted (Harmer and Howe, 2003; Rédei et al., 2007; Starr, 2008; Stajic et al., 2009; Stähr, 2013).

In low coppice, trees are usually cut close to the ground, the stools being given a sloping surface (up to 20°) to prevent water from settling. By cutting this way, the shoots appear at ground level in order to form independent roots. In drier areas of Europe, trees such as Holm oak (*Quercus ilex*) are cut below ground level, the stools are covered with earth and roots produce suckers. This system is also used for black locust in countries such as Hungary and Romania.

In temperate regions, late winter–early spring before the beginning of growing season is considered the best time for low coppice cut (Evelyn, 1664; Hartig, 1805; Lorentz and Parade, 1867). The only major exceptions to this optimum period are the oak tan-bark coppice (cutting in May–early June, after the growing season has commenced – Hartig, 1805; Lorentz and Parade, 1867; Schwappach, 1904) and alder, willow and poplar coppices in swampy sites (cut in winter, when the ground is firm – Schwappach, 1904; Schwappach et al., 1914).

Between the two coppice cuts, tending operations such as cleaning-respacing and thinning are important; they target the removal of unwanted species or individuals, improvement of the quality and quicker growth of final crop, and production of small- and medium-sized material that may increase the financial return (Matthews, 1991). The number of these operations depends chiefly on the rotation length, competition among shoots, and the wood market. For instance, in the black locust coppice stands of Hungary and Romania with rotations of 25–35 years, there are one–two cleaning-respacing and one–two thinnings common (Rédei et al., 2007; Anon., 2000), compared to only two thinnings in France (Carbonnière et al., 2007). In sweet chestnut coppices, the number of tending operations ranges from none in Britain (Crowther and Evans, 1984) to three in Greece (Bourgeois, 1991).

Since long ago, low coppices reaching the rotation age have been worked by the method of annual coupes by area (already recommended by the Royal French Ordinance of 1545, according to Huffel, 1907), after deciding the rotation on the basis of the size of material required. The total area treated as low coppice is divided into annual coupes equal to the number of years in the rotation; each year, one coupe is coppiced. All material should be removed from the cutting area before flushing begins, so as to avoid damage to the young shoots, which are very fragile (Cotta, 1841; Lorentz and Parade, 1867; Schwappach et al., 1914).

After repeated coppicing, stools start to die and show a gradual decline in yield, the potential of producing young and vital shoots decreasing with increasing age and shoot diameter (Nyland, 1996; Hamilton, 2000; Harmer and Howe, 2003; Matula et al., 2012). Consequently, the stools should be replaced after two–three coppice cycles in temperate regions (Lorentz and Parade, 1867; Troup, 1928; Nyland, 1996) or three–four coppice cycles in tropical regions (Evans, 1992; Evans and Turnbull, 2004).

The productivity of simple coppices is very variable but can reach 7 m^{-3} ha^{-1} yr^{-1} in black locust stands of France (Carbonnière et al., 2007), 8–10 m^{-3} ha^{-1} yr^{-1} in sweet chestnut coppices of France and Great Britain (Bacchetta, 1984; Rollinson and Evans, 1987), and over 10 m^{-3} ha^{-1} yr^{-1} in the poplar and willow coppices of Germany (Stähr, 2013). At global level, the most productive coppices are those of eucalypts, yielding 5–10 m^{-3} ha^{-1} yr^{-1} on marginal sites, 20–30 m^{-3} ha^{-1} yr^{-1} on more favourable sites, and reaching 40–50 m^{-3} ha^{-1} yr^{-1} where the combination of site, species and cultural conditions is exceptional (Cannell, 1979; NSA, 1980, both cited in Matthews, 1991).

7.2.2 Short-Rotation Coppice

Short-rotation coppice (SRC) is a special form of low coppice, of major economic importance for the production of *bioenergy* (chips, pellets, splinters) or *biomass* (pulpwood, paper, cardboard, fibreboard, chipboards, etc.) in the last decades (CREFF, 2012). The 'engine' of this form of coppice was the energy crisis in the 1970s, when willow species started to be used as an alternative source of renewable energy in order to replace fossil fuels, especially in North European countries (Savill and Evans, 1986; Hart, 1994; Dimitriou, 2009; Verwijst et al., 2013). In Europe and North America, these cultures have been considered as a *viable alternative* for the surplus of agricultural land, possibly leading to agricultural over-production (Dawson, 1991; Ford-Robertson et al., 1991; Maryan, 1991; Christersson, 1998; Cannell, 2004), as well as for reducing the greenhouse effect, as they act as a major carbon sink (Bewa and Bouvot-Mauduit, 1999).

Figure 7.2. Short rotation willow coppice (Covasna County, Romania) (photo V.-N. Nicolescu).

In this system, the most used species are willow (*Salix* spp., see Figure 7.2), poplars and eucalypts; alders and black locust are also used on a smaller scale (CREFF, 2012).

They are established using 18–20 cm long bare-rooted cuttings (Tubby and Armstrong, 2002; Verwijst et al., 2013). Planting distances, using double rows of cuttings, are 0.75 m between these rows and 1.5 m between each pair of rows. The planting distance along the rows is 0.4–1.0 m so that the number of plants per ha lies between 10,000 and 25,000, with ca. 15,000 plants ha^{-1} on average (CREFF, 2012; Guidi et al., 2013).

In all species treated as SRC, rotation age is maximum 10 years. This ranges between two and four years (black locust – Paris et al., 2006), two and five years (willows – Hinton-Jones and Valentine, 2008; Hochbichler et al., 2011; Stolarski et al., 2011; Schweier and Becker, 2012; Verwijst et al., 2013), four and eight years (eucalypts – Cannell, 2004), and up to 8–10 years (poplars – Cannell, 2004; Johnson, 1999; Tubby and Armstrong, 2002; Hochbichler et al., 2011). When reaching the rotation, the shoots are harvested fully mechanically in autumn or winter time, after the leaf fall. After four to five coppice cycles, at age 20–25 (seldom 30) years, the old stools show reducing yield and are replaced with cuttings so the SRC is re-established (Dimitriou, 2009; CREFF, 2012; Savill, 2013).

The productivity of SRC reaches, in general, (8) 10–15 (20) t dry matter ha^{-1} yr^{-1}, depending on species or clone, site, region and planting scheme (Hochbichler et al., 2011; CREFF, 2012; Verwijst et al., 2013). However, SRC should produce at least 10 t dry matter ha^{-1} yr^{-1} to be economically viable (Tubby and Armstrong, 2002).

Figure 7.3. White willow (*Salix alba*) repeatedly pollarded (photo V.-N. Nicolescu).

Currently 'short rotation poplar and willow coppices are becoming more common in Europe as a renewable energy source' (FOREST EUROPE, 2015), being considered as a *strategic resource of wood products* (FAO, 2008, cited in Schweier and Becker, 2012).

7.2.3 High Coppice (Pollarding)

High coppice (HC) is, in fact, a simple coppice (Chivulescu, 1886), and consists of cutting the tops of trees so as to stimulate production of numerous straight shoots on the top of the cut stem (Matthews, 1991), as shown in Figure 7.3. These shoots (most used species: oak, willow, poplar, linden, elm, ash, black locust, hornbeam, plane tree) are trimmed off periodically to provide fodder but also material for basket-making, fencing hurdles, fascines, pulpwood and even sawn timber. After this series of trimmings, the upper part of the trunk looks like a reversed stump (with roots above the stump), sometimes called 'chair' (Drăcea, 1942).

Pollarding is used for park alley and garden trees, those along the streets and roadsides, around fields and hop gardens. The system is also used in areas with long pastoral traditions (Basque Regions of France and Spain) or with large-scale silvo-pastoral systems (Spain, Portugal) (Bastien, 1999). In these cases, pollarding is done at heights of 2.5 to 3 (3.5) m, well out of the reach of cattle and sheep (Schwappach et al., 1914; Dengler, 1935; Drăcea, 1942; Matthews, 1991).

The most important forestry use of HC is in the case of willows and poplars along the sides of rivers, streams and ditches. This is the case along the Danube River in Romania, where the pollarding is done at heights between (1) 2 and 3 m – *above the highest flooding levels over a long chronosequence* – to avoid any damage to the high stump caused by the flooding waters (Drăcea, 1923).

The cutting of shoots is carried out as in the case of simple coppice, especially during the winter (legal period 15 September to 31 March – Anon., 2011).

In time, after two–three cycles of cuts, pollards start to deteriorate (often becoming hollow), the coppicing potential and vigour of shoots being more and more reduced; consequently, pollards are replaced with seedlings or long cuttings, (1) 2 m long and 3–5 cm thick, which will be treated subsequently as pollards (Rădulescu and Vlad, 1955; Anon., 2000).

7.2.4 Coppice Selection System

The coppice selection system (CSS) is similar to the selection system in high forest. A target diameter is fixed according to the size of aimed wood product, followed by an estimate of the age at which material of this size is produced. This age determines the rotation, which is divided into a number of felling cycles, and the total area of forest under CSS is divided into annual coupes equal in number to the number of years in the felling cycle. Each year coppice fellings are carried out in one of the annual coupes (Matthews, 1991). Shoots of one to three (seldom four) ages, depending on the number of felling cycles in the rotation, coexist on the same stool. Only shoots reaching the target diameter are cut while the others are thinned.

The coppice selection system has been historically used for European beech in the Pyrenees (rotation of 30 years, with two felling cycles of 15 years or three of 10 years), Morvan Massif of southern France (rotation of 36 years, with four felling cycles of nine years) (Matthews, 1991), Apennine Massif in Italy (rotation of 27–36 years, with three felling cycles of 9 to 12 years) (Coppini and Hermanin, 2007) as well throughout the Balkans. It has many *advantages* (the young shoots are better protected by the cover of the older shoots; the soil remains permanently covered – it is a system appropriate in erosion prone and watershed protection areas) but also *disadvantages* (cutting of the large shoots is more difficult and tedious, and is apt to damage the smaller ones; cutting at ground level is generally impossible as the stool becomes taller after each cut; development of the shoots is poorer than in the case of simple coppice) (Lorentz and Parade, 1867; Boppe, 1889; Cochet, 1971; Matthews, 1991; Stajic et al., 2009).

Based on these facts it has for a long time been considered that this silvicultural system has not many advantages and too many disadvantages so it cannot be expanded outside the area where it was born (Lorentz and Parade, 1867).

7.2.5 Coppice with Standards

Coppice with standards (CWS) is a silvicultural system in which selected stems are retained as standards (see Figure 7.4) at each coppice harvest to form an uneven-aged overstorey,

Figure 7.4. Coppice-with-standards stand in Austria (photo Viktor Bruckman). (A black and white version of this figure will appear in some formats. For the colour version, please refer to the plate section.)

which is removed selectively on a rotation consisting of a multiple of the coppice cycle (IUFRO, 2005). In such a stand, which is 'the oldest form of irregular forest' (Garfitt, 1995), two distinct elements are found (Cotta, 1841; Lorentz and Parade, 1867; Boppe, 1889; Jolyet, 1916):

(a) A lower even-aged storey (*underwood*), originating from shoots and treated as coppice. This storey plays *economic* (produces small- and medium-sized timber, used especially as firewood) and *cultural* roles (protects the soil as well as trunks of standards in the upper storey).
(b) An upper uneven-aged storey (*overwood*) composed of taller but scattered trees, originating from both shoots and seeds, distributed as uniformly as possible and treated as high forest. It also has *economic* (produces a certain proportion of large timber) and *cultural* roles (provides seeds for natural regeneration) (Chivulescu, 1886; Drăcea, 1942; Rădulescu and Vlad, 1955).

In order to establish a CWS stand, after fixing the rotation of coppice, the following operations are carried out (Dengler, 1935; Drăcea, 1942):

1. The coppice stand reaching the rotation age *r* (usually 20–25 years) is clear cut as simple coppice and a certain number of coppice trees (standards), of desired species, with good forms and increments, are reserved.

2. After another coppice rotation of 20–25 years, the great majority of standards of **2r** (40–50 years) are reserved, extracting those that have deteriorated or are slow growing. Among the coppice storey, the majority of individuals are removed while a certain number of trees are reserved as a second cohort of standards **r**.
3. The same operation is repeated regularly for several coppice rotations of **r** years so the coupe about to be felled consists of coppice aged **r** years together with standards aged **2r**, **3r**, **4r** … years, and a number of young prospective standards aged **r** years.

Standards should originate from seed or, if not possible, from young and vigorous shoots, already individualized from the stool, or from root suckers. The trees reserved as standards originate from valuable and light-demanding species, with tall, large, balanced and open crowns, wind-firm and scattered as regularly as possible (Cotta, 1841; Lorentz and Parade, 1867; Bagneris, 1878; Boppe, 1889). In CWS, standards are tall, with weak natural pruning and so have shorter boles than high forest trees, as well as large crowns, with a high proportion of branches (Drăcea, 1942; Savill, 1993, Starr, 2008).

The most recommended species as standards are oak (Figure 7.5), elm and ash. Other important species are sycamore (*Acer pseudoplatanus*), Norway maple (*Acer platanoides*), wild cherry (*Prunus avium*), wild service tree (*Sorbus torminalis*), service tree (*S. domestica*), black walnut (*Juglans nigra*), as well as conifers such as larch (*Larix* spp.) and pine (*Pinus* spp.) (Lorentz and Parade, 1867; Muel, 1884; Chivulescu, 1886; Boppe, 1889; Schwappach, 1904; Jolyet, 1916). European beech is not well suited as a standard mainly because of its sensitivity to sun scorch when isolated as well as densely foliaged crowns, affecting the growth of the coppice storey (Boppe, 1889; Schwappach, 1904; Troup, 1928).

The number of standards existing in a CWS at any given moment has evolved from a minimum of 30 trees/ha in Britain in 1543 (Crowther and Evans, 1984) to 40–50 (France – Forest Law of 1827 – Bastien, 1999) or even 100 trees/ha (Germany – Cotta, 1841).

Nowadays, the proposed number of standards is 50–100 trees/ha, of all age classes; the number of standards in each age class should be about half of the number in the age class immediately younger. For instance there can be fifty standards in age class I (youngest), 30 in age class II, 13 in age class III, and seven in age class IV (oldest) (Harmer and Howe, 2003). Hochbichler (2008, 2009) developed stem number guidelines for different overwood cover percentages. The number of standards ranges between 82 and 163 trees ha^{-1} before cut in relation to an overwood canopy cover of 33% and 66%, respectively (target diameter of 60 cm; moderate sites; height of the overwood: 18–20 m; rotation period: 30 years).

The rotations adopted for standards, 'that should be reserved as long as they are healthy, vigorous, and growing sustainably' (Muel, 1884), reaches: oak, 100–130 years (Crowther and Evans, 1984; Garfitt, 1995; Hochbichler, 2008); ash, elm, *Acer* spp., 75–(90) 100 years (Muel, 1884; Drăcea, 1942; Hochbichler, 2008); larch, pine, 80 years (Fankhauser, 1921); wild cherry, *Sorbus* spp., 50–70 years (Muel, 1884; Drăcea, 1942); birch, wild cherry, 40–60 years; *Sorbus* spp. 80–120 years (Hochbichler, 2008).

Figure 7.5. Oak standard trees in (a) France (Forêt-de-Brin, and (b) Romania (Târgu-Mureş area) (photos V.-N. Nicolescu). (A black and white version of this figure will appear in some formats. For the colour version, please refer to the plate section.)

The **underwood** (coppice storey) in CWS consists of a mixture of species coppicing vigorously, able to withstand the shadow of standards (at least semi-shade tolerant species), and producing firewood of good quality (Schwappach, 1904; Jolyet, 1916). The most recommended species for underwood are hornbeam, field maple (*Acer campestre*), European beech, linden, sweet chestnut and hazel (*Corylus avellana*) (Hartig, 1877; Schwappach, 1904; Schwappach et al., 1914; Jolyet, 1916; Drăcea, 1923; Troup, 1928; Dengler, 1935).

The rotations of underwood are nowadays 20–30 years (Boudru, 1989; Dubourdieu, 1997; Hochbichler, 2008), the age of 20 years being considered as 'minimum minimorum' (Lafouge, 1964).

In CWS, the silvicultural operations to be carried out depend on the stand storey:

(a) Release cutting, cleaning-respacing and 1–2 thinnings are performed in the underwood, only if considered necessary, the latter operations in order to prepare the standards for their life after the cutting of the coppice storey (Rădulescu and Vlad, 1955).
(b) Removal of epicormic branches along the stems of standards (especially pedunculate oak) receiving a surplus of light after the cutting of coppice storey (Lorentz and Parade, 1867; Bagneris, 1878; Broilliard, 1881).

The production of CWS located on rich sites can be higher than that of high forests (Machar, 2009), while the coppice storey in a CWS is less vigorous than that of simple coppice, its production being a bit higher than half of the latter (Boppe, 1889; Savill, 1993).

7.3 Historical Background and Current Significance of Coppice Forests in Europe

Simple (low) coppice is considered as *the oldest silvicultural system* known (Troup and Jones, 1952; AFOCEL, 1982), its first traces in Great Britain dating back to Neolithic times (Crowther and Evans, 1984; Peterken, 2001). It was the only form of forestry systematically practised by the Greeks and Romans, the latter preferring it over high forest (Cato the Elder, 234–149 BC, cited in AFOCEL, 1982). The Romans were aware of the coppicing potential of broadleaves and used coppice forests (so-called *sylvae caeduae* or *sylvae minutae*) for the production of vine stakes or firewood (Huffel, 1907). Pliny the Elder (AD 23–79), in his *Naturalis historia* (vol. III – Botany), mentioned rotations of seven years in sweet chestnut for vine stakes and 10 years for Hungarian oak (*Quercus frainetto*).

During the Middle Ages (from AD 337 to AD 1500), simple coppice was practised on a large scale in Central Europe. In France, the rotation age of simple coppices was shorter than 10 years (five, four, three or even two years in communal or private forests – Huffel, 1907) until the eighteenth century, even though all royal ordinances released since the sixteenth century (in 1563, 1573, 1587, 1588 and 1669) sought to impose rotations of minimum 10 years in order to produce larger diameter trees not only firewood (Huffel, 1907; Mathey, 1929). Rotations over 10 years were used only in state forests, reaching 15–20 years in the eighteenth century (Huffel, 1907). In Germany, the rotation age of simple coppices during the Middle Ages was also short (7–12 years) but grew up to 20 or 30 years at the end of the sixteenth century and during the seventeenth century (Huffel, 1907). In this period, but also in the eighteenth century, coppice continued to supply domestic firewood, building material, fencing and, in addition, increasing quantities of fuel for industry and oak bark for tanning (Matthews, 1991).

The decline of European simple coppice forests started as soon as coal began to supplement wood as a fuel (mid-nineteenth century – Savill, 1993). This decline in the coppice system accelerated in Britain after the First World War as rural electrification spread (Crowther and Evans, 1984), and again after the Second World War as oil and gas became readily available (Rădulescu and Vlad, 1955; Crowther and Evans, 1984; Matthews, 1991).

Even though many simple coppice forests have been converted to high forest since the mid-nineteenth century, this system is still used on a very large scale in Central European countries. This is the case in France (2.4 million ha – Bastien, 1999) and Italy (0.915 million ha – Diaci et al., 2012).

The first traces of **coppice with standards** in Britain also date from Neolithic times when standard trees were kept occasionally for more rotations in order to produce construction wood (Rackham, 1990, cited in Rietbergen, 2001). The system has been applied in Europe since around AD 600 in Germany, before AD 1200 in France and since AD 1295 in Britain (Garfitt, 1995). Its application was legally authorized in Europe based on:

- France: Royal Ordinance (1376) on the preservation of standards in royal coppice forests, this prescription being expanded to all French forests in the second half of the fifteenth century. All subsequent Royal Ordinances (1523, 1544, but especially Colbert's from 1669) have included the retention of standards (Lafouge, 1964).

- Germany: instructions released at the end of the fifteenth century and during the sixteenth century on the number of standard trees and the age when they should be selected (Dengler, 1935).
- Britain: the *Act for the Preservation of Woods* (1543), indicating the way in which the CWS should be applied (James, 1991), with a minimum of 30 standard trees retained per ha (Troup, 1928).
- Austria: instructions for management of coppice with reserves/standards were found in an urbarium of the lordship Wolkersdorf from 1553 (cited by Frank, 1937, and Schöfberger, 2006). The villagers were allowed to cut underwood trees for firewood, but a defined number of trees (reserves, standards) had to be left. Old trees (standards) and fruit trees may not be cut in principle.

However, the current name of this silvicultural system (*taillis-sous-futaie* or *taillis composé* in French) was used in the forestry literature only in the first half of nineteenth century, as part of Lorentz and Parade's Handbook of Silviculture (*Cours élémentaire de culture des bois*), published in 1837.

In France, CWS reached its greatest extent during the eighteenth and early nineteenth centuries, followed by its conversion to high forest starting in 1824 (Bernard Lorentz – Forêt d'Amance, close to Nancy). This process was based on Lorentz's belief that CWS 'has an essential defect, to bring together two diametrically opposed elements' (Bastien, 1999).

CWS was the most important silvicultural system applied to broadleaves in Britain until the mid-nineteenth century. With the industrial revolution/development, its importance diminished and since the end of the Second World War an important share of derelict CWS has been converted to high forest (Matthews, 1991).

In Germany, the process of conversion started during the same period so that the area of coppice with standards decreased from 40% of national forests in 1800 down to 1.5% in 1961 (Aubry and Druelle, 1987).

Although under conditions, coppice with standards is still practised in large areas, especially in the south and centre of Europe. This is the case in France (4.2 million ha, of which over two million ha is in private forests – Anon., 2006), Italy (3.2 million ha – Diaci et al., 2012), and Greece and Spain (Kuusela, 1994).

Other important European countries in terms of coppice forests are Albania (405,000 ha), Bulgaria (1,750,000 ha), Croatia (512,000 ha), Macedonia (557,000 ha), Hungary (501,000 ha), Montenegro (298,000 ha), Romania (369,000 ha) and Serbia (1,456,000 ha) (Stajic et al., 2009).

In Europe, coppices of all kinds cover an area of about 23 million ha (UN/ECE-FAO, 2000). In some European countries, specifically Montenegro, Hungary, Portugal, Turkey, Ukraine, France and Bulgaria, the forest area regenerated through coppicing represented around 10% of the forest area and accounted for 8.2 million ha in 2010 (FOREST EUROPE, 2015).

7.4 Ecology of Coppice Forests

Apart from their rediscovered *economic (utilisation)* importance, coppice forests have high *ecological* significance. For instance, in many parts of the world, the preservation of coppices is considered as necessary taking into account their important role in *biodiversity conservation* (especially of rare or endangered species of bird and plant) in rural areas, coppices being considered today as a rare and relevant element in the sense of forest diversity (Vacik et al., 2009). As shown by Sjölund and Jump (2013), the habitat heterogeneity of coppice and pollard systems is a key factor promoting high floral and faunal diversity. In this respect, numerous taxa (e.g. small mammals, breeding birds, understorey plants, saproxylic invertebrates and epiphytes) benefit from the abundance of microhabitats arising from the multi-stemmed growth form of the trees and the continuity of ages within an area (Mitchell, 1992, cited in Sjölund and Jump, 2013). In addition, coppice forests have a lower volume of deadwood than high forests and here, late successional species will benefit from more mature trees and deadwood as found in abandoned coppice (Greatorex-Davies and Marrs, 1992, cited in Sjölund and Jump, 2013).

The conservation of coppices is also necessary for their high *habitat value* as well as *historical value* (Savill, 1993; Nagaike et al., 2005; Starr, 2008). The *genetic resource value* of coppice forests is also to be considered, even though repeated vegetative reproduction from stump shoots or root suckers might eventually lead to genetic 'fixing' or 'freezing' of variation in coppices. Such 'fixed' coppice forests may have a reduced capacity to adapt to accelerating rates of climate change compared with sexually regenerating populations (high forests) (Buckley and Mills, 2015). However, even though the neglected/abandoned coppice populations can harbour high genetic diversity, they can display an unexpected excess of homozygotes, usually a sign of inbreeding (Sjölund and Jump, 2013).

The management of coppice forests involves the *mechanical harvesting* of shoots, meaning more potential damage to ground vegetation and shrub layers. In addition, the greater frequency of harvesting in coppice forests means that the soil carbon levels tend to be lower than under high forest, but considerably more than in arable land. SRC can, as a result, increase soil carbon stocks compared with the farmland it replaces (Buckley and Mills, 2015).

In coppice forests, the soil is subject to high nutrient removal as the trees are in the juvenile stage and show high increment. In such forests, the nutrient input into the ecosystem is low, and soil degradation, grassiness and unproductive shrub-rich softwood stocking can result (Stähr, 2013). However, European soils are able to keep their productivity in spite of high removal of nutrients – sometimes up to 100 years or more (Dengler, 1935; Dohrenbusch, 1982, both cited in Stähr, 2013). In addition, coppice forests can play an important and positive role in soil conservation/erosion control, especially on steep slopes.

Under coppice conditions, prudent biomass harvesting (by carefully selecting the tree compartments to harvest and/or by reducing the harvesting intensity) is required, especially in poor sites with low nutrient reserves. Another solution is the shift from simple coppice to coppice with standards, leaving large trees, containing high nutrient amounts, standing for many coppice rotations (Pyttel et al., 2015).

7.5 Case Studies (CWR and CWS Austria)

Due to site conditions, low coppice, coppice with reserves and coppice with standards systems have been a widespread silvicultural practice in the eastern part of Austria for centuries. Seventy-five thousand hectares belong to the 'land-coppice system' and 25,000 ha are part of coppice forests in the alluvial plains. The case studies are located in the Weinviertel in Lower Austria. The thermophile oak-hornbeam forests are the dominating forest vegetation type in the main growth zone 'Sommerwarmer Osten' (Kilian et al., 1994). The average yearly rainfall is 450–600 mm and the average annual temperature is 9.3°C. Dryer periods occur in spring and autumn. In this climate region, the trees have a high potential for sprouting. Therefore, the landowners used the two regeneration tactics (seed and sprouting) within the various coppice systems (low coppice, coppice with reserves and coppice with standards systems). Periodic changes of forest management objectives, influenced by the purpose of optimization of the performance of forestry systems (coppice vs. high forest) and decreasing demand for firewood, caused a decreasing relevance of coppice systems up to 1990 (Krissl and Müller, 1989; Tiefenbacher, 1996; Hochbichler, 1997).

7.5.1 Carbon Stock of a Coppice With Reserves System

Coppice with reserves (CWR), a silvicultural system where widely spaced selected trees are allowed to grow to large size for a second rotation (*2r*), is typical for Austria and widespread, showing different forest structures (Krissl and Müller, 1989). Calculations based on a production model and biomass functions (tree and shrub specific) were developed from stand inventories and single tree biomass analysis (Hochbichler, 2008). Figure 7.6 shows the development of carbon stocks in a CWR (*Acer campestre*, *Quercus* spp., *Carpinus betulus*) in relation to age and percentage of shrubs (*Cornus* spp., *Ligustrum vulgare*). The total carbon stock ranges between 61 and 89 Mg ha^{-1} at the end of rotation. That implies an annual increment of 2–3 Mg ha^{-1} y^{-1} at 30 years of age. In relation to an increasing percentage of shrubs it shows a decreasing carbon stock, therefore mixture regulation for young growth phase in CWR is recommended to optimize productivity.

7.5.2 Carbon Stocks of a CWS System

A CWS system consists of standards (overwood) and coppice (underwood), as mentioned above. The share of standards in the total biomass is flexible and depends on local silvicultural aims determined by environmental and economic conditions. CWS is still commonly used as a local management practice for producing high-quality logs as well as biomass for energy production in the same stand. Figure 7.7 illustrates the C stocks in an Austrian *Quercus-Carpinus* CWS system. Stand location, sampling procedures and analysis details can be obtained from Bruckman et al. (2011). While the age of the standards (*Quercus petraea*) lies between 100 and 130 years, we used the chronosequence approach (time-for-space substitution) to illustrate a potential temporal development of a typical CWS stand with *Carpinus betulus*-dominated coppice underwood.

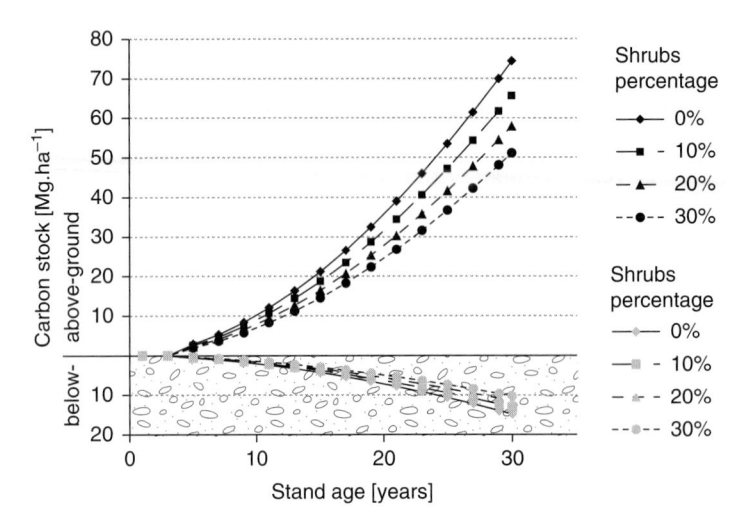

Figure 7.6. Above- and below-ground biomass C stocks (Mg ha⁻¹) in relation to age [y] and percentage of basal area at the height of 10 cm of shrubs for 0, 10, 20 and 30% (shrubs [10%] = shrubs percentage 10%) for coppice with reserves system in Ebenthal (Hochbichler, 2008).

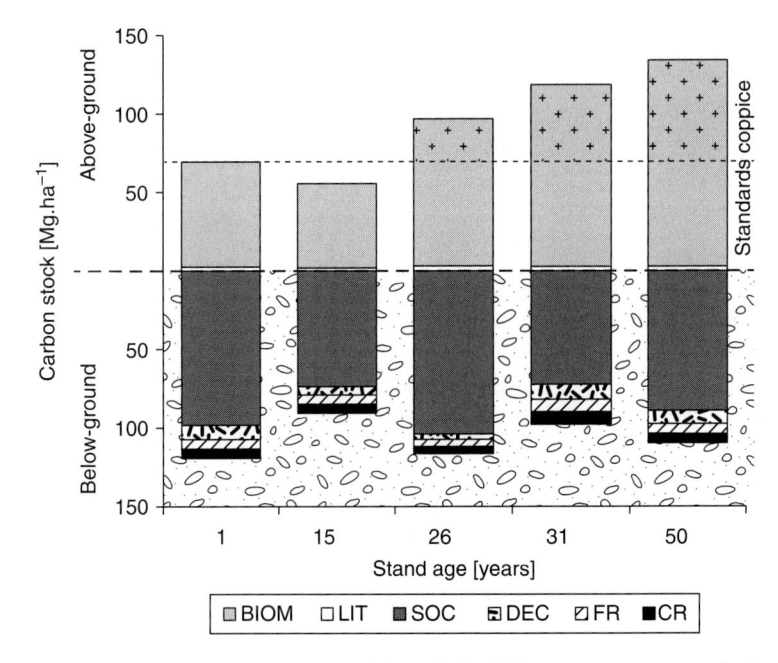

Figure 7.7. The share and temporal dynamics of C stocks in different compartments obtained from a CWS chronosequence study (Bruckman et al., 2011). BIOM = above-ground biomass, LIT = litter, SOC = soil organic carbon, DEC = decay, FR = fine roots, CR = coarse roots. Stand age is defined as the age of the coppice (underwood), while the age of the standards lies between 100 and 130 years in all five stands.

The chronosequence results demonstrate that the below-ground C stocks up to 50 cm soil depth are relatively constant, with a mean value of 87 Mg ha^{-1}. Depending on stand age, above-ground biomass excluding litter contributed to values between 54 and 131 Mg ha^{-1} in the 15 and 50 years old stands respectively. The fact that total above-ground biomass was lower in the 15 years old stand than in the freshly coppiced stand shows the dominant role of standards in terms of stand volume and subsequent C stocks. In our chronosequence model, we observed an increase of above-ground stocks up to 50 years, which represents an over-mature stand. In this stand, the coppice C stocks represent nearly 50% of the entire above-ground biomass, or 95 m^{-3} ha^{-1}. The standards (stocking = 32 individuals ha^{-1}, basal area = 10.2 m^{-2} ha^{-1}) contributed to 90 m^{-3} ha^{-1}.

7.6 Conclusions

Ongoing demand for biomass (energy wood) and valuable hardwood has increased interest in these traditional silvicultural systems in Austria. Restoration, conversion and transformation strategies are discussed in relation to site conditions (e.g. coppice with standards on moderate sites; coppice with reserves and/or simple coppice on drier and less productive sites) and in order to improve the natural and economic performances (Hochbichler, 1993). The silvicultural systems CWR and CWS in particular allow for flexible, sustainable production of firewood and valuable sawlogs in small forest areas. The share of biomass (energy wood) production decreases from 100% in simple coppices to 60–70% in stands treated as coppice with standards. Due to increasing interest in fuelwood and other forms of biomass (woodchips, pellets) the main goal of current forest management is to restructure degraded coppice stands (e.g. over-mature coppice stands; 'false' high forest stands; unbalanced coppice with standards stands) in contrast to favouring conversion to high forests. The two case studies demonstrate clearly the site- and tree- (shrub-)specific opportunities to combine and optimize biomass and sawlog production. In this respect, the use of coppice biomass not only for firewood but also for pyrolysis processes is feasible and can represent an important end use of coppice products in the future.

These facts highlight the landowners' flexibility to respond quickly to environmental and market constraints within coppice management.

References

AFOCEL (1982). *Culture de biomasse ligneuse – taillis à courte rotation*. Nangis: AFOCEL (Association Forêt Cellulose).
Anon. (2000). *Norme tehnice privind alegerea şi aplicarea tratamentelor 3*. Bucureşti: Ministerul Apelor, Pădurilor şi Protecţiei Mediului.
Anon. (2006). *Indicators for the Sustainable Management of French Forests 2005*. Nogent-sur-Vernisson: Inventaire Forestier National.
Anon. (2011). *Ordin nr. 1540 din 3 iunie 2011 pentru aprobarea Instrucţiunilor privind termenele, modalităţile şi perioadele de colectare, scoatere şi transport al materialului lemnos*. Bucureşti: Monitorul Oficial, no. 430/20.06.2011.

Aubry, S. and Druelle, P. (1987). *Histoire du taillis-sous-futaie ou La résurection d'un 'mort qui reste à tuer'*. Nogent-sur-Vernisson: ENITEF (Ecole Nationale des Ingénieurs des Travaux des Eaux et Forêts).

Bacchetta, R. (1984). Le châtaignier à bois et la mise en valeur des taillis de châtaignier. *Forêt-entreprise*, 22, pp. 12–23.

Bagneris, G. (1878). *Éléments de Sylviculture. 2ème édition*. Nancy: Imprimerie Berger-Levrault et Cie.

Bastien, Y. (1999). *Les modes de traitement des forêts*. Nancy: Ecole Nationale du Génie Rural, des Eaux et des Forêts.

Berg, Å. (2002). Breeding birds in short-rotation coppices on farmland in central Sweden – the importance of Salix height and adjacent habitats. *Agriculture, Ecosystems & Environment*, 90, pp. 265–276.

Bewa, H. and Bouvot-Mauduit, S. (1999). Short rotation crops in France. In: Christersson, L. and Ledin, S. (eds) *Short Rotation Crops for Energy Purposes. Proceedings of the First Meeting of IEA, Bioenergy, Task 17, Uppsala, Sweden, 4–6 June 1998*. Uppsala: Department of Short Rotation Forestry, pp. 31–36.

Boppe, L. (1889). *Traité de Sylviculture*. Paris and Nancy: Berger-Levrault et Cie, Libraires-Éditeurs.

Boudru, M. (1989). *Forêt et sylviculture: traitement des forêts*. Gembloux: Les Presses Agronomiques de Gembloux.

Bourgeois, C. (1991). Le châtaignier de la montagne sacrée. *Forêt-entreprise*, 4, p. 40.

Broilliard, Ch. (1881). *Le traitement des bois en France à l'usage des particuliers*. Paris and Nancy: Berger-Levrault et Cie, Libraires-Éditeurs.

Bruckman, V. J., Yan, S., Hochbichler, E. and Glatzel, G. (2011). Carbon pools and temporal dynamics along a rotation period in *Quercus* dominated high forest and coppice with standards stands. *Forest Ecology and Management*, 262, pp. 1853–1862.

Bruckman, V. J., Terada, T., Fukuda, K., Yamamoto, H. and Hochbichler, E. (2016). Overmature periurban *Quercus-Carpinus* coppice forests in Austria and Japan: a comparison of carbon stocks, stand characteristics and conversion to high forest. European Journal of Forest Research (online first). doi:10.1007/s10342-016-0979-2.

Buckley, G. P. (1992). *Ecology and Management of Coppice Woodlands*. London and New York: Chapman & Hall.

Buckley, P. and Mills, J. (2015). Coppice silviculture: from the Mesolithic to the 21st century. In: Kirby, K. J. and Watkins, C. (eds) *Europe's Changing Woods and Forests: From Wildwood to Managed Landscapes*. Wallingford and Boston: The Centre for Agriculture and Bioscience International, pp. 77–92.

Cannell, M. G. R. (2004). Short rotation forestry for biomass production. In: Burley, J., Evans, J. and Youngquist, Y. A. (eds) *Encyclopedia of Forest Sciences, vol. 2*. Amsterdam, Boston, Heidelberg: Elsevier and Academic Press, pp. 872–877.

Carbonnière, T., Debenne, J-N., Merzeau, D. and Rault, M. (2007). Le robinier en Aquitaine. *Forêt-entreprise*, 177, pp. 13–17.

Chivulescu, T. (1886). *Catehismul silvicultorului. Noțiuni de silvicultură (generalități)*. București: Tipografia proprietari F. Göbl și Fii.

Christersson, L. (1998). *Theoretical background to and practical utilization of short-rotation and energy forestry*. In: Christersson, L. and Ledin, S. (eds) *IEA, Bioenergy Task 17 Short Rotation Crops for Energy Purposes. Proceedings of the First Meeting of IEA, Bioenergy Task 17, Uppsala, Sweden, 4–6 June 1998. Rapport 64*. Uppsala: Swedish University of Agricultural Sciences, pp. 5–19.

Cochet, P. (1971). *Étude et culture de la forêt. 3ème édition*. Nancy: Ecole Nationale du Génie Rural, des Eaux et des Forêts.

Coppini, M. and Hermanin, L. (2007). Restoration of selective beech coppices: a case study in the Appenines (Italy). *Forest Ecology and Management*, 249, pp. 18–27.

Cotta, H. (1841). *Principes fundamentaux de la science forestière. 2ème édition corrigée.* Paris: Bouchard-Huzard and Nancy: George-Grimblot, Thomas et Raybois.

CREFF (Cost Reduction and Efficiency Improvement of Short Rotation Coppice) (2012). *Technical Guide. Short rotation coppice.* [online] Available at: www.creff.eu [Accessed 10 July 2013]

Crowther, R. E. and Evans, J. (1984). *Coppice.* Forestry Commission Leaflet 83. London: Her Majesty's Stationery Office (HMSO).

Dawson, W. M. (1991). Short rotation coppice willow: the Northern Ireland experience. In: Aldhous, J. R. (ed.) *Wood for Energy: The Implications for Harvesting, Utilisation and Marketing.* Edinburgh: Institute of Chartered Foresters, pp. 235–247.

Dengler, A. (1935). *Waldbau auf ökologischer Grundlage. Ein Lehr- und Handbuch.* Berlin: Verlag von Julius Springer.

Diaci, J., Govedar, Z., Krstić, M. and Motta, R. (2012). Importance and perspectives of silviculture for the science and practice of forestry. In: Govedar, Z. and Dukić, V. (eds) *Proceedings, International Conference 'Forestry Science and Practice for the Purpose of Sustainable Development of Forestry – 20 Years of the Faculty of Forestry in Banja Luka'.* Banja Luka, 1–4 November 2012, pp. 23–40.

Dimitriou, I. (2009). Short rotation coppice with willow for energy and phytoremediation in Sweden. In: *Proceedings, International Conference 'Forestry in Achieving Millenium Goals'.* Novi Sad, pp. 87–92.

Drăcea, M. (1923). *Silvicultură.* Bucharest: Şcoala Politehnică.

Drăcea, M. D. (1942). *Curs de Silvicultură. Vol. I. Regime şi tratamente.* Bucharest: Editura Politehnicei.

Dubourdieu, J. (1997). *Manuel d'aménagement forestier. Technique & Documentation*, Paris: Lavoisier.

Evans, J. (1992). Coppice forestry – an overview. In: Buckley, G. P. (ed.) *Ecology and Management of Coppice Woodlands.* New York, Tokyo, Melbourne, Madras: Chapman & Hall, pp. 18–27.

Evans, J. and Turnbull, J. (2004). *Plantation Forestry in the Tropics.* 3rd Edition. Oxford: Oxford University Press.

Evelyn, J. (1664). *Silva, or a Discourse of Forest Trees and the Propagation of Timber in His Majesty's Dominions.* Ed. Guy de la Bédoyère (Internet Edition, 1995).

Fankhauser, F. (1921). *Guide pratique de Sylviculture. Troisième édition.* Lausanne and Geneva: Librairie Payot et Cie.

Ford-Robertson, J. B., Watters, M. P. and Mitchell, C. P. (1991). Short rotation coppice willows for energy. In: Aldhous, J. R. (ed.) *Wood for Energy: The Implications for Harvesting, Utilisation and Marketing.* Edinburgh: Institute of Chartered Foresters, pp. 218–234.

FOREST EUROPE (2015). *State of Europe's Forests 2015. Ministerial Conference on the Protection of Forests in Europe.* Madrid: FOREST EUROPE Liaison Unit Madrid.

Frank, J. (1937). *Der Hochleithenwald, Einführung zur Wälderschau des NÖ Forstvereins in das Rudolf Graf von Abensperg – Traun'sche Forstrevier Wolkersdorf.* St. Pölten: Selbstverlag NÖ Forstverein.

Garfitt, J. E. (1995). *Natural Management of Woods: Continuous Cover Forestry.* New York, Chichester, Toronto, Brisbane, Singapore: Research Studies Ltd. Taunton, England: John Wiley & Sons.

Guidi, W., Pitre, F. E. and Labreque, M. (2013). Short-rotation coppice of willows for the production of biomass in eastern Canada. In: Matovic, M. D. (ed.) *Biomass Now – Sustainable Growth and Use.* Rijeka: InTech. doi: 10.5772/2583, pp. 421–448.

Hamilton, L. (2000). *Managing Coppice in Eucalypt Plantations*. Agriculture Notes no. 0814. Kingston: State of Victoria, Department of Primary Industries.
Hamm, J. (1900). Leitsätze für den Mittelwaldbetrieb. *Forstwissenschaftliches Centralblatt*, 8, pp. 392–404.
Harmer, R. (1995). *Management of Coppice Stools*. Research Information Note 259. Wrecclesham, Alice Holt Lodge: The Forestry Authority Research Division.
Harmer, R. (2004). Coppice silviculture practiced in temperate regions. In: Burley, J., Evans, J. and Youngquist, Y. A. (eds) *Encyclopedia of Forest Sciences, vol. 3*. Amsterdam, Boston, Heidelberg: Elsevier and Academic Press, pp. 1045–1052.
Harmer, R., and Howe, J. (2003). *The Silviculture and Management of Coppice Woodlands*. Edinburgh: Forestry Commission.
Hart, C. (1994). *Practical Forestry for the Agent and Surveyor*. Stroud: Alan Sutton Publishing.
Hartig, G. L. (1805). *Instruction sur la culture du bois à l'usage des forestiers*. Paris: L'Imprimerie de C.F. Patris.
Hartig, G. (1877). *Lehrbuch für Förster. II Band*. Stuttgart: J.G. Cott'sche Buchhandlung.
Hinton-Jones, M. and Valentine, J. (2008). Variety and altitude effects on yield and other characters of SRC willow in Wales. *Aspects of Applied Biology*, 90, pp. 67–73.
Hochbichler, E. (1993). Methods of oak silviculture in Austria. *Annales des Sciences Forestières*, 50, pp. 591–593.
Hochbichler, E. (1997). Waldbaustrategien und betriebswirtschaftliche Aspekte für die Waldbewirtschaftung im sommerwarmen Osten Österreichs. In: Müller, F. (ed.) *Waldbau an der unteren Waldgrenze*. FBVA – Berichte 95, pp. 105–110.
Hochbichler, E. (2008). *Fallstudien zur Struktur, Produktion und Bewirtschaftung von Mittelwäldern im Osten Österreichs (Weinviertel)*. Forstliche Schriftenreihe, 20. Vienna: Österr. Gesellschaft für Waldökosystemforschung und experimentelle Baumforschung an der Universität für Bodenkultur.
Hochbichler, E. (2009). Coppice forestry in Austria. In: Marusak, R., Kratochvilova, Z., Trnkova, E. and Hajnala, M. (eds) *Forest, Wildlife and Wood Sciences for Society Development – Conference Proceedings*. Prague: Czech University of Life Sciences, Faculty of Forestry and Wood Sciences, pp. 19–35.
Hochbichler, E., Bellos, N., Diwold, G., Hofmann, H., Zeitlhofer, C. and Liebhard, P. (2011). Produktionsmodelle und Bewirtschaftung – Biomassefunktionen für Pappel und Weide zur Ermittlung von Ertragsleitung und Produktivität von Kurzumtriebsflächen [16. Österreichischer Biomasse – Tag – Kurzumtriebstagung, Wieselburg, 16.-18.11.2011]. In: Schuster K. (ed.) *Innovative Energiepflanzen – Erzeugung und Verwendung von Kurzumtriebsholz*. Wieselburg: NÖ-Landeslandwirtschaftskammer, BLT Francisco Josephinum.
Huffel, G. (1907). *Economie forestière. Tome troisième*. Paris: Lucien Laveur, Éditeur.
IUFRO (2005). *Multilingual Pocket Glossary of Forest Terms and Definitions*. IUFRO SilvaVoc Terminology Project. Vienna: IUFRO (International Union of Forest Research Organizations).
James, N. D. G. (1991). *An Historical Dictionary of Forestry and Woodland Terms*. Oxford, UK and Cambridge, MA: Basil Blackwell.
Jansen, P. and Kuiper, L. (2004). Double green energy from traditional coppice stands in the Netherlands. *Biomass and Bioenergy*, 26, pp. 401–402.
Johnson, J. D. (1999). Hybrid poplar production in the Pacific Northwest and the United States. In: *Proceedings, Joint Meeting on Short Rotation Forestry, IUFRO and International Energy Agency (IEA)*, University of Philippines Los Baños College, Laguna, Philippines, 3–7 March 1999, pp. 103–108.

Jolyet, A. (1916). *Traité pratique de Sylviculture*. 2nd Edition. Paris: Librairie J.-B. Baillière et Fils.

Kilian, W., Müller, F. and Starlinger, F. (1994). Die forstlichen Wuchsgebiete Österreichs – Eine Naturraumgliederung nach waldökologischen Gesichtspunkten. FBVA Berichte 82, Forstliche Bundesversuchsanstalt, Vienna.

Kneifl, M., Kadavy, J. and Knott, R. (2011). Gross value yield potential of coppice, high forest and model conversion of high forest to coppice on best sites. *Journal of Forest Science*, 57, pp. 536–546.

Krausmann, F. (2013). A city and its hinterland: Vienna's energy metabolism 1800–2006. In: Singh, S. J., Haberl, H., Chertow, M., Mirtl, M. and Schmid, M. (eds.) *Long Term Socio-Ecological Research: Studies in Society-Nature Interactions Across Spatial and Temporal Scales*. New York: Springer, pp. 247–268.

Krissl, W. and Müller, F. (1989). Waldbauliche Bewirtschaftungsrichtlinien für das Eichenmischwaldgebiet Österreichs. *FBVA-Berichte*, 40.

Kuusela, K. (1994). *Forest Resources in Europe*. EFI Research Report 1, Cambridge: Cambridge University Press.

Lafouge, R. (1964). *Les taillis-sous-futaie et ses problèmes. Ameliorations et transformations*. Nancy: Ecole Nationale des Eaux et Forêts.

Lorentz, B. and Parade, A. (1867). *Cours élémentaire de culture des bois*. 5th Edition. Paris: Mme Ve Bouchard-Huzard and Nancy: Nicolas Grosjean.

Machar, I. (2009). Coppice-with-standards in floodplain forests – a new subject for nature protection. *Journal of Forest Science*, 55, pp. 306–311.

Maryan, P. S. (1991). The potential for short rotation fuel wood crops. In: Aldhous, J. R. (ed.) *Wood for Energy: The Implications for Harvesting, Utilisation and Marketing*. Edinburgh: Institute of Chartered Foresters, pp. 178–182.

Mathey, A. (1929). *Traité théorique & pratique des taillis*. Le Mans: Imprimerie M. Vilaire.

Matthews, J. D. (1991). *Silvicultural Systems*. Oxford: Clarendon Press.

Matula, R., Svátek, M., Kůrová, J., Úradniček, L., Kadavý, J. and Kneifl., M. (2012). The sprouting ability of the main tree species in Central European coppices: implications for coppice restoration. *European Journal of Forest Research*, 131, pp. 1501–1511.

Muel, E. (1884). *Notions de Sylviculture*. Paris: Ducher et Cie, Editeurs.

Nagaike, T., Yoshida, T., Miguchi, H., Kamitani, T. and Nakashizuka, T. (2005). Rehabilitation for species enrichment in abandoned coppice forests in Japan. In: Stanturf, J. A. and Madsen, P. (eds) *Restoration of Boreal and Temperate Forests*. Boca Raton, FL, London, New York, Washington, DC: CRC Press, pp. 371–381.

Nyland, R. D. (1996). *Silviculture: Concepts and Applications*. New York: McGraw-Hill.

Nyland, R. D. (2002). *Silviculture: Concepts and Applications*. 2nd Edition. New York: McGraw-Hill.

Paris, P., Todaro, L., Saccheti, R., Mugnozza, G. S., Pisanelli, A. and Cannata, P. (2006). La robinia per le piantagioni da biomasa in zone marginali. *Alberi e Territorio*, 10/11, pp. 22–27.

Peterken, G. F. (2001). *Natural Woodland. Ecology and Conservation in Northern Temperate Regions*. New York: Cambridge University Press.

Pliny the Elder (2002). *Naturalis Historia. Enciclopedia cunoștințelor din antichitate. Vol. III Botanica*. Iași: Editura Polirom.

Pyttel, P. L., Köhn, M. and Bauhus, J. (2015). Effects of different harvesting intensities on the macro nutrient pools in aged oak coppice forests. *Forest Ecology and Management*, 349, pp. 94–105.

Rădulescu, A. and Vlad, I. (1955). Regime și tratamente. In: *Manualul Inginerului Forestier 80 – Cultura pădurilor și Bazele naturalistice*. Bucharest: Editura Tehnică, pp. 471–511.
Rédei, K., Veperdi, I., Osváth-Bujtás, Z., Bagaméry, G. and Barna, T. (2007). La gestion du robinier en Hongrie. *Forêt-entreprise*, 177, pp. 44–49.
Rietbergen, S. (2001). The history and impact of forest management. In: Evans, J. (ed.) *The Forests Handbook. Vol. 2. Applying Forest Science for Sustainable Management.* Oxford, London: Blackwell Science, pp. 1–25.
Rollinson, T. J. D. and Evans, J. (1987). *The Yield of Sweet Chestnut Coppice*. Forestry Commission Bulletin, 64. London: HMSO.
Savill, P. (2013). *The Silviculture of Trees used in British Forestry*. 2nd Edition. Wallingford and Boston: CAB International.
Savill, P. S. (1993). *Coppice and Coppice-with-Standards*. Oxford: Oxford Forestry Institute.
Savill, P. S. and Evans, J. (1986). *Plantation Silviculture in Temperate Regions*. Oxford: Clarendon Press.
Schöfberger, H. (2006). *Waldbauliche Untersuchungenim Mittelwald unter Berücksichtigung der energetischen Verwertung von Biomasse*. Vienna: Diplomarbeit Universität für Bodenkultur. (Unpublished.)
Schwappach, A. (1904). *Forestry*. London: The Temple Primers.
Schwappach, A., Eckstein, K., Herrmann, E. and Borgmann, W. (1914). *Manual silvic. Partea a V-a Cultura pădurilor*. Bucharest: Alfred Baer.
Schweier, J. and Becker, G. (2012). Harvesting of short-rotation coppice – harvesting trials with a cut and storage system in Germany. *Silva Fennica*, 46, pp. 287–299.
Sjölund, M. J. and Jump, A. S. (2013). The benefits and hazards of exploiting vegetative regeneration for forest conservation management in a warming world. *Forestry*, 86, pp. 503–513.
Spinelli, R., Ebone, A. and Gianella, M. (2014). Biomass production from traditional coppice management in northern Italy. *Biomass and Bioenergy*, 62, pp. 68–73.
Stähr, F. (2013). *Renaissance and global utilisation of the coppice system – Is the historical silvicultural system 'coppice forest' topical again?* [online] Available at: www.hnee.de/_obj/FB563809-E403-4728-84C1-D15C000385A9/outline/StaehrVortrag_EKonferenz_englisch_Konferenzband.pdf [Accessed 10 September 2013]
Stajic, B., Zlatanov, T., Velichkov, I., Dubravac, T. and Trajkov, P. (2009). Past and recent coppice forest management in some regions of south eastern Europe. *Silva Balcanica*, 10, pp. 9–19.
Starr, C. (2008). *Woodland Management. A Practical Guide*. Ramsbury: The Crowood Press.
Stolarski, M. J., Szczukowski, S., Tworkowski, J., Wróblewska, H. and Krzyżaniak, M. (2011). Short rotation willow coppice biomass as an industrial and energy feedstock. *Industrial Crops and Products*, 1, pp. 217–223.
Terada, T., Yokohari, M., Bolthouse, J. and Tanaka, N. (2010). Refueling Satoyama woodland restoration in Japan: enhancing restoration practice and experiences through woodfuel utilization. *Nature and Culture*, 5, pp. 251–276.
Tiefenbacher, H. (1996). Waldbau – Strategie angesichts eines 'Eichensterbens'. *Centralblatt für das gesamte Forstwesen*, 113, pp. 83–96.
Troup, R. S. (1928). *Silvicultural Systems*. Oxford: Clarendon Press.
Troup, R. S. and Jones, E. W. (1952). *Silvicultural Systems*. 2nd Edition. Oxford: Clarendon Press.

Tubby, I. and Armstrong, A. (2002). *Establishment and Management of Short Rotation Coppice*. Forestry Commisison Practice Note no. 7. Edinburgh: Forestry Commission.

UN ECE-FAO (United Nations Economic Commission for Europe/Food and Agriculture Organization of the United Nations) (2000). Forest resources of Europe, CIS, North America, Australia, Japan and New Zealand. Main Report. *Geneva Timber and Forest Study Papers*, 17.

Vacik, H., Zlatanov, T., Trajkov, P. and Dekanic, S. (2009). Role of coppice forest in maintaining forest biodiversity. *Silva Balcanica*, 10, pp. 35–45.

Verwijst, T., Lundkvist, A., Edelfeldt, S. and Albertsson, J. (2013). Development of sustainable willow short rotation forestry in northern Europe. In: Matovic, M. D. (ed.) *Biomass Now – Sustainable Growth and Use*. Rijeka: InTech. doi: 10.5772/2583, pp. 479–502.

Zlatanov, T. and Lexer, M. J. (2009). Coppice forestry in south-eastern Europe: problems and future prospects. *Silva Balcanica*, 10, pp. 5–8.

8

Towards Environmental and Economic Sustainability via the Biomass Industry: the Malaysian Case Study

KOK MUN TANG, WAN ASMA IBRAHIM
AND WAN RASHIDAH KADIR

> **Abstract**
>
> Despite the rapid industrialization of Malaysia, the agriculture sector remains one of the major economic contributors. Important crops include oil palm, rubber, short rotation forestry and paddy. Biomass has always been used for the upstream generation of steam and power at the milling and primary processing stage in Malaysia. The production and utilization of biochar from agro-based biomass is one of the major research areas in Malaysian universities and research institutions. The biomass sector in Malaysia has the potential to satisfy the three key pillars of sustainable development, that is environment, business and social benefits. Biochar can be used as a soil enhancer where it can be mixed into compost with other ingredients to produce high performance biofertilizers. The use of biochar in the urban landscape and green building sector is also being explored as it has been successfully applied for tropical rooftop greening as far back as 2004. Nevertheless, there are still challenges and gaps that need to be addressed via a combination of policy and legal instruments as well as actions and strategies with a focus on the biomass sources.

8.1 Biomass in Malaysia

Despite Malaysia's heavy focus on the industrialization of its economy throughout the 1980s and 1990s, the agriculture sector remains one of the main contributors to the overall economy. Approximately 11% of Malaysia's gross national income (GNI) is generated from the agriculture sector. Out of a total cultivable area estimated at 14.2 million ha in Malaysia, there are about 5.1 million ha of oil palm plantation, 1.0 million ha of rubber estates and 0.7 million ha of paddy fields (AIM, 2011).

One of the byproducts from the agriculture sector is the large amount of biomass solid and effluent waste, generated in the fields during harvesting as well as from the upstream processing facilities. These include palm wastes (empty fruit bunches, kernel shell, mill

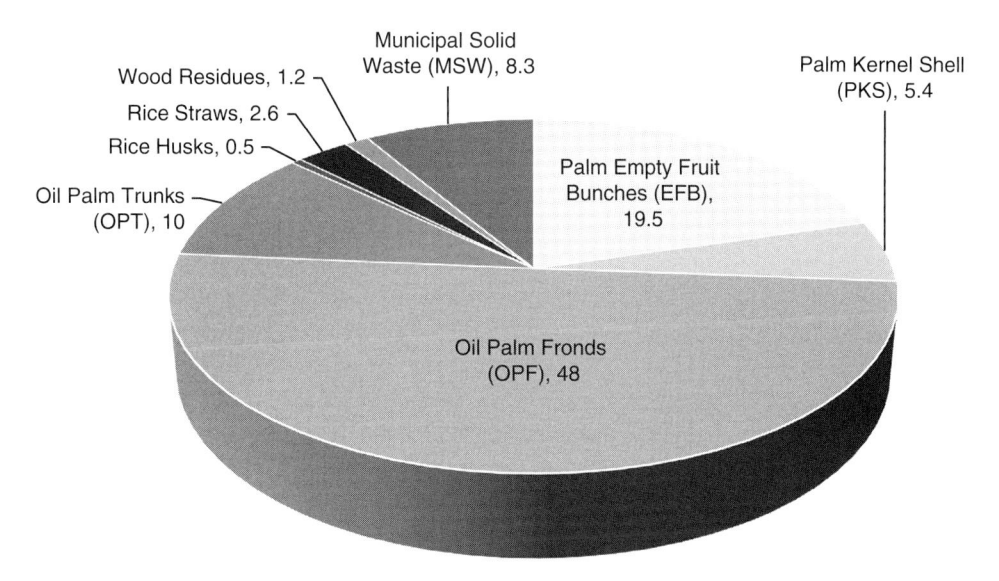

Figure 8.1. Annual generation of biomass in Malaysia (million tons).
Source: MBIC (2015).

effluent, palm fronds, palm trunks from replanting), rice straws and husks, wood off-cuts and sawdust. Figure 8.1 provides the typical quantities of biomass generated annually based on 2013 and 2014 data (MBIC, 2015) in Malaysia. The largest amounts of biomass were generated from the palm oil industry in the forms of palm fronds (48 million tons) followed by empty fruit bunches (19.5 million tons) and palm kernel shells (5.4 million tons). In the case of Malaysia, municipal solid waste (MSW) is also categorized as a form of biomass waste due to its high content of organic matter, but certainly needs special treatment (Arpah, 2013).

8.1.1 Wood-Based Biomass

In 2012, Malaysia processed a total of 7,800 million m^3 of wood, generating wood waste in the form of sawdust and offcuts (MIGHT, 2013). The majority of this waste is utilized as fuel to power sawmills, as feedstock for the production of medium density fiber (MDF) boards or for mushroom cultivation media. Another source of wood biomass is rubber trees (*Hevea brasiliensis*) that are felled for replanting after approximately 25 years as they are no longer producing economical yields for latex production. It is estimated that annually around 2 million m^3 of *Hevea* logs are harvested and utilized for the production of furniture and furniture components. The wood waste generated in the form of sawdust and offcuts is either utilized for the manufacturing of particleboards and fiberboards or as fuel for the drying of rubber sheets and timber wood, for the brick stone industry, and for curing of tobacco leaves (LGM, 2016).

8.1.2 Palm-Based Biomass

Presently, the majority of biomass residues in Malaysia are generated by the palm oil industry. The palm oil industry generates residues from its two main activities: plantation management and oil milling. Major residues from the planting activities are the old palm trunks (OPT) and oil palm fronds (OPF) generated from the replanting and pruning activities. Biomass generated from the mills can be classified into empty fruit bunches (EFB), palm kernel shells (PKS), mesocarp fibers, palm kernel cakes and palm oil mill effluent (POME). All of these, except POME, are generated in solid form. Some of these residues, such as mesocarp fibers and PKS, are typically fully used for energy generation within the mills (Wan Asma et al., 2012).

The disposal and management of palm waste was a major concern of the industry and authorities in the 1980s and 1990s due to the high biochemical oxygen demand (BOD) and methane emissions of these wastes (Wu et al., 2010). Just one decade ago, owners of palm oil mills had to pay contractors to dispose of the palm wastes. Various methods were also assessed to dispose or treat these wastes, including incineration, mulching and composting. Results showed that incineration of biomass waste created the problem of air pollution due to the high moisture content of biomass and was later abandoned. Currently, mulching and composting are popular methods used in oil palm estates, but the end products are not able to generate sufficient additional revenue to offset the investment and operational costs.

The paradigm has now changed and these materials are no longer considered as waste but byproducts, demanded mainly for generation of renewable energy and downstream production of a range of products, such as bio-fertilizers, packaging materials, fibermats and palm timber. Mill owners are now being paid by external parties to purchase their palm biomass such as EFB and PKS.

8.1.3 Rice-Based Biomass

From the rice fields, main biomass residues are in the form of rice husks and rice straw, with about 500,000 and 2,600,000 tons per year being generated respectively. Rice straw is left in the field after harvest and usually burned off to sterilize the topsoil and to return the carbon of charred biomass and mineral nutrients to the soil (Chan and Cho, 2012). Rice husks are generated in the rice milling process and therefore utilized as a fuel source in rice mills to generate heat and power. However, the traditional burning of rice straw in the fields produces smoke and particulates, forming haze. The haze limits visibility and is dangerous to motorists along highways, in addition to various health risks and environmental pollution (Gadde et al., 2009).

8.1.4 Biomass Availability vs. Accessibility

While it appears that there is a large amount of biomass available in Malaysia, the amount that is accessible for downstream utilization is considerably lower. Many constraints and barriers still exist that impede the accessibility of the material as feedstock to the biomass industry.

The combination of the high moisture content in biomass and warm weather in tropical countries such as Malaysia promotes rapid decomposition of biomass, resulting in material loss and reduced quality. In the case of palm biomass, the remote locations of many mills add logistical challenges to the transportation of the biomass to downstream manufacturers. For example, untreated EFB contains about 60% moisture, making the transportation of this resource in raw form highly inefficient. Consequently, there is a need to perform in situ pretreatment of the biomass to remove the high water content and therefore improve the efficiency of transportation (Chan, 2009).

The lack of common technical specifications and standards to promote the trading of biomass along the value chain represents another barrier to accessing potential biomass sources. Since 2009, Malaysia has developed common product standards to facilitate the production and trading of palm fibers (SIRIM, 1997). This effort should be extended to other types of biomass in order to further facilitate trading and hence downstream utilization of biomass.

In addition, the biomass industry in Malaysia is in its infancy and is lacking a formal trading platform and associated trading instruments to facilitate buying and selling of biomass. The unavailability of market information on biomass volume, spot and forward pricing leads to potential risks for investors. This is in contrast to other regions such as the European Union (EU) and North America with mature markets to trade biomass fuel pellets as a form of commodity, for instance.

In the specific case of the oil palm sector, the current practice in the plantations is to recycle the palm EFB back to the field, usually in the form of mulch or compost, as mentioned above. This has become a standard practice under the zero waste policy in the oil palm industry. It was shown that proper recycling of EFB contributes to sustaining the soil fertility via adding soil organic matter (SOM), while increasing crop yield. It was reported that one ton of EFB would have a fertilizer equivalent of 7 kg urea, 2.8 kg rock phosphate, 19.3 kg muriate of potash and 4.4 kg kieserite (Singh et al., 2009). Managers of oil palm plantations are therefore concerned about unknown potentially negative impacts from the removal of the EFB and other biomass from the plantations.

8.2 Status of the Biomass Industry in Malaysia

In the early decades of the commercial agriculture sector, biomass waste generated from harvesting, pre-treatment and processing of agricultural products such as rice, timber, rubber and palm oil was mainly used for the generation of heat and power. The excess biomass that cannot be utilized in this way is typically left on site to recycle nutrients and organic matter in order to be available for the subsequent crop season. Using palm EFB for mulching in the field improves the water holding capacity of the soil, reduces leaching, and enhances the soil pH and nutrient availability for plant growth (Singh et. al, 2009). On the other hand, OPF from pruning also release essential nutrients during the decomposition process on site and should be therefore left at the plantation site.

Downstream utilization of biomass waste was initiated in the 1980s when intensive research and promotion by Malaysian researchers resulted in the treatment and use of rubberwood for manufacturing high-quality furniture as well as sawdust as feedstock for MDF composite boards and for mushroom cultivation (Gan et al., 2013). Rubberwood products are now exported to well over 20 countries, with export values reaching USD 600 million in 2012.

The next era of industrial utilization of biomass occurred in the 1990s within the expanding oil palm sector to address increasing air (biomass burning) and water pollution from the EFB and POME waste. The supporting palm equipment industry has successfully developed the processes and machinery required to process EFB biomass in order to produce fibers (Mahmudin et al., 2007). This opened up a range of possibilities to utilize EFB fibers as feedstock for downstream manufacturing. In the early years of this breakthrough, most of the fibers were baled and exported mainly to China to replace coconut fibers for various uses such as manufacturing of fiber-based mattresses. Some of the fibers were also compressed to fibermats and used for various applications such as for mulching (Wan Rashidah and Wan Asma, 2007), topsoil protection, erosion control and planting media.

The participation of Malaysia in the Clean Development Mechanism (CDM) under the Kyoto Protocol since 2002 has further opened up commercial opportunities to utilize primarily palm biomass waste for green power generation and compost production. As of June 2013, Green Technology Corporation Malaysia reported on its website that there were 86 biomass and biogas projects in Malaysia with a total of 4.59 million carbon emission reductions (CERs) or carbon credits issued. The benefit for the local biomass industry is the increasing capacity development that leads to locally adapted solutions and know-how that helps to strengthen Malaysia's green technology industry, which can benefit from exporting its services, technology and equipment to the entire ASEAN (Association of Southeast Asian Nations) region.

The commitment of Annex I countries to reduce their national carbon emissions under the Kyoto Protocol has also benefited the local biomass industry via the creation of export demand for biomass fuel pellets. Some early players in this industry source their feedstock for pellet production directly from the mills, while others form joint ventures to produce the pellets in situ, taking advantage of the excess power from mills. Pelletizing equipment is typically imported from China and Taiwan with modifications carried out locally to suit the physical and chemical characteristics of the local biomass feedstock. Major markets for the biomass fuel pellets are China, South Korea and Japan.

Nevertheless, Malaysia is still a relatively small exporter as compared to neighboring countries with a large timber sector such as Thailand, Indonesia and Vietnam. Due to the accessibility issue highlighted earlier, local players are not able to secure sufficient amounts of feedstock on a long-term basis to supply the large volumes demanded by the overseas market. Moreover, the use of EFB fuel for large-scale power generation to directly substitute coal or wood chips still faces a number of limitations in meeting the required technical specifications, specifically regarding the ash and chloride content. Therefore, innovative industry players use pre-treatment processes such as EFB washing to remove unwanted minerals or blend different types of biomass (palm, wood etc.) in order to achieve the required specifications.

In contrast, the use of PKS as a fuel source in boilers and cement kilns for heat and power generation is in high demand, as it has similar characteristics and energy density to coal (~20–25 MJ kg^{-1}). There is currently one cement plant in Malaysia utilizing up to 68,000 tons of PKS per annum as a renewable energy source for its cement production to partially substitute fossil fuel. PKS is also used as feedstock for the production of activated carbon for various applications, such as filter materials (Puad et al., 2007).

The last decade has seen exciting developments in the biomass industry in Malaysia, especially in regard to biotechnology. Biomass is now seen as a potential feedstock for the production of not only second-generation biofuels, but also industrial feedstock such as bio-chemicals and bio-polymers. The main component of interest is the cellulose content in the biomass, while the lignin can be used to generate energy for processing or as a co-product. Several pilot projects are underway in various parts of Malaysia to demonstrate the technical and commercial viability of these new technologies of biomass utilization. Many of these projects are jointly funded by the major plantation players in Malaysia, where access to large amounts of biomass feedstock is critical to ensure the viability of such ventures.

The Malaysian biomass industry also acknowledges the participation of numerous smaller players such as small- to medium-sized enterprises (SMEs) and technology start-ups. Even though limited in funding and access to high-end technologies, these small players excel in coming up with innovative products and finding niche markets. A number of these players are collaborating with local universities and research institutes to benefit from available expertise and lab facilities, as well as supporting research and development (R&D) output commercialization.

8.2.1 Potentials of the Biomass Industry in Malaysia

The biomass industry may be able to deliver great potential for Malaysia's economy as well as for enhancing the nation's environmental sustainability. As a renewable source of energy, biomass can help to reduce Malaysia's dependency on fossil-based energy sources and hence, exposure to fluctuating prices caused by global incidents.

Table 8.1 provides an overview of the types of potential biomass resources for both energy and non-energy uses in Malaysia. The majority of the available biomass has the potential to be converted into solid fuel pellets. Many biomass players in Malaysia are currently tapping into immediate local and overseas markets with minimal start-up investment costs. However, some companies are beginning to shift into higher-value forms of utilization of the biomass feedstock by developing in-house technologies, commercializing R&D outputs from local universities or research institutions, and working with foreign technology providers. This development is crucial for ensuring a long-lasting benefit for Malaysia's green development.

Table 8.1. Overview of biomass sources and further processing in Malaysia. Biochar is already produced from a range of feedstock materials.

Product type	Solid fuels	Liquid fuels	Gaseous fuels	Green chemicals	Bio-fertilizers	Biochar	Biocomposites	Others
Empty fruit bunches	Fuel pellets	Bioethanol bio-oil	Syngas	Bio-sugars	Mulching, compost	Biochar powder, fibers	Fiberboard	Pulp, celluose, eco-products
Palm kernel shell	Fuel pellets					Biochar powder		Activated carbon
Oil palm trunks	Fuel pellets	Bioethanol	Syngas	Bio-sugars	Mulching, compost	Biochar powder	Engineered lumber	Animal feed
Oil palm fronds	Fuel pellets	Bioethanol		Bio-sugars	Mulching, compost			Animal feed phyto-nutrients
Palm kernel cake	Fuel pellets				Compost			Animal feed
Palm effluent		Bio-diesel	Biogas	Bio-polymers	Compost			
Sawdust	Fuel pellets	Bioethanol	Syngas	Bio-sugars		Biochar powder	Fiberboard	Mushroom cultivation
Rice husk	Fuel pellets						Fiberboard	Silica aerogel
Rice straw	Fuel pellets				Mulching			Silica animal feed
Kenaf							Composite panel	
Sago waste	Fuel pellets	Bioethanol	Biogas	Bio-polymers				Animal feed supplement
MSW	Fuel pellets		Biogas		Compost			

Source: MIGHT (2013).

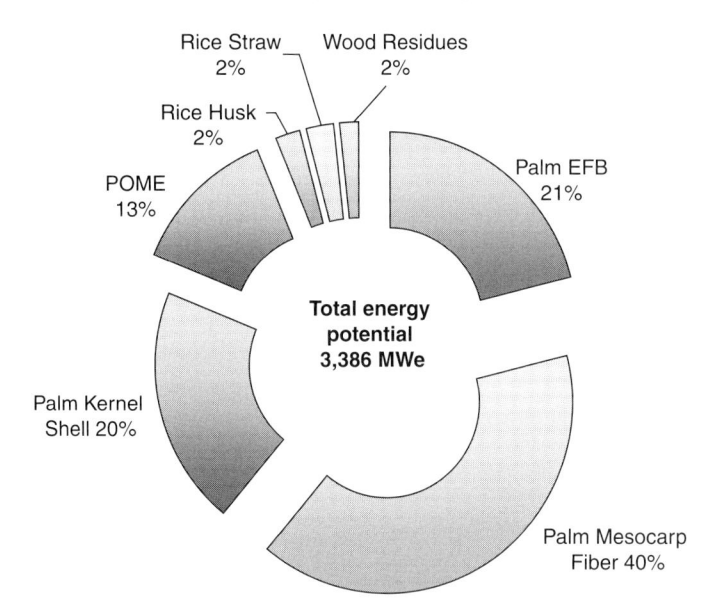

Figure 8.2. The electricity generation potential from biomass (MWe) in Malaysia, 2007. *Source*: Milbrandt and Overend (2008).

The Biomass Industry Action Plan 2020 (MIGHT, 2013) projected that 30% of total biomass available in Malaysia can be made accessible for downstream utilization. This amount of resource is estimated to be able to generate an annual GNI of USD 4.4 billion per annum, with the figure reaching USD 10 billion by 2020. The National Biomass Strategy 2020 target is to mobilize the majority of biomass from the palm sector, estimated at about 25 million tons made accessible annually. According to this strategy paper, the growth of this industry is also projected to create about 66,000 new green jobs by 2020 (MIGHT, 2013).

From the perspective of renewable energy (RE), the total electricity generation potential of accessible biomass resources in Malaysia is estimated to be 3,386 MWe, as shown in Figure 8.2.

RE can be generated from these biomass resources by utilizing solid biomass fuels, liquid bio-fuels and biogas, depending on the technical and economic factors of local supply and demand structures.

The use of biomass for RE generation is also a promising prospect to enable Malaysia to contribute to the global effort to mitigate the impacts of greenhouse gas (GHG) emissions. For example, a life cycle assessment by SIRIM determined that the use of MSW for electricity generation will only have carbon footprints for all the processes involved of 0.138 kg CO_2e kWh^{-1}, as opposed to the grid electricity carbon footprint of 0.741 kg CO_2e kWh^{-1} in Peninsular Malaysia as of 2012 (Chen et al., 2011).

Based on 2014 figures, the existing number of grid-connected RE projects in Malaysia utilizing biomass and biogas will result in CO_2e avoidance of about 466,000 tons per year (SEDA, 2015). There are already a total of 102 projects registered as CDM projects under the United Nations Framework Convention on Climate Change (UNFCCC). Out of these,

31 are biomass-based projects and another 55 are biogas-based projects with the combined issuance of 4,950,000 tons of CER credits.

Other than RE generation, the utilization of palm biomass waste (specifically EFB and POME) for the in situ production of bio-fertilizers also has the potential to partially substitute the use of mineral-based fertilizers or to increase their efficiency. Malaysia imports chemical fertilizers worth about USD 1.0–1.6 billion annually for its commercial agricultural sector, whereby the oil palm sector consumes about 75%. The wider production and use of compost and bio-fertilizers is expected to reduce the dependency of the nation's agricultural sector on external sources of chemical fertilizers as well as the associated water pollution from leaching that is a major problem in tropical countries. In addition, some of this biomass can be converted into biochar products that can be used as soil conditioner to be mixed into compost and bio-fertilizers to enhance the performance of the final products.

8.3 Government Policies and Actions

Over the last ten years, Malaysia has implemented a number of policies and specific actions that assist the promotion of the biomass industry. These policies aim at creating positive momentum along the entire supply chain of the industry, including creation of market demand, development of technologies, investment and financing, as well as administrative support by selected government agencies.

8.3.1 National Biotechnology Policy

The National Biotechnology Policy was announced in 2005 with a 15-year action plan to promote and develop the biotechnology industry in Malaysia with the aim of making the nation a key player in the global biotechnology sector. The key argument for this policy is that Malaysia's rich biological resources and diversity can serve as a comparative advantage to excel in this sector globally if proper technologies, human capital and a suitable environment can be provided. The Malaysia Biotechnology Corporation has been set up under this policy to oversee and coordinate the implementation of actions that include human capacity building, commercialization of bio-based technologies, promotion of biotechnology investment into Malaysia and setting up of the Bio-XCell Biotechnology Industrial Park in the southern state of Johor in Peninsular Malaysia. One of the key aims, however, is the positioning of Malaysia as favorable location of investment to attract foreign technology providers to set up production plants and research facilities to add value to locally available biomass and create products for the global market.

8.3.2 National Green Technology Policy

This policy was established in 2009 with the aim to develop and promote the implementation of green technologies in four major sectors: energy, building, waste & water

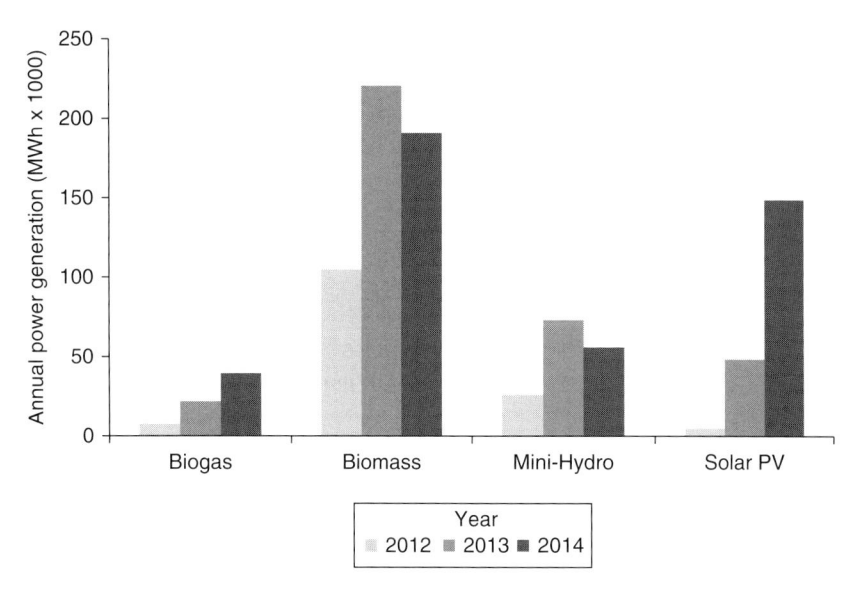

Figure 8.3. Annual power generation of commissioned RE installations until September 2014 (MWh). *Source*: SEDA (2015).

management and transportation. The main objective of this policy is to promote green technology as a new driver for the nation's future economic growth as well as contributing to sustainable development in Malaysia.

Under this policy, investment and tax incentives to encourage the production of green products and services are provided. In addition, the utilization of green technologies that aim to reduce the consumption of energy, water and raw materials is promoted. The government has allocated a total of USD 1 billion in the form of loan subsidy and guarantees under the Green Technology Financing Scheme to companies undertaking qualifying green business ventures and projects.

8.3.3 *National Renewable Energy Policy and Action Plan*

Also in 2009, the Malaysian government launched the National Renewable Energy Policy and Action Plan with the main objective of enhancing the utilization of local RE resources to ensure energy security. This policy has five strategic thrusts to guide its implementation of specific actions: (1) formulation of RE law; (2) creation of conducive environment for RE business; (3) human capital development; (4) R&D action plan; and (5) development of RE advocacy programs to support the policymaking process. This plan was further strengthened in 2011 by implementing the Renewable Energy Act, which oversees the implementation of a feed-in-tariff (FiT) mechanism, setting up an RE fund as well as the formation of the Sustainable Energy Development Authority (SEDA). The FiT applies to both biomass and non-biomass renewable sources for RE generation including biomass, biogas (including landfills), solar (photovoltaic, PV), small-scale hydro, wind and geothermal resources.

8.3.4 Biomass Industry Strategic Action Plan

From 2010 to 2013, a joint program was implemented between the EU and Malaysia to assist SMEs in Malaysia to utilize local biomass resources for high-value utilization. The Biomass Sustainable Production Initiative (Biomass-SP) focuses on gaps in the value chain of the Malaysian biomass industry and develops sub-programs to address them.

One of the major outcomes is the formulation of the Biomass Industry Strategic Action Plan 2020 to develop the entire biomass industry in Malaysia. Three primary strategies are proposed under this action plan:

1. Unlocking biomass feedstock potential via optimizing the efficiencies of resource utilization upstream at the plantation and milling stage. Upstream, biomass is used for both energy generation and nutrient recycling to the field. If this upstream utilization can be made more efficient, that is less biomass for the same outputs, more biomass will be available for downstream utilization.
2. Smart utilization of biomass for high-value production via commercialization and scaling-up of local know-how and expertise and setting up Biomass Smart Hubs. Each of these Smart Hubs will focus on a specific high-value biomass sub-sector whereby resources such as feedstock access, conversion technologies, research expertise, testing and certification, as well as marketing and branding, can be pooled to facilitate the commercialization process.
3. Positioning Malaysia as a regional and international biomass hub via establishing the nation as the focal point for internal and external biomass stakeholders in aspects such as trading, logistics, technology development, engineering and equipment.

Four sub-industries – bio-energy, bio-chemicals, bio-fertilizers and bio-composites/materials – have been identified as potential contributors for the nation's development towards biomass expertise.

In addition, the program also facilitated the formation of the Malaysia Biomass Industry Confederation (MBIC), initiated by the biomass players in the industry, especially the SMEs. Via MBIC, the industry players such as biomass SMEs will have a collective voice to engage the Malaysian government, its relevant agencies and other stakeholders to develop this industry.

8.3.5 National Biomass Strategy 2020

Another key action for the development of the biomass industry in Malaysia is the 2011 launch of the National Biomass Strategy 2020 (AIM, 2011). A second follow-up version was later published in 2013 (AIM, 2013). The main focus of this strategic document is the utilization of oil palm-based biomass ranging from EFB to OPT and OPF. The main issues to be addressed in this strategic plan are the economics for the mobilization of these biomass resources from the field and the impact of the removal of biomass on soil fertility and therefore future yields.

The plan adopts the short-term strategy of promoting the export of palm biomass pellets for the global market, which may generate immediate benefits. The long-term strategy promotes development of bio-based liquid fuels and the bio-chemicals sector via engagement of key stakeholders such as large plantation companies, technology providers and oil palm industry clusters. This should be achieved by installing the proposed Oil Palm Technology Centre to consolidate the required resources.

8.4 R&D on Biomass Conversion Technologies

R&D activities on biomass utilization in Malaysia are focused on the potential of biomass to substitute fossil-based products such as fuels, chemicals, fertilizers and manufactured goods (Wan Asma et al., 2012). There are three main biomass conversion technology platforms currently being explored for the production of bio-based fuels and chemicals. These are the thermal, the biological and the chemical conversions, but several intermediate methods combining these platforms exist as well. Thermal conversion breaks biomass down into syngas (gasification, see Chapters 10 and 12) using heat and a low oxygen environment with subsequent application of thermo-chemical processes to generate fuels and chemicals. Biological conversion involves treatment with chemicals and enzymes to produce sugars that are subsequently fermented into desired fuels or chemicals. Chemical conversion involves treatment of biomass and sugars using chemicals only in a variety of methods to produce fuels or to extract valuable chemical compounds. Being lignocellulosic and fibrous in nature, biomass residues are also a very promising resource for conversion into value-added goods and products such as pulp for paper production, bio-composites, packaging materials and further eco-friendly products.

8.4.1 Wood-Based Composite Products

The R&D conducted on woody biomass resources from forestry and forest-based industries includes technology development in using alternative sources to the increasingly depleting stocks of valuable tropical timber. Fast-growing species (e.g. *Paraserianthes falcataria*, *Gmelina aborea*, *Acacia mangium*), agriculture residues (e.g. from rubberwood, *Hevea brasiliensis*) and non-wood resources such as oil palm plantations, bamboo and rattan were identified to serve as an alternative resource for the conventional wood processing industries. The utilization of these new materials for products, such as for pulp and paper, furniture manufacturing and engineered wood composites (particle board, MDF, oriented strand board, cement board etc.) is well developed. Concerted R&D efforts with various government agencies, for instance, has resulted in the commercial utilization of rubberwood and oil palm biomass for use in furniture industries and the final products now present global commodities. One good example is the rubberwood furniture industry, which has evolved into a leading industry in Malaysia. Forestry Research Institute Malaysia (FRIM) provides standard testing services for quality assurance of the products. It has 11 ISO 17025 accredited testing laboratories servicing the forest-based industries and is therefore positioned at the forefront of sustainable biomass market development. FRIM recently focused on life cycle analysis (LCA) of forest products manufactured in Malaysia. The aim was to decrease the carbon footprint of existing technologies towards a greener environment (Gan et al., 2013). Other agencies involved in forestry and forest-based industry R&D include the Malaysian Timber Industry Board (MTIB) and tertiary education institutes such as UPM (Universiti Putra Malaysia), USM (Universiti Sains Malaysia), UM (Uinversiti Malaya) and UiTM (Universiti Teknologi Mara Malaysia), among others.

8.4.2 Pulp and Paper

The R&D activities of FRIM during the 1970s were focused on conversion of biomass into pulp and paper. This included the development of fast-growing species plantations and the utilization of oil palm biomass to obtain raw materials in sufficient quality and quantity. The work on mechanical and chemical pulping of oil palm biomass has been a success. From the point of view of logistics and cost, EFB offers the best prospects for commercial exploitation. Back in 2007, FRIM and Eko Pulp Sdn. Bhd. embarked on the commercial set up of a pulp and paper plant in Tawau, Sabah for the production of a high-grade printing paper from 100% EFB fibers with a production capacity of 30,000 tons per year (Wan Asma et al., 2012).

8.4.3 Biomass Energy

Malaysia's largest potential in renewable energy lies in the development of biomass resources. The palm oil industry is a perfect example of biomass that can lead the country to be a forerunner for a green bio-based economy in the region. Higher biomass production and utilization will contribute a significant USD 9 billion to GNI by 2020, reduce carbon emissions and create about 66,000 jobs as highlighted in the National Biomass Strategy 2020 (AIM, 2011, 2013).

Biomass is traditionally used to generate thermal energy. While the principle of biomass combustion still plays a major role, a large number of methods were developed in the past decades to meet specific demands and to increase efficiency. In general fuels for biomass energy can be classified as solid or liquid fuels.

Charcoal is a solid biofuel with a long tradition in Malaysia as well as neighboring countries (see Chapter 13, for instance). Carbonization of woody biomass has been one of the older research and technology transfer activities in FRIM, where more efficient and flexible solutions, such as the development of mobile kilns, were recently topics of interest to enable production of charcoal on site by local entrepreneurs. The traditional charcoal production technology using beehive kilns is still widely utilized in Malaysia. Commonly *Rhizophora* species are used as feedstock material. Charcoal is locally used for cooking purposes either in stoves or for barbequing. Charcoal manufacturing industries in Malaysia are thriving due to high demand from Japanese customers. The small-scale industries in Malaysia make about USD 808,300 a month (ITN Source, 2009).

The charcoal industry soon adopted technologies to utilize cheap and easily available feedstock materials such as sawdust, which is an interesting feedstock also in terms of its low water content. The sawdust is compacted and subsequently carbonized to produce charcoal briquettes, which are being used for energy generation and heating purposes in overseas markets, such as South Korea, Japan and the Middle East. The Malaysian wood pellet industry is also expanding at a steady pace. Most of the wood pellets are directed to the industrial energy sector, especially for export to Japan, South Korea and China.

Second-generation biofuels, such as bio-oil, are actively being pursued by various universities and research institutes using different types of palm biomass as precursor materials. Most activities are refining the pyrolysis process, where the optimization of temperature and process time is being investigated to produce the desirable yields and properties of bio-oil, syngas and biochar for further downstream use. Bio-oil is of particular commercial interest due to its convenience of transport, handling and storage as a liquid fuel. In addition, bio-oil also contains a wide range of high-value organic compounds that can be extracted and used for industrial processes.

8.4.4 Bio-Based Chemicals

A wide range of bio-based chemicals can be produced from biomass, such as bio-sugars, biopolymers, bioethanol, amino acids, poly-lactic acids, glutamic acids and so on. The utilization of local biomass from sustainable resources for bio-based chemicals would promote green technology and enhance environmental sustainability due to the decreasing dependency on fossil resources. Most of the conversion technologies using local biomass are currently still at the R&D stage. Initiatives towards commercial uptake are being carried out by a consortium of oil palm plantation companies and governmental institutions of Malaysia. Such initiatives include the MYBiomass initiative under the Malaysian Industry–Government Group for High Technology (MIGHT) as well as the setting up of the Oil Palm Biomass Centre (OPBC) proposed in the National Biomass Strategy 2020 (AIM, 2013).

8.5 Biochar Activities in Malaysia

Biochar is gaining popularity as a soil amendment to improve fertilizer efficiency in Malaysian agriculture. Endowed with high-temperature weather throughout the year, the weathering processes of soil minerals are comparatively fast in tropical conditions, leading to highly weathered soils, with low cation exchange capacities and a high acidity. Moreover, high rates of precipitation lead to the loss of nutrients through leaching. Incorporation of biochar into such soils has showed potential to improve soil health, crop yield and fertilizer efficiency (Rosenani et al., 2012).

Local research on biochar production and application is extensive, with a wide array of publications available. University Kuala Lumpur (UniKL) and UPM are the two leading tertiary education institutions that are actively involved in biochar research. The Malaysian Agriculture Research and Development Institute (MARDI) is one of the leading governmental counterparts working in this area. Several types of biomass feedstock were examined for biochar production, including oil palm biomass, pineapple waste, rice husk and stalk, and other agricultural waste materials.

At the FRIM, laboratory-scale production of biochar was carried out using sludge waste generated from paper production mills. Transformation of paper mill bio-sludge (PMS) into biochar could be one of the promising solutions addressing the high costs of waste disposal. The pyrolysis process was adopted and the initial moisture content of PMS has been found to significantly influence the biochar yield.

8.5.1 Benefits from Biochar Applications

Over recent years, there has been increased attention on the effect of biochar addition to agricultural soils to ascertain benefits on soil quality and yield of crops (see Part 4 of this book). Biochar can be used as a soil amendment for various growth media. It acts as a plant nutrient source as well as a nutrient buffer that slowly releases them into the soil solution. Regarding physical soil properties, biochar creates more aeration, loosens soil pores and improves water retention (Smith et al., 2010). It also influences soil organisms and to some extent regulates effective mycorrhiza association (Noguera et al., 2010).

The industry is currently exploring the potentials of converting biomass into biochar for soil improvement. These companies are typically already in the biomass business such as producing biomass fuel pellets, briquettes or activated carbon. Biochar will serve as an additional product stream for their business using their existing equipment and manufacturing facilities.

8.5.2 Commercial Applications

Commercial ventures into biochar production and applications for agriculture use are relatively new in Malaysia even though the slash-and-burn practice has been traditionally practiced for generations. There are a number of favorable developments on both the supply and demand side that can make such ventures commercially viable.

On the supply side, the standard agriculture practice of open burning of biomass waste to return nutrients is no longer acceptable due to air pollution, risk of uncontrolled fire hazards and decreasing buffer zones between agricultural and residential areas. The other option is to compost biomass on site, however, this is a relatively slow process that potentially attracts unwanted pests and pathogens. Therefore, large amounts of biomass are now available to be collected from the field, aggregated and further processed, for example to produce biochar or bio-energy as well as other valuable products as detailed above.

The demand for biochar is also increasing especially with greater awareness among the potential end users and entrepreneurs due to the R&D efforts of researchers on its benefits to crop planting. The adoption of more sustainable practices in both large- and small-scale agriculture via the use of bio-fertilizers to reduce the use of chemical fertilizers has been picking up pace in Malaysia over the last decade. In the oil palm sector, a significant number of mills are building composting facilities to produce compost from the mix of EFB and effluent waste. The compost can be further enhanced by the addition of biochar as well as beneficial microbes to produce higher-value soil improvement products (Kong et al., 2014).

Another favorable development on the demand side is in the green property development sector. A recent trend of architecture in the ASEAN region, specifically Malaysia and Singapore, is the inclusion of vegetation as elements of commercial and residential buildings, such as roof-top gardens, green façades and so on. Due to the low-weight density and water-draining capabilities of biochar, it is possible to use either pure biochar or clay/biochar mixture to plant trees and vegetation in high-rise buildings as well as to use them as elements of urban landscaping.

One of the first projects of this nature has been in place in Malaysia since 2004 in a retail property development project called 1 Utama within the Klang Valley (see Figure 1.2). This project created 2,800 m^2 of green space on top of the shopping center and a rain forest integrated in the building structure, consisting of altogether up to 600 species of tropical trees and other plants. Depending on the types of species, the growth medium mixture ranges from pure biochar to a mix of biochar from charcoal briquettes, coconut fiber and clayey subsoil at different ratios (30:70, 50:50, 70:30). The function of the biochar is to reduce the weight of the substrate and to prevent the clay from lumping together, therefore improving the water draining and aeration properties of the final mixture (Richards, 2015). In addition, FRIM is currently testing the use of biochar on a sandy 5-acre plot joint afforestation project with Universiti Malaysia Pahang in Pekan, Pahang.

A plant in Sekinchan, Selangor, which has been in operation for 13 years, produces biochar from rice husks at a capacity of 8 tons per day. This plant operates in the vicinity of paddy fields, giving easy access to raw material. The biochar produced is used as a rice plant germination medium. About 1 kg of biochar is placed in a tray measuring 3 × 2 ft, and the seeds germinate well (Figure 8.4). The tray eases the mechanical transplanting of the seedlings to the fields by enabling the seedlings to be stacked onto the mechanical transplanter. No other mixtures are added; only fertilizer is sprayed 10 days after germination. Adopting this technology, the owner provides planting material and assistance to local farmers, charging USD 300 for a 3-acre planting plot. This biochar material also sells at USD 1.10 per 12 kg bag.

The commercial use of biochar in Malaysia is still in the early stage, due to the small supply and demand market as well as low awareness of its applications and potential. Production of biochar is also expensive with currently very few players in this part of the value chain. Additionally, the use of biochar for agriculture applications has to compete with other applications such as bio-energy use (charcoal and pellets) as well as activated carbon. Therefore, it is foreseeable that the use of biochar in Malaysia in the immediate future will be confined to niche applications such as urban landscaping and flora cultivation in green building, where it has several advantages over normal soil.

Figure 8.4. (a) Rice seedlings germinated on biochar from rice husk and (b) stacked on a rice seedlings transplanter. (A black and white version of this figure will appear in some formats. For the colour version, please refer to the plate section.)

8.6 Gaps and Challenges

This section describes some of the gaps and challenges in two categories – technical and non-technical – faced by the biomass industry in Malaysia.

8.6.1 Technical Challenges

(a) **Stability of biomass**: Due to the tropical climate, biomass degrades rapidly, therefore compromising the quality of the feedstock for downstream utilization. The distances between the sources of biomass and the utilization facilities also pose logistical challenges. It is therefore not feasible to transport and handle biomass in its raw form. Some form of pre-treatment such as drying, size reduction and compaction is needed to stabilize the biomass for downstream uses.

(b) **Common technical standards**: In a mature biomass industry, trading of biomass feedstock is an important component of the value chain to ensure stability in supply and pricing. For biomass to be easily traded as a form of commodity, common technical standards that are accepted by the industry stakeholders are needed to facilitate trading. Currently, technical standards for biomass in Malaysia are limited to long and short palm EFB fibers under SIRIM Malaysian Standard 1408:1997 (SIRIM, 1997). More extensive technical standards will need to be developed for other types of biomass as well as for catering to different applications of the biomass.

(c) **Appropriate conversion technologies**: Palm EFB contains more chloride, ash and oil than wood-based biomass, which leads to a number of operational and maintenance issues when it is used directly to substitute the latter whether as a renewable fuel source or feedstock for downstream processes. Imported technologies therefore need to be properly evaluated and adapted to ensure that they can work successfully with tropical-based biomass resources.

8.6.2 Non-Technical Challenges

(a) ***Access to biomass feedstock***: Biomass in Malaysia, while available in large amounts, is still not easily accessible to many industry players. Conventional agricultural practices, such as in the oil palm and rice sectors, dictate that much of the biomass needs to be returned to the planting fields to recycle nutrients and maintain soil fertility (Chew et al., 2009). This makes the plantation managers and farmers reluctant to sell off biomass as it may affect the yield of the next planting cycle.

(b) ***Access to financing***: For investors and financial institutions, biomass ventures are being perceived as high-risk ventures due to a number of factors. These include the non-guaranteed supply of biomass feedstock, lack of any pricing stability mechanism, unproven technologies, and the overall green market being mostly driven by government interventions rather than real economic needs.

(c) ***Low-value utilization of biomass***: In the early years of the biomass industry in Malaysia, much of the accessible biomass was used to produce low-value intermediate products (e.g. palm fibers, compost and fuel pellets) using relatively low-tech conversion technologies. The risk of new competitors into the same market space was therefore very high. There is a need for biomass industry stakeholders to adopt new technologies and a long-term strategic view to move into high-value utilization of biomass in order to protect their ventures from competitors. Moreover, the shift to high-value utilization is beneficial for further development of a green economy in terms of higher investment and income from a generally limited amount of biomass resources (Wan Asma et al., 2007, 2012).

(d) ***Sustainability certification***: One of the main hurdles faced by Malaysia in biomass utilization is the issue of the sustainability of its palm-based biomass, which constitutes up to 90% of the available biomass. As most of it is converted into products for the global green economy, most overseas buyers require that the biomass feedstock must be of sustainable origin. As an example, previous exports of palm biodiesel from Malaysia to major markets in Europe have led to legislative barriers due to the sustainability requirements imposed under the Renewable Energy Directive established in 2009.

8.6.3 Challenges to the Commercialization of Biochar

In addition to the general above mentioned challenges faced by the biomass industry, the commercialization of biochar for use in the agriculture sector needs to overcome more specific challenges.

Higher local acceptance and demand for biochar needs to be created by demonstrating the material's unique advantages in large-scale commercial plantings, especially in taking

Malaysia's agriculture sector towards a higher level of sustainability and profitability. Practices, standards and assessment tools need to be developed and promoted to farmers, plantation companies and other stakeholders to quantify the benefits of biochar application. Further research is still needed in order to estimate under which conditions the use of biochar would be beneficial.

Existing pyrolysis technologies used in Malaysia for biochar production are still expensive and the use of the still common traditional charring methods poses great risks, especially to kiln operators (also see Chapter 13) due to considerable pollution. This will limit the production facilities of biochar to sites that may be far away from locations where it is utilized, therefore increasing the cost of logistics and transportation. The ability of facilities to produce multiple streams of products in addition to biochar is also desirable as it will diversify investment risks.

8.7 Recommendations and Way Forward

Both technology and economics play an important role in ensuring that the biomass industry is able to grow and sustain itself in Malaysia. While technology development will ensure that biomass can be used for various applications and increasingly substitute fossil resources, the economics will determine whether it is feasible to sustain the industry over the long run. The challenges listed in the previous section must be addressed in a comprehensive and integrated manner to support the industry. Therefore, we propose the following recommendations:

(a) *Installation of a biomass commodity exchange*: Access to biomass feedstock at stable prices, quantity and quality is important for biomass industry players to sustain their businesses. A commodity exchange would be able to provide a trading platform to stabilize the pricing and supply of the feedstock. One of the benefits of such an exchange is that it would lower the risks of biomass ventures and make them more attractive for financing and funding by investors. Such an exchange could be either institutionally or privately driven and it could also be regional at the ASEAN or Asia Pacific level.

(b) *Issuing sustainable consumption policies*: More efforts and actions should be made to create a market pull for biomass-based products at the consumption side. In the case of Malaysia, most if not all ventures in the production of biomass fuel pellets are export-oriented due to the favorable pricing offered by the importing countries because of their transition towards renewable energy. On the other hand, Malaysia is still highly dependent on fossil fuels such as coal, oil and gas for its economic growth. The same scenario applies to other biomass-based products where the export market is more favorable than local markets. Such mismatch in demand and supply needs to be addressed in order to support the local biomass industry. Sustainable biomass products will also need to be certified and promoted to consumers to drive the demand for sustainable consumption.

(c) *Commercialization of local technologies for high-value utilization*: Technologies and know-how are the key drivers for moving the national biomass industry up the global value chain. The challenge to commercialization is of course to determine which technologies are commercially viable. Therefore, the development of biomass technologies must be industry-driven and research-assisted. One of the components missing in the biomass industry value chain is the establishment of commercial-based technology development centers to fully serve the needs of the industry. Expertise in these centres must also be multidisciplinary in nature in order to provide comprehensive technological solutions to industry players.

(d) **ASEAN-wide biomass industry**: Member countries of ASEAN should work towards a regional biomass industry, with each country making use of its respective competitive advantages to fit into the regional value chain. For example, Malaysia, Singapore, Brunei and Thailand have a strong base in manufacturing, technical expertise and financing, while Indonesia, the Philippines and Vietnam are growing economies with strong demand for energy and materials, and Cambodia, Laos and Myanmar are agriculture-based economies which could be significant suppliers of biomass feedstock. As the region moves towards an ASEAN Economic Community (AEC), the biomass industry could be one of the industries promoted under this regional integration initiative.

The recommendations above apply for the production and consumption of biochar just as well as for all other biomass products in general. Having a biomass commodity exchange would ensure the biochar entrepreneurs can secure consistent biomass feedstock and hence, financing for their ventures. The same exchange could also serve as the platform to market and trade biochar products.

Specific sustainable consumption policies could be formulated to promote the wider use of biochar in commercial agriculture and reduce the consumption of chemical fertilizers due to higher efficiencies. Local technologies need to be promoted and commercialized to develop cleaner production methods of biochar as well as add value to biochar. Last but not least, an ASEAN-wide biochar sub-industry could be one of the key drivers to shift the regional agriculture sector towards more sustainable practices to meet the increasing food demand of the growing ASEAN population while having lower impacts on the environment.

References

AIM (2011). *National Biomass Strategy 2020: New Wealth Creation for Malaysia's Palm Oil Industry*. Selangor: Agensi Inovasi Malaysia (AIM).

AIM (2013). *National Biomass Strategy 2020: New Wealth Creation for Malaysia's Biomass Industry*. Selangor: Agensi Inovasi Malaysia (AIM).

Arpah, A. R. (2013). Solid waste management in Malaysia: a way forward. Paper presented at the International Solid Waste Association World Congress, Vienna. Austria: Vienna, October 2013.

Chan, C. W. and Cho, M. C. (2012). Country report: Malaysia. Paper presented at the Asia Pacific Economic Cooperation (APEC) Workshop on Food Security, Tokyo. Japan: Tokyo, 17–19 January 2012.

Chan, K. W. (2009). Biomass production and uses in oil palm industry. In: Singh, G. et al. (eds.) *Sustainable Production of Palm Oil: A Malaysian Perspective*. Kuala Lumpur: Malaysian Palm Oil Association, pp. 133–161.

Chen, S. S., Isnazunita, I., Abdul Nasir, A. and Puvaneswari R. (2011). Refuse derived fuel – case study of waste as renewable resource. *International Journal for Sustainable Innovations*, 1, pp. 81–86.

Chew, P. S., Kee, K. K. and Goh, K. J. (2009). Cultural practices and their impact. In: Singh, G. et al. (eds.) *Sustainable Production of Palm Oil: A Malaysian Perspective*. Kuala Lumpur: Malaysian Palm Oil Association, pp. 163–198.

Gadde, B., Bonnet, S., Menke, C. and Garivait, S. (2009). Air pollutant emissions from rice straw open field burning in India, Thailand and the Philippines. *Environmental Pollution*, 157, pp. 1554–1558.

Gan, K. S, Lim, S. C. and Rahim, S. (2013). *Forest Products R&D at FRIM: Yesterday and Today*. Kepong: Forest Research Institute Malaysia (FRIM).

ITN Source (2009). Malaysian charcoal industry thrives as Japan demand for it rises. [online] Available at: www.itnsource.com/shotlist//RTV/2009/03/09/RTV412009/ [Accessed 2 November 2015].

Kong, S. H., Loh, S. K., Bachmann, R. T., Rahim, S. A. and Salimon, J. (2014). Biochar from oil palm biomass: A review of its potential and challenges. *Renewable and Sustainable Energy Reviews*, 39, pp. 729–739.

LGM (2016). Natural rubber statistics 2016. Lembaga Getah Malaysia (LGM, Malaysian Rubber Board). [online] Available at: www.lgm.gov.my/nrstat/nrstats.pdf [Accessed 15 November 2015].

Mahmudin, S., Wan Asma, I. and Puad, E. (2007). Processing of oil palm lignocellulosic residues. In: Wan Asma, I. et al. (eds.) *Turning Oil Palm Residues into Products*. FRIM Research Pamphlet, 127, pp 11–25.

MBIC (2015). *Malaysia Biomass Industries Review 2015/2016*. Putrajava: Malaysia Biomass Industries Confederation Putrajava.

MIGHT (2013). *Malaysian Biomass Industry Action Plan 2020: Driving SMEs Towards Sustainable Future*. Selangor: Malaysian Industry–Government Group for High Technology (MIGHT).

Milbrandt, A. and Overend, R. P. (2008). *Survey of Biomass Resource Assessments and Assessment Capabilities in APEC Economies. Report of the Asia-Pacific Economic Cooperation (APEC) Energy Working Group*. Singapore: APEC Secretariat.

Noguera, D., Rondon, M., Laossi, K. R. et al. (2010). Contrasted effect of biochar and earthworms on rice growth and resource allocation in different soils. *Soil Biology and Biochemistry*, 42, pp. 1017–1027.

Puad, E., Mahanim, S. and Rafidah, J. (2007). Opportunities and challenges in the charcoal industry in Malaysia and South East Asia. In: *Proceedings of the Seminar on Energy from Biomass 2006*, Kuala Lumpur: FRIM.

Richards, T. (2015). Profile: Bandar Utama – A history of biochar application in Malaysia. [online] Available at: www.biochar-international.org/profile_Malaysia. [Accessed 2 November 2015].

Rosenani, A. B., Ahmad, S. H, Nurul, A. S., Khairunissa, M. Y., Tan, W. I. and Lee, S. C. K. (2012). Biochar as a soil amendment to improve crop yield, soil health and carbon sequestration for climate change mitigation. In: Wan Rasidah, K. et al. (eds.) *Proceedings of the Soil Science Conference of Malaysia 2012. Soil Quality Towards Sustainable Agriculture Production*. 10–12 April 2012, Kota Bharu, Kelantan. Kuala Lumpur: Malaysian Society of Soil Science, pp. 5–14.

SEDA (2015). Statistics and monitoring – CO_2 avoidance. [online] Available at: www.seda.gov.my [Accessed 2 November 2015].

Singh, G., Low, D. L., Lee, K. H., Lim, K. C. and Loong, S. G. (2009). Empty fruit bunches as mulch. In: Singh, G. et al. (eds.) *Sustainable Production of Palm Oil: A Malaysian Perspective.* Kuala Lumpur: Malaysian Palm Oil Association, pp. 301–315.

SIRIM (1997). *Specification for Oil Palm Empty Fruit Bunch Fibre. Malaysian Standard MS1408:1997.* Kuala Lumpur: Institut Piawaian dan Penyelidikan Perindustrian Malaysia (SIRIM).

Smith, J. L., Collins, H. P. and Bailey, V. L. (2010). The effect of young biochar on soil respiration. *Soil Biology and Biochemistry*, 42, pp. 2345–2347.

Wan Asma, I., Wan Rasidah, K. and Mohd Nor, M. Y. (2007). *Turning Oil Palm Residues into Products.* FRIM Reasearch Pamphlet No. 127. Kepong: Forest Research Institute Malaysia (FRIM).

Wan Asma I., Wan Rasidah K., Rafidah, J. and Khairul, A. J. (2012). *As Good as Wood.* Kepong: Forest Research Institute Malaysia (FRIM).

Wan Rasidah, K. and Wan Asma, I. (2007). Oil palm residues as organic mulch and soil conditioner. In: Wan Asma, I. et al. (eds.) *Turning Oil Palm Residues into Products.* FRIM Research Pamphlet, 127. Kepong: Forest Research Institute Malaysia (FRIM), pp. 99–111.

Wu, T. Y., Mohammad, A. W., Jahim, J. M. and Anuar, N. (2010). Pollution control technologies for the treatment of palm oil mill effluent (POME) through end-of-pipe processes. *Journal of Environmental Management*, 91, pp. 1467–1490.

9
Carbon Sequestration Potential of Forest Biomass in Turkey

BETÜL UYGUR AND YUSUF SERENGIL

Abstract

The role of Turkish forests in mitigating GHG emissions under the present policy incentives is evaluated in this chapter. Currently, the forestry sector is an increasing sink whereas the GHG emissions of the energy, industry, agriculture and waste sectors are increasing faster. The main drivers of GHG removals by forests are the difference between wood increment and harvest rate, and ambitious afforestation/reforestation programs. Potentially, wood biomass could supply a significant part of the energy demand and could be a substitute for fossil fuels. However, the current policy framework does not reward the implementation of such initiatives. Renewable resources constitute some 9% of Turkey's primary energy supply, and almost half of it is provided from biomass resources. Mostly agricultural and animal wastes are collected and converted into energy in this way but the amount decreases every year in contrast with the global trend. The use of forest biomass as an alternative energy source is very limited in Turkey and there is not much support from the government on this. The biomass energy sources are emphasized in the Climate Change Action Plan of Turkey prepared in 2011 but the government has failed to incentivize this with efficient policy approaches yet.

9.1 Introduction

Renewable energy sources received attention as the war against climate change intensified during the last decade (Christiansen, 2002; Anandarajah and Strachan, 2010; Sapkota et al., 2015). The Kyoto Protocol (KP) has obligated Annex 1 Parties to prepare their annual national inventory reports (NIR) of greenhouse gas (GHG) emissions by source and removals by sink every year for submission under the United Nations Framework Convention on Climate Change (UNFCCC). The accounting rules were defined by guidelines issued by the Intergovernmental Panel on Climate Change (IPCC, 1995) and include the sector land use, land-use change and forestry (LULUCF). A new international climate change agreement that will have legal force and be applicable to all countries is being negotiated under the auspices of the UNFCCC. The pathway towards a new agreement for the post-2020 period may bring out several positive outcomes as the increased global energy demand causes an upward trend in GHG emissions.

Forest ecosystems have significant potential to reduce the concentration of atmospheric CO_2 by storing carbon in the above- and below-ground biomass, litter, deadwood and soil. Many studies have been focused on the role of forest ecosystems in the global carbon cycle around the world and also in Turkey (Dixon et al., 1994; Brown, 1996; Kaygusuz and Türker, 2002; Kaygusuz and Sarı, 2003; Demirbaş, 2004; Swift and Cuzner, 2012; Ahmad et al., 2014). The future sink strength of CO_2 of forest ecosystems depends on the implemented management strategies (Paoletti et al., 2013). In the case of carbon sequestration by differently managed forest ecosystems, the effects of forest management on C balance in forestry has been reviewed in many studies (Johnson and Curtis, 2001; Johnson et al., 2002; Jandl et al., 2007; Morrison et al., 2012) pointing out that management is a key determinant. Forest management has two main effects on the C sequestration process: first is the potential to increase the terrestrial C pool and second is the influence on the soil C pool because of disturbance of soil dynamics (Jandl et al., 2007).

The most important forest management activity that the studies mainly have focused on was harvesting. Johnson et al. (2002) suggested that harvesting causes little lasting effect on soil C after 15–16 years. Another suggestion of that research was that the differences caused by harvesting treatments in ecosystem C content at a later time can be significantly related to the differences in biomass C and how they increase or decrease over time. Another study emphasizes that the long-term balance depends on the extent of soil disturbance and defines the effects of harvesting in two ways: leaving harvest residues on the soil surface causes an increment of C stock of forest floor and soil C loss because of disturbance of the soil structure (Jandl et al., 2007). According to these suggestions, a key factor of harvesting operations is that soil disturbance affects C sequestration of forest ecosystems negatively. Soil disturbances during forest management should therefore be minimized.

A long-term C sink can be generated by improving the soil fertility by application of biomass-derived C (biochar) to soil (Lehmann et. al., 2006). The benefits of adding biochar to soil are increased C sequestration, crop yield and retention of nutrients and chemicals (H. Wang, 2010; J. Wang, 2010). Also monitoring the conversion of biomass into biochar and usage of it as a soil amendment tool is easier and more economic than other sequestration approaches (Lehmann, 2007).

Our objective is to discuss the position of Turkey in the UNFCCC process, analyze the carbon stock potential of forests and to identify possibilities for biomass resources.

9.2 Position of Turkey in UNFCCC

The forest biomass utilization capacity of a country depends on its biomass resources together with the economic, social and legislative regulations that the carbon economy is highly dependent on.

Turkey ratified the UNFCCC in 2004 and is one of the country parties that has been listed as Annex 1 but with "specific circumstances". Its name was deleted from Annex 2 and was not included in Annex B of the KP1 (first commitment period of the Kyoto Protocol 2008–2012) and KP2 (Kyoto Protocol 2013–2020). Annex 1 Parties have to report their GHG emissions by source and removals by sink every year. LULUCF is one of the six sectors controlled by the UNFCCC and KP and the only one where removals have the potential to partly compensate for the emissions of a party.

Turkey is not expected to contribute to the climate financing regime, but rather to benefit from it. Turkey does not have the binding GHG emission reduction commitment inscribed in Annex B to the KP. The current status of Turkey gives few if any incentives for the implementation of a green economy and curbs motivation for the uses of renewable resources.

9.3 Carbon Stocks of Forests in Turkey

Turkey's contribution to global GHG emissions reached 439.9 Gg CO_2 equivalents in 2012. This represented an increase of 139% since 1990 as the economy and population grew by around 4.5% (Beyazıt, 2004), and 3.1% per year (TUIK, 2014), respectively. Like most developing countries the energy sector represents the largest share of overall emissions (Figure 9.1).

The percentage increase in emissions represents the highest increase among the Annex 1 countries compared to 1990. However this high emission is caused by a high population because the per capita emission of Turkey is 5.06 Mg CO_2 equivalents per capita per year, which is considerably below the average of Annex 1 countries (9.31 Mg CO_2 equivalents per capita per year) (UNFCCC, 2015).

LULUCF is the only sector that has been a sink for GHGs. The removals in this sector are increasing but not as much as emitting sectors.

In Turkey, degradation of forests occurs commonly through conversion to agriculture, pasture and urban areas as in Mediterranean forests (Kaya and Raynal, 2001; Biringer, 2003). In addition forest fires are a major threat in Turkey. According to the General Directorate of Forestry (GDF) approximately 60% of the forests have been specified as very sensitive to fires.

In this regard, the historical changes in forest areas have a significant role in the annual exchange of carbon between the atmosphere and forests. To define the changes, the forests have been grouped based on three criteria and identified by the GDF under six sub-categories:

i) coniferous (around 76% of the area of pure high forest) vs deciduous forest (around 24%);

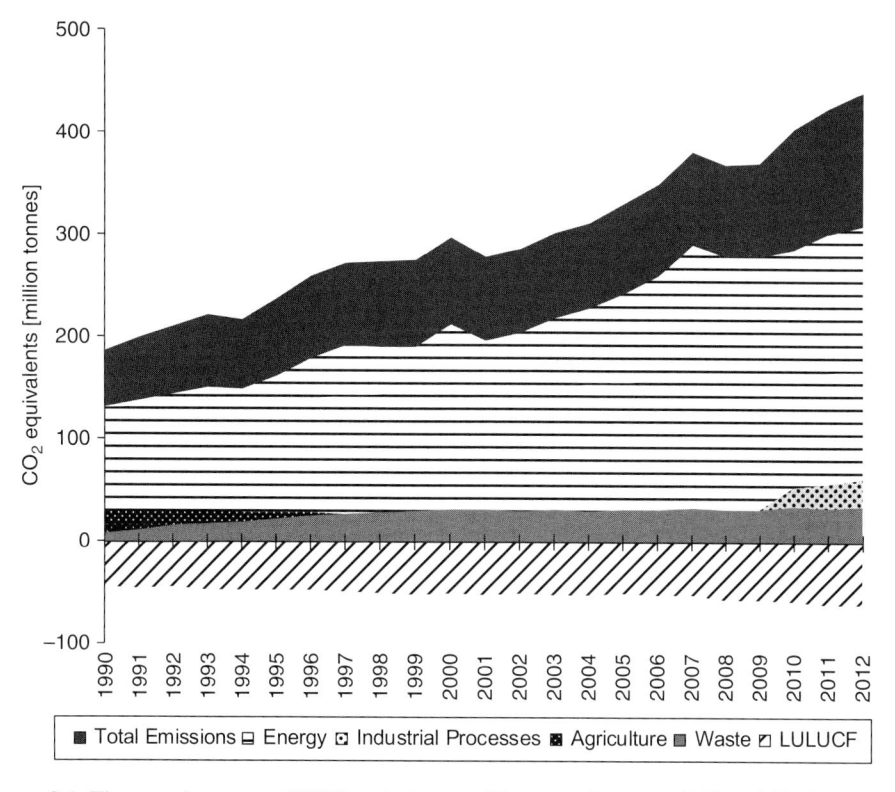

Figure 9.1. The greenhouse gas (GHG) emission trend by sector between 1990 and 2012 (modified from NIR, 2014).

ii) productive (more than 10% forest cover) vs degraded (between 1% and 10% forest cover);
iii) and high forests (80% of the total forest area) vs coppices (20% of the total forest area).

The national forest definition has some differences from the FAO (Food and Agriculture Organization of the UN) definition of forest land (area > 0.5 ha; tree > 5 m; tree canopy cover > 10%; no inclusion of land predominantly under agricultural or urban land use). This forest definition is equivalent to the national definition of "productive forest" (which can be high forest or coppice); the FAO definition of Other Wood Land (OWL) is partially equivalent to the national definition of "degraded forest" (which can be high forest or coppice): as the definition of degraded forest captures land with 1–10% tree cover, the area of degraded forest is larger than the area of OWL (with tree cover between 5% and 10%).

According to the national definition, there is around 21.9 M ha of forest (27% of Turkey), 54% considered "productive" (above 10% of crown closure) and 46% considered "degraded" (between 1% and 10% of crown closure) (Table 9.1). These figures show that almost half of the forests in the country still have crown closure of less than 10% or in other words classed as degraded.

Table 9.1 *Shares of productive vs degraded forest area*

	Mixed high forest (M ha)	Coppices (M ha)	Total (M ha)
Productive	10.86	1.08	11.94
Degraded	7.15	2.76	9.91
Total	18.01	3.84	21.86

Source: calculated based on ENVANIS.

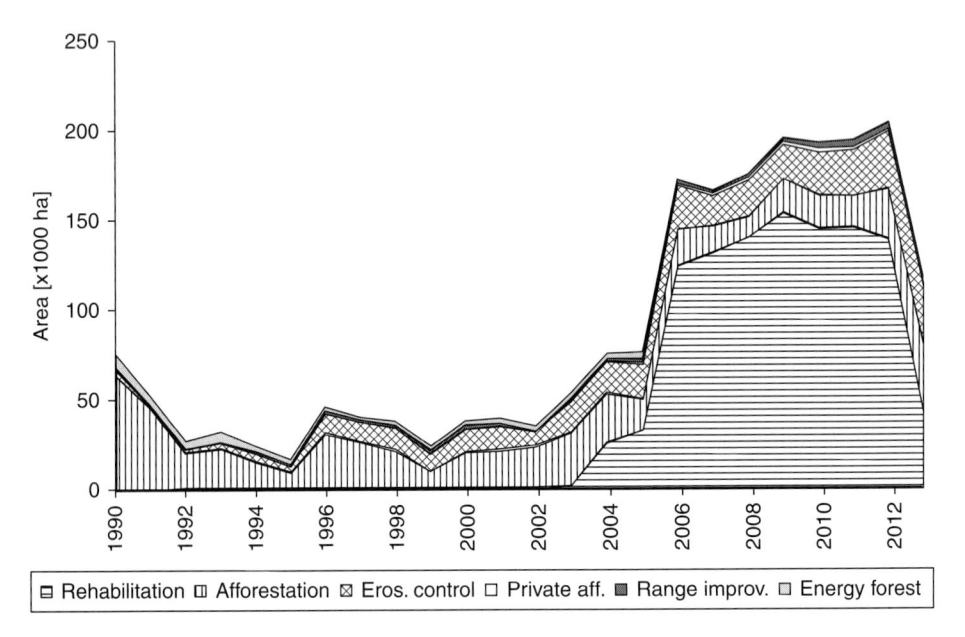

Figure 9.2. Changes in various forestry activities (ha/yr) from 1990 to 2013 (modified after Bouyer and Serengil, 2014).

National Forest Inventories (NFI) were carried out in Turkey in 1972 and 2004. After 2004, a system called ENVANIS was created, which compiles data from forest management units and classifies stands using three criteria: species mix, crown closure and age classes. Therefore, it allows the calculation of area changes, volume increment changes and stock changes year by year.

In order to increase the forest area of Turkey, the National Afforestation and Erosion Control Action Plan was initiated in 2008. The GDF has conducted various forestry activities such as afforestation, rehabilitation, erosion control activities, and artificial regeneration in degraded forests (Figure 9.2).

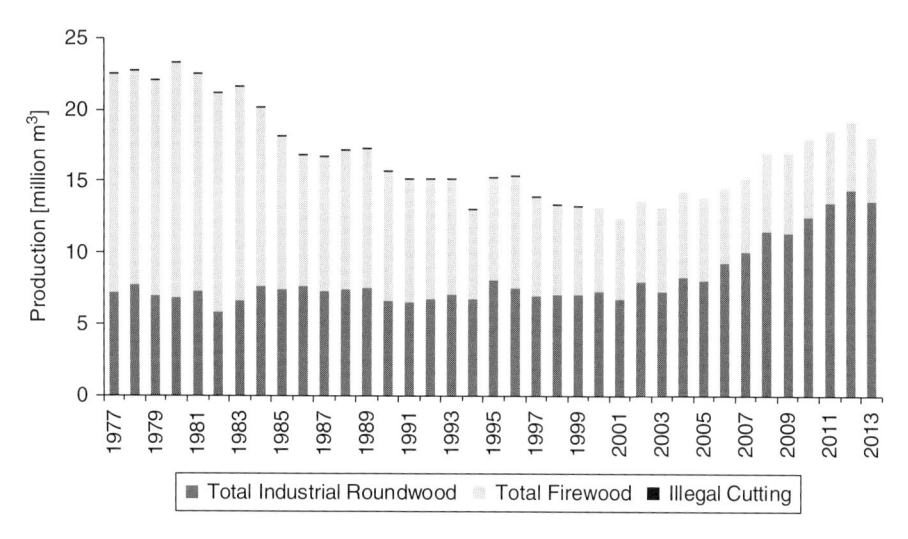

Figure 9.3. Annual harvest rate in Turkish forests from 1977 to 2013 (updated from Bouyer and Serengil, 2014).

The demand for industrial wood in Turkey is steadily increasing, mainly to meet the needs of the construction industry (Figure 9.3). This increase in industrial roundwood and decrease in firewood caused a gain in carbon stocks in the Harvested Wood Products (HWP) pool of the country. The carbon stock in HWP has been estimated by Bouyer and Serengil (2016) to be around 7.5 Gg CO_2 equivalents in 2013, representing around 10% of the whole LULUCF sector removals.

9.4 The Role of Biomass in Carbon Management in Turkey

Biomass is a cost-effective and often sustainable energy source that can substitute fossil fuels and consequently mitigate GHG emissions. According to 2012 data from the Ministry of Energy and Natural Resources, fossil fuels constitute 90% while renewable energy sources constitute 10% of Turkey's primary energy supply (MENR, 2014). Wood and biomass constitute 47% of the renewable energy sources but this figure decreases every year in contrast with the global trend (MENR, 2011).

According to the latest NIR of Turkey the C stock increase in productive forests was 19.32 Mt by 2012. An additional 2.57 Mt C were added from dead organic matter stocks mainly because of conversion to forestland. Carbon losses were due to commercial cutting (5.43 Mt C), fuel wood gathering (0.90 Mt C) and forest fires (0.08 Mt C). The net C sequestration has been 15.47 Mt C. The net C sequestration from degraded forests was lower than this (1.11 Mt C) (NIR, 2014). These figures show that the C stocks in forests in Turkey are increasing mainly due to increased standing volume per hectare and an increase in

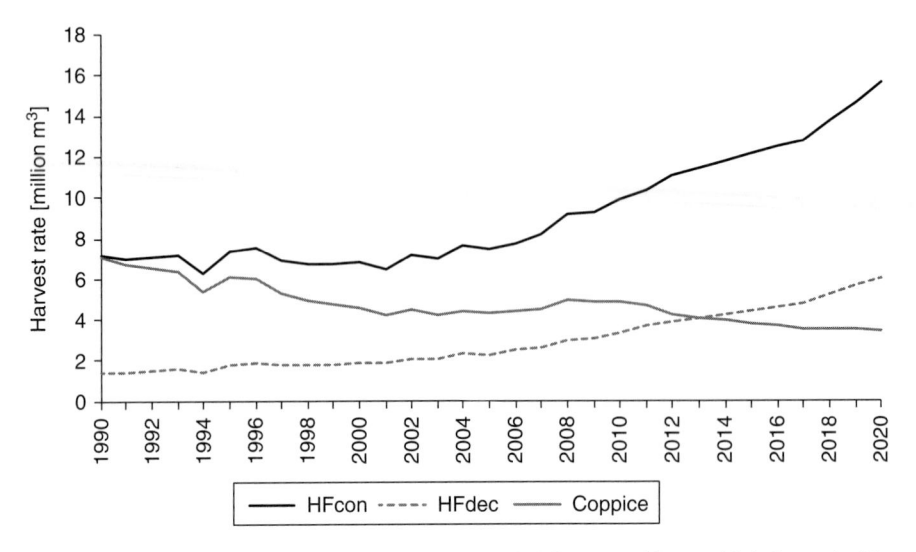

Figure 9.4. 2020 projections of harvest per forest type (high forest coniferous, high forest deciduous and coppice) in the extensive scenario (modified after Bouyer and Serengil, 2014).

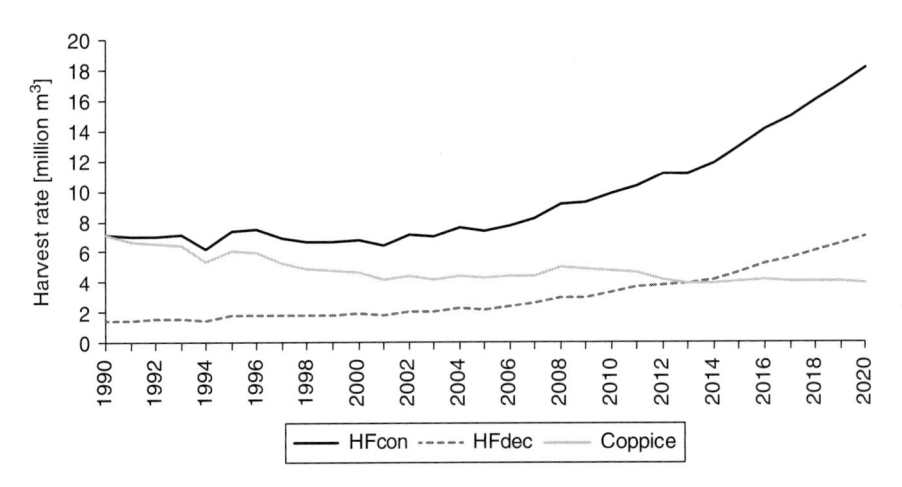

Figure 9.5. 2020 projections of harvest per forest type (high forest coniferous, high forest deciduous and coppice) in the intensive scenario (modified after Bouyer and Serengil, 2014).

the cover of productive forests. These increases are expected to further increase the annual increment and therefore carbon intensity of forests in the coming years.

The biomass stock has a significant potential to contribute to reducing GHG emissions by storing a large amount of CO_2. However, in the 2014 NIR of Turkey, the LULUCF sector net removals reached 60.787 Gg CO_2 eq. This shows that there is a large mitigation potential in this sector and it can be further enlarged with some cost-effective activities that may include grassland rehabilitation, afforestation/reforestation, pest control measures and so on. Bouyer and Serengil (2014) identified two alternative scenarios based on the GDF Strategic Plan for forest management: 90 m m³ of roundwood harvest between 2013 and 2017 (intensive harvest) and 25 m m³/yr of felling (industrial roundwood) harvest by 2020 (extensive harvest). The corresponding volumes of firewood, felling and total roundwood were forecasted accordingly from 2013 to 2020. Having estimated two 1990–2020 data series of roundwood production for GDF, an extensive one (25 m m³/yr by 2020) and an intensive one (29 m m³/yr by 2020, + 4 m m³/yr compared to the other), this harvest has been allocated to the three main forest types (coniferous, deciduous, coppice) (Figures 9.4, 9.5).

9.5 Biomass Potential as a Renewable Energy Source in Turkey

Using biomass as a renewable source for energy has been an option in respect of climate change mitigation in Turkey. Also, meeting the demand for energy in an economic way has been very important in the case of spending a significant amount of money for energy. The forest area in Turkey is 21.7 million hectares with 9.6 m ha subject to timber production according to official numbers provided by GDF. The increment of forests in Turkey is estimated as 33.6 m m³ while production remains at a level of 17 m m³. This means that half of the increment is added to the biomass stock and there is some more potential for harvest. However this may disturb sustainability of the managed forests considering the other functions of the forests.

Fuel wood gathering decreased in Turkey for the last couple of decades. The conversion of coppices to high forests was a significant reason for this. Nevertheless, the amount in 2012 was 4.49 M m³ (= 0.9 M t C), offers promising potential for biomass resources. The harvested wood in Turkey was used in the following major sub-sectors (excluding fuel wood) in 2013 in m m³: 3.47 as roundwood, 0.41 as mine pole, 0.53 industrial wood, 1.65 pulp wood, 4.16 fiber and particle wood and 10.67 as pole (all data from the ENVANIS System).

There are different estimations on the annual biomass energy potential in Turkey. Ediger and Kentel (1999) estimated 17.2 Mtoe (million tonnes of oil equivalent) while Demirbaş et al. (2006) and Demirbaş (2008) estimated the potential as 32 Mtoe (Saraçoğlu, 2008). The difference between the estimations suggests that the potential almost doubled between these years. However the contribution of forest residues has been given as 64% of the total biomass energy (Taşkıran, 2009). According to the FAO, some countries that are members of the International Energy Agency (IEA) have the potential to meet their energy requirements by using biomass energy, mentioning a percentage of 22 in Finland, 18 in Sweden and 14 in Austria. Also Turkey has the potential to increase the use of renewable energy by up to 12% by using the forest residues; however, the actual value is approximately 1.5% (GDF, 2009). The difference between the IEA countries and Turkey is derived from the classic and modern energy forest. For a long time the classic energy forest approach has been applied in Turkey, especially in degraded forests (Saraçoğlu, 2008; Geray and Okan, 2008). In this approach coppices are utilized by forest villagers for fuel wood on an annual basis. The eighth five-year

Development Plan stated that modern energy forest establishments should be promoted in Turkey (DPT, 2001). Based on GDF data, the amount of annual average biomass related to forest residues which can be utilized for bioenergy in Turkish forests is 5–7 million tons. This is a considerable amount both economically and environmentally for Turkey. In this regard, the GDF has been working on the determination of biomass potentials in Turkish forests. On the other hand, the GDF has been involved in many projects that have had the aim of using forest biomass for heating. The organization's goal is making forest biomass a renewable energy source by setting up bioenergy systems in Turkey (GDF, 2009). The GDF expects to have official regulations put in place in respect of forest biomass utilization as a bioenergy source, taking place in the sector that is supported by the government and producing bioenergy more economically in Turkey.

Although not recommended in most cases, the litter layer in forests is a potential resource for biomass in Turkey. The southern regions of the country where Calabrian pine forests prevail provide suitable conditions for this purpose. Litter layer removal along the roads is already done as a measure to diminish fires hazard in this region. There are some estimates for litter potentials but considering the diversity of forest types and climate variations it is hard to estimate a reliable amount for this. The C stored on the forest floor is estimated in a range of 1.70–7.46 Mg per hectare by Tolunay (2011). The C stock of productive deciduous forests in a warm-dry region (Istanbul) has been given as 2.86 Mg per hectare (std dev. = 1.65) for deciduous forests, 4.43 Mg per hectare (std dev. = 3.27) for coniferous forests, and 4.02 Mg per hectare (std dev. = 1.77) for mixed forests by Serengil et al. (as cited in NIR, 2014). The litter layer stock is considerably low or ignorable for grasslands (0.06 Mg ha^{-1}) and croplands (0.27 Mg ha^{-1}) according to the same study. The amount of litter in forests is considerable in some ecosystems and regions in Turkey. However the aim of forest management in Turkey is not carbon storage or carbon conservation in forest ecosystems. In recent years, GDF has tried to determine new strategies and policies towards adaptive forest management. Biomass has an important role in respect of GHG mitigation and being a new aspect of renewable energy sources (GDF, 2009).

9.6 Conclusions

Turkey is a developing country experiencing a fast urbanization trend. One of the reasons for increasing forest cover is migration to the cities from rural regions. This is advantageous in terms of sustainable natural resource management, in particular for forests. Illegal cutting, deforestation and wildfires have decreased substantially during the last decade as demands from forests have diminished. This situation provides less problematic forestry due to fewer inhabitants around the forests. However, wood demand is increasing and other ecosystem services are demanded more severely. Urban forestry becomes a new focus under these new conditions. All these changes in society cause changes in forestry policy and management.

The forest cover in Turkey is not sufficient to satisfy the demand of all ecosystem services required for a large population. Half of the forest area is already degraded or has a crown cover less than 10%. Thus, per hectare litter and dead wood biomass stocks that can be subject to biomass energy are low. The increasing demands of the construction industry cannot be compensated by national resources and therefore an increase in importation or a more intensive harvest can be anticipated for the coming decades. The strategic document of the GDF puts this forward and estimates increased production. This is reflected also by the economic point of view as the import of industrial roundwood is expected to reach more than 1 billion US dollars by 2012.

The possibility to use forest biomass as an energy resource is not likely to happen to a significant extent in these conditions. In other words we do not expect any significant investment by the government for the next five years. However, this may change if the UNFCCC process adds some motivation on GHG mitigation investments in the country. A more realistic scenario is the strengthening of the carbon economy in Turkey by projects funded by bilateral or multilateral agencies.

Litter layer and dead wood in a forest are vital to sustain the water and nutrient cycles. However, woody residues left in forests after forest operations can raise the greenhouse gas emissions because of causing wildfires in Turkey (GDF, 2009). In many parts of the country residues along the roads are cleared and carried away to disable fire occurrence. These residues if left on site are carbon stocks ready for mineralization or for oxidation by fire. Every once in a while, for example every ten years, they can be used as a resource for biomass projects.

The possibility of biochar as a mitigation strategy should be compared with other mitigation options such as afforestation or range rehabilitation also taking the total economic values into account. The use of biochar as a soil rehabilitation option seems to be more feasible especially to mitigate soil and water pollution.

References

Ahmad, A., Mirza, S. N. and Nizami, S. M. (2014). Assessment of biomass and carbon stocks in coniferous forest of Dir Kohistan, KPK. *Pakistan Journal of Agricultural Sciences*, 51, pp. 335–340.

Anandarajah, G. and Strachan, N. (2010). Interactions and implications of renewable and climate change policy on UK energy scenarios. *Energy Policy*, 38, pp. 6724–6735.

Beyazıt, M. F. (2004). Türkiye Ekonomisi ve Büyüme Oranının Sürdürülebilirliği (Turkish economy and sustainability of growth rate). *Doğuş Üniversitesi Dergisi*, 5, pp. 89–99.

Biringer, J. (2003). Forest ecosystems threatened by climate change: promoting long-term forest resilience. In: Hansen, L. J., Biringer, J. L. and Hoffman J. R. (eds.) *Buying Time: A User's Manual for Building Resistance and Resilience to Climate Change in Natural Systems.* WWF Climate Change Program.

Bouyer, O. and Serengil, Y. (2014). *Cost and Benefit Assessment of Implementing LULUCF Accounting Rules in Turkey.* Istanbul: OGM.

Bouyer, O. and Serengil, Y. (2016). Carbon stored in harvested wood products in Turkey and projections for 2020. *Journal of the Faculty of Forestry Istanbul University (JFFIU)*, 66, pp. 295–302.

Brown, S. (1996). Present and potential roles of forests in the global climate change debate. [online] Available at: www.fao.org/docrep/w0312e/w0312e03.htm#TopOfPage [Accessed 1 January 2015]

Christiansen, A. C. (2002). New renewable energy developments and the climate change issue: a case study of Norwegian politics. *Energy Policy*, 30, pp. 235–243.

Demirbaş, A. (2004). The importance of biomass. *Energy Sources*, 26, pp. 361–366. doi. 10.1080/0090831049077406.

Demirbaş, A. (2008). Importance of biomass energy sources for Turkey. *Energy Policy*, 36, pp. 834–842.

Demirbaş, A., Pehlivan, E. and Turkan, A. (2006). Potential evolution of Turkish agricultural residues as bio-gas, bio-char and bio-oil sources. *International Journal of Hydrogen Energy*, 31, pp. 613–620.

Dixon, R. K., Solomon, A. M., Brown, S., Houghton, R. A., Trexier, M. C. and Wisniewski, J. (1994). Carbon pools and flux of global forest ecosystems. *Science*, 263, pp. 185–190. doi:10.1126/science.263.5144.185.

DPT, (2001). VIII. Bes yıllık kalkınma planı, ormancılık özel ihtisas komisyonu raporu, Silvikültür ve Enerji Ormanları. *DPT: 2531-ÖİK*, 547, pp. 255–278.

Ediger, V. S. and Kentel, E. (1999). Renewable energy potential as an alternative to fossil fuels in Turkey. *Energy Conservation and Management*, 40, pp. 743–755.

Geray, A. U. and Okan, T. (2008). Yenilenebilir enerji kaynağı olarak ormanlar. [online] Available at: www.foresteconomics.org/ugeray2.pdf [Accessed 17 February 2015]

IPCC. (1995). *Climate change 1995: IPCC second assessment report. A report of the Intergovernmental Panel on Climate Change*. [online] Available at: www.ipcc.ch/pdf/climate-changes-1995/ipcc-2nd-assessment/2nd-assessment-en.pdf [Accessed 07 January 2015]

Jandl, R., Lindner, M., Vesterdal, L., et al. (2007a). How strongly can forest management influence soil carbon sequestration? *Geoderma*, 137, pp. 253–268.

Jandl, R., Vesterdal, L., Olsson, M., Bens, O., Badeck, F. and Rock, J. (2007b). Carbon sequestration and forest management. *CAB Reviews: Perspectives in Agriculture, Veterinary Science, Nutrition and Natural Resources*, 2, doi: 10.1079/ PAVSNNR20072017.

Johnson, D. W. and Curtis, P. S. (2001). Effects of forest management on soil C and N storage: meta analysis. *Forest Ecology and Management*, 140, pp. 227–238.

Johnson, D. W., Knoepp, J. D., Swank, W. T. et al. (2002). Effects of forest management on soil carbon: results of some long-term resampling studies. *Environmental Pollution*, 116, pp. 201–208.

Kaya, Z. and Raynal, D. J. (2001). Biodiversity and conservation of Turkish forests. *Biological Conservation*, 97, pp. 131–141.

Kaygusuz, K. and Türker, A. (2002). Biomass energy potential in Turkey. *Renewable Energy*, 26, pp. 661–678.

Kaygusuz, K. and Sarı, A. (2003). Renewable energy potential and utilization in Turkey. *Energy Conversion and Management*, 44, pp. 459–478.

Lehmann, J., Gaunt, J. and Rondon, M. (2006). Bio-char sequestration in terrestrial ecosystems – a review. *Mitigation and Adaptation Strategies for Global Change*, 11, pp. 403–427. doi: 10.1007/s11027-005-9006-5.

Lehmann, J. (2007). A handful of carbon. *Nature*, 447, pp. 143–144.

Ministry of Energy and Natural Resources (2014). Türkiye Ulusal Yenilenebilir Enerji Eylem Planı. [online] Available at: www.eie.gov.tr/duyurular_haberler/document/Turkiye_Ulusal_Yenilenebilir_Enerji_Eylem_Plani.PDF [Accessed 8 August 2016]

Morrison, T., Matthews, R., Miller, G. et al. (2012). Understanding the carbon and greenhouse gas balance of forests in Britain. *Forestry Commission Research Report*, pp. 68–69. [online] Available at: www.forestry.gov.uk/pdf/FCRP018.pdf/$FILE/FCRP018.pdf [Accessed 4 July 2016]

NIR (2014). *National Greenhouse Gas Inventory Report 1990–2012*. Annual report submission under the framework convention on climate change. Ankara: TurkStat (Turkish Statistical Institute).

Paoletti, E., De Vries, W., Mikkelsen, T. N. et al. (2013). Key indicators of air pollution and climate change impacts at forest supersites. Climate change, air pollution and global changes: understanding and perspectives from forest research. In: Matyssek, R., Clarke, N., Cudlin, P., Mikkelsen, T. N., Tuovinen, J. P., Wieser, G. and Paoletti, E. (eds.) *Developments in Environmental Science*, 13, pp. 497–518.

Sapkota, A., Lu, Z., Yang, H. and Wang, J. (2015). Role of renewable energy technologies in rural communities' adaptation to climate change in Nepal. *Renewable Energy*, 68, pp. 793–800.

Saraçoğlu, N. (2008). Biyokütleden enerji üretiminde enerji ormancılığının önemi. VII. Ulusal Temiz Enerji Sempozyumu, UTES. 17–19 Aralık 2008, İstanbul.

Swift, K. and Cuzner, D. (2012). Natural disturbance effects on forest carbon dynamics, and the role of forest management in the process. Climate-induced changes to natural disturbance regimes and management responses in British Columbia: impacts on natural and human systems. *FORREX Series* 28.

Taşkıran, D. (2009). *Yenilenebilir enerjide orman biyokütlesinin durumu ve Orman Genel Müdürlüğü'nde biyoenerji konusunda yapılan çalışmalar*. [online] Available at: www.slideserve.com/alton/ekim-2009-ogm-tr# [Accessed 22 December 2014]

Tolunay, D. (2011). Total carbon stocks and carbon accumulation in living biomass in forest ecosystems of Turkey. *Turkish Journal of Agriculture and Forestry*, 35, pp. 265–279. doi: 10.3906/tar-0909-369.

TUIK (Türkiye İstatistik Kurumu, Turkish Statistical Institute) (2014) [online] Available at: www.tuik.gov.tr/UstMenu.do?metod=temelist [Accessed 27 November 2014]

UNFCCC (2015). *Greenhouse Gas Inventory Data*. [online] Available at: http://unfccc.int/ghg_data/items/3800.php [Accessed 05 January 2015]

Wang, H. (2010). *Biochar for climate change mitigation: the role of forest industry*. [online] Available at: www.biochar-international.org/sites/default/files/Hailong_Wang.pdf [Accessed 21 January 2015]

Part III
Biochar Production

10
Biochar Production

FREDERIK RONSSE

Abstract

This chapter gives an overview of the key technologies to produce biochar. First, an introduction will be given to the different thermochemical conversion techniques of dry biomass (including pyrolysis) which result in char as one of the product fractions. A second part of this chapter is devoted to the discussion on how the biochar physicochemical properties result from the type of biomass feedstock used, as well as from the prevailing process conditions applied during thermochemical conversion – as some of these physicochemical properties in biochar have a major impact on the functionality and stability of biochar in soil. A major challenge for the successful deployment of biochar systems is to render its production economically profitable. Hence, this chapter concludes with an economic assessment of biochar. This last part of the chapter also emphasizes the potential increase in value creation in the biochar production process by identifying potential economic uses of co-products, including bio-oil and producer gas.

10.1 Introduction to Thermochemical Conversion Processes

Biochar can be defined as the carbon-rich solid residue resulting from the thermal treatment of organic matter (Lehmann and Joseph, 2009; Masek et al., 2013b; Qian et al., 2015). This thermal treatment is better known as thermochemical conversion, which is an umbrella term that encompasses all processes in which heat is applied to decompose or convert biomass into more useful products. Traditionally, three main processes are distinguished within thermochemical conversion, in which biochar is either the main product or a co-product: slow and fast pyrolysis, and gasification – as shown in Table 10.1.

In pyrolysis, the thermal decomposition is carried out under oxygen-free or oxygen-limiting conditions, resulting in the organic material being broken down into three distinct product fractions: (a) vapors, which after condensation result in a liquid product,

Table 10.1. *Typical process conditions and product yields in wt%, obtained from different types of (dry and wet) thermochemical conversion processes relevant to biochar production*

	\multicolumn{5}{c}{Thermochemical conversion process}
	\multicolumn{4}{c}{Dry processes}	Wet processes
	Fast pyrolysis	Carbonisation (slow pyrolysis)	Gasification	Torrefaction (partial slow pyrolysis)	Hydrothermal carbonization (HTC)
Temperature	~ 500°C	> 400°C	600–1800°C	< 300°C	180–260°C
Heating rate	Fast, up to 1000°C/min	< 80°C/min	–	–	5–10°C/min
Reaction time	Few seconds	Hours to days	–	< 2h	5 min–12 hours
Pressure	Atmospheric (and vacuum)	Atmospheric (or elevated up to 1 MPa)	Atmospheric, pressurized up to 8 MPa	Atmospheric	Corresponding water vapor pressure, i.e. 1.0–4.7 MPa
Medium	Oxygen-free	Oxygen-free or oxygen-limited	Oxygen-limited (air or steam/oxygen)	Oxygen-free	Pressurized water
Bio-oil	75%	30%	5%	5%	5–25%
Permanent gases	13%	35%	85%	15%	2–5%
Char (solids)	12%	35%	10%	80%	45–70%

Source: data from Bridgwater (2012); van der Stelt et al. (2011); Williams and Besler (1996); Bain and Broer (2011); Nachenius et al. (2013); and Kambo and Dutta (2015).

(b) non-condensable or so-called permanent gases, and (c) the solid residue, being char (Mohan et al., 2006). The liquid condensation product of pyrolysis is often referred to as 'pyrolysis oil,' 'bio-oil' or 'tar,' although the latter is more commonly used to designate the liquid product obtained in biomass gasification. The liquid product obtained through slow pyrolysis or carbonization is also referred to as 'pyroligneous acids' or 'wood vinegar.' Depending on the feedstock type as well as on the pyrolysis process conditions different distributions and yields of the aforementioned pyrolysis product fractions can be obtained (Bridgwater and Peacocke, 2000; Lu et al., 2009).

Pyrolysis processes are further differentiated according to the heating rate the biomass particle is subjected to. In fast pyrolysis, high heating rates (i.e. several hundred °C/min) are selected to maximize the yield of bio-oil while the residence time of the produced vapors should be kept as low as possible to minimize unwanted secondary vapor cracking (Czernik and Bridgwater, 2004; Balat et al., 2009; Bridgwater, 2012). There is no universal minimum heating rate criterion as a requirement to fast pyrolysis. Various authors have given different interpretations to the required heating rate due to differences in reactor type and feedstock size used in fast pyrolysis research. A more appropriate quantitative particle heating criterion to define fast pyrolysis is the heat transfer coefficient external to the biomass particle surface (Lédé, 2013). Typically, in fast pyrolysis of low-ash lignocellulosic biomass feedstock, bio-oil yields range between 60 and 70 *wt%* (on feedstock basis), while char yields are between 12 and 15 *wt%* (Isahak et al., 2012).

Slow pyrolysis is characterized by lower heating rates and long vapor/solids contact times, which incur maximum yields in char. Often, the term 'carbonization' is used to denote slow pyrolysis, however the term is more broad: carbonization refers to complete pyrolytic conversion of biomass into a char material and is the basis of the production of charcoal. Slow pyrolysis covers, besides carbonization, more recent process developments such as torrefaction. In essence being a partial slow pyrolysis process, torrefaction uses mild temperatures (up to 300°C) to pretreat biomass as a solid fuel (Basu, 2013). Torrefaction could increase energy density, hydrophobicity and grindability, and reduce its biodegradability in comparison with the untreated biomass feedstock (Nachenius et al., 2013).

Gasification differs from both pyrolysis processes in its use of higher temperatures (> 700°C) and the use of sub-stoichiometric amounts of oxidant, being pure O_2, air, CO_2, steam or mixtures thereof. The primary aim of gasification is the production of a combustible mixture of permanent gases (CO, H_2, CO_2, CH_4 and lighter hydrocarbons), called producer gas, which can then be further used to power internal combustion engines and gas turbines, or used as a feedstock in the production of liquid fuels and chemical intermediates (McKendry, 2002). Char yields are generally low, less than 10 *wt%*.

All three major thermochemical conversion processes (slow and fast pyrolysis, gasification) require relatively dry feedstock, with moisture content below 30 *wt%*, but moisture contents in the vicinity of 10 *wt%* are preferred. As the moisture in the biomass feed increases, more energy will be consumed to account for the increasing required heat of vaporization during the heating of the biomass towards the pyrolysis reaction temperature (Antal and Grønli, 2003), greatly increasing the energy requirements of the thermochemical conversion reactor. Furthermore, the gases and vapors resulting from conversion processes using a high moisture feedstock are diluted with steam and thus have a lower calorific content, which could potentially hamper downstream usage of these gas and vapor streams. Hence, biomass feeds with a high moisture content (70 *wt%* or more) are preferably processed in an aqueous environment at elevated temperatures and elevated pressures (i.e. the saturation vapor pressure corresponding to the temperature operated at). These processes, collectively called hydrothermal processes, do not require a drying step prior to conversion and water plays an active role, not only as a solvent, but also as a reactant (Kruse and Dahmen, 2015). Similar to dry thermochemical processing, categorization can be made into three key processes based on the process conditions used: at low temperatures (around 200°C), mostly carbonaceous solids are formed in a process known as hydrothermal carbonization (HTC). Between 300 and 350°C hydrothermal liquefaction (HTL) takes place, in which the biomass feed is preferably converted into a liquid product (bio-oil, biocrude). Near and above the critical point of water (i.e. 374°C; 22 MPa), biomass is mostly converted into a mixture of non-condensable gases (H_2, CO, CO_2 and CH_4), a process better known as supercritical water gasification (SCWG) (Kruse, 2009). More details regarding HTC for biochar production purposes are given in Section 10.3.4.

10.2 Pyrolysis Chemistry and Char Formation

Typically, plant-based biomass is a complex matrix consisting, apart from water, mainly of structural cell wall components including hemicellulose, cellulose and lignin. Non-structural compounds are also present in the form of ash, proteins, soluble sugars, starches, waxes and lipids; these are collectively known as extractives (Mohan et al., 2006). The different chemical structure of each of the biomass constituents is reflected in the thermal stability and the temperature at which these biomass constituents decompose. When investigating the chemical changes brought about by heating a biomass particle in an inert atmosphere, the first constituent to be removed is water, both free and bound, to temperatures up to 160°C (Grønli and Melaaen, 2000; Zhou et al., 2013). Heating further up to 220°C, the extractives are devolatilized or decomposed (Yang et al., 2014). The first biopolymer to decompose is hemicellulose, at temperatures between 220 and 315°C (Stefanidis et al., 2014; Yang et al., 2007; Zhou et al., 2013; Yang et al., 2014). Hemicellulose is the least stable of the three major plant-based biopolymers, as it consists of heteropolysaccharides built from various C5 and C6 sugars (Saha, 2003) with a low degree of polymerization and being amorphous. Cellulose, on the other hand, is a polymer solely consisting of glucose units and has a high degree of polymerization, in the order of 10,000 (White et al., 2011). Van der Waals forces and hydrogen bonds stabilize adjoining linear cellulose molecules into a predominant crystalline structure (Vanholme et al., 2013). Consequently, cellulose exhibits higher thermal stability and decomposes at higher temperatures, that is between 315 and 400°C (Stefanidis et al., 2014; Yang et al., 2007). Finally, lignin exhibits the most complex decomposition behavior: its decomposition starts early on at 160°C, but continues gradually until complete conversion at around 900 to 1000°C (Zhou et al., 2013). This complex decomposition behavior stems from lignin's intricate chemical structure: it is a complex polymer consisting of phenylpropanoid monomers (i.e. so-called monolignols) that are linked through a variety of chemical bonds with different dissociation enthalpies (Zakzeski et al., 2010; Kim et al., 2011).

The reactions in which the plant biopolymers are broken down, with the accompanying release of gases and volatiles, are collectively known as 'primary reactions' and are comprised of depolymerization, fragmentation and rearrangement reactions (Di Blasi, 2008; Collard and Blin, 2014) (see Figure 10.1). The char formed directly out of the plant cell wall biopolymers through primary reactions (i.e. rearrangement reactions) is considered

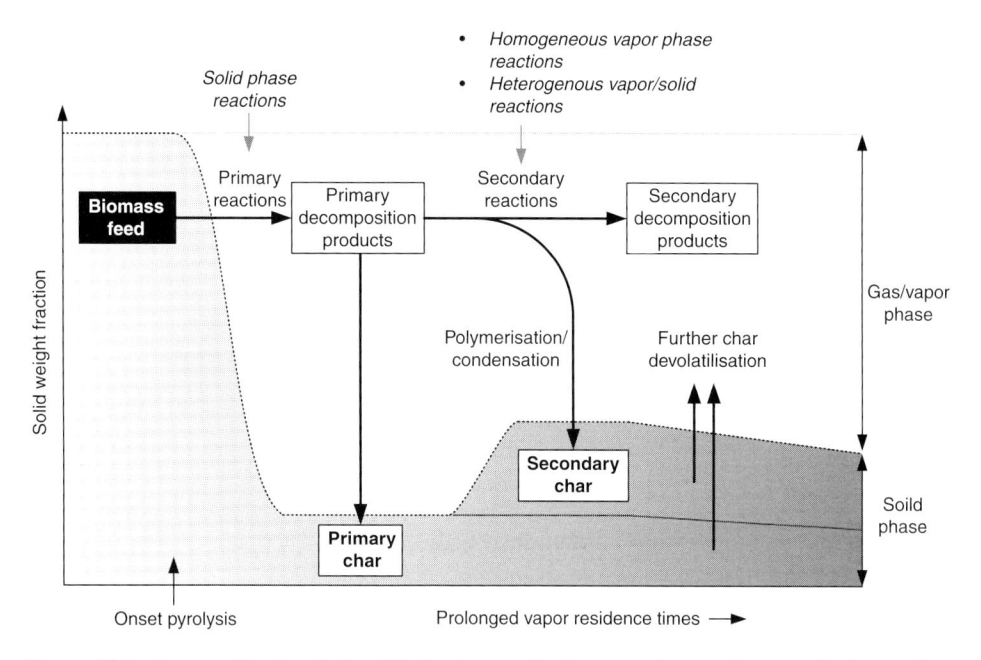

Figure 10.1. A generalized and simplified scheme of primary and secondary pyrolysis reactions occurring in biomass constituents (i.e. hemicellulose, cellulose and lignin).

as 'primary char' and consists of an aromatic polycyclic structure (Hajaligol et al., 2001; Azargohar et al., 2014). Furthermore, it is widely accepted that these primary reactions are highly endothermic (Antal and Grønli, 2003; Yang et al., 2007). The produced primary vapor-phase reaction products are not stable and, given sufficient vapor residence time, they will undergo secondary reactions that are typically exothermic. These reactions include cracking as well as recombination or repolymerization reactions (Wei et al., 2006; Collard and Blin, 2014), the latter of which result in the formation of 'secondary char.' Secondary reactions, thus yielding additional char, can be catalyzed at the surface of the primary char or by minerals and ash components contained therein (Ronsse et al., 2012). The chemical nature of secondary char, formed through homogeneous reactions from wood pyrolysis tarry vapors, has also been demonstrated to be highly aromatic in nature (Morf et al., 2002).

The distinction into primary and secondary reactions occurring in thermal decomposition of biomass also highlights the fundamental differences between fast and slow pyrolysis. Fast pyrolysis aims at maximizing the yield of condensable vapors (bio-oil) by preventing secondary decomposition reactions from taking place, which is achieved by sustaining high biomass heating rates and by rapid removal and quenching of the condensable vapors (Venderbosch and Prins, 2010). By eliminating secondary reactions, char yield is generally low (10 to 15 $wt\%$), as only primary char is being produced and the overall fast pyrolysis process is highly endothermic, requiring up to 1.8 MJ/kg biomass (Venderbosch and Prins, 2010). On the other hand, slow pyrolysis promotes secondary reactions by imposing long vapor residence times, thereby maximizing the yield of char, which consists of both primary and secondary char. Due to the extensive occurrence of secondary reactions, the overall slow pyrolysis reaction is generally exothermic in nature (Park et al., 2010; Manya et al., 2013).

10.3 Biochar Production Systems and Reactor Development

From the above description of char formation in thermochemical conversion reactions, it is clear that a wide range of technologies in which char (biochar) is either obtained as the main product, or as a co-product, exist. The following section will give a brief overview of the different key reactor systems.

10.3.1 Slow Pyrolysis Reactors

Slow pyrolysis or carbonization systems are a well-established technology and have been used for centuries in the production of charcoal. However, woody biomass, typically in the form of logs or cordwood, has relatively simple process requirements. The increased focus on agricultural residues and other mixed waste feedstock for the production of biochar with a consistent composition and quality, and on the coproduction of energy (heat, electricity) or chemicals, pose additional challenges in slow pyrolysis reactor development.

10.3.1.1 Kilns

The oldest and simplest form of slow pyrolysis systems are pit and mound kilns. A major feature of kilns is that the heat required for pyrolysis is generated by the internal combustion of a part of the supplied biomass (Schenkel et al., 1998). These kilns are operated by setting fire to a stack of wooden logs, either on flat terrain (mound kiln, Figure 10.2a) or in a pit (pit kiln) and once sufficient heat has been generated, the pit or mound is covered with a layer of soil to starve the kiln of oxygen (air) and to allow the pyrolysis reactions to complete (FAO, 1983). The main advantages of these types of kilns, and the reasons for their continued use in developing countries, is that they are easy and cheap to set up with minimal infrastructural requirements, they can be deployed near or in the biomass collection area and are able to handle relatively large batches (Brown, 2009; Duku et al., 2011; Gwezi et al., 2015). However, they suffer from numerous disadvantages as well: process control is absent, char yields are generally low (Antal and Grønli, 2003) and the heterogeneous conditions with respect to pyrolysis temperature, heat transfer and gas flow give rise to heterogeneity in composition and quality of the char from a single batch (i.e. some parts of the biomass will not be carbonized completely, while other parts have been burnt completely, only yielding pure ash) (de Oliveira Vilela et al., 2014). Finally, the pyrolysis vapors and gases are directly emitted into the environment without additional gas treatment. These emissions not only contain gases such as CO_2, CH_4 and N_2O, which are greenhouse gases

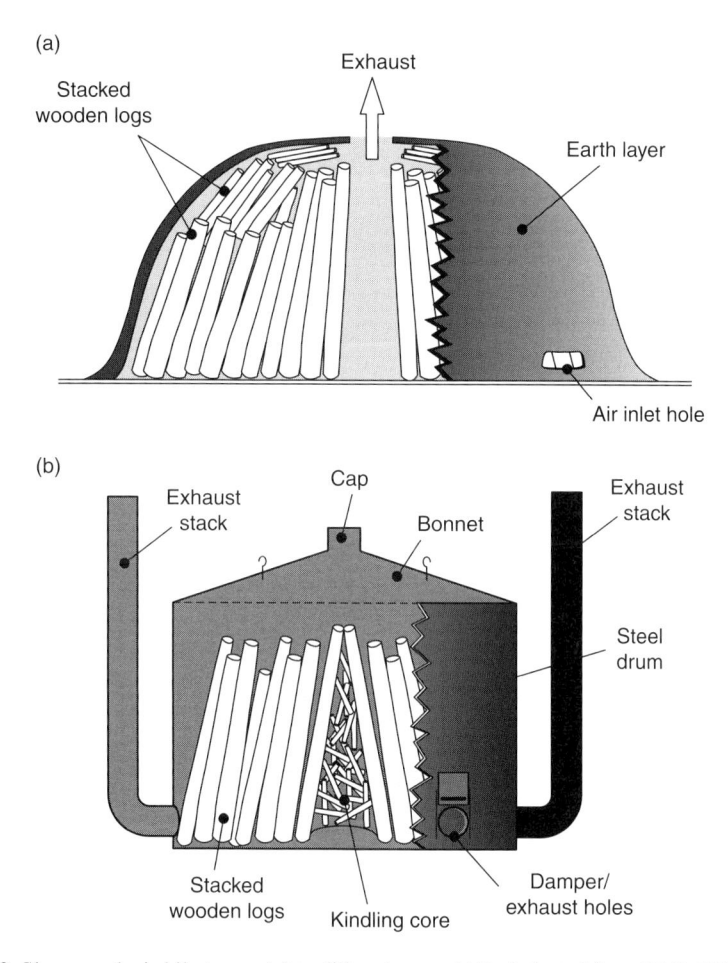

Figure 10.2. Slow pyrolysis kiln types: (a) traditional mound kiln (adapted from FAO, 1983), (b) steel kiln, New Hampshire type (adapted from Baldwin, 1958).

(Pennise et al., 2001; Adam, 2009; Chidumayo and Gumbo, 2013), but also particulate matter (black carbon) and toxic volatile organic compounds that affect the local environment and pose a health hazard to the workers operating these kilns (Pennise et al., 2001).

Some of the traditional kiln's drawbacks, including poor gas flow and high heating losses, can be overcome by constructing the kiln out of refractory brick (e.g. beehive kilns), reinforced concrete (e.g. Missouri kiln) or steel. In these more advanced designs, temperature can be monitored using thermocouples and process control is provided by varying the air flow rate using valves and dampers. An example of a transportable steel kiln, the New Hampshire Kiln, is given in Figure 10.2b. Recirculating or thermally oxidizing the exhaust gases can be included to reduce hazardous emissions of pyrolysis gases and vapors (Adam, 2009). These kilns are in essence batch reactors and are operated using a sequence consisting of manual loading, heating, carbonization, cooling down and manual unloading.

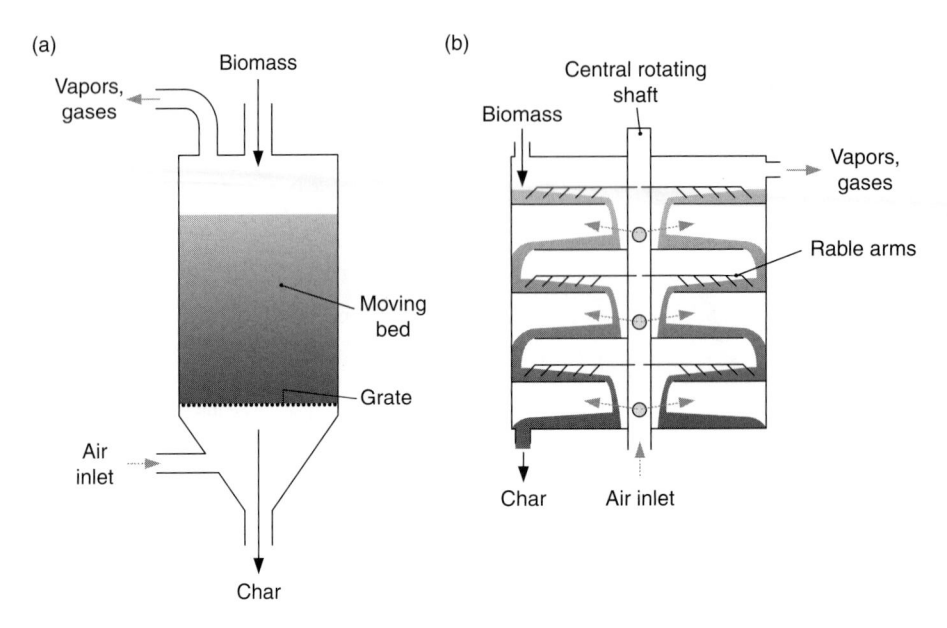

Figure 10.3. Slow pyrolysis continuous kilns: (a) the Schottdorf kiln, (b) Hereshoff multiple hearth furnace (adapted from Thomas et al., 2009).

For high-volume, industrial charcoal or biochar production, continuous kilns have been designed, including the Schottdorf kiln used by Carbon Terra (Germany) (Figure 10.3a) and the Hereshoff multiple hearth furnace (Figure 10.3b) (Thomas et al., 2009; Wiedner et al., 2013). These continuous kilns operate on the principle of a vertically downward moving bed. By having the biomass flowing countercurrently to the gas, improved thermal efficiency can be obtained: the upward flowing pyrolysis gases and vapors will heat and dry the freshly added biomass in the upper segments of the reactor. An additional benefit of continuous processing is that the reactor's exhaust gases have a consistent composition, thereby facilitating gas treatment or the recovery of valuable vapor-phase constituents (i.e. condensates). As these continuous reactors have more strict requirements with respect to the flowability of the biomass feed, size reduction (i.e. chipping) is usually needed (Nachenius et al., 2013).

10.3.1.2 Retorts

Retorts differ from kilns in their method of heating: whereas in kilns the heat is generated in the reactor by partial combustion of the biomass feed, retorts rely on externally supplied heat. In most cases, this heat is generated by combustion of the mixture of pyrolysis gases and vapors, although an external fuel (natural gas, additional biomass) may be used as well (Grønli, 2005). The heat is either supplied directly, by bringing the combustion gases into contact with the biomass in the reactor, or indirectly using a heat exchange surface (i.e. the reactor wall is heated). The environment in the reactor is completely free from oxygen

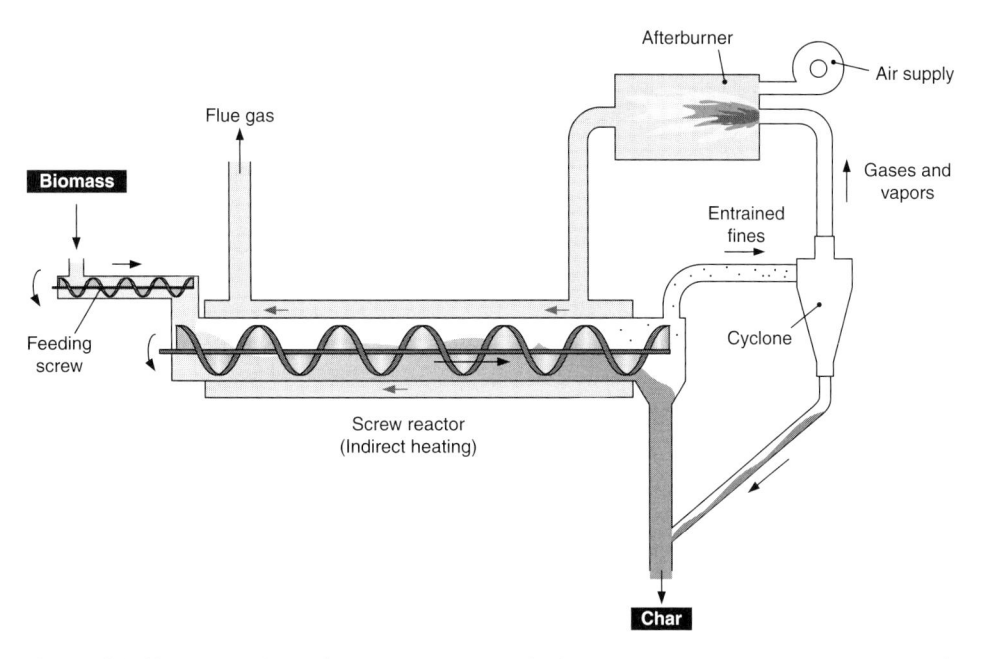

Figure 10.4. Slow pyrolysis continuous screw reactor, indirectly heated with the combusted pyrolysis gases and vapors. (A black and white version of this figure will appear in some formats. For the colour version, please refer to the plate section.)

and no biomass is sacrificed in the retort for heat production, consequently the overall char yields are higher compared to kilns. Batch operated retorts have been scaled to large industrial sizes, examples of which are the Reichert retort, developed by Evonik (formerly Degussa, Germany) with a reactor volume of 100 m³ and the Arkansas retort in which the biomass is stacked onto steel wagons to easily move it in and out of the retort (FAO, 1985). Continuously operated industrial-scale retorts exist as well, such as the Lambiotte retort, which is a downward moving bed reactor similar to the ones in Figure 10.3 but supplied with the recirculated, combusted pyrolysis gases instead of air. The largest Lambiotte retorts are capable of an annual production of 13,500 tons of charcoal per year (Grønli, 2005).

The large-scale industrial charcoal retorts, like the Reichert and the Lambiotte designs, cannot cope with biomass feedstock with particle dimensions smaller than those of logs (cordwood) due to the excessive pressure drop that otherwise would be created across the fixed or moving bed of biomass. In order to handle feedstock with smaller particle sizes and differing morphologies (i.e. sticky, powdery or fibrous material), which are typical of the agricultural and forestry residues of interest in biochar production, modern continuous pyrolysis retorts have been developed, including the rotating drum and screw (or auger) pyrolyzers. Rotating drum pyrolysis units consist of a horizontal, or slightly inclined, rotating drum which is either directly or indirectly heated using combusted pyrolysis gases. Biomass is fed at one end of the drum and paddles or baffles are mounted on the inside of the drum to ensure proper mixing of the biomass as it travels to the other end of the barrel.

Screw pyrolysis units for the production of char are similar in construction to those for fast pyrolysis, with the exception that the inert heat carrier is omitted and heating is usually performed through the barrel only, as the rapid heating conditions typical to fast pyrolysis are not needed in slow pyrolysis or carbonization. A well-known example of a screw-based slow pyrolysis system, indirectly heated by the combustion of the pyrolysis off-gases, is the reactor system developed by Pyreg (Germany), which operates similar to the principle detailed in Figure 10.4. Both the rotating drum and screw reactor offer good control over the particle residence time and can handle a wide range of particle morphologies (Klose and Wiest, 1999; Nachenius et al., 2015).

Some retorts have unconventional methods of heating: in the case of **microwave pyrolysis**, the heat to drive the pyrolysis reactions inside the particles is transferred by means of microwave radiation. Conventional heating methods always transfer heat to the particle surface, mostly through convection and conduction, and subsequently within the particle by means of conduction. Consequently, the pyrolysis reactions initiate at the particle surface and then move gradually towards the particle's center. Microwave pyrolysis, on the other hand, can heat the particle from within and initiate pyrolysis reactions from the inside of the particle. This highly efficient heat transfer is beneficial in fast pyrolysis applications for the production of bio-oil (Domínguez et al., 2006; Lei et al., 2009: Salena and Ani, 2011), however for biochar production in slow pyrolysis, lower char yields have been reported compared to conventionally heated retorts (Masek et al., 2013a). Some retorts, more specifically screw reactors, can also be heated using **induction heating**: by applying a high-frequency external magnetic field, electric currents are induced within the helical screw which in turn heats up by Joule heating. An example of an induction-heated pyrolysis screw retort is Biogreen's Spirajoule (ETIA, France) (Antoniou et al., 2014). The disadvantage of both microwave and induction heating is that they require electricity, a more costly and higher quality energy source compared to heat resulting from the combustion of pyrolysis gases and vapors.

10.3.1.3 Other Dry Carbonization Techniques

Flash carbonization is a technique developed by Antal and coworkers (Antal et al., 2003). Basically being a pressurized, oxygen blown and batch operated kiln, the high pressure (1 megapascal) on the outside of the biomass particles reduces the net pressure gradient in the particle (i.e. pressure generated by the generated pyrolysis vapors and gases). Consequently, the vapors take a longer time to flow out of the porous particle, thereby increasing the extent of secondary char formation. High char yields, approaching that of the thermodynamic equilibrium, have been reported and reaction rates are generally higher than those in slow pyrolysis at atmospheric pressure.

10.3.2 Fast Pyrolysis Reactors

Typical fast pyrolysis conditions include moderate temperatures (400–600°C) and rapid heating rates (>100°C min^{-1}) combined with short residence times of the biomass particles (0.5–2 s) (Demirbas, 2001). To date, several reactor designs have been developed and

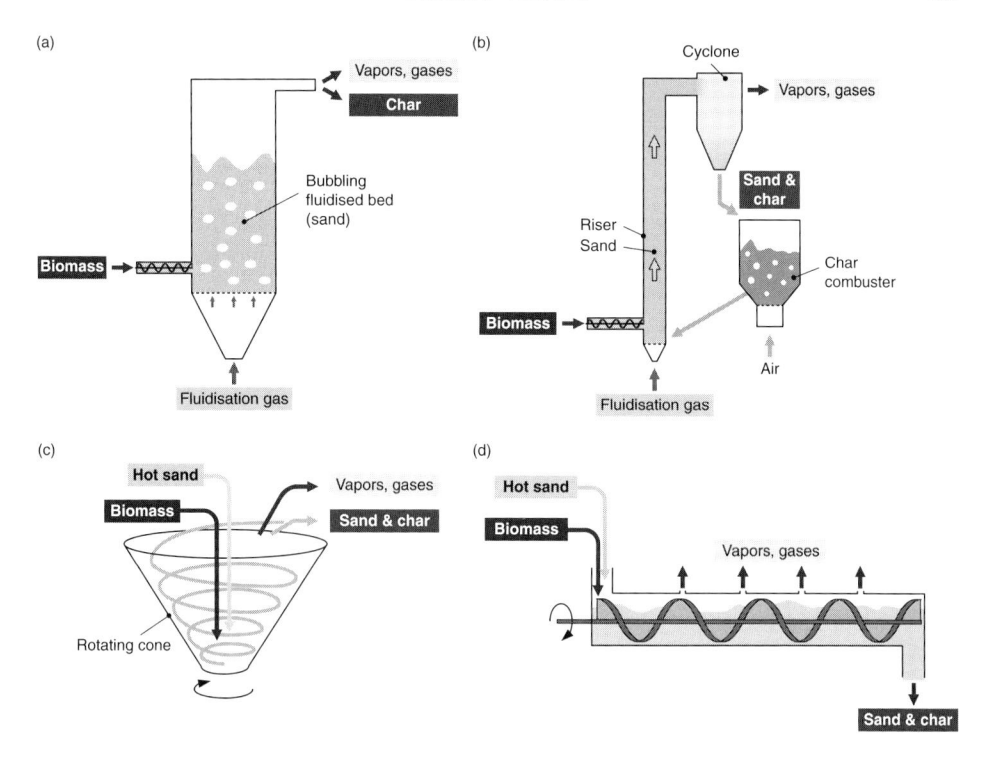

Figure 10.5. Fast pyrolysis reactor types: (a) bubbling fluidized bed, (b) circulating fluidized bed, (c) rotating cone reactor and (d) screw or auger reactor. (A black and white version of this figure will appear in some formats. For the colour version, please refer to the plate section.)

tested, including the entrained flow, ablative, vacuum moving bed, fluidized bed, auger and rotating cone reactors, which have been covered in more recent reviews by Venderbosch and Prins (2010), Bridgwater (2012) and Nachenius et al. (2013) – only the more common designs will be discussed below.

The most popular reactor design is the fluidized bed (Figure 10.5a), in which a solid heat carrier, usually sand, is suspended in a stream of hot gas. The vigorous motion of the fluidized heat carrier particles ensures optimal mixing behavior and rapid heat transfer to the biomass particles that are fed to the bed. Typically, the fed biomass particles are fairly small (2–3 mm). The fluidizing gas flow ensures rapid evacuation of the produced vapors thus keeping the vapor residence times sufficiently low. Char particles are entrained in the gas stream, requiring cyclones to separate them from the mixture of gases and vapors. This type of fluidized bed is commonly known as the bubbling fluidized bed. A second type, the circulating fluidized bed (Figure 10.5b), does not separate the char particles, but instead relies on higher fluidization gas velocities to completely entrain the biomass/heat carrier mixture upwards through a narrow riser (i.e. turbulent fluidization regime). At the top of the riser, the vapors are separated from the char/heat carrier mixture using a cyclone. To reheat the heat carrier, char is combusted in a secondary air or oxygen blown reactor after which the heat carrier is recirculated to the bottom of the riser.

More recent reactor designs include the rotating cone reactor, developed by Twente University (The Netherlands) (Wagenaar et al., 1994) and Biomass Technology Group (The Netherlands), in which mechanical mixing of the heat carrier with the biomass particles is achieved by centrifugal forces generated on the inside of a rotating, inverted cone. Biomass particles and heat carrier are fed at the bottom of the cone and the mixture is spun upwards (Figure 10.5c). Screw or auger reactors (Figure 10.5d) consist of one or two counter-rotating helical screws which transport a mixture of biomass and heat carrier through a cylindrical shell (barrel). Screw reactors can accommodate a wider range in morphology of the biomass feed, but suffer from a minor penalty in bio-oil yields (typically 10% lower) due to longer vapor residence times (5 to 30 s) compared to fluidized bed fast pyrolysis reactors (Liaw et al., 2012). Both the rotating cone and the screw reactor can operate in the absence of an inert carrier gas, thereby simplifying the process of bio-oil condensation.

10.3.3 Gasification

In gasification, the aim is to convert the biomass into a gaseous mixture (mainly composed of CO, CO_2, H_2, CH_4 and some light hydrocarbons) by the sub-stoichiometric addition of an oxidizer (i.e. oxygen, steam) under a high temperature (> 700°C) (McKendry, 2002; Qian et al., 2015). The primary aim of gasification is to produce a combustible gas, called producer gas, which can be used in heat and power applications, or the production of chemicals or liquid transportation fuels. As already detailed in Table 10.1, gasification is a highly effective conversion process, with gas yields typically exceeding 85 *wt*% and carbon efficiencies up to 95%. Consequently, gasification yields rather small amounts of char, and depending on the biomass feedstock ash content, with limited amounts of carbon. As such, gasification appears to be of little interest to char production. However, given the maturity of the gasification process, and the number of pilot and commercial scale gasification systems, significant quantities of char may be sourced from gasification systems for biochar purposes.

Gasification on a large scale is typically performed in a fluidized bed reactor (see Section 3.1), and is air, pure oxygen or steam blown. Smaller scale moving bed reactors are in use as well and are air blown only, either with the air (and gases/vapors) moving concurrent to the biomass (downdraft gasifier) or the air (and gases/vapors) moving countercurrent to the biomass (updraft gasifier), the latter similar to the design in Figure 10.5a.

10.3.4 Hydrothermal Carbonization

When biomass is mixed with water and heated to high temperatures (up to 300°C) and under high pressures to retain the water in a liquid form (i.e. using the saturation vapor pressure corresponding to the applied temperature), it can be converted into a solid, carbonaceous product (Hu et al., 2010). This charcoal-like product, obtained in the process better known as hydrothermal carbonization (or in short, HTC), is often designated as 'hydrochar' or

HTC char. The unique selling point of HTC is its ability to handle high moisture-containing feedstock without requiring a drying step. As the HTC solid product is more hydrophobic in nature compared to its original feedstock, dewatering the solid product is straightforward and requires less energy (Titirici et al., 2007). As such, many of the agricultural, food industry and biorefinery residues (including manure, sewage sludge and digestate from anaerobic digestion) may be preferably and more cost-effectively treated using HTC to yield a char-like product.

HTC char is not a new technology, and has been performed mainly for the production of solid fuels (Kruse et al., 2013). However, the use of HTC char as a biochar in soil amendments is relatively new. The chemical nature of HTC char as well as the transformations taking place during hydrothermal processing are not well known, especially when comparing against char resulting from slow pyrolysis or carbonization. Nevertheless, differences in composition and behavior in HTC char have been reported as clear examples that it is a different product altogether, when comparing against slow pyrolysis char: one such difference is that HTC char usually contains potential phytotoxic and genotoxic compounds (Busch et al., 2013), which may require an additional post-treatment step to render it safe for soil applications. HTC char also appears to be less stable in soils: its projected mean residence time is at least one order of magnitude lower than comparable slow pyrolysis biochar (Gajic et al., 2012). Consequently, HTC char's long-term soil sequestration potential appears to be limited.

10.4 Relevant Biochar Properties

Biochars present themselves with a wide range of physicochemical properties, depending on the feedstock as well as the applied conditions in thermochemical conversion. The next section will briefly discuss how some of these major properties, which in turn impact biochar's performance and stability in soil, relate to both the feedstock as well as the pyrolysis process.

10.4.1 Proximate Analysis

One of the simplest biochar compositional essays is the so-called proximate analysis: by heating a biochar sample to certain threshold temperatures under reductive and/or oxidative environments, distinction can be made between moisture, ash, volatile matter (VM) and fixed carbon (FC) present in the biochar. Typically, FC is defined as the amount of organic (ash-free) material that remains after heating in an oxygen-less environment up to 950°C according to the ASTM D 1762–82 standard for charcoal analysis (ASTM, 2007), while VM is calculated by subtracting the FC weight from the biochar's dry and ash-free weight. The term 'volatile matter' might be a bit of a misnomer: it simply implies all organic compounds that are still volatilizable in a high (950°C) temperature treatment. For example, an intact cellulose polymer molecule is a near 100% VM, while small oxygenated

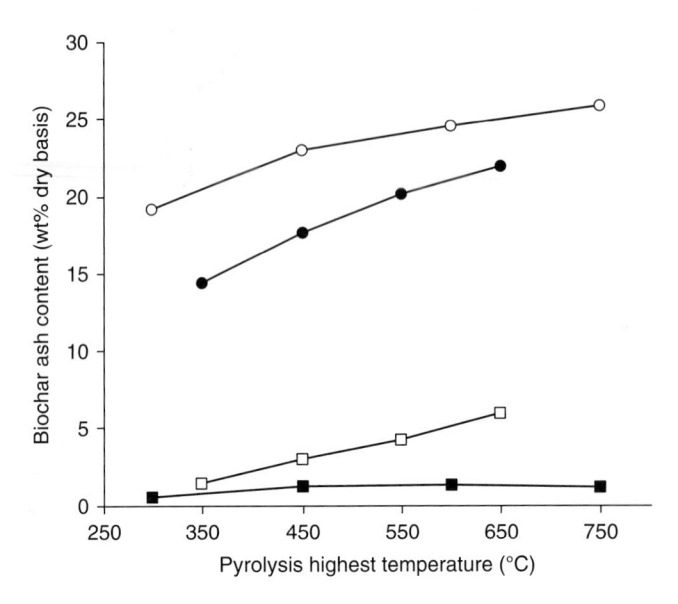

Figure 10.6. Influence of pyrolysis temperature on biochar ash content, data pooled from Ronsse et al., 2013 (■: pine wood and ●: wheat straw, original biomass ash content 0.2 and 7.9 wt%, respectively) and Crombie et al., 2015 (□: pine wood chips and ○: wheat straw, original biomass ash content 0.2 and 7.9 wt%, respectively).

pyrolytic compounds that may absorb onto the char structure and are truly 'volatile' (i.e. having an equilibrium vapor pressure) are also counted amongst the VM.

The term 'ash' designates all inorganic material, including alkali and heavy metals, as well as chlorine, phosphorus and sulfur present in the material. Biochars having a high ash content have been shown to be beneficial to soil productivity, an effect which was attributed to the release of minerals and the liming effect (see also Section 10.4.3) (Smider and Singh, 2014). However, the higher the ash content, the lower the organic carbon content and thus, the lower the carbon sequestration potential (per weight unit of biochar). Most ash constituents exhibit relatively low volatility under typical conditions applied in either slow or fast pyrolysis (Du et al., 2014). Consequently, the ash content in the resulting biochar is determined by the initial ash content in the biomass feedstock and the biochar yield as determined by the prevailing pyrolysis conditions. Typically, the more severe the pyrolysis or carbonization conditions (i.e. a combination of temperature and residence time), the more concentrated the ash in the biochar. For example, Figure 10.6 illustrates the biochar ash content of which data has been pooled from Ronsse et al. (2013) and Crombie et al. (2015) for the slow pyrolysis of wheat straw and pine wood: with the exception of high temperatures (750°C) at which certain ash constituents may volatilize, there is a clear relationship between ash content and pyrolysis temperature.

The presence of ash in the biomass feedstock also affects the thermochemical conversion process: it has been demonstrated that alkali and earth alkaline metals catalyze primary

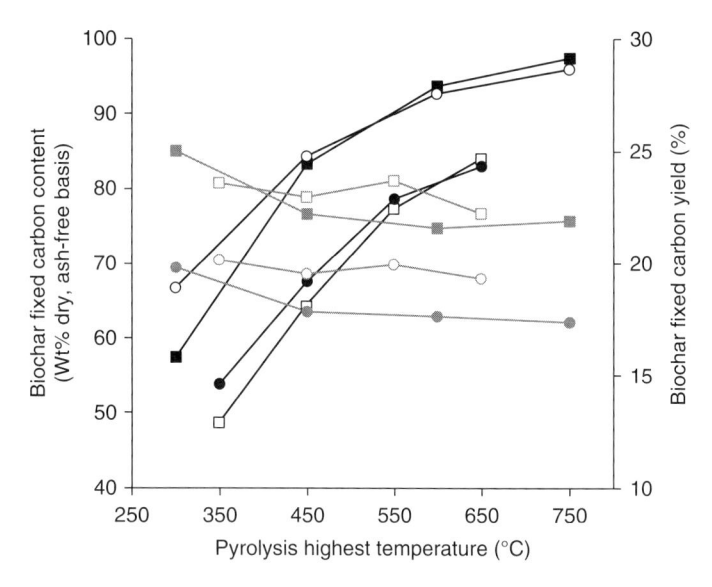

Figure 10.7. Influence of pyrolysis temperature on biochar fixed carbon content, expressed on dry and ash-free basis (━) and biochar fixed carbon yield (━), data pooled from Ronsse et al., 2013 (■,■: pine wood and ●,●: wheat straw) and Crombie et al., 2015 (□, □: pine wood chips and ○,○: wheat straw, original).

pyrolysis vapors, thereby yielding additional gas and char (Patwardhan et al., 2010). Certain ash constituents (i.e. alkali metals in the presence of chlorine) can pose considerable challenges, due to their tendency to cause equipment fouling and slagging. Chlorine and sulfur lead to increased corrosion as well. Pyrolysis not only retains most of the mineral nutrients in the char, but also if the feedstock is contaminated with heavy metals – with the exception of the volatile mercury – these will accumulate in the char as well, potentially limiting its use for soil applications (Van Wesenbeeck et al., 2014).

Originally, proximate analysis and the distinction of the organic fraction into VM and FC was made to assess the properties of solid fuels, including coal and charcoal. A large presence of VM gives rise to a lower ignition temperature and larger smoke production during combustion. Domestically used charcoal (i.e. for fuel purposes) typically has VM content between 20 and 30 *wt%* (Antal and Grønli, 2003). As the severity of the carbonization process increases, that is at higher pyrolysis temperatures and with longer biomass residence times, the FC content in the produced char increases while the VM content inherently decreases (Cordero et al., 2001), an example of which is given in Figure 10.7. However, with increasing pyrolysis temperature, the overall char yield diminishes: a so-called FC yield, expressing the FC in the char relative to the biomass feedstock weight, has also been presented in Figure 10.7. It can be observed that at temperatures above 450°C, the FC yield is invariant to further increases in pyrolysis temperature, which has also been reported by Antal and Grønli (2003), Masek et al. (2013b) and Crombie et al. (2013).

Despite the method's origins in solid fuel quality assessment, the quantification of FC and VM also has merits in soil applications: Zimmerman (2010), Cross and Sohi (2011) and Ronsse et al., (2013) found that biochars produced at low pyrolysis temperatures (i.e. below 400~450°C) encountered higher mineralization rates in char-soil incubation experiments, which correlated well with the chars' VM content. It is hypothesized that the organic compounds, identified as VM in proximate analysis, are in an amorphous phase and contain a significant fraction of hetero-atoms (O as well as N) and can therefore be easily metabolized by soil microbiota. Some of the effects beneficial or detrimental to crop productivity have been attributed to the presence of VM in biochar (Deenik et al., 2010; Graber et al., 2010; Spokas et al., 2011).

Fixed carbon content is closely correlated to the aromaticity of the char, that is the extent to which carbon is bound into polycyclic aromatic ring structures (Manya et al., 2014). The increasing FC content in chars produced at higher pyrolysis temperatures indicates that aromaticity is strongly correlated to pyrolysis severity. As these aromatic polycyclic structures are believed to be highly resistant to microbial decay and chemical degradation occurring in soils, the FC content has been demonstrated to be a relatively good indicator for biochar stability. However, it lacks universal prediction capability, probably due to differences in feedstock composition such as ash content (Enders et al., 2012; Crombie et al., 2013) and as such, it can be used – within limits – to compare or predict the stability of various char samples sourced from a single feedstock and produced using a specific conversion technique, but using varying process conditions.

10.4.2 Ultimate Composition

As the biomass is undergoing devolatilization during pyrolysis, the remaining solid product is progressively enriched in carbon. Major constituents of pyrolysis vapors are water and non-condensable gases including CO, CO_2 and H_2, while the organic compounds in the pyrolysis vapors have O/C and H/C ratios similar to those of virgin biomass. Consequently, H and O are preferably removed over C, and the O/C and H/C ratios of the biomass undergoing its transformation into char tend to decrease during pyrolysis (Krull et al., 2009). Typically, for fuel applications H/C and O/C ratios are used to assess the degree of aromaticity and maturation (Krull et al., 2009) and can be plotted against each other in a so-called 'van Krevelen' diagram – an example of which has been given in Figure 10.8. Please note that only organic carbon is being referred to and that carbonate-carbon is omitted from this discussion. Chars produced through low temperature pyrolysis have high H/C and O/C ratios, and lie in close proximity to the original biomass in the upper right hand corner of the diagram. The higher the pyrolysis temperature or more generally, the more severe the pyrolysis conditions, the closer the char approaches coal in the lower left hand corner of the van Krevelen diagram.

Hydrochars or chars produced in hydrothermal carbonization deviate from this behavior: they have a higher H/C ratio, typically ranging between 0.6 to 1.3, and a higher O/C ratio, between 0.2 and 0.4 (Schimmelpfennig and Glaser, 2012). This clearly indicates

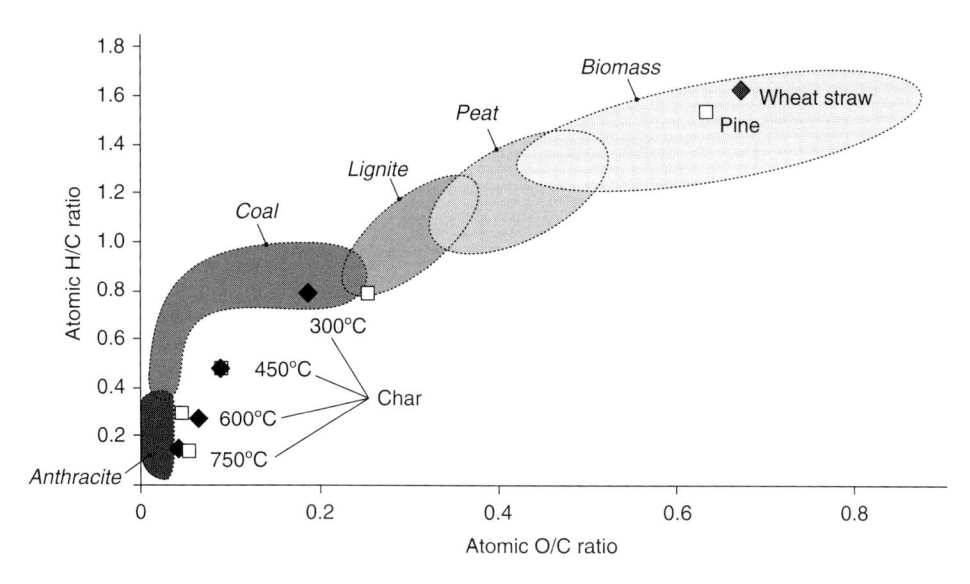

Figure 10.8. The van Krevelen diagram, adapted from van der Stelt et al. (2011) and with the pine (♦) and wheat pellet (□) based chars produced by slow pyrolysis in Ronsse et al. (2013) indicated. Temperatures indicate highest pyrolysis temperatures.

that in HTC the selective removal of O and H through dehydration, decarboxylation and decarbonylation reactions (reactions characteristic to pyrolysis) only occur to a smaller extent.

Elemental ratios have also been used extensively as a proxy for biochar stability. Both the O/C and the H/C ratio negatively correlate with char stability in soil (Spokas, 2010; Cross et al., 2013), and the latter is a common required stability criterion found in several voluntary biochar certifications schemes or guidelines (IBI Guidelines, 2014; EBC Certificate, 2012), where the cut-off value is 0.7 (i.e. charred materials with a H/C > 0.7 are not considered as biochar). However, as both Enders et al. (2012) and Cross et al. (2013) pointed out and similar to FC content as a stability proxy, there is a relative spread when correlating between H/C ratio and char stability, the latter determined by methods including wet oxidation or soil incubation. These findings somehow partially invalidate the H/C (as well as O/C) ratio as a 'true' universal stability criterion, and care must be taken when comparing samples produced from different feedstocks, as well as from different thermochemical conversion techniques.

10.4.3 Char pH

The pH of freshly produced biochar is highly variable, ranging from below pH 4 to above pH 12 (Lehmann et al., 2011). Two major factors influence the biochar's pH value: the temperature at which biochar has been produced and the ash content, especially the presence of alkali in the ash (Ronsse et al., 2013). The higher the pyrolysis temperature, the more selective the removal of oxygen from the char and consequently the lower the concentration of functional groups, including COOH, on the char surface. This effect can be observed by the lowering of the O/C ratio as the severity of the pyrolysis process increases (see also Section 10.4.2). Furthermore, as the pyrolysis temperature increases, so does the ash content in the resulting char (see Section 10.4.1), which further amplifies the pH increase in high-temperature chars. Because of the smaller extent of removal of O and O-containing functional groups in hydrothermal carbonization, HTC chars have typically an acidic pH, between 3 and 5 (Busch et al., 2013).

10.4.4 BET (Brunauer-Emmett-Teller) Surface Area

Along with the chemical transformation of biomass into char, physical properties such as the specific surface area are altered as well. Biochar is characterized as a highly porous structure and has a large specific surface area, which can be as large as 400 m²g⁻¹, depending on the biomass feedstock and the pyrolysis conditions used (Downie et al., 2009; Van Zwieten et al., 2009; Bruun et al., 2012; Chen et al., 2012; Ronsse et al., 2013; Schimmelpfennig and Glaser, 2012; Song and Guo, 2012). A large specific surface area in biochars is a preferable trait, as it indicates a high porosity and consequently, the char may be suitable for colonization by beneficial soil microorganisms (Lehmann et al., 2011).

The porosity and surface area in biochar is caused by two factors: the microstructure of the original biomass used to produce the biochar (e.g. capillary structures such as xylem in woody biomass), and the physical cracking due to mass loss and shrinkage during the pyrolysis process. High specific surface areas (i.e. > 100 m²/g) can be attained when a lignocellulosic (esp. woody based), low ash containing feedstock is used (Ronsse et al., 2013). With respect to pyrolysis temperature, a clear correlation can be seen between temperature and specific surface area in the resulting char. However, the surface development has a temperature optimum: at higher temperatures (≥ 750°C), rearrangement of the carbon structure or ash melting causes a reduction in specific surface area (Schimmelpfennig and Glaser, 2012).

10.5 Techno-economic Evaluation

10.5.1 Use of Liquid and Gaseous Co-products

The success of commercial biochar production depends, in part, on the valorization potential of co-products, including the non-condensable gases and the vapors (in condensed form: bio-oil). Pyrolysis gases (i.e. non-condensable gases) are mainly made up of CO, CO_2, H_2, CH_4 and some lighter hydrocarbons. For safe disposal, these are typically combusted on-site in a thermal oxidizer (afterburner). The heat released during gas treatment is commonly supplied to the pyrolysis reactor. In the case of fluidized bed reactors, the combusted gases can also be used as a fluidizing medium.

The use of pyrolysis liquids obtained from fast pyrolysis processes is still heavily vested upon research-wise, as pyrolysis oils (bio-oils) prove to be a rather challenging product to valorize. Fast pyrolysis oils have a rather high content in water (between 15 and 30 *wt%*) and mainly consist of oxygenated organic compounds (Bridgwater, 2012). The high concentration of water in bio-oil is not solely the result of water being present in the feedstock; the majority of water in bio-oil is produced in chemical dehydration reactions taking place during pyrolysis. As such, the heating value of pyrolysis oils ranges between 16 and 19 MJ/kg, which is considerably lower than that of conventional petroleum-based fuels, which have higher heating rates of between 42 and 44 MJ/kg (Venderbosch and Prins, 2010). Furthermore, the presence of compounds like carboxylic acids gives rise to a low, acidic pH (< 2) and causes the bio-oil to be corrosive. Finally, some of the compounds present in bio-oil, including aldehydes and carboxylic acids, might react over prolonged periods of time. Typically, certain compounds will polymerize yielding a bio-oil with a higher viscosity and a higher water content, and phase separation into an aqueous and organic phase may occur as well. This process of quality deterioration in bio-oils is also referred to as 'ageing.'

As fast pyrolysis oils are combustible liquids, albeit immiscible with petroleum, they can serve as fuel in a variety of stationary energy applications, including boilers, turbines and compression ignition engines (i.e. diesel engines). However, some modifications are required to cope with the bio-oil's corrosiveness, lower calorific value, higher viscosity and poor ignition properties (Bridgwater, 2012).

Currently, a lot of research in fast pyrolysis is devoted to the so-called upgrading of bio-oil. By using a catalyst, either directly in the pyrolysis process or in a post-pyrolysis upgrading process known as hydrodeoxygenation, the aim is to reduce the content of oxygen in the pyrolysis oil and to obtain a liquid whose composition is more similar to the hydrocarbon composition of refinery blendstocks. As such, these upgraded bio-oils can be co-fed in existing petrochemical refineries or be blended in today's transportation fuels. Unfortunately, some of these upgrading processes are material as well as energy intensive, and – in their current status – are not cost competitive with fossil crude oil refining.

With respect to condensates obtained from slow pyrolysis, the energy content is much reduced compared to fast pyrolysis bio-oil because the process of slow pyrolysis stabilizes a higher fraction of the biomass feedstock C in the char. For instance, in Figure 10.9 a plot is made using literature data sets reporting char yield, char proximate and elemental composition. By plotting the fraction of feedstock energy that is recoverable in the pyrolysis gas and vapor phase against the C-yield of the char (i.e. the percentage of feedstock carbon retained in the char), it is clear that if one aims for high carbon retention in the char, little energy will be available in the combined pyrolysis gases and vapors and their heating value will be relatively low (Cantrell and Martin, 2012; Ronsse et al., 2013; Crombie et al., 2015; Crombie and Masek, 2014). Fast pyrolysis is characterized by rather low C-yields in the char (i.e. below 30%), thereby retaining the largest portion of feedstock energy in the gases and vapors. By keeping the majority of the biomass carbon in the liquid pyrolysis product, fast pyrolysis is able to produce a liquid having sufficient heating value for heating and/or fuel purpose at a cost which is competitive to the cost, expressed on an energy basis (i.e.

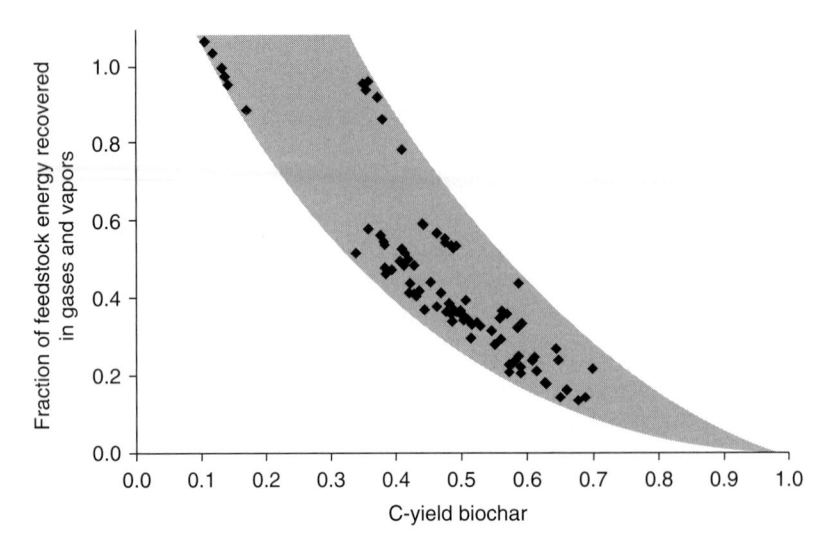

Figure 10.9. The fraction of biomass feedstock energy recovered in the gas and vapor phase versus the carbon yield in biochar. Data pooled from Cantrell and Martin (2012); Ronsse et al. (2013); Crombie et al. (2015); and Crombie and Masek (2014). If heating values were not reported, Dulong's formula was used to estimate heating value based on reported elemental composition.

in $/GJ), of heating fuel oil (HFO) (Rogers and Brammer, 2012). For instance, Wright et al. (2010) found that fast pyrolysis oil could commercially be produced from corn stover at a projected cost of 0.21 USD per liter. However, compared to crude oil, upgrading of pyrolysis oil into drop-in transportation fuels is a costly process, requiring the input of hydrogen gas to deoxygenate the pyrolysis oil. According to the same study, the projected cost of upgraded pyrolysis oils ranged between 0.56 and 0.82 $/l, depending on whether the hydrogen gas was produced internally or bought externally. Arbogast et al. (2013) determined that this cost estimate needs to be lowered by up to 40% in order to be competitive with today's petroleum-based fuel production.

Slow pyrolysis, however, aims at producing a char for carbon sequestration purposes and thus maximizes the retention of the biomass carbon in a larger yielding fraction of char (C-yield > 50%). Consequently, the energy content of the vapors will be low and water will be present, up to 80 *wt%* in the condensate (Crombie and Masek, 2014). Therefore and in lieu of the non-existence of a market for slow pyrolysis liquids, current-day slow pyrolysis units do not resort to vapor condensation systems. Instead, both vapors and gases are combusted on-site and the heat may be recovered for feedstock drying and for process energy. If excess thermal energy is being produced, electrical power generation and grid supply could be considered. However, for smaller scale biochar production units, the relatively low temperature of the flue gases, after having been used for supplying heat to the pyrolysis reactor, may require advanced thermal recovery solutions, including Stirling engines and ORC systems (Organic Rankine Cycle), which warrant further research. Another benefit of direct combustion is the elimination of the need to install complex and specialized liquid collection systems.

10.5.2 Biochar Production Cost

With the emergence of biochar for agricultural soil applications and the growing interest from commercial ventures, a careful cost analysis is at hand. Such production cost analysis has to take into account the long-term evolution of feedstock supply and cost, the development of production techniques, the incentivization of bio-energy coproduction and the carbon markets (Roberts et al., 2010).

In their study, Dickinson et al. (2015) calculated the production cost of two hypothetical scenarios: the first being biochar production in the Sub-Saharan region, by processing local agricultural residues in traditional kilns. The second scenario was an annual 100,000 ton production facility in Northwest Europe using forestry residues as feedstock and where excess heat was used for electricity production. Electricity not consumed on-site was supposed to be delivered to the grid at average 2013 feed-in tariffs. From their results, average biochar production costs were estimated to be 123 and 207 USD/t (2013 USD currency) for the Sub-Saharan and Northwestern Europe scenario, respectively. These results were in line with previous reported estimates on forestry-derived biochar: Kumar and Sarkar (2009) computed a production cost of 173 USD/t, the study of McCarl et al. (2009) yielded a 258 USD/t estimate and finally, Shackley et al. (2011) estimated forestry biochar production costs to be between 166 and 230 GBP/t. The breakdown of the cost estimates in these studies clearly pinpoints the cost at which the feedstock is procured to be the most determining factor for a typical Western, mid- to large-sized production facility. Switching to waste stream biomasses could further lower the production costs. However, operators face significant legal (Van Laer et al., 2015) and technical challenges, including the sourcing of a steady supply of waste material, consistent in composition. In the cost estimates by Dickinson et al. (2015), the revenue generated by the supply of surplus electricity to the grid could dampen the biochar unit production cost by 109 USD/t, which was already factored in the 207 USD/t estimate.

From these cost estimates, it is clear that agronomic benefits or the added value of biochar in any given application need to be significantly favorable to warrant market prices that permit break-even investments. One also needs to be aware of the boundaries to the biochar market: if biochar's market value drops below that of solid fuels like coal, lignite and charcoal in terms of cost per equivalent heating value (i.e. USD/GJ), then the agricultural value of biochar will be overshadowed by its use as a solid fuel (Field et al., 2013; Dickinson et al. 2015). Similarly, some biochars have been demonstrated to be valuable absorbents, with a high specific surface area, and could therefore provide an alternative to activated carbon, which is currently mostly sourced from fossil resources. From the above remarks, it may be suggested that biochars unfit for combustion (i.e. high ash containing chars) or unfit for absorption applications (i.e. low BET surface area) will be readily available for agricultural uses and free from competitive use models. However, the performance of said biochars is currently not fully clear.

References

Adam, J. C. (2009). Improved and more environmentally friendly charcoal production system using a low-cost retort-kiln (eco-charcoal). *Renewable Energy*, 34, pp. 1923–1925.

Antal, M. J. and Grønli, M. (2003). The art, science, and technology of charcoal production. *Industrial and Engineering Chemistry Research*, 42, pp. 1619–1640.

Antal, M. J., Mochidzuki, K. and Paredes, L. S. (2003). Flash carbonization of biomass. *Industrial Engineering Chemistry Research*, 42, pp. 3690–3699.

Antoniou, N., Stavropoulos, G. and Zabaniotou, A. (2014). Activation of end of life tyres pyrolytic char for enhanching viability of pyrolysis – critical review, analysis and recommendations for a hybrid dual system. *Renewable and Sustainable Reviews*, 39, pp. 1053–1073.

Arbogast, S., Bellman, D., Paynter, J. D. and Wykowski, J. (2013). Advanced biofuels from pyrolysis oil … opportunities for cost reduction. *Fuel Processing Technology*, 106, pp. 518–525.

ASTM (2007). *D1762-84: Standard Method for Chemical Analysis of Wood Charcoal*. American Society for Testing and Materials international.

Azargohar, R., Nanda, S., Kozinski, J. A., Dalai, A. K. and Sutarto, R. (2014). Effects of temperature on the physicochemical characteristics of fast pyrolysis bio-chars derived from Canadian waste biomass. *Fuel*, 125, pp. 90–10.

Bain, R. and Broer, K. (2011). Gasification. In: Brown, R. C. and Stevens, C. (eds.) *Thermochemical Processing of Biomass – Conversion into Fuels, Chemicals and Power*. London: John Wiley and Sons, pp. 47–77.

Balat, M., Balat, M., Kirtay, E. and Balat, H. (2009). Main routes for the thermo-conversion of biomass into fuels and chemicals. Part 1: pyrolysis systems. *Energy Conversion and Management*, 50, pp. 3147–3157.

Baldwin, H. I. (1958). *The New Hampshire Charcoal Kiln*. Concord, NH: New Hampshire Forestry and Recreation Commission.

Basu, P. (2013). *Biomass Gasification, Pyrolysis and Torrefaction: Practical Design and Theory*. 2nd Edition. Burlington, MA: Academic Press.

Bridgwater, A. V. (2012). Review of fast pyrolysis of biomass and product upgrading. *Biomass and Bioenergy*, 38, pp. 68–94.

Bridgwater, A. V. and Peacocke, G. V. C. (2000). Fast pyrolysis processes for biomass. *Renewable Sustainable Energy Reviews*, 4, pp. 1–73.

Brown, R. C. (2009). Biochar production technology. In: Lehmann, J. and Joseph, S. (eds.) *Biochar for Environmental Management: Science and Technology*. London: Earthscan, pp. 127–146.

Bruun, E. W., Ambus, P., Egsgaard, H. and Hauggaard-Nielsen, H. (2012). Effects of slow and fast pyrolysis biochar on soil C and N turnover dynamics. *Soil Biology and Biochemistry*, 46, pp. 73–79.

Busch, D., Stark, A., Kammann, C. I. and Glaser, B. (2013). Genotoxic and phytotoxic risk assessment of fresh and treated hydrochar from hydrothermal carbonization compared to biochar from pyrolysis. *Ecotoxicology and Environmental Safety*, 97, pp. 59–66.

Cantrell, K. B. and Martin, J. H. (2012). Stochastic state-space temperature regulation of biochar production. Part II: application to manure processing via pyrolysis. *Journal of the Science of Food and Agriculture*, 92, pp. 490–495.

Chen, Y., Yang, H., Wang, X., Zhang, S. and Chen, H. (2012). Biomass-based pyrolytic polygeneration system on cotton stalk pyrolysis: influence of temperature. *Bioresource Technology*, 107, pp. 411–418.

Chidumayo, E. N. and Gumbo, D. J. (2013). The environmental impacts of charcoal production in tropical ecosystems of the world: a synthesis. *Energy for Sustainable Development*, 17, pp. 86–94.

Collard, F.-X. and Blin, J. (2014). A review on pyrolysis of biomass constituents: mechanisms and composition of the products obtained from the conversion of cellulose, hemicelluloses and lignin. *Renewable and Sustainable Energy Reviews*, 38, pp. 594–608.

Cordero, T., Marquez, F., Rodriguez-Mirasol, J. and Rodriguez, J. J. (2001). Predicting heating values of lignocellulosics and carbonaceous materials from proximate analysis. *Fuel*, 80, pp. 1567–1571.

Crombie, K., Masek, O., Sohi, S. P., Brownsort, P. and Cross, A. (2013). The effect of pyrolysis conditions on biochar stability as determined by three methods. *Global Change Biology Bioenergy*, 5, pp. 122–131.

Crombie, K. and Masek, O. (2014). Pyrolysis biochar systems, balance between bioenergy and carbon sequestration. *Global Change Biology Bioenergy*, 7, pp. 349–361.

Crombie, K., Masek, O., Cross, A. and Sohi, S. (2015). Biochar – synergies and trade-offs between soil enhancing properties and C sequestration potential. *Global Change Biology Bioenergy*, 7, 1161–1175.

Cross, A. and Sohi, S. P. (2011). The priming potential of biochar products in relation to labile carbon contents and soil organic matter status. *Soil Biology and Biochemistry*, 43, pp. 2127–2134.

Cross, A. and Sohi, S. P. (2013). A method for screening the relative long-term stability of biochar. *GCB Bioenergy*, 5, pp. 215–220.

Czernik, S. and Bridgwater, A. V. (2004). Overview of applications of biomass fast pyrolysis oil. *Energy & Fuels*, 18, pp. 590–598.

De Oliveira Vilela, A., Lora, E. S., Quintero, Q. R., Vicintin, R. A. and da Silva e Souza, R. P. (2014). A new technology for the combined production of charcoal and electricity through cogeneration. *Biomass & Bioenergy*, 69, pp. 222–240.

Deenik, J. L., McClellan, T., Uehara, G., Antal, M. J. and Campbell, S. (2010). Charcoal volatile matter influences plant growth and soil nitrogen transformations. *Soil Science Society of America Journal*, 74, pp. 1259–1270.

Demirbas, A. (2001). Biomass resource facilities and biomass conversion processing for fuels and chemicals. *Energy Conversion Management*, 42, pp. 1357–1378.

Di Blasi, C. (2008). Modeling chemical and physical processes of wood and biomass pyrolysis. *Progress in Energy and Combustion Science*, 34, pp. 47–90.

Dickinson, D., Balduccio, L., Buysse, J., Ronsse, F., Van Huylenbroeck, G. and Prins, W. (2015). Cost-benefit analysis of using biochar to improve cereals agriculture. *GCB Bioenergy*, 7, pp. 850–864.

Domínguez, A., Menéndez, J. A, Inguanzo, M. and Pís, J. J. (2006). Production of bio-fuels by high temperature pyrolysis of sewage sludge using conventional and microwave heating. *Bioresource Technology*, 97, pp. 1185–1193.

Downie, A., Crosky, A. and Munroe, P. (2009). Physical properties of biochar. In: Lehmann, J. and Joseph, S. (eds.) *Biochar for Environmental Management: Science and Technology*. London: Earthscan, pp. 13–32.

Du, S., Yang, H., Qian, K., Wang, X. and Chen, H. (2014). Fusion and transformation properties of the inorganic components in biomass ash. *Fuel*, 117, pp. 1281–1287.

Duku, M. H., Gu, S. and Hagan, E. B. (2011). Biochar production potential in Ghana – a review. *Renewable and Sustainable Energy Reviews*, 15, pp. 3539–3551.

EBC (2012). European Biochar Certificate – Guidelines for a sustainable production of biochar. [online] Available at: www.european-biochar.org/en/download. European

Biochar Foundation (EBC), Arbaz, Switzerland, Version 5 of 1 January 2015. [Accessed 29 December 2014]

Enders, A., Hanley, K., Whitman, T., Joseph, S. and Lehmann, J. (2012). Characterization of biochars to evaluate recalcitrance and agronomic performance. *Bioresource Technology*, 114, pp. 644–653.

FAO, Food and Agriculture Organization of the United Nations (1983). *Simple Technologies for Charcoal Making*. FAO Forestry Paper 41.

FAO, Food and Agriculture Organization of the United Nations (1985). *Industrial Charcoal Making*. FAO Forestry Paper 63.

Field, J. L., Keske, C. M. H., Birch, G. L., Defoort, M. W. and Cotrufo, M. F. (2013). Distributed biochar and bioenergy coproduction: a regionally specific case study of environmental benefits and economic impacts. *Global Change Biology Bioenergy*, 5, pp. 177–191.

Gajic, A., Ramke, H.-G., Hendricks, A. and Koch, H.-J. (2012). Microcosm study on the decomposability of hydrochars in a Cambisol. *Biomass & Bioenergy*, 47, pp. 250–259.

Graber, E. R., Harel, Y. M., Kolton, M., Cytryn, E., Silber, A., David, D. R., Tsechansky, L., Borenshtein, M. and Yigal, E. (2010). Biochar impact on development and productivity of pepper and tomato grown in fertigated soilless media. *Plant and Soil*, 337, pp. 481–496.

Grønli, M. (2005). *Industrial Production of Charcoal*. Sintef Energy Research Paper, N-7465. Norway: Trondheim.

Grønli, M. and Melaaen, M. C. (2000). Mathematical model for wood pyrolysis – comparison of experimental measurements with model predictions. *Energy & Fuels*, 14, pp. 791–800.

Gwezi, W., Chaukura, N., Mukome, F. N. D., Machado, S. and Nyamasoka, B. (2015). Biochar production and applications in sub-Saharan Africa: opportunities, constraints, risks and uncertainties. *Journal of Environmental Management*, 150, pp. 250–261.

Hajaligol, M., Waymack, B. and Kellogg, D. (2001). Low temperature formation of aromatic hydrocarbon from pyrolysis of cellulosic materials. *Fuel*, 80, pp. 1799–1807.

Hu, B., Wang, K., Wu, L., Yu, S.-H., Antonietti, M. and Titirici, M.-M. (2010). Engineering carbon materials from the hydrothermal carbonization process of biomass. *Advanced Materials*, 22, pp. 813–823.

International Biochar Initiative (IBI) Guidelines (2014). *Standardized product definition and product testing guidelines for biochar that used in soil. Internation Biochar Initiative, Westeville (OH), US. Version 2 as of 27 October 2014.* [online] Available at: www.biochar-international.org. [Accessed 29 December 2014]

Isahak, W. N. R. W., Hisham, M. W. M., Yarmo, M. A. and Hin, T. Y. Y. (2012). A review on bio-oil production from biomass by using pyrolysis method. *Renewable & Sustainable Energy Reviews*, 16, pp. 5910–5923.

Kambo, H. and Dutta, A. (2015). A comparative review of biochar and hydrochar in terms of production, physico-chemical properties and applications. *Renewable and Sustainable Energy Reviews*, 45, pp. 359–378.

Kim, S., Chmely, S. C., Nimlos, M. R., Bomble, Y. J., Foust, T. D., Paton, R. S. and Beckham, G. T. (2011). Computational study of bond dissociation enthalpies for a large range of native and modified lignins. *The Journal of Physical Chemistry Letters*, 2, pp. 2846–2852.

Klose, W. and Wiest, W. (1999). Experiments and mathematical modeling of maize pyrolysis in a rotary kiln. *Fuel*, 78, pp. 65–72.

Krull, E. S., Baldock, J. A., Skjemstad, J. O. and Smernik, R. J. (2009). Characteristics of biochar: organo-chemical properties. In: Lehmann, J. and Joseph, S. (eds.) *Biochar for Environmental Management: Science and Technology*. London: Earthscan. pp. 53–66.

Kruse, A. (2009). Hydrothermal biomass gasification. *The Journal of Supercritical Fluids*, 47, pp. 391–399.

Kruse, A., Funke, A. and Titirici, M.-M. (2013). Hydrothermal conversion of biomass to fuels and energetic materials. *Current Opinion in Chemical Biology*, 17, pp. 515–521.

Kruse, A. and Dahmen, N. (2015). Water – a magic solvent for biomass conversion. *The Journal of Supercritical Fluids*, 96, pp. 36–45.

Kumar, A. and Sarkar, S. (2009). *Techno-economic Assessment of Biomass Conversion to Charcoal for Carbon Sequestration*. Edmonton (Canada): University of Alberta.

Lédé, J. (2013). Biomass fast pyrolysis reactors: a review of a few scientific challenges and of related recommended research topics. *Oil & Gas Science and Technology – Revue de IFP Energies Nouvelles*, 68, pp. 801–814.

Lehmann, J. and Joseph, S. (2009). Biochar for environmental management: an introduction. In: Lehmann, J. and Joseph, S. (eds.) *Biochar for Environmental Management: Science and Technology*. London: Earthscan, pp. 1–12.

Lehmann, J., Rillig, M. C., Thies, J., Masiello, C. A., Hockaday, W. C and Crowley, D. (2011). Biochar effects on soil biota – a review. *Soil Biology and Biochemistry*, 43, pp. 1812–1836.

Lei, H., Ren, S. and Julson, J. (2009). The effects of reaction temperature and time and particle size of corn stover on microwave pyrolysis. *Fuel*, 23, pp. 3254–3261.

Liaw, S.-S., Wang, Z., Ndegwa, P., Frear, C., Hu, S., Li, C.-Z. and Garcia-Perez, M. (2012). Effect of pyrolysis temperature on the yield and properties of bio-oils obtained from the auger pyrolysis of Douglas Fir wood. *Journal of Analytical and Applied Pyrolysis*, 93, pp. 52–62.

Lu, Q., Li, W.-Z. and Zhu, X.-F. (2009). Overview of fuel properties of biomass fast pyrolysis oils. *Energy Conversion and Management*, 50, pp. 1376–1383.

Manya, J. J., Laguarta, S. and Ortigosa, M. A. (2013). Study on the biochar yield and heat required during pyrolysis of two-phase olive mill waste. *Energy & Fuels*, 27, pp. 5931–5939.

Manya, J. J., Ortigosa, M. A., Laguarta, S. and Manso, J. A. (2014). Experimental study on the effect of pyrolysis pressure, peak temperature, and particle size on the potential stability of vine shoots-derived biochar. *Fuel*, 133, pp. 163–172.

Masek, O., Budarin, V., Gronnow, M., Crombie, K., Brownsort, P., Fitzpatrick, E. and Hurst, P. (2013a). Microwave and slow pyrolysis biochar – comparison of physical and functional properties. *Journal of Analytical and Applied Pyrolysis*, 100, pp. 41–48.

Masek, O., Brownsort, P., Cross, A. and Sohi, S. (2013b). Influence of production conditions on the yield and environmental stability of biochar. *Fuel*, 103, pp. 151–155.

McCarl, B. A., Peacocke, C., Chrisman, R., Chih-chun, K. and Sands, R. D. (2009). Economics of biochar production, utilization and greenhouse gas offsets. In: Lehmann, J. and Joseph, S. (eds.) *Biochar for Environmental Management: Science and Technology*. London: Earthscan, pp. 341–358.

McKendry, P. (2002). Energy production from biomass (part 3): gasification technologies. *Bioresource Technology*, 83, pp. 55–63.

Mohan, D., Pittman, C. U. and Steel, P. H. (2006). Pyrolysis of wood/biomass for bio-oil: a critical review. *Energy & Fuels*, 20, pp. 848–889.

Morf, P., Hasler, P. and Nussbaumer, T. (2002). Mechanisms and kinetics of homogeneous secondary reactions of tar from continuous pyrolysis of wood chips. *Fuel*, 81, pp. 843–883.

Nachenius, R. W., Ronsse, F., Venderbosch, R. H. and Prins, W. (2013). Biomass pyrolysis. In: Murzin, D. Y. (ed.) *Advances in Chemical Engineering*. Burlington: Academic Press, pp. 75–139.

Nachenius, R. W., Van de Wardt, T. A., Ronsse, F. and Prins, W. (2015). Residence time distributions of coarse biomass particles in a screw conveyer reactor. *Fuel Processing Technology*, 130, pp. 87–85.

Park, W. C., Atreya, A. and Baum, H. R. (2010). Experimental and theoretical investigation of heat and mass transfer processes during wood pyrolysis. *Combustion and Flame*, 157, pp. 481–494.

Patwardhan, P. R., Satrio, J. A., Brown, R. C. and Shanks, B. H. (2010). Influence of inorganic salts on the primary pyrolysis products of cellulose. *Bioresource Technology*, 101, pp. 4646–4655.

Pennise, D. M., Smith, K. R., Kithinji, J. P., Rezende, M. E., Raad, T. J., Zhang, J. and Fan, C. (2001). Emissions of greenhouse gases and other airborne pollutants from charcoal making in Kenya and Brazil. *Journal of Geophysical Research*, 106, pp. 24143–24155.

Qian, K., Kumar, A., Zhang, H., Bellmer, D. and Huhnke, R. (2015). Recent advances in utilization of biochar. *Renewable and Sustainable Energy Reviews*, 42, pp. 1055–1064.

Roberts, K. G., Gloy, B. A., Joseph, S., Scott, N. R. and Lehmann, J. (2010). Life cycle assessment of biochar systems: estimating the energetic, economic, and climate change potential. *Environmental Science & Technology*, 44, pp. 827–833.

Rogers, J. G. and Brammer, J. G. (2012). Estimation of the production cost of fast pyrolysis bio-oil. *Biomass and Bioenergy*, 36, pp. 208–217.

Ronsse, F., Bai, X., Prins, W. and Brown, R. C. (2012). Secondary reactions of levoglucosan and char in the fast pyrolysis of cellulose. *Environmental Progress and Sustainable Energy*, 31, pp. 256–260.

Ronsse, F., Van Hecke, S., Dickinson, D. and Prins, W. (2013). Production and characterization of slow pyrolysis biochar: influence of feedstock type and pyrolysis conditions. *Global Change Biology Bioenergy*, 5, pp.104–115.

Saha, B. (2003). Hemicellulose bioconversion. *Journal of Industrial Microbiology and Biotechnology*, 30, pp. 279–291.

Salena, A. and Ani, F. N. (2011). Microwave induced pyrolysis of oil palm biomass. *Bioresource Technology*, 102, pp. 3388–3395.

Schenkel, Y., Bertaux, P., Vanwijnsberghe, S. and Carre, J. (1998). An evaluation of the mound kiln carbonization technique. *Biomass & Bioenergy*, 14, pp. 505–516.

Schimmelpfennig, S. and Glaser, B. (2012). One step forward toward characterization: some important material properties to distinguish biochars. *Journal of Environmental Quality*, 41, pp. 1001–1013.

Shackley, S., Hammond, J., Gaunt, J. and Ibarrola, R. (2011). The feasibility and costs of biochar deployment in the UK. *Carbon Management*, 2, pp. 335–356.

Smider, B. and Singh, B. (2014). Agronomic performance of a higher ash biochar in two contrasting soils. *Agriculture, Ecosystems & Environment*, 191, pp. 99–107.

Song, W. and Guo, M. (2012). Quality variations of poultry litter biochar generated at different pyrolysis temperatures. *Journal of Analytical and Applied Pyrolysis*, 94, pp. 138–145.

Spokas, K. A. (2010). Review of the stability of biochar in soils: predictability of O:C molar ratios. *Carbon Management*, 1, pp. 289–303.

Spokas, K. A., Novak, J. M., Stewart, C. E., Cantrell, K. B., Uchiyama, M., DuSaire, M. G. and Ro, K. S. (2011). Qualitative analysis of volatile organic compounds on biochar. *Chemosphere*, 85, pp. 869–882.

Stefanidis, S. D., Kalogiannis, K. G., Iliopoulou, E. F., Michailof, C. M., Pilavachi, P. A. and Lappas, A. A. (2014). A study of lignocellulosic biomass pyrolysis via the pyrolysis of cellulose, hemicellulose and lignin. *Journal of Analytical and Applied Pyrolysis*, 105, pp. 143–150.

Thomas, R., Grose, A., Obaje, G., Taylor, R., Rownson, N. and Blackburn, S. (2009). Residence time investigation of a multiple hearth kiln using mineral tracers. *Chemical Engineering and Processing: Process Intensification*, 48, pp. 950–954.

Titirici, M.-M., Thomas, A., Yu, S.-H., Müller, J.-O. and Antonietti, M. (2007). A direct synthesis of mesoporous carbons with bicontinuous pore morphology from crude plant material by hydrothermal carbonization. *Chemistry of Materials*, 19, pp. 4205–4212.

Van der Stelt, M. J. C., Gerhauser, H., Kiel, J. H. A. and Ptasinski, K. J. (2011). Biomass upgrading by torrefaction for the production of biofuels: a review. *Biomass and Bioenergy*, 35, pp. 3748–3762.

Vanholme, B., Desmet, T., Ronsse, F., Rabaey, K., Van Breusegem, F., De Mey, M., Soetaert, W. and Boerjan, W. (2013). Towards a carbon-negative sustainable bioeconomy. *Frontiers in Plant Science*, 4, p. 174.

Van Laer, T., De Smedt, P., Ronsse, F., Ruysschaert, G., Boeckx, P., Verstraete, W., Buysse, J. and Lavrysen, L. J. (2015). Legal constraints and opportunities for biochar: a case analysis of EU law. *Global Change Biology*, 7, pp. 14–25.

Van Wesenbeeck, S., Prins, W., Ronsse, F. and Antal, M. J. (2014). Sewage sludge carbonization for biochar applications. Fate of heavy metals. *Energy & Fuels*, 28, pp. 5318–5326.

Van Zwieten, L., Kimber, S., Morris, S., Chan, K. Y., Downie, A., Rust, J., Joseph, D. and Cowie, A. (2009). Effects of biochar from slow pyrolysis of papermill waste on agronomic performance and soil fertility. *Plant and Soil*, 327, pp. 235–246.

Venderbosch, R. H. and Prins, W. (2010). Fast pyrolysis technology development. *Biofuels, Bioproducts & Biorefining*, 4, pp. 178–208.

Wagenaar, B. M., Prins, W. and Van Swaaij, W. P. M. (1994). Pyrolysis of biomass in the rotating cone reactor: modelling and experimental justification. *Chemical Engineering Science*, 49, pp. 5109–5126.

Wei, L., Xu, S., Zhang, L., Zhang, H., Liu, C., Zhu, H. and Lio, S. (2006). Characteristics of fast pyrolysis of biomass in a free fall reactor. *Fuel Processing Technology*, 87, pp. 863–871.

White, J. E., Catallo, W. J. and Legendre, B. L. (2011). Biomass pyrolysis kinetics: a comparative critical review with relevant agricultural residue case studies. *Journal of Analytical and Applied Pyrolysis*, 91, pp. 1–33.

Wiedner, K., Rumpel, C., Steiner, C., Pozzi, A., Maas, R. and Glaser, B. (2013). Chemical evaluation of chars produced by thermochemical conversion (gasification, pyrolysis and hydrothermal carbonization) of agro-industrial biomass on a commercial scale. *Biomass & Bioenergy*, 59, pp. 264–278.

Williams, P. T. and Besler, S. (1996). The influence of temperature and heating rate on the slow pyrolysis of biomass. *Renewable Energy*, 7, pp. 233–250.

Wright, M. M., Daugaard, D. E., Satrio, J. A. and Brown, R. C. (2010). Techno-economic analysis of biomass fast pyrolysis to transportation fuels. *Fuel*, 89, pp. S2-S10.

Yang, H., Yan, R., Chen, H., Lee, D.H. and Zheng, C. (2007). Characteristics of hemicellulose, cellulose and lignin pyrolysis. *Fuel*, 86, pp. 1781–1788.

Yang, S. I., Wu, M. S. and Wu, C. Y. (2014). Application of biomass fast pyrolysis part I: pyrolysis characteristics and products. *Energy*, 66, pp. 162–171.

Zakzeski, J., Bruijninck, P. C. A., Jongerius, A. L. and Weckhuysen, B. M. (2010). The catalytic valorization of lignin for the production of renewable chemicals. *Chemical Reviews*, 110, pp. 3552–3599.

Zhou, H., Long, Y. Q., Meng, A. H., Li, Q. H. and Zhang, Y. G. (2013). The pyrolysis simulation of five biomass species by hemi-cellulose, cellulose and lignin based on thermogravimetric curves. *Thermochimica Acta*, 566, pp. 36–43.

Zimmermann, A. R. (2010). Abiotic and microbial oxidation of laboratory-produced black biochar (biochar). *Environmental Science and Technology*, 44, pp. 1295–1301.

11

Biomass Pyrolysis for Biochar Production: Kinetics, Energetics and Economics

BYUNGHO SONG

Abstract

This chapter briefly introduces the pyrolysis process in terms of kinetics, energetics and economics. In this scope, firstly simple kinetic models are described for biomass pyrolysis and then the product distribution is discussed. The yields of gas, oil and char are predicted by using chemical kinetics, however it is very difficult to predict the yields of each gas species. Therefore, in this chapter, empirical correlations are used together with the elemental balances to predict the yields of gas species for reactor simulation. Also, a detailed application of the empirical relationships from Neves et al. (2011), elemental balances, and energy balance on the prediction of pyrolysis products are given for the simulation of the devolatilization stage in a biomass gasification process. The production of biochar through the pyrolysis process is simply described. Finally the results from several studies on energy balances and the economic feasibility of the addition of biochar application to the normal bio-energy system are summarized.

11.1 Background of Pyrolysis

Pyrolysis is the thermal decomposition of organic matter in an oxygen-free atmosphere. Biomass is converted into the following products through pyrolysis:

(1) a gas product called syngas or biogas consisting of CO, CO_2, H_2, CH_4 and higher hydrocarbons;
(2) a liquid product called bio-oil or bio-crude;
(3) a solid product that can be used as a soil additive (biochar) or as a solid fuel (char).

Most of the biochar is produced by utilizing two types of pyrolysis reactions. These are slow and fast pyrolysis respectively, where the differences depend on their heating rate and duration (see Chapters 10 and 12). Slow pyrolysis generally yields more biochar and less bio-oil, while the yields are conversely related in the case of fast pyrolysis (Lehmann and Joseph, 2009; Sohi et al., 2010). When the temperatures are lower and vapor residence times are longer, more charcoal can be produced. The proportions of the three products can be controlled by the process parameters. Typical product distribution obtained at various conditions of wood pyrolysis can be found in the literature (Bridgwater, 2003; Ringer et al., 2006; Wright et al., 2008).

Various types of biomass may be utilized for biochar production, including crop- and forestry waste products, such as logging residues, urban green wastes, industrial biomass byproducts, municipal sewage sludge and animal manures. The use of biochar as a soil amendment has been proposed to simultaneously mitigate anthropogenic climate change, improve soil fertility and consequently enhance crop production (Woolf et al., 2010). The potential benefits of biochar as a soil amendment are, for example, mitigation of global warming through application of stable C into soil, waste management, production of bioenergy, improving soil health and yielding productivity benefits. Specifically, biochar as a soil additive has certain greenhouse gas implications by improving the efficiency of nitrogen fertilizer and by sequestering carbon. In addition to environmental advantages, biochar may realize economic benefits. Already very early work, for example by Adams (1991), has observed the increase of soil fertility after soil amendment with biochar. Many studies reported a potential to increase crop yield, reduce irrigation needs and enhance fertilizer efficiency when biochar is incorporated into the soil matrix (Steiner et al., 2007). These effects have economic implications as production costs can be reduced (increased fertilizer efficiency) while income might be increased (yields as a consequence of improvements of soil fertility).

11.2 Temperature Effect on Pyrolysis Product Yields

Various process conditions like feedstock type, temperature, heating rate and pressure influence the product quality. The effects of pyrolysis conditions on the products were summarized for a range of biomass types (Yaman, 2004). Typical product yields from different types of feedstock are presented in Figure 11.1, which shows the simulated temperature dependence of the four main products from flax biomass. Most biomass gives similar results and the maximum yield of bio-oil can occur between 480 and 600°C.

11.3 Modeling Biomass Pyrolysis

Biomass is primarily decomposed into char, bio-oil, light gases and water through the devolatilization process. A model can be developed to predict the yields of the species released, as well as the relative amounts of bio-oil and char. The elemental composition of char and bio-oil needs to be known. The composition of bio-oil is more important as the bio-oil contains a range of hydrocarbon species.

The decomposition is a complex reaction, that is, it involves various physical and chemical transformations. Thus, the release rates of the various species cannot be explained by

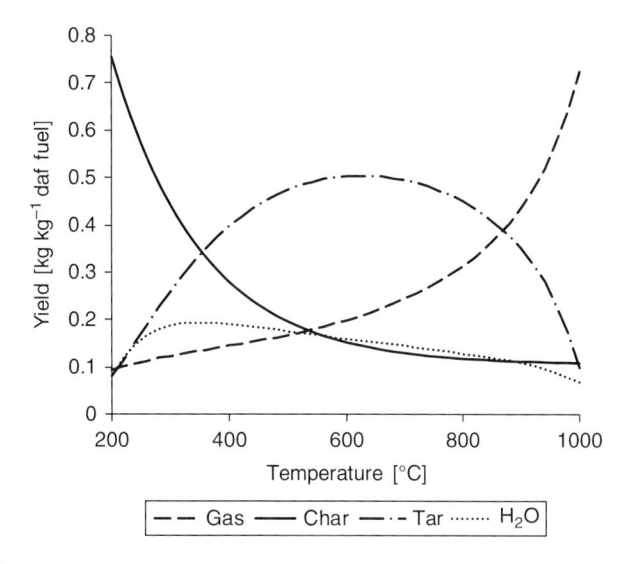

Figure 11.1. Variation of products from flax with temperature. Daf = dry and ash free

a single reaction. For this reason, kinetic models, in which the prediction is based on a set of chemical reactions, are needed. Although there are particle models considering physical aspects, only kinetic models will be introduced in this chapter.

11.3.1 Kinetic models

The kinetic models trying to describe the chemical transformation can be categorized into basic, distributed activation energy and structural models (Gomez-Barea and Leckner, 2010). A basic model (global model) describes pyrolysis by either a single reaction (which is insufficient to estimate release rates as mentioned above) or by a combination of series and parallel reactions, which are treated as first-order and independent reactions (Di Blasi, 2008). The global model for pyrolysis can be described as:

$$\text{biomass} \xrightarrow{k_{py}} v_{py}(\text{volatile}) + (1 - v_{py})(\text{char}) \tag{1}$$

The kinetic constant k_{py} and the coefficient of distribution between the primary volatiles and the char, v_{py} are the parameters to be determined by experiments. Many studies have been devoted to the calculation of kinetic parameters, for example k_{py} for various fuels, but the parameters can vary widely between cases. Despite these variations, this type of equation is preferred because of its simplicity. Series and parallel reactions can be still combined for an improvement. The kinetic scheme of pyrolysis includes primary and secondary conversion processes as shown in Figure 11.2. As shown in this figure, the prediction for the primary process comprises reactions of (A) biomass to volatiles and char, (B) volatiles to gas 1 and bio-oil 1, (C) bio-oil 1 to char 1 and (D) bio-oil 1 to gas 1.

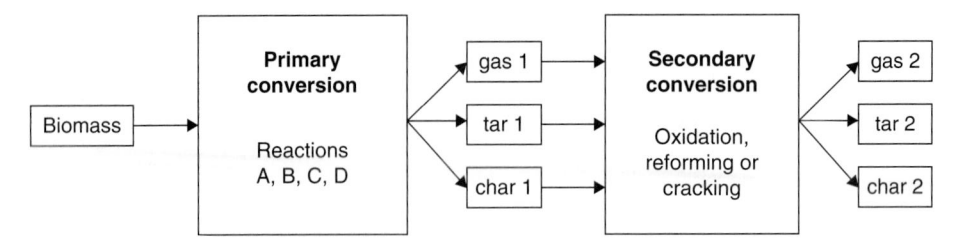

Figure 11.2. Simplified pyrolysis concept.

The method of distributed activation energies was mentioned for the application to devolatilization (Pitt, 1962; de Diego et al., 2002; Gomez-Barea and Leckner, 2010). It is assumed that the activation can be described by a continuous distribution function due to the large number of reactions. When the distribution function is selected, the kinetic information of first-order reactions, such as the pre-exponential factor, can be determined from the experimental thermal gravimetric analysis (TGA) curves of volatile release.

Structural models were developed by considering the chemical constitution of the solid fuels: (1) functional groups, (2) aromatic nuclei and bridges, (3) aromatic clusters connected by weaker aliphatic and ether bridges and so on. Among the models predicting not only the yields of gas, tar and char, but also the composition of the gas, are the Bio-FLASHCHAIN model (Niksa, 2000), the Bio-FG model (de Jong et al., 2007) and the Ranzi model (Ranzi et al., 2008).

11.3.2 Chemical Kinetics of Pyrolysis

Understanding of pyrolysis kinetics coupled with transport phenomena can provide tools for the design of pyrolysis reactors. In addition, knowledge of fuel reactivity is required. Several reviews on pyrolysis kinetics of lignocellulosic biomass were published within the last 20 years (Varhegyi et al., 1996; Burnham and Braun, 1999; Conesa et al., 2001; Di Blasi, 2008; White et al., 2011).

Primary conversion in Figure 11.2 can be considered as a one-component reaction as expressed in Equation (2), where Y is solid mass fraction and k [1/s] is the global rate constant. Shafizadeh and Chin (1977) proposed the above concept through their wood pyrolysis experiments. Under such conditions, a kinetic mechanism consists of three parallel reactions for the formation of the main products: gas, bio-oil and char. Subsequently, the global rate constant can be expressed as shown in Equation (3), where k_C, k_L and k_G are the rate constant for the formation of char, liquid and gas, respectively.

$$\frac{dY}{dt} = -kY \qquad (2)$$

$$k = k_C + k_L + k_G \qquad (3)$$

To estimate the global rate constant, k, the mass conservation equations can be integrated over the entire duration of the pyrolysis process (Di Blasi and Branca, 2001). The common

Arrhenius plot, describing the effect of temperature on the rates of chemical reactions, and a least square analysis provide the activation energy and pre-exponential factor of the global degradation kinetics. Di Blasi (2008) summarized these parameters for the formation rates of gas, liquid and char products from the literature. It is apparent from the summary that the activation energy of the global reaction rate widely varied between 56 and 174 kJ/mol. With the kinetic information available, prediction of the product yields can be obtained under the given conditions. The prediction also indicates that the char yield decreases as the temperature increases. The use of thick oak particles larger than 0.65 mm (Thurner and Mann, 1981) clearly leads to heat and mass transfer limitations, ultimately providing higher char yields.

11.4 The Prediction of Yields of Gas Species, Bio-oil and Biochar

Mathematical modeling is a powerful tool for process design, evaluation of reactor performance, and analysis of process transient behavior, and also it provides strategies for effective reactor control. Prediction of product yields from pyrolysis is very important for process simulation. However, the models mentioned above cannot predict the yields of the main gas species, although they can predict the total amount of gas, bio-oil and char produced. Therefore, empirical correlations have been developed for coal (Loison and Chauvin, 1964; Goyal and Rehmat, 1993; Song and Watkinson, 2004) and biomass (Nunn et al., 1985; Boroson et al., 1989). These were successfully implemented to obtain gas yields in reactor simulations. However, these correlations have been obtained for a particular set of operating conditions and type of fuel, which are likely not similar to those being simulated. Song and Watkinson (2000) developed a multi-stage, well-mixed reactor model for a spouted bed coal gasifier, in which they evaluated yields of gas species and bio-oil composition by empirical equations of temperature given by Goyal and Rehmat (1993). Prasad and Kuester (1988) proposed an empirical model obtained from regression to predict the composition of pyrolysis gas in a fluidized bed biomass gasifier where each gas component was represented as a function of pyrolysis temperature and steam to fuel ratio.

Sadaka et al. (2002) estimated the gas composition in their fluidized bed biomass gasifier using the elemental balance under the assumption that the volatiles in the biomass are considered to be completely converted to the five gaseous components: CO, CO_2, CH_4, H_2 and H_2O.

$$n_C = n_{CO} + n_{CO2} + n_{CH4} \tag{4}$$

$$n_H = 4n_{CH4} + 2n_{H2} + 2n_{H2O} \tag{5}$$

$$n_O = n_{CO} + 2n_{CO2} + n_{H2O} \tag{6}$$

where n is the number of moles. Nitrogen in the biomass feedstock is assumed to form only inert nitrogen gas, consequently the elemental balance for nitrogen becomes

$$n_N = 2n_{N2} \tag{7}$$

In order to have zero degree of freedom in the above equations, two more equations are required. The fraction of CO and CH_4 formed may be defined as:

$$\phi_{CO} = \frac{n_{CO}}{n_{CO2}} \tag{8}$$

$$\phi_{CH4} = \frac{n_{CH4}}{n_{H2}} \tag{9}$$

These fractions can be determined from pyrolysis experiments. Nguyen et al. (2012) examined the pyrolysis data of wood, coconut shell and straw in Fagbemi et al. (2001) and they described the above two fractions as a function of temperature for the simulation of a biomass gasification reactor. However, the above approach predicts only gas yield and the type of biomass seems to be somewhat limited.

A mathematical model of a biomass gasifier can be developed as shown in Figure 11.3. Assumptions are as follows:

1st stage: Drying and devolatilization of biomass instantaneously occur at the reactor inlet.
2nd stage: Oxygen is completely consumed in a zone of negligible thickness close to the inlet.
3rd stage: Carbon gasification takes place in a well-mixed isothermal reactor and the corresponding kinetic information can be used. Entrainment, abrasion or fragmentation of the bed particles do not affect the reaction kinetics.
4th stage: Water gas shift reaction is considered as near to its equilibrium state.

At the first stage, the given biomass feed is being pyrolyzed to gas, liquid and char. Pyrolysis products can be predicted through elemental balances, energy balance and empirical relationships. Neves et al. (2011) have developed a very useful and relatively simple empirical relationship from the collected data over the wide temperature range for biomass pyrolysis. It should be noticed that the ash content (inorganics) is not considered throughout this calculation.

$$\text{Input} = \text{output}$$

$$\sum_j Y_{j,F} = Y_{ch,F} + Y_{tar,F} + Y_{H2O,F} + Y_{G,F} \tag{10}$$

$$Y_{G,F} = Y_{CH4,F} + Y_{CO,F} + Y_{CO2,F} + Y_{H2,F} \tag{11}$$

$$Y_{tar,F} + Y_{H2O,F} + Y_{M,F} = \text{total liquid (bio-oil)} \tag{12}$$

$Y_{j,F}$: mass fraction of jth element in fuel, dry and ash free (daf) basis (kg j/kg daf fuel)
$Y_{i,F}$: yield of ith product, daf basis (kg i/kg daf fuel)
$Y_{ash,F}$: ash content of fuel, daf basis (kg ash/kg daf fuel)
$Y_{M,F}$: moisture content of fuel, on daf fuel basis (kg moisture/kg daf fuel)

The relationship for char yield is given as:

$$Y_{ch,F} = 0.106 + 2.43 \cdot \exp(-0.66 \times 10^{-2} T) \tag{13}$$

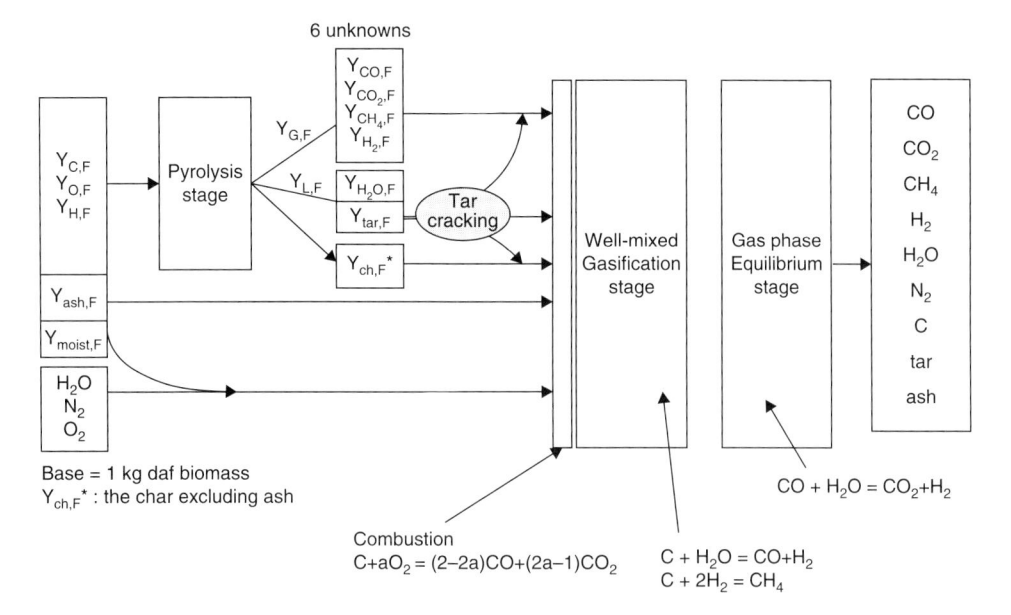

Figure 11.3. The well-mixed reactor model for a biomass gasifier.

The elemental composition of char:

$$Y_{C,ch} = 0.93 - 0.92 \cdot \exp(-0.42 \times 10^{-2} T) \tag{14}$$

$$Y_{O,ch} = 0.07 + 0.85 \cdot \exp(-0.48 \times 10^{-2} T) \tag{15}$$

$$Y_{H,ch} = -0.41 \times 10^{-2} + 0.10 \cdot \exp(-0.24 \times 10^{-2} T) \tag{16}$$

The CHO composition of bio-oil is relatively close to that of parent fuel:

$$Y_{C,tar} / Y_{C,F} = 1.14 \tag{17}$$

$$Y_{H,tar} / Y_{H,F} = 1.13 \tag{18}$$

$$Y_{O,tar} / Y_{O,F} = 0.8 \tag{19}$$

The elemental carbon balance at first stage:

$$Y_{C,F} = Y_{C,ch} Y_{ch,F} + Y_{C,tar} Y_{tar,F} + Y_{C,G} Y_{G,F} \tag{20}$$

This can be rearranged as:

$$Y_{C,F} - Y_{C,ch} Y_{ch,F} = Y_{C,tar} Y_{tar,F} + Y_{C,CH4} Y_{CH4,F} + Y_{C,CO} Y_{CO,F} + Y_{C,CO2} Y_{CO2,F} \tag{21}$$

The oxygen balance:

$$Y_{O,F} - Y_{O,ch} Y_{ch,F} = Y_{O,tar} Y_{tar,F} + Y_{O,CO} Y_{CO,F} + Y_{O,CO2} Y_{CO2,F} + Y_{O,H2O} Y_{H2O,F} \tag{22}$$

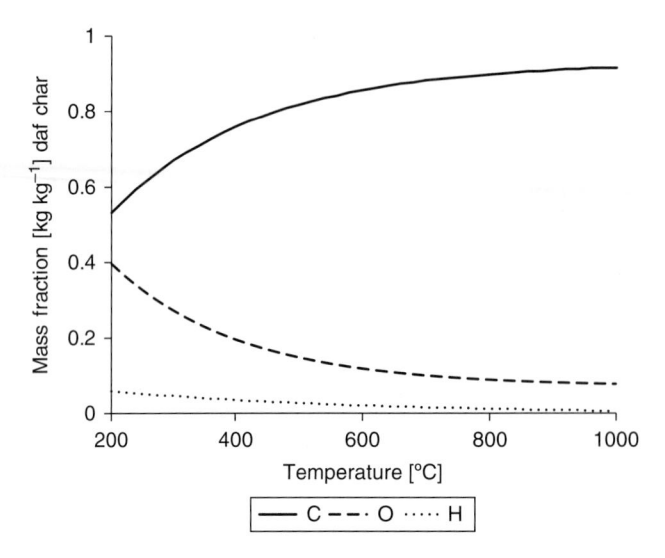

Figure 11.4. The predicted product gas composition from the pyrolysis of flax (C = 46.71%, H = 5.77%, N = 0.82%, S = 0.2%, O = 46.5%).

Hydrogen balance:

$$Y_{H,F} - Y_{H,ch}Y_{ch,F} = Y_{H,tar}Y_{tar,F} + Y_{H,CH4}Y_{CH4,F} + Y_{H,H2}Y_{H2,F} + Y_{H,H2O}Y_{H2O,F} \tag{23}$$

$$Y_{H2,F} = 1.145\left[1 - \exp(-0.11 \times 10^{-2} T)\right]^{9.384} \tag{24}$$

$$Y_{H2,F} = \left[3 \times 10^{-4} + \frac{0.0429}{1 + (T/632)^{-7.23}}\right] Y_{CO,F} \tag{25}$$

$$Y_{CH4,F} = 0.146 \cdot Y_{CO,F} - 2.18 \times 10^{-4} \tag{26}$$

The above set of six simultaneous equations ((21) to (26)) could be solved as a linear system A*X=B ('linsolve' in MATLAB). In addition, Neves et al. (2011) fitted the collected data on the lower heating value of total pyrolysis gas by equation 27:

$$LHV_G = -6.23 + 2.47 \times 10^{-2} T \tag{27}$$

Not only can the yields of char, bio-oil and gas be derived from the above equations, they allow in addition the estimation of product gas composition with the variation of reactor temperature (200–1000°C). The calculation result with flax data is shown in Figure 11.4 for the composition of char, and Figure 11.5 for the composition of the product gas. Figure 11.1 clearly presents the decreasing char yield with increasing temperature and it finally reaches 0.2 kg/kg daf fuel at 500°C from an initial value of approx. 0.7 kg/kg daf fuel. As the temperature increases, the bio-oil yield increases up to a maximum and subsequently decreases because of secondary bio-oil cracking reactions. The fraction of bio-oil cracked to gas may be used as an adjustable parameter in the reactor model, however.

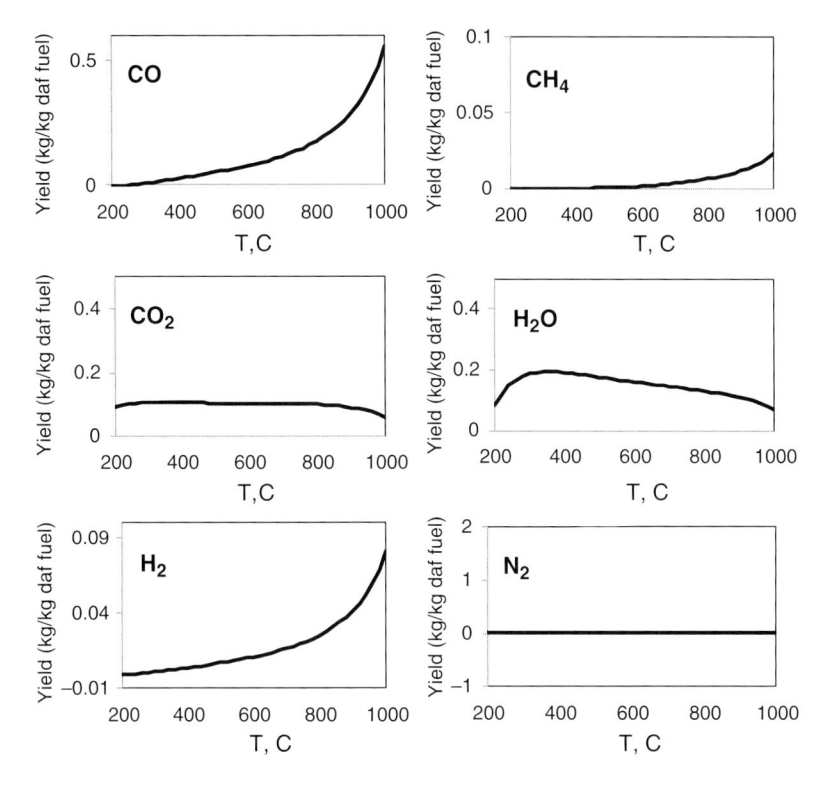

Figure 11.5. The predicted char composition with temperature variation.

11.5 The Production of Biochar in Favor of Maximum Char Yields

The energy required by the pyrolysis process can be supplied either (1) by the heat of reaction itself, or (2) by flue gases from combustion of byproducts and additional feedstock if necessary. The characteristics of biochar, such as composition, density, particle size distribution and pH, depend on the feedstock conditions (type, nature, origin and so on) as well as pyrolysis conditions (Zhang et al., 2008). For example, Demirbas (2004) reports that biomass with high lignin content, such as olive husks, produces high biochar yields due to the stability of lignin to thermal degradation (see Chapter 12 for more details).

Depending on pyrolysis conditions, it can be categorized into slow and fast pyrolysis, among others, as detailed in Chapters 10 and 12. Reviews on maximizing biochar yield by altering pyrolysis conditions and reactor configurations can be found in previously published work, for example Antal and Gronli (2003) or Zhang et al. (2008). Reviews on commercial pyrolysis processes can also be found (Meier and Faix, 1999; Bridgwater, 2003). These reviews cover the design of the entire process, pyrolysis reactors and the global status of pyrolysis plants. Important conditions for maximum char yield are long residence times of the vapor products and slow heating rates with a generally low temperature peak. Demirbas (2004) summarized favorable pyrolysis conditions for high biochar yield as follows: (i) high lignin, ash and nitrogen contents in the biomass; (ii) low pyrolysis temperature (< 400°C); (iii) high pressure; (iv) long vapor residence time; (v) extended vapor/solid contact; (vi) low heating rate; and (vii) large feedstock particle size. These conditions enhance cracking reactions that consequently decrease the bio-oil yield and increase the biochar yield (see Chapter 12).

11.6 Energetic and Economic Potential of Pyrolysis and Biochar

Kung et al. (2015) recently studied the economic feasibility of biochar production and utilization for different purposes. Their aim was to understand and compare issues about bioenergy generation, agricultural cost savings and enhancement of atmospheric quality. The study investigates the net economic and environmental costs associated with pyrolysis and biochar application using multiple feedstock and agricultural wastes at The Poyang Lake, China. Unfortunately, benefits were offset by costs of feedstock collection and manipulation and increased GHG (greenhouse gas) emissions during production. Both fast and slow pyrolysis can be profitable under certain assumptions, except for the cases of rice straw and corn stover. Poplar used for fast pyrolysis profits the most while rice straw used for slow pyrolysis generated a loss. The authors conclude therefore, if the focus lies solely on environmental benefits such as carbon sequestration, animal wastes may be a better alternative than other agricultural residuals.

Gaunt and Lehmann (2008) optimized a slow pyrolysis system for biochar and energy production and compared the result with that of a system solely for energy production. Comparison tables and the detailed energy calculation can be found in their work. For energy crops and crop wastes as feedstock, they calculated energy inputs (field production, transportation and processing) and outputs, and compared two slow pyrolysis systems that were optimized for energy production and biochar production, respectively. Two to five times greater emission reductions could be achieved when biochar was used for agricultural land than when the biochar was used solely for fossil energy offset. About a half of these emission reductions are related to the retention of C in biochar.

Roberts et al. (2010) studied the economic assessment of a drum kiln slow-pyrolysis plant of 10 ton feedstock per hour capacity. This plant is operated at 450°C and a drum residence time of several minutes. The objective of the process is to produce biochar for soil application and syngas for heat supply. According to this study, 16 USD per ton dry feedstock can be earned by biochar production, taking into account the sales of the biochar, heat produced from the synthesis gas, an income of disposal fee of yard waste, and the sales of GHG offset certificates. The authors mentioned that the revenues from the sales of GHG offset certificates are an important factor in their calculation.

References

Adams, M. D. (1991). The mechanisms of adsorption of Hg(CN)$_2$ and HgCl$_2$ on to activated carbon. *Hydrometallurgy*, 26, pp. 201–210.

Antal, M. J. and Gronli, M. (2003). The art, science and technology of charcoal production. *Journal of the American Chemical Society*, 42, pp. 1619–1640.

Boroson, M. L., Howard, J. B., Longwell, J. P. and Peters, W. A. (1989). Product yields and kinetics from the vapor phase cracking of wood pyrolysis tars. *AIChE Journal*, 35, pp. 120–128.

Bridgwater, A. V. (2003). Renewable fuels and chemicals by thermal processing of biomass. *Chemical Engineering Journal*, 91, pp. 87–102.

Burnham, A. K. and Braun, L. R. (1999). Global kinetic analysis of complex materials. *Energy Fuels*, 13, pp. 1–22.

Conesa, J. A., Marcilla, A., Caballero, J. A. and Font, R. (2001). Comments on the validity and utility of the different methods for kinetic analysis of thermogravimetric data. *Journal of Analytical and Applied Pyrolysis*, 58–59, pp. 617–633.

Demirbas, A. (2004). Effect of temperature and particle size on biochar yield from pyrolysis of agricultural residues. *Journal of Analytical and Applied Pyrolysis*, 721, pp. 243–248.

de Diego, L. F., García-Labiano, F., Abad, A., Gayán P. and Adánez, J. (2002). Modeling of the devolatilization of nonspherical wet pine wood particles in fluidized beds. *Industrial & Engineering Chemical Research*, 41, pp. 3642–3650.

de Jong, W., Di Nola, G., Venneker, B. C. H., Spliethoff, H. and Wojtowicz, M. A. (2007). TG-FTIR pyrolysis of coal and secondary biomass fuels: determination of pyrolysis kinetic parameters for main species and NOx precursors. *Fuel*, 86, pp. 2367–2376.

Di Blasi, C. (2008). Modeling chemical and physical processes of wood and biomass pyrolysis. *Progress in Energy and Combustion Science*, 34, pp. 47–90.

Di Blasi, C. and Branca, C. (2001). Kinetics of primary product formation from wood pyrolysis. *Industrial & Engineering Chemical Research*, 40, pp. 5547–5556.

Duku, M. H., Gu, S. and Hagan, E. B. (2011). Biochar production potantial in Ghana – a review. *Renewable and Sustainable Energy Reviews*, 15, 3539–3551.

Fagbemi, L., Khezami, L. and Capart, R. (2001). Pyrolysis products from different biomasses: application to the thermal cracking of tar. *Applied Energy*, 69, pp. 293–306.

Gaunt, J. L. and Lehmann, J. (2008). Energy balance and emissions associated with biochar sequestration and pyrolysis bioenergy production. *Environmental Science & Technology*, 42, pp. 4152–4158.

Gomez-Barea, A. and Leckner, B. (2010). Modeling of biomass gasification in fluidized bed. *Progress in Energy & Combustion Science*, 36, pp. 444–509.

Goyal, A. and Rehmat, A. (1993). Modelling of a fluidized-bed coal carbonizer. *Industrial & Engineering Chemical Research*, 32, pp. 1396–1410.

Kung, C. C., Kong, F. and Choi, Y. (2015). Pyrolysis and biochar potential using crop residues and agricultural wastes in China. *Ecological Indicators*, 51, pp. 139–145.

Lehmann, J. and Joseph, S. (eds.) (2009). *Biochar for Environmental Management: Science and Technology*. London: Earthscan.

Loison, R. and Chauvin, R. (1964). Pyrolyse Rapide du Charbon. *Chimie et Industrie*, 91, pp. 269–275.

Meier, D. and Faix, O. (1999). State of the applied fast pyrolysis of lignocellulosic materials. *Bioresource Technology*, 68, pp. 71–77.

Neves, D., Thunman, H., Matos, A., Tarelho, L. and Gomez-Barea, A. (2011). Characterization and prediction of biomass pyrolysis products. *Progress in Energy & Combustion Science*, 37, pp. 611–630.

Nguyen, T. D. B., Ngo, S. I., Lim, Y. I., Lee, J. W., Lee, U. D. and Song, B. H. (2012). Three-stage steady-state model for biomass gasification in a dual circulating fluidized-bed. *Energy Conversion and Management*, 54, pp. 100–112.

Niksa, S. (2000). Predicting the rapid devolatilization of diverse forms of biomass with bio-flashchain. *Proceedings of the Combustion Institute*, 8, pp. 2727–2733.

Nunn, T. R., Howard, J. B., Longwell, J. P. and Peters, W. A. (1985). Product compositions and kinetics in the rapid pyrolysis of sweet gum hardwood. *Industrial & Engineering Chemistry Process Design and Development*, 24, pp. 836–844.

Pitt, G. J. (1962). The kinetics of the evolution of volatile products from coal. *Fuel*, 41, pp. 267–274.

Prasad, B. V. R. K. and Kuester, J. L. (1988). Process analysis of a dual fluidized bed biomass gasification system. *Industrial & Engineering Chemical Research*, 27, pp. 304–10.

Ranzi, E., Cuoci, A., Faravelli, T., Frassoldati, A., Migliavacca, G. and Pierucci, S. (2008). Chemical kinetics of biomass pyrolysis. *Energy Fuels*, 22, pp. 4292–4300.

Ringer, M., Putsche, V. and Scahill, J. (2006). *Large-scale pyrolysis oil production: a technology assessment and economic analysis*. Technical Report NERL/TP-51037779, doi: 10.2172/894989.

Roberts, K. G., Gloy, B. A., Joseph, S., Scott, N. R. and Lehmann, J. (2010). Life cycle assessment of biochar systems: estimating the energetic, economic, and climate change potential. *Environmental Science & Technology*, 44, pp. 827–833.

Sadaka, S. S., Ghaly, A. E. and Sabbah, M. A. (2002). Two phase biomass air-steam gasification model for fluidized bed reactors: part I – model development. *Biomass Bioenergy*, 22, pp. 439–462.

Shafizadeh, F. and Chin, P. P. S. (1977). Thermal deterioration of wood. *ACS Symposium Series*, 43, pp. 57–81.

Sohi, S. P., Krull, E., Lopez-Capel, E. and Bol, R. (2010). A review of biochar and its use and function in soil. *Advances in Agronomy*, 105, pp. 47–82.

Song, B. H. and Watkinson, A. P. (2000). Three-stage well-mixed reactor model for a pressurized coal gasifier. *Canadian Journal of Chemical Engineering*, 78, pp. 143–155.

Song, B. H. and Watkinson, A. P. (2004). Effect of temperature on the gas yield from flash pyrolysis of bituminous coals. *Journal of Industrial and Engineering Chemistry*, 10, pp. 460–467.

Steiner, T., Mosenthin, R., Zimmermann, B., Greiner, R. and Roth, S. (2007). Distribution of phytase activity, total phosphorus and phytate phosphorus in legume seeds, cereals and cereal products as influenced by harvest year and cultivar. *Animal Feed Science and Technology*, 133, pp. 320–334.

Thurner, F. and Mann, U. (1981). Kinetic investigation of wood pyrolysis. *Industrial & Engineering Chemistry Process Design and Development*, 20, pp. 482–488.

Varhegyi, G., Antal, M. J., Jakab, E. and Szabo, P. (1996). Kinetic modeling of biomass pyrolysis. *Journal of Analytical and Applied Pyrolysis*, 42, pp. 73–87.

White, J. E., Catallo, W. J., Legendre, B. L. (2011). Biomass pyrolysis kinetics: a comparative critical review with relevant agricultural residue case studies. *Journal of Analytical and Applied Pyrolysis*, 91, pp. 1–33.

Woolf, D., Amonette, J. E., Street-Perrott, F. A., Lehmann, J. and Joseph, S. (2010). Sustainable biochar to mitigate global climate change. *Nature Communications*, 1, pp. 1–9.

Wright, M. M., Brown, R. C. and Boateng, A. A. (2008). Distributed processing of biomass to bio-oil for subsequent production of Fischer-Tropsch liquids. *Biofuels, Bioproducts and Biorefining*, 2, pp. 229–238.

Yaman, S. (2004). Pyrolysis of biomass to produce fuels and chemical feedstocks. *Energy Conversion and Management*, 45, pp. 651–671.

Zhang, Q., Yang, Z. and Wu, W. (2008). Role of crop residue management in sustainable agricultural development in the North China Plain. *Journal of Sustainable Agriculture*, 32, pp. 137–148.

Figure 1.2 The "Rain forest" (a) and the "Secret garden" (b) in the 1 Utama shopping centre in Kuala Lumpur both use a mixture of biochar, coconut fibre and clayey subsoil as a growth medium. The ratio of the three compartments was adjusted according to the actual vegetation. Charcoal briquettes from sawdust, produced for barbecue purposes, were used to provide biochar. Photographs by V.J. Bruckman.

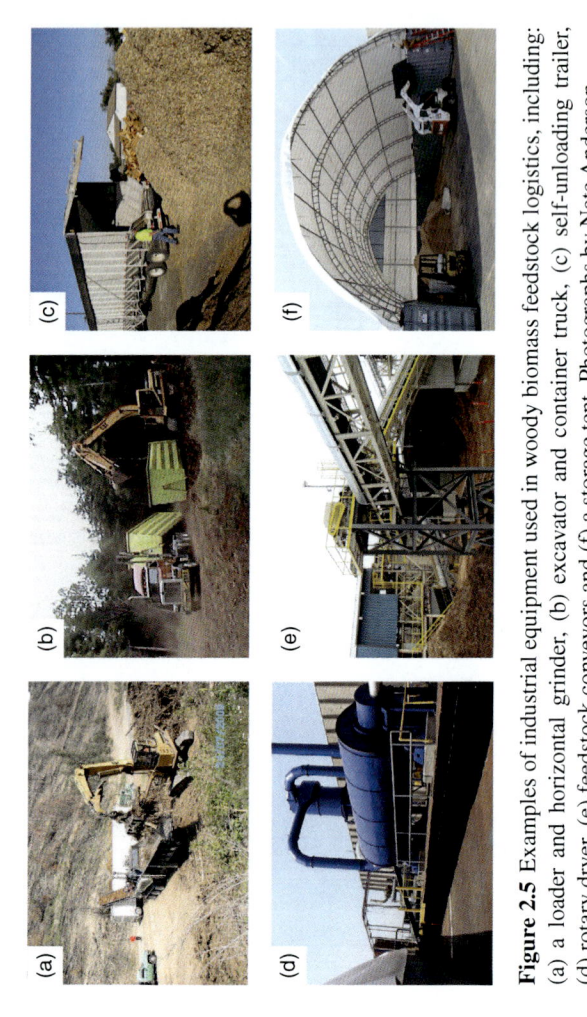

Figure 2.5 Examples of industrial equipment used in woody biomass feedstock logistics, including: (a) a loader and horizontal grinder, (b) excavator and container truck, (c) self-unloading trailer, (d) rotary dryer, (e) feedstock conveyors and (f) a storage tent. Photographs by Nate Anderson.

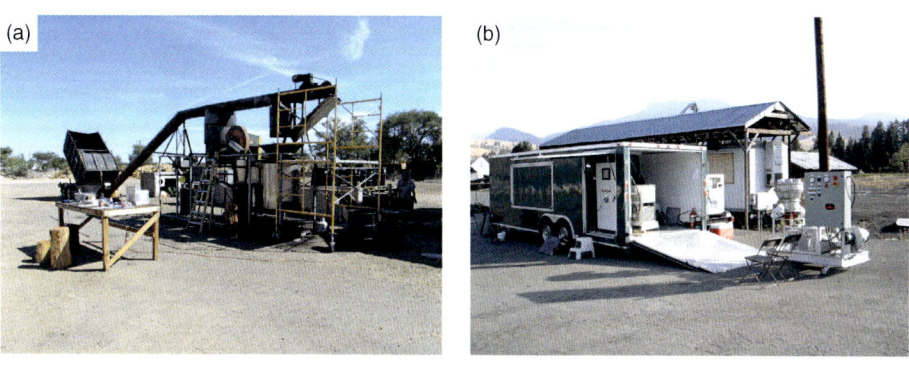

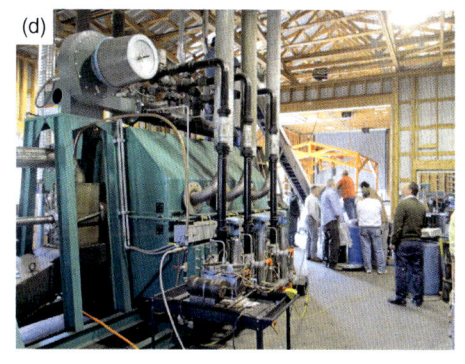

Figure 2.7 Examples of mobile and distributed-scale pyrolysis conversion systems producing co-products with biochar: (a) biochar and heat, (b) biochar with low-energy gas and bio-oil, (c) biochar with low-energy gas and bio-oil and (d) biochar with medium-energy gas. Photographs by Nate Anderson.

Figure 2.8 A six-wheeled forwarder, normally configured to carry logs, here mounted with a modified pellet spreader to apply biochar pellets on forested sites developed by the Missoula Technology Development Center, Missoula, MT. Photo by Han-Sup Han.

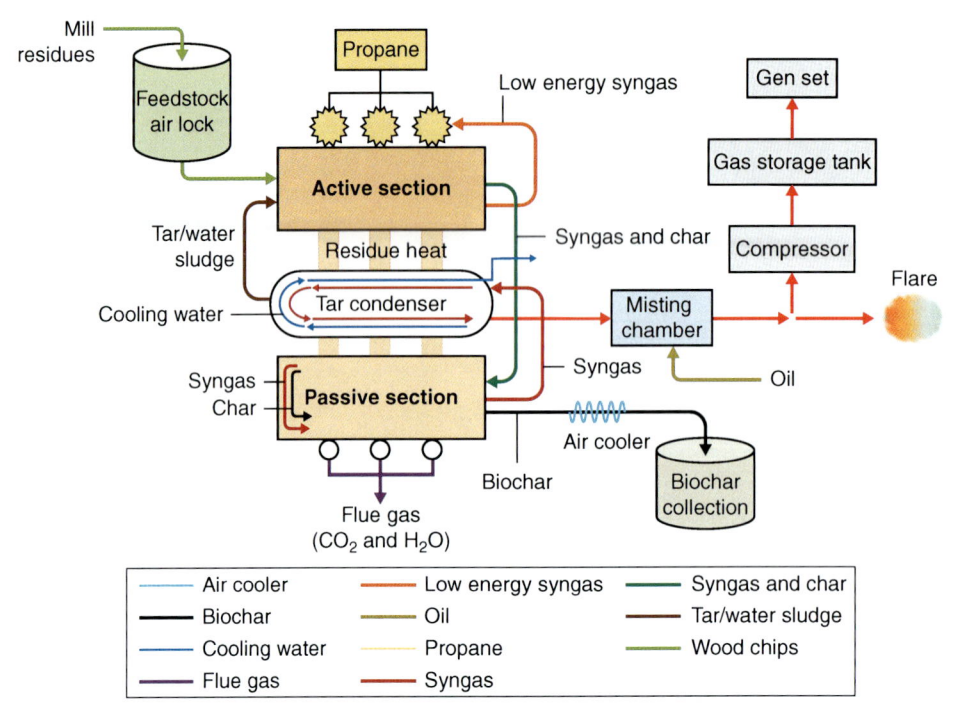

Figure 3.4 Process diagram for the Tucker Renewable Natural Gas unit.

Figure 7.4 Coppice-with-standards stand in Austria (photo Viktor Bruckman).

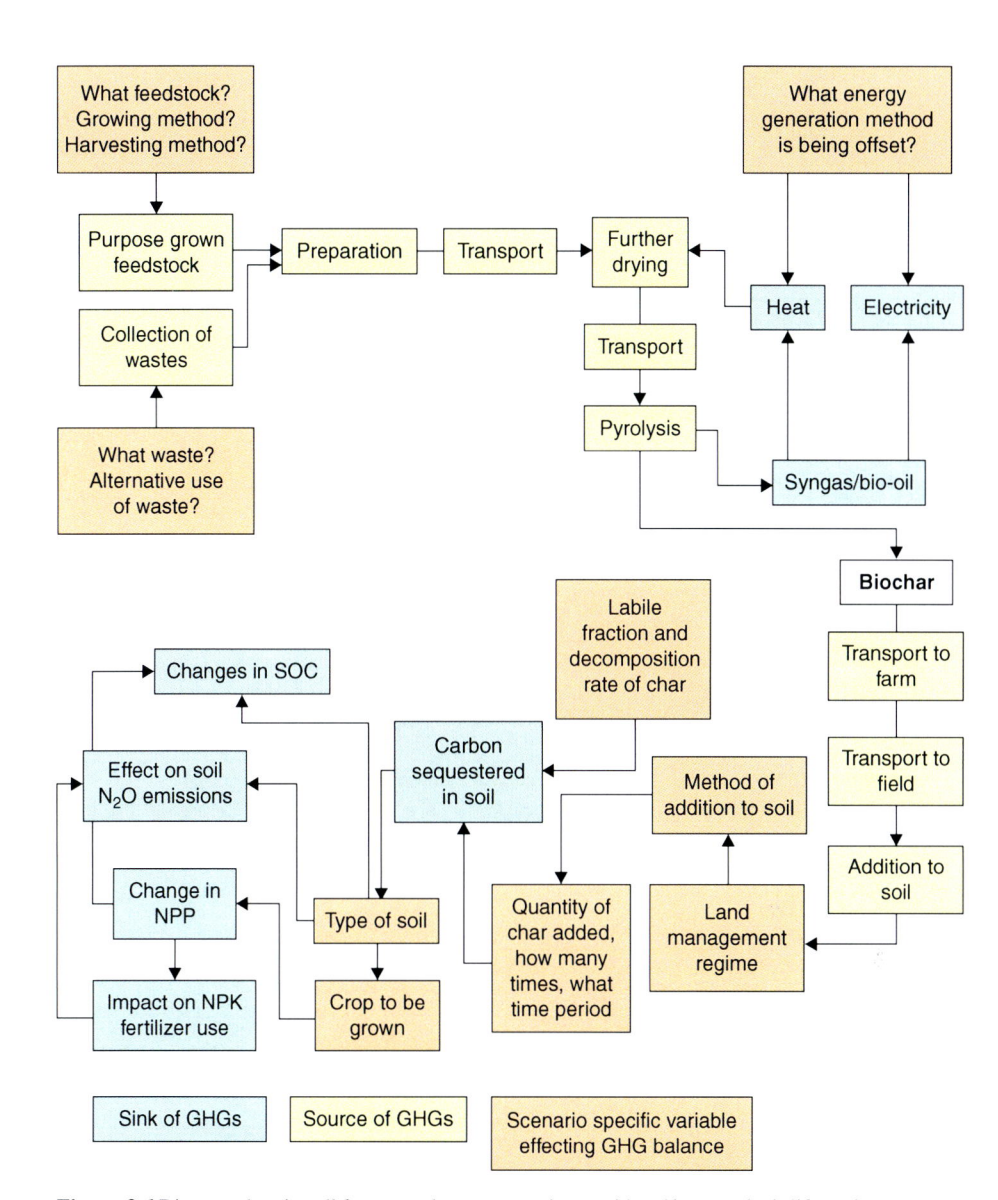

Figure 3.6 Diagram showing all factors and processes to be considered in a pyrolysis life cycle assessment.
Source: Hammond et al. (2011): permission to use granted by Elsevier through Copyright Clearance Center's Rightslink service.

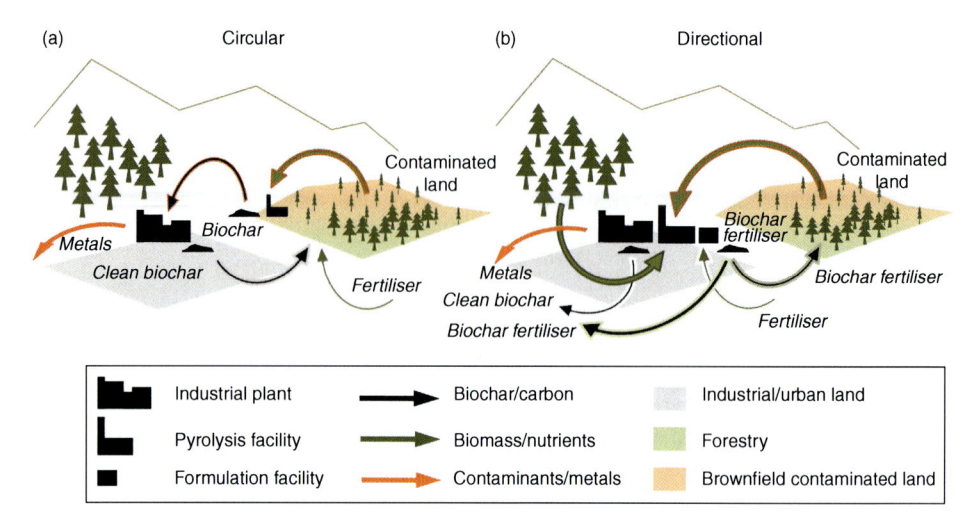

Figure 4.1 Biochar in scenarios for the remediation of contaminated brownfield land: (a) biomass of wood hyper-accumulator species grown on contaminated sites is pyrolysed with the biochar extracted for metals recovery, amended with nutrients or microbial inoculums and used to support further phyto-extraction-/localisation (or facilitated degradation) of organic contaminants where present (circular system); (b) remediation of land accelerated by optimising biochar using other forest sources of biomass, with cleaned biochar from hyper-accumulators used in other fertiliser products (directional system).

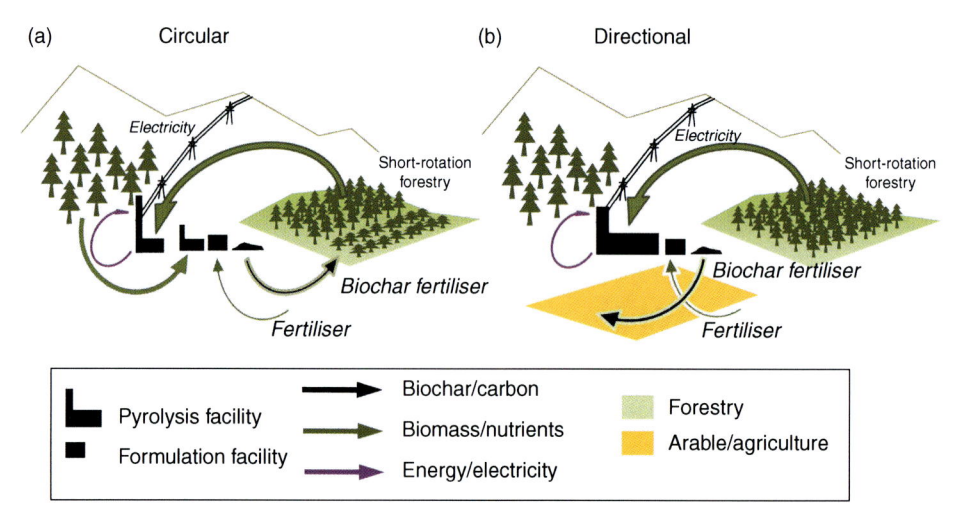

Figure 4.2 Biochar in scenarios involving short-rotation forestry: (a) processing of minor ash-rich fractions of trees grown mainly for bioenergy using pyrolysis (possibly augmented by the same from other forestry), with biochar used to convey and slowly recycle minerals to newly established stands (circular system); (b) ash-rich biochar used in horticulture as a slow-release, nutrient-enhanced ingredient in growing media, substituting peat and chemical nutrients (directional system).

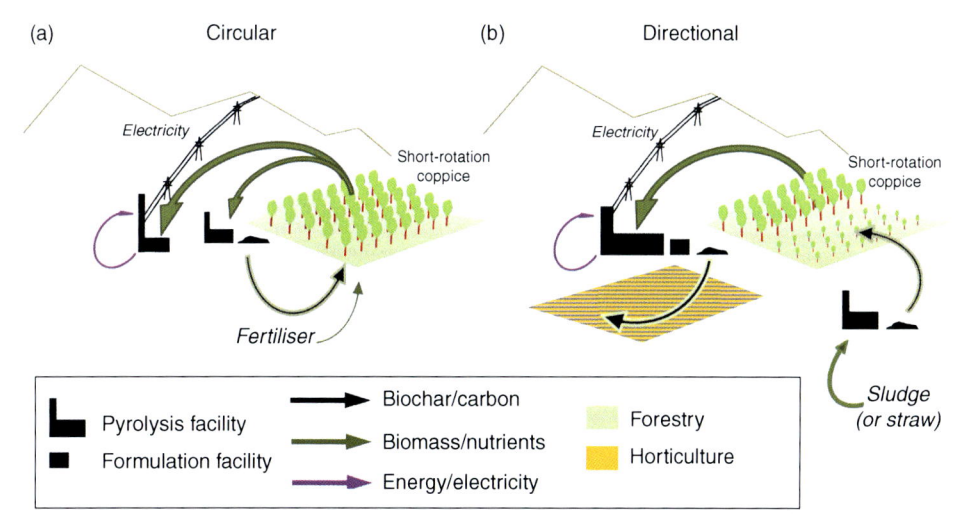

Figure 4.3 Biochar in scenarios involving short-rotation coppice: (a) part of the harvested biomass is pyrolysed rather than combusted, the biochar used to compensate for any carbon loss in land conversion, return mineral nutrients and improve nitrogen cycling/use efficiency (co-applied with chemical fertiliser at each harvest) (circular system); (b) harvest biomass pyrolysed in large facilities with lower electricity yield than the combustion alternative, but biochar used as a nitrogen carrier in distributed mainstream fertiliser; and coppice establishment supported by biochar from alternative feedstock (directional system).

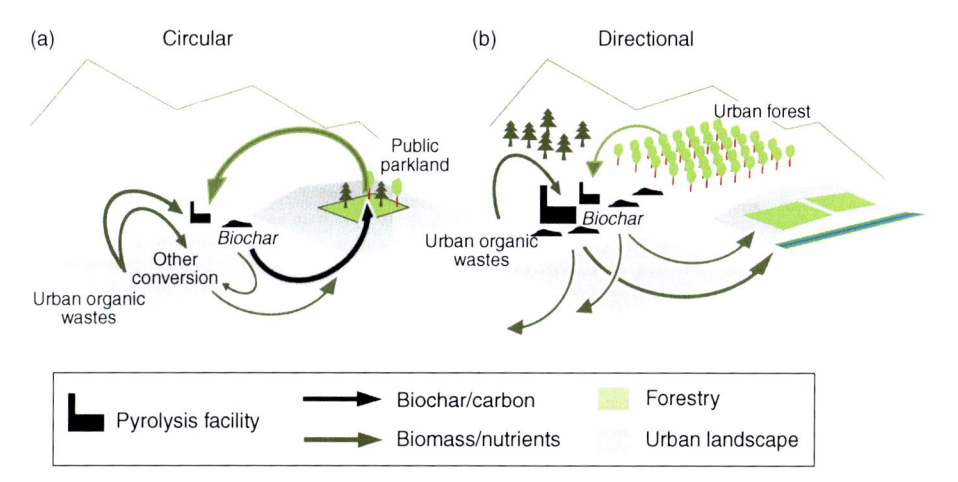

Figure 4.4 Biochar in scenarios involving urban forestry and green waste: (a) waste biomass from parkland management pyrolysed to create biochar for use in park landscaping in synergy with compost etc., produced also using biochar as an ingredient (circular system); (b) community pyrolysis facility used to convert urban forest and green waste, biochar used in enhancing services provided by urban environment (directional system).

Figure 7.5 Oak standard trees in (a) France (Forêt-de-Brin, and (b) Romania (Târgu-Mureş area) (photos V.-N. Nicolescu).

Figure 8.4 (a) Rice seedlings germinated on biochar from rice husk and (b) stacked on a rice seedlings transplanter.

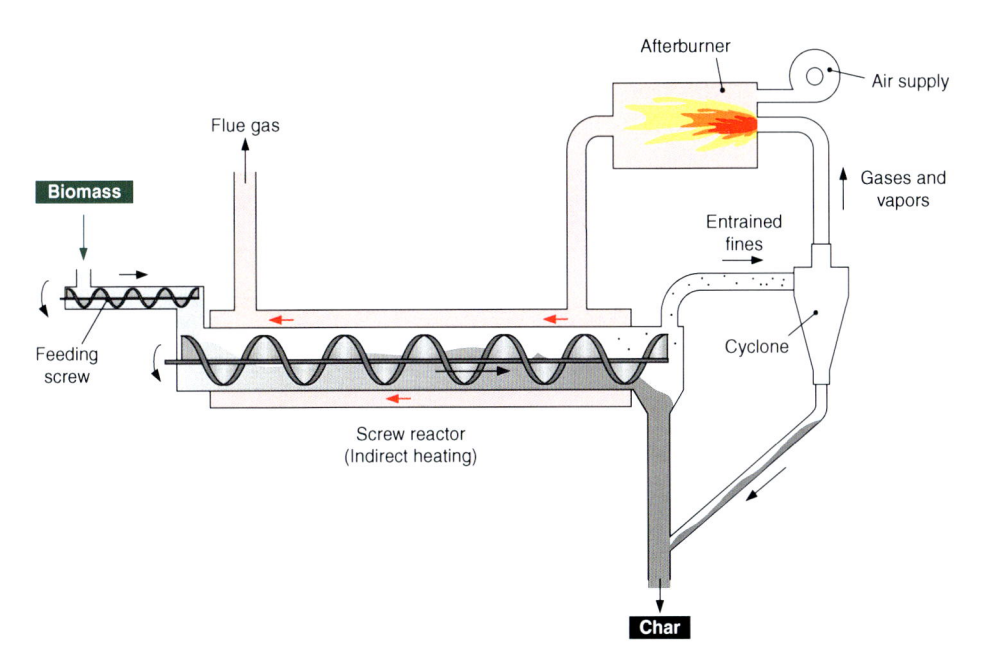

Figure 10.4 Slow pyrolysis continuous screw reactor, indirectly heated with the combusted pyrolysis gases and vapors.

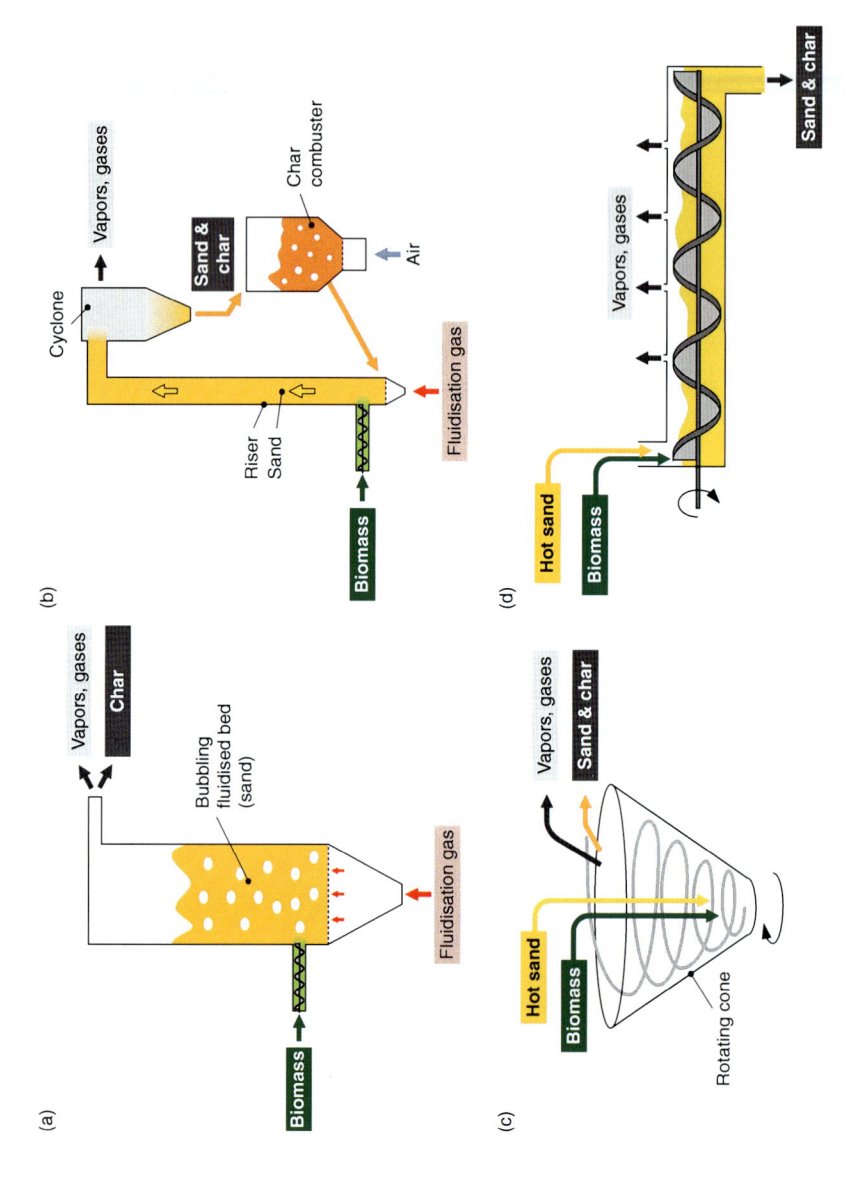

Figure 10.5 Fast pyrolysis reactor types: (a) bubbling fluidised bed, (b) circulating fluidised bed, (c) rotating cone reactor and (d) screw or auger reactor.

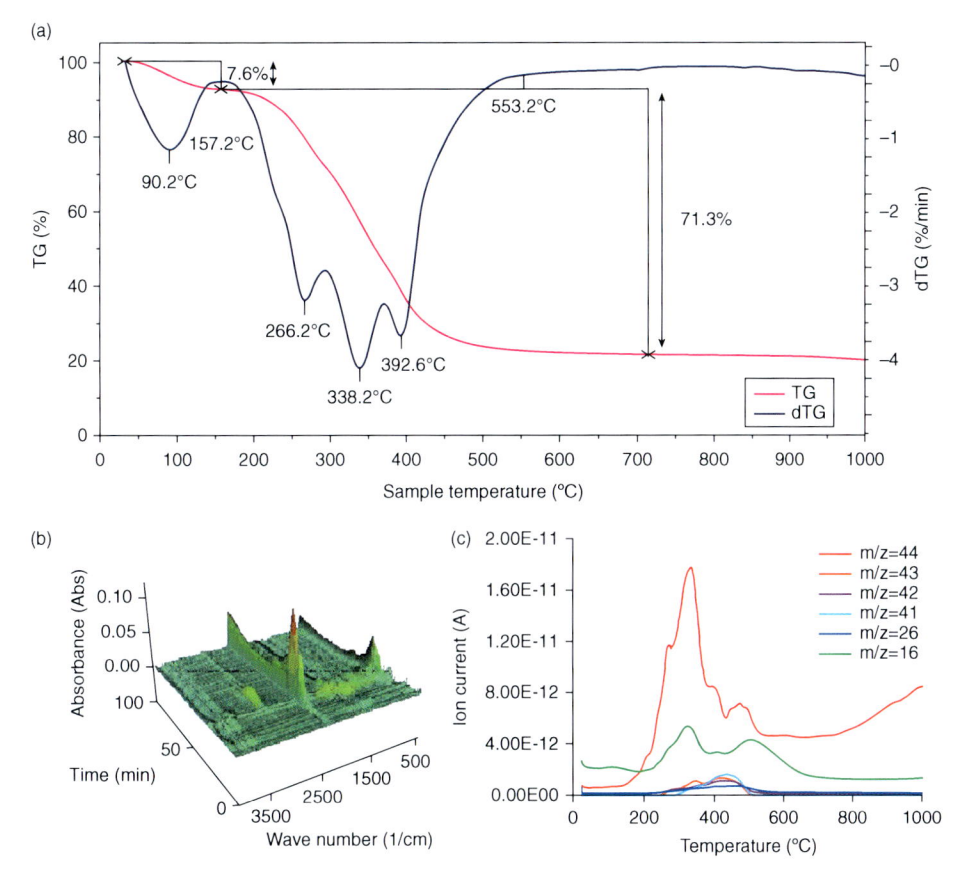

Figure 12.2 Thermal decomposition behavior of a biomass: cottonseed (a) TG-dTG curves, (b) 3d FT-IR spectra, (c) single ion current curves for the selected fragments. [Reprinted from the *Journal of Analytical and Applied Pyrolysis*, 105, E. Apaydın-Varol, B. B. Uzun, E. Önal, A. E. Pütün, Synthetic fuel production from cottonseed: Fast pyrolysis and a TGA/FT-IR/MS study, 83–90, Copyright (2014), with permission from Elsevier.]

Figure 13.4 The Thai-Iwate kiln is based on Japanese technology and it is widely used in Asia. This particular kiln type allows efficient temperature control. (a) Detailed view of a freshly constructed Thai-Iwate kiln with bamboo sticks supporting the dome during cement curing. (b) The Thai-Iwate kiln can be operated with various dimensions of feedstock material. (c) The front opening is tightly closed prior to pyrolysis with two rows of bricks and a hollow space in between to exactly control the airflow from a vent. Photo credit (a–c): Maliwan Haruthaithanasan.

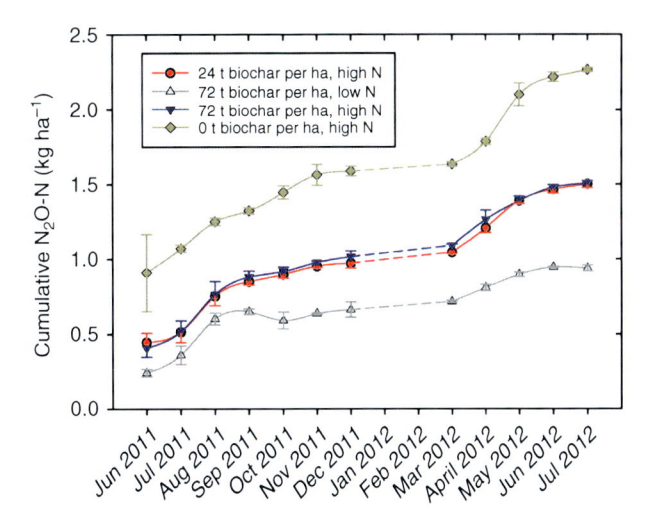

Figure 14.4 Cumulative nitrous oxide fluxes from differently biochar-amended plots (see graph legend) of a field experiment on a Chernozem in the period July 2011–July 2012.

Source: Kitzler et al., in preparation.

Figure 16.2 Growth response to mycorrhizal inoculation and different biochar application rates (10 and 20 Mg ha^{-1}) of sour orange (*Citrus aurantium*) seedlings.

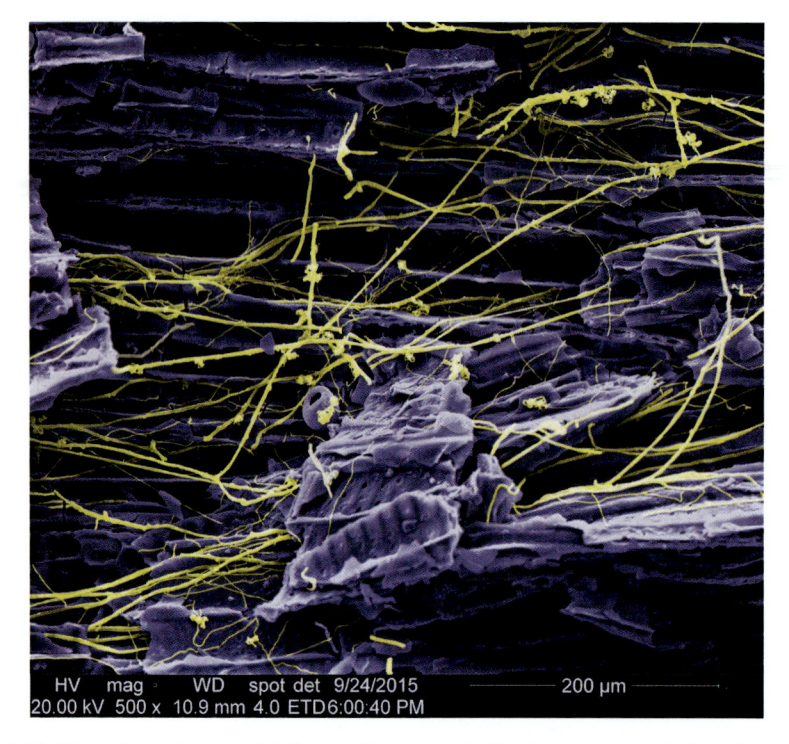

Figure 16.4 Charred woody material from a forest site in Northern Austria, derived from ~10 cm cm soil depth. The material was identified as 110 years old *Picea abies* charcoal from slash burning. Intensive colonization with mycorrhiza and the frequent occurrence of testaste amoebae (round object in the center) suggests an active role in maintaining soil functions.

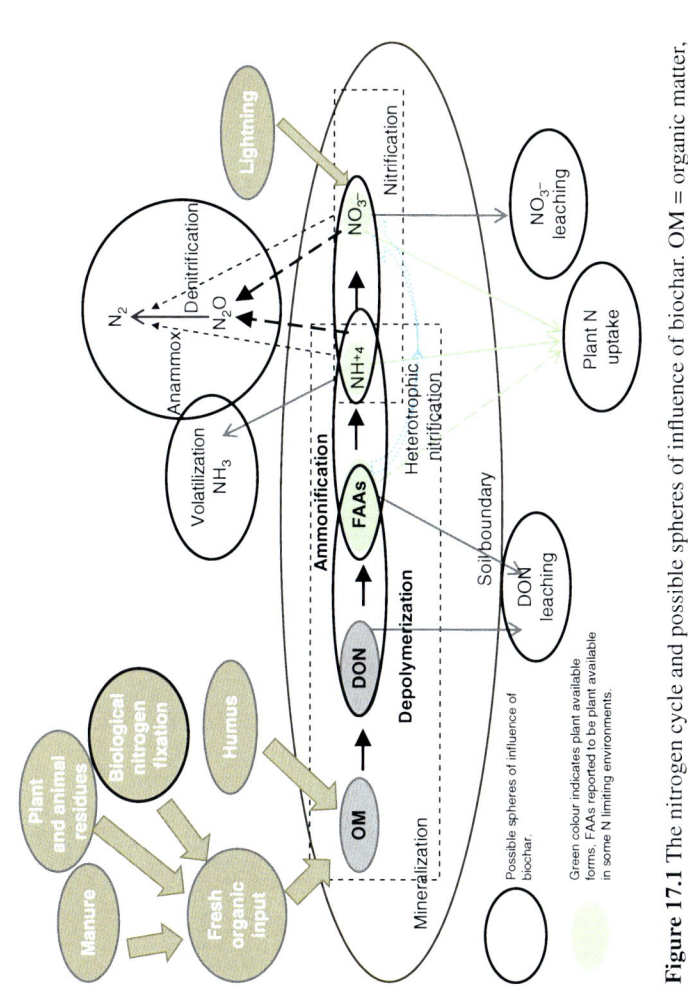

Figure 17.1 The nitrogen cycle and possible spheres of influence of biochar. OM = organic matter, FAAs = free amino acids, DON = dissolved organic nitrogen. Black arrows indicate microbially mediated processes.

Figure 18.4 Biochar production process: (a) = arranging biomass vertically in the retort for producing biochar; (b) = fuelwood is being placed horizontally in the space between the retort and the furnace; (c) = biomass is being heated to a temperature of about 450–600°C in the retort, after ignition and closing of the furnace; (d) = the retort can be opened after a cooling down phase of about three hours.

12

Pyrolysis: a Sustainable Way From Biomass to Biofuels and Biochar

BAŞAK B. UZUN, ESIN APAYDIN VAROL AND
ERSAN PÜTÜN

Abstract

Biomass provides 14% of the world's primary energy production, but it is largely wasted by inefficient and unsustainable use. To exploit the full potential of this energy source, new approaches and modern technologies such as pyrolysis and gasification are needed. Pyrolysis is the most promising thermal decomposition method for the conversion of biomass into valuable bio-products. The process produces a solid fraction (biochar), a liquid fraction (bio-oil) and a mixture of gases. Depending on pyrolysis conditions, biochar for soil amendment, activated carbon, carbon fibers, bio-fuels, value-added chemicals (PF type adhesives, phenolics, levoglucosan, octane enhancers, fertilizers) and gas products (hydrogen, methane, ethane and propane) could be achieved. The ratio of the products varies with the chemical composition of the biomass and operating conditions such as pyrolysis temperature, heating rate, reactor configuration, pyrolysis atmosphere, reaction time, particle size and so on. In the scope of this issue, this chapter covers the definition and sources of biomass, thermal behavior of biomass and its components, fundamentals of the pyrolysis process, and effects of the process parameters on yields and composition of products. Moreover, properties of bio-oil and biochar are explained according to their utilization areas.

12.1 Introduction

Energy is the major requirement of present societies for their sustainable development. The world's population will increase from today's 7.3 billion people to 9.7 billion in 2050 and 11.2 billion by the year 2100, and it is estimated that the demand for energy will increase by 2.8 times the present demand (American Statistical Association, 2015; IPCC, 2015). Thus, significant consumption and depletion of fossil fuels and the evident global warming lead countries to find alternative energy sources to fossil fuels and to develop new technologies. Among the renewables, biomass can be considered to be one of the most popular and diverse. It can also be named as "green waste" and it is an attractive source due to its ease of availability, high carbon content, low moisture and ash content, low or no sulfur content, low or even no cost, low or no conflict arising from alternative usage, solving solid waste disposal problems and keeping the environment clean (Islam et al., 2013).

Biomass is an extremely abundant resource, which is produced in agriculture and forestry or could also be collected as waste residue. It is renewable and environmentally friendly and can be converted into biofuels as an efficient alternative substitute for fossil fuels. Biomass comprising 47% of the total renewable energy consumption is the single largest renewable energy resource currently being used (Mohan et al., 2006). In contrast to fossil fuels, the use of biomass as an energy source provides significant environmental benefits. The most important advantage is that biomass fuels are carbon neutral or even carbon negative. During plant growth, atmospheric carbon dioxide is captured by photosynthesis and thus carbon dioxide emissions from the thermal degradation processes are balanced. Currently, there is no commercially viable method to offset the carbon dioxide added to the atmosphere (and the resultant greenhouse effect) this way. The climate change effects of carbon dioxide from fossil fuels are now generally recognized as a potential serious environmental problem. To meet the goals of the Kyoto agreement, 15 member countries of the European Union were committed to reducing their collective greenhouse gas (GHG) emissions in the period 2008–2012 to 8% below levels in 1990. The United States must reduce GHG emissions to a level 7% below the 1990 emissions in the same period. Japan established the Kyoto Protocol in 2002, and has committed to reducing its GHG emissions by 6% below its 1990 levels. Carbon dioxide is one of the dominant contributors to the increased concentration of GHGs. Two-thirds of the total anthropogenic CO_2 emissions originate from the combustion of fossil fuels, with the balance attributed to land use changes (Mohan et al., 2006).

Pyrolysis is the thermal decomposition of any organic material into liquids, gases and solid residue (char) in the absence of oxygen. Pyrolytic products can be used as fuels with or without prior upgrading and utilized as feedstock for chemical/material industries. Because of the nature of the process, yield of useful products is higher as compared to other processes (Sadaka, 2015).

In this context, this chapter covers the definition of biomass and its thermal decomposition behavior in terms of its constituents, pyrolysis process and its products. Liquid product (bio-oil) and solid product (biochar) will be discussed in detail. The effect of various pyrolysis parameters on product yields and characteristics will also be illustrated with examples from the literature.

12.2 Biomass

The term biomass is ascribed to biological materials derived from living, or recently living, organisms. The chemical composition of biomass is complex and very different from that of fossil fuels. The major constituents can be listed as cellulose, hemicellulose, lignin, organic extractives and inorganic minerals, where the composition depends on the type of biomass. Cellulose is the most abundant renewable organic high molecular weight (300,000 and 500,000 g/gmol) molecule in nature and can be symbolized as $(C_6H_{10}O_5)_n$. Cellulose fibers provide wood's strength and comprise approximately 40–50 wt% of dry wood. Hemicellulose, being a mixture of various polymerized monosaccharides, is the second major chemical constituent comprising approximately 25–35% of the mass of dry wood. Hemicelluloses are chemically related to cellulose and generally heterogeneous, built up of different hexoses (C6-sugars) and pentoses (C5-sugars). The third major component of biomass is lignin, which accounts for 23–33% of the mass of softwoods and 16–25% of the mass of hardwoods. It is an amorphous cross-linked resin with no exact structure deposited by woody plants for the purpose of providing the growing plant with mechanical support. Benzene rings, methoxy-, hydroxy- and propyl- groups are joined together to form the chemical structure of lignin. Biomass also contains inorganics, mainly K, Na, Ca, Mg and P, which end up in the pyrolytic ash. Organic extractives are the last group of constituents of biomass, including fats, waxes, alkaloids, proteins, phenolics, simple sugars, pectins, mucilages, gums, resins, terpenes, starches, glycosides, saponins and essential oils. Also, biomass has a substantial amount of water both in free and bound forms. The lignocellulosic complex holds water in the form of fibers, vessels and other anatomical parts (Theander, 1985; Glasser, 1985; Lipinsky, 1985; Klass, 1998; McKendry, 2002; Mohan et al., 2006; Apaydın-Varol and Mutlu, 2013).

Biomass can be found in many different forms, and is generally categorized as follows:

i) forest biomass: from forestry, arboricultural activities or from wood processing;
ii) energy crops: high yield crops grown specifically for energy applications;
iii) agricultural residues: residues from agriculture harvesting or processing (stalks, branches, leaves, straw, waste from pruning and so on) and biomass from the byproducts of the processing of agricultural products (residue from cotton ginning, olive pits, fruit pits etc.);
iv) industrial waste and co-products: from manufacturing and industrial processes;
v) animal waste: from intensive livestock operations, from poultry farms, pig farms, cattle farms and slaughterhouses;
vi) aquatic crops: such as algae, giant kelp, other seaweed and marine microflora.

When biomass is considered to be a renewable energy source, in the long term, diverse technologies are needed to make use of these potential energy sources. The most common technologies involve biochemical, chemical and thermochemical conversion processes. As seen in Figure 12.1, thermochemical technologies are combustion, pyrolysis, liquefaction and gasification. Among them, pyrolysis is the most promising route for the production of bio-oils and biochar because it is possible to produce three different product yields depending on the process conditions via this method (McKendry, 2002).

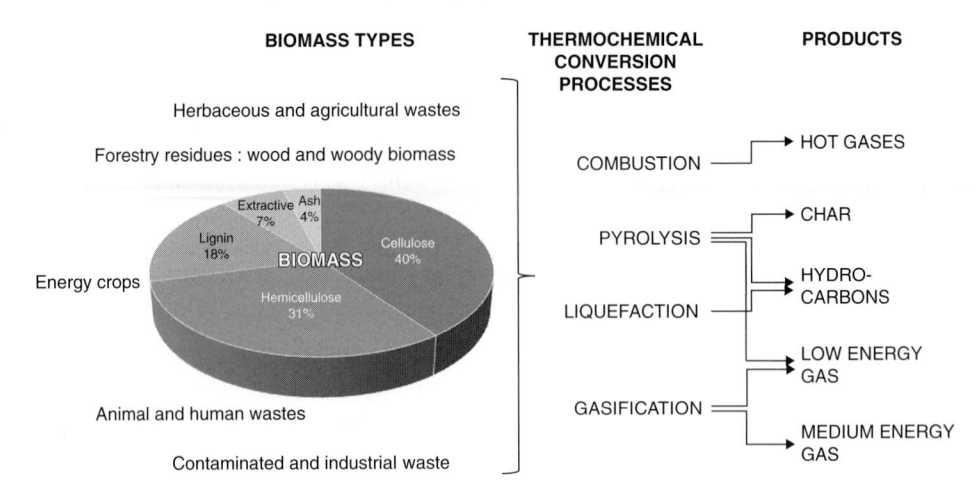

Figure 12.1. Thermochemical conversion technologies: from biomass to valuable products.

12.2.1 Thermal Decomposition Behavior of Biomass

Understanding the thermal behavior of biomass is fundamental to thermochemical conversions and mostly to pyrolysis. It is known that the contents in biomass such as cellulose, hemicellulose, lignin, ash and extractives are some of the important parameters for the evaluation of pyrolysis characteristics (Lv et al., 2010).

Thermogravimetric analysis (TGA) has been widely used for pyrolysis research studies. Thermogravimetry involves measuring the mass of a sample as its temperature increases. Suitable samples for thermogravimetry are solids (reactants) that undergo one of the two general types of reactions:

$$\text{Reactant(s)} \rightarrow \text{Product(s)} + \text{Gas}$$

$$\text{Gas} + \text{Reactant(s)} \rightarrow \text{Product(s)}$$

The first process involves a mass loss, whereas the second involves a mass gain. Pyrolysis of biomass, as shown below, resembles the first type of reaction.

$$\text{Biomass} \rightarrow \text{Solid char} + \text{Liquid bio-oil} + \text{Gaseous products}$$

Biomass and biomass constituents like cellulose, hemicellulose and lignin have been pyrolyzed in thermogravimetric analyzers under different conditions to investigate the reaction parameters and kinetics of pyrolysis reactions. TGA, differential thermal analysis (DTA) and differential scanning calorimetry (DSC) or combinations of these are commonly used for this purpose. Recently, the combination of TGA with Fourier transform infrared spectrometry (FT-IR) and mass spectrometry (MS) has attracted attention for both the determination of thermal degradation profiles and the detection of evolved gases during pyrolysis (Seebauer et al., 1997; Sorum et al., 2001; Suarez-Garcia et al., 2002; Apaydın et al., 2003; Sanchez-Silva et al., 2012; Apaydın-Varol et al., 2014).

Several studies investigated the pyrolysis mechanism of biomass in terms of its three main components: cellulose, hemicellulose and lignin. To better understand the mechanism of biomass pyrolysis either different biomass feedstocks containing various amounts of these constituents or the constituents individually have been examined. Due to their larger proportions, cellulose and hemicellulose are the dominant constituents in typical biomass and hence their effect on the pyrolysis mechanism is significant. Depending on the heating rate, the main weight loss of hemicellulose starts at temperatures of about 220°C, whereas cellulose decomposition starts at about 310°C. The decompositon of hemicellulose and cellulose is completed around temperatures of 315 and 400°C, respectively (Shen and Gu, 2009). Lignin thermally decomposes over a broad temperature range between 160 and 900°C. Since various functional groups of oxygen from its structure have different thermal stabilities, their scission occurs within a wide temperature range. As a result, lignin generates high char yields at typical pyrolysis temperatures (Brebu and Vasile, 2010; Dorez et al. 2014).

Figure 12.2a shows the thermogravimetric and derivative thermogravimetric curves for a biomass sample, cotton seed, when heated to 1000°C with a heating rate of 10°C min^{-1} under nitrogen atmosphere using a thermogravimetric analyzer (Setaram-Labsys). The temperature at which the decomposition appears to start and end, as well as the shape of the curve, depends upon many factors. Some of these are heating rate, heat of reaction, furnace atmosphere, the amount of the sample, nature of the sample container, particle size and packing of the sample. However, typical biomass decomposition shows four main peaks for moisture release and decompositions of hemicellulose, cellulose and lignin. Several studies showed that the superposition of the last three peaks is commonly observed (Dorez et al., 2014). In Figure 12.2a, it can be observed from the dTG curve that there are four peaks at different temperature ranges. The first peak, around 90°C, is obviously responsible for the moisture retained in the biomass. The following three peaks are due to main decomposition reactions indicating the presence of three major biomass constituents, namely hemicellulose, cellulose and lignin, and also the presence of a significant amount of oil in the cotton seed. On the other hand, lignin decomposes in a broad range of temperatures, completing its decomposition around 550°C. From the TG curve, total weight loss for the decomposition of cotton seed between temperatures of 25–550°C is observed to be 77.9% (Apaydın-Varol et al., 2014).

As mentioned earlier, recent studies include not only thermal behavior but also the detection of evolved gases during pyrolysis. For this purpose, TGA is coupled with an FT-IR for functional group analysis and MS. Figures 12.2b and 12.2c present the FT-IR and MS results for the evolved gases during the thermal decomposition of cotton seed. Pyrolytic reactions start at about 200°C and simultaneously related products such as CO_2, CO, CH_4, H_2O, ketones, aldehydes, acids and other hydrocarbons are formed (Figure 12.2b). Three-dimensional FT-IR data show that C=O stretching vibrations related with CO_2 (band between 2210–2400 cm^{-1}) have their maximum at 345°C, whereas the C-H stretching vibration band (2840–3061 cm^{-1}) assigned to the presence of hydrocarbon gases, mainly CH_4, has its maximum absorbance at 460°C. H_2O peaks (O-H stretching vibration band between 3100 and

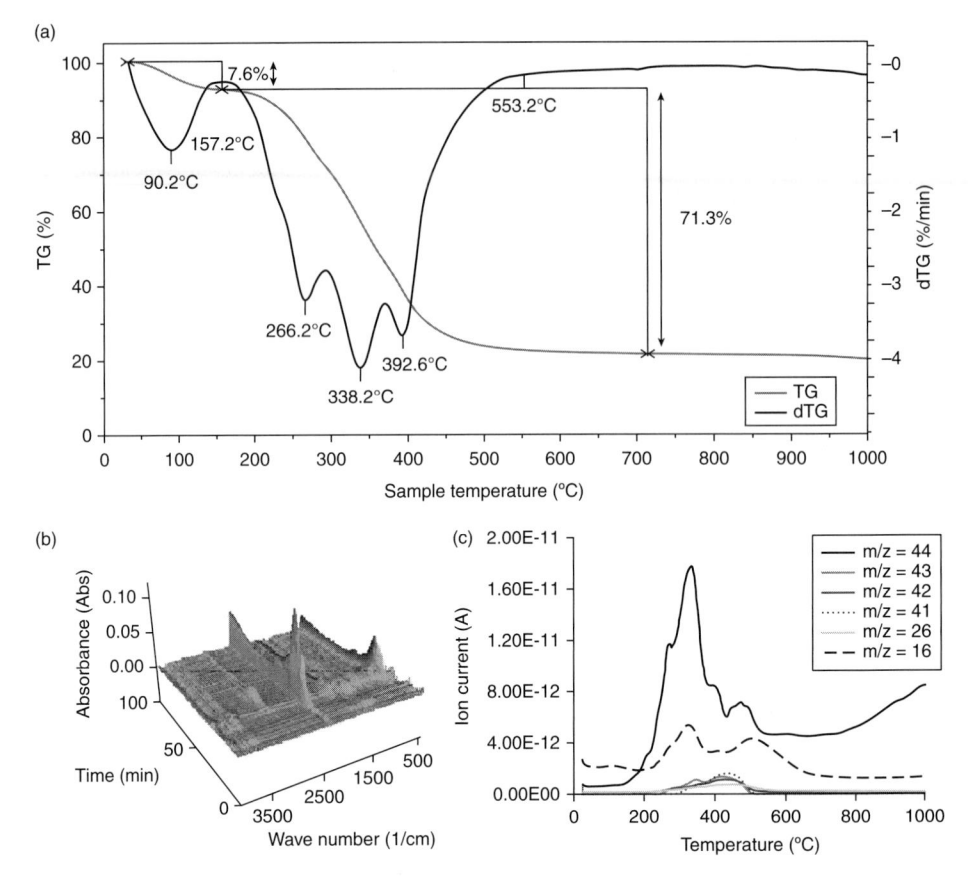

Figure 12.2. Thermal decomposition behavior of a biomass: cottonseed (a) TG-dTG curves, (b) 3d FT-IR spectra, (c) single ion current curves for the selected fragments. [Reprinted from the *Journal of Analytical and Applied Pyrolysis*, 105, E. Apaydın-Varol, B. B. Uzun, E. Önal, A. E. Putun, Synthetic fuel production from cottonseed: Fast pyrolysis and a TGA/FT-IR/MS study, 83–90, Copyright (2014), with permission from Elsevier.] (A black and white version of this figure will appear in some formats. For the colour version, please refer to the plate section.)

3500 cm^{-1}) are seen in two sections indicating i) the moisture release from the biomass (temperatures between 85 and 140°C) and ii) water formed as a result of many parallel and consecutive reactions (temperatures between 185 and 330°C). Carbonyl and C=C stretch vibration bands seen between 1690 and 1850 cm^{-1} have two maximum peaks at temperatures of 290 and 353°C indicating the release of aldehydes and acids.

Functional group information about the evolved gases during pyrolysis can be detected via FT-IR. However, mass spectroscopy has an important role in the determination of exact composition. Figure 12.2c represents the single ion current curves for the main pyrolysis products obtained during the pyrolysis of cotton seed. Intensive signals of the evolved gases appear within the temperature range of 187 and 523°C, being consistent with

TGA measurements. Carbon dioxide with a mass to charge ratio (m/z) of 44 has the highest concentration among the others and its maximum value is obtained at 330°C. Formation of methane can be confirmed by the m/z of 16, and its peak in Figure 12.2c starts at 216°C, where pyrolytic reactions mainly due to the decomposition of cellulose and hemicellulose start. The signals of m/z 18 can be assigned to H_2O and the moisture release and water formation due to dehydration reactions. The formation of hydrocarbon gases such as C_2H_5 and CHO (m/z = 29), C_3H_5 (m/z = 41), C_3H_6 and C_2H_2O (m/z = 42), C_3H_7 and CH_3CO (m/z = 43) can be also identified from MS data (Apaydın-Varol et al., 2014).

12.3 Pyrolysis

The definition of pyrolysis is generally expressed as the thermal decomposition of any organic material in the absence of oxygen. Since ancient Egyptian times, this technology has been used for the production of various products, for example pitch for embalming purposes and for waterproofing boats, acetic acid for vinegar production, acetone for smokeless gun powder, methanol and turpentine from wood pyrolysis and so on. Pyrolysis processes have been improved and are widely used for coke and charcoal production. As explained in previous chapters, among the various pyrolysis technologies two of them gain attention as the main pyrolysis processes depending on the temperature range and residence times. These are i) slow and ii) fast pyrolysis. Slow pyrolysis or carbonization is mainly used for production of char. In recent decades, it is also used for energy production, such as heat and electricity. In the 1980s, researchers found that the pyrolysis liquid yield could be increased using fast pyrolysis technologies, where a biomass feedstock is heated at a rapid rate and the vapors produced are also condensed rapidly. The bio-oil obtained via fast pyrolysis can be upgraded for subsequent use in diesel engines and turbines, blending with other fuels as a fuel or for heating purposes in boilers or value-added chemicals (Carpenter, 2014).

12.3.1 Slow Pyrolysis

Conventional slow pyrolysis has been applied for thousands of years and mainly used for the production of charcoal at low temperatures and low heating rates. In this process, typically biomass (wood) is heated slowly with a heating rate lower than 10°C min^{-1} to temperatures generally between 350 and 500°C, and the vapor residence time varies between five and 30 minutes. Vapors should be continuously removed as they are formed to produce high yields of bio-oil but in this case vapors do not escape as rapidly as they do in fast pyrolysis. Thus, components in the vapor phase continue to react with each other to form high yields of the solid char, especially at low pyrolysis temperatures (below 500°C). Depending on the type of feedstock, 20–40 wt% liquid oil, 20–35 wt% solid char and 30–40 wt% gaseous products are achieved typically as a result of slow pyrolysis. However, slow pyrolysis has some limitations in terms of high-quality bio-oil production. Due to high residence times (5 to 30 min), cracking of the primary volatiles occurs and hence bio-oil yield and quality is affected adversely. Moreover, long residence times and low heat transfer rates demand extra energy input (Bridgwater and Meier, 1999; Bridgwater et al., 2001).

Kilns are generally used in conventional slow pyrolysis processes. In a simple process, liquid and gas products are often not collected. Developments through the end of the nineteenth century led to industrial scale processes using large retorts, operated either in batch or continuous modes to allow recovery of organic liquid products and recirculation of gases to supply process heat recovery. Modern pyrolyzers are designed to capture the volatiles for the production of bio-oil to generate important organic liquid products, especially acetic acid and methanol. In the late twentieth century, slow pyrolysis technologies for biochar production based on horizontal tubular kilns, agitated drum kilns, rotary kilns and screw pyrolyzers were introduced (See Chapter 10). Although some of these technologies have well-established commercial applications, the development of integrated systems that produce energy and biochar at high efficiency is still mainly at the research scale (Antal and Grønli, 2003; Sohi et. al., 2010).

12.3.2 Fast Pyrolysis

Fast pyrolysis is a relatively high-temperature and rapid process in which biomass is heated in the absence of oxygen. The essential features of a fast pyrolysis process are: i) very high heating and heat transfer rates, which usually requires a finely ground biomass feed; ii) carefully controlled pyrolysis reaction temperature with short vapor residence times; and iii) rapid cooling of the pyrolysis vapors to give the bio-oil product (Bridgwater and Meier, 1999).

Typical temperatures between 500–550°C are chosen for fast pyrolysis processes, with a heating rate of about 100–300°C sec^{-1} and a very short residence time (1–5 seconds) to reduce the formation of intermediate products and increase the yield of bio-oil. During this fast heating, biomass decomposes into vapors, aerosols and some charcoal-like char. After cooling and condensation of the vapors and aerosols, a dark brown mobile liquid (so-called bio-oil) is formed with a lower heating value than conventional fuels (IEA Bioenergy Task 34, 2015). During fast pyrolysis, no waste is generated, because the bio-oil and solid char can each be utilized in various areas and the gas can be recycled back into the process. Depending on the biomass used, typically fast pyrolysis processes produce 60–75 wt% of bio-oil, 15–25 wt% of char and 10–20 wt% of noncondensable gases if the temperature is kept around 500°C in order to limit the reaction of vapor cracking, facilitating condensation and the subsequent formation of liquid. Intermediate pyrolysis liquid products are obtained by higher heating rates and rapid quenching. Also, char formation is minimized with higher reaction rates. Under some proper conditions, no char is observed. When the pyrolysis temperature is increased, gas product is obtained with high yields (Mohan et al., 2006; Uzun et al., 2007).

12.4 Pyrolysis Products

The three primary products obtained from the pyrolysis of biomass are solid char, permanent gases and condensable vapors as bio-oil. Under proper conditions of slow pyrolysis at around

500°C final temperature, approximately equal amounts of these products are obtained, as well as water that should be removed to improve the quality of bio-oil. In this section, bio-oil and biochar yields and properties will be discussed (Jahirul and Rasul, 2012).

12.4.1 Bio-oil

Fast pyrolysis is a common process for high yields of bio-oil production. However, significant amounts of bio-oil are also produced in slow pyrolysis, as described earlier. Liquid products obtained from pyrolysis are generally called bio-oils, pyrolytic oils, biocrude oil, wood oil, liquid wood, pyrolysis liquids, pyrolytic tar and so on (Czernik and Bridgwater, 2004; Mohan et al., 2006). Utilization of bio-oil in ancient Egyptian times was to produce tar for caulking boats and embalming agents by pyrolysis. Today, bio-oil seems to be a good renewable energy source that may replace conventional fuels due to its negligible sulfur content, no net CO_2 emission to the atmosphere and easy transportation and storage. Bio-oil has moderately higher heating values in the range of 15 to 38 MJ kg^{-1}, and hence it is useful as a fuel either for adding to petroleum refinery as feedstock or by upgrading to produce petroleum-grade refined fuels (Mohan et al., 2006; Islam et al., 2010; Uzun and Sarioğlu, 2009).

The chemical composition of bio-oil is very complex due to the heterogeneity arising from the feedstocks, that is due to the presence of different molecules derived from the depolymerization and fragmentation of cellulose, hemicellulose and lignin. More than 300 types of major and minor organic compounds can be detected in bio-oil, which include acids, alcohols, ketones, aldehydes, phenolics, hydroxyaldehydes, carboxylic acids, hydroxyketones, ethers, esters, sugars, furans, nitrogen compounds and multifunctional compounds. Basically, bio-oil is comprised of water, guaiacols, catecols, syringols, vanillins, furancarboxaldehydes, phenolics, isoeugenol, pyrones, acetic acid, formic acid and other carboxylic acids (Mohan et al., 2006; Islam et al., 2010).

12.4.1.1 Bio-oil Yields and Effect of Various Parameters on Bio-oil Production

Bio-oil yields strongly depend on the type of raw material and on the applied pyrolysis conditions. Moisture content, cellulose/lignin ratio, particle size and presence of inorganics can be listed as the feedstock-related effects, whereas pyrolysis temperature, heating rate, pyrolysis atmosphere and type, and reactor geometry are the process-related effects. Bio-oil yield can be calculated in different ways, however generally it is expressed in dry ash-free (daf) terms, and hence it is the ratio of bio-oil obtained from pyrolysis over daf weight of the feedstock. Yields of bio-oil from various biomass sources are in the range of 30–95 wt%, depending on the factors listed above.

The composition of biomass is the primary factor affecting bio-oil yield. It was demonstrated that high lignin contents in a biomass have a tendency to give lower bio-oil yields, while the presence of cellulose, hemicellulose and hexane-soluble extractives in high amounts increases the bio-oil yield (Pütün et al., 2005; Mohan et al., 2006). Table 12.1

demonstrates a list of bio-oil yields from different feedstocks with a variety of compositions. Raw materials with higher hexane-soluble fractions such as olive oil residue and soybean cake are known to be advantageous since they produce more bio-oil with a higher quality, that is higher calorific value. Inorganic content of the raw material increases the formation of solid products by accelerating the solidification and dehydration reactions, and reduces the tar formation. Therefore, ash content of biomass is another parameter affecting the bio-oil yield (Bonelli et al., 2001; Apaydın-Varol and Pütün, 2012).

The effect of feedstock used in the pyrolysis is not limited to its composition. The heterogeneous physical structure of biomass, such as its moisture content and particle size, also has an important role. The particle size can be homogenized by pre-grinding and screening processes prior to pyrolysis. Likewise, the moisture content can be lowered to the desired value by drying the feedstock. Removal of moisture from the raw material is also essential for further steps since it requires less energy to reach the targeted pyrolysis temperature with a dry feedstock (Zaror and Pyle, 1982).

The effect of pyrolysis temperature on bio-oil yield might be considered as the secondary factor. During the thermal degradation of biomass up to temperatures of 150°C, feedstocks seem to be stable and only the moisture on the surface is released. The decomposition of hemicellulose and cellulose starts at about 220 and 310°C respectively and after 450°C decomposition of lignin becomes dominant. Therefore, choosing pyrolysis temperatures higher than 450°C is preferred when bio-oil is the target product. As seen in Table 12.1, previous studies showed that for various biomass samples a temperature between 500–550°C gives almost maximum yields of bio-oil. Higher heating rates also accelerate bio-oil formation. Although it gives low amounts of biochar and gas products, fast pyrolysis is the most suitable process for bio-oil production.

Pyrolysis atmosphere is an important factor for product distribution. Especially, fast pyrolysis is operated under nitrogen atmosphere. It is known that the presence of a sweeping gas during pyrolysis enhances bio-oil formation by minimizing the secondary reactions such as recondensation, repolymerization and char formation (Pütün et al., 2004). Some studies showed that using steam as the sweeping gas favors the formation of bio-oil significantly. Steam not only sweeps the primary products from pyrolysis reactions, but also reacts with the products and helps the evaporation of oils from biomass and this reduces the formation of solid char (Pütün et al., 2004; Pütün et al., 2005; Özbay et al., 2008). Table 12.1 also shows that bio-oil yields of more than 30% were achieved from slow pyrolysis of agricultural waste (rice straw) and industrial waste (apricot pulp) when steam was used as the pyrolysis atmosphere.

The pyrolysis process is composed of many parallel and consecutive reactions and hence the presence of catalysts might influence the power of these reactions. Pütün et. al. (2009) studied the effect of various zeolites, either natural or synthetic, on fast pyrolysis of olive oil residue. It was reported in the same study that catalytic treatment with synthetic zeolites provided enhancement of the gas yields and reduction of the bio-oil yields. However, it was also noticed that after application of the catalytic treatment, the long chains of alkanes and alkenes present in the bio-oil were converted

Table 12.1. *Bio-oil yields for various biomass samples via pyrolysis*

Biomass	Cellulose + hemicellulose (wt%)	Lignin (wt%)	Hexane soluble oil (wt%)	Pyrolysis temperature (°C)	Reactor type	Heating rate (°C/min)	Pyrolysis atmosphere	Catalyst	Maximum bio-oil yield (wt%)	Heating value (MJ/kg)	Reference
Pinewood sawdust	–	–	–	500	Conical spouted bed reactor	Fast pyrolysis	Nitrogen	–	75.0	14.6 (Lower heating value)	Amutio et al., 2012
Hard wood or soft wood feedstocks	–	–	–	450	Tubular vacuum pyrolysis reactor	Slow pyrolysis	Nitrogen	–	50.0–55.0	–	Ortega et al., 2011
Rice straw	60.27	14.07	1.9	550	Fixed-bed reactor	5	Steam	–	35.86	32.58	Pütün et al., 2004
Apricot pulp	–	–	–	550	Fixed-bed reactor	5	Steam	–	27.7	35.63	Özbay et al., 2008
Olive residue	58.83	34.98	5.46	500	well-swept and high-speed heated fixed-bed batch reactor	500	Nitrogen	–	46.72	29.6	Pütün et al., 2009
Euphorbia rigida	–	–	–	550	Fixed-bed reactor	7	Static	Criterion-534	30.98	–	Ateş et al., 2005

Table 12.2. *Comparison of bio-oil with light and heavy fuel oils in terms of some typical physical properties*

Property	Bio-oil	Light fuel oil	Heavy fuel oil
Moisture content (wt%)	15–30	–	0.1
Ash content (wt%)	<0.02	<0.01	0.03
Specific gravity	1.2	–	0.94
pH	2–3	–	–
Viscosity at 50°C (centistokes)	7	4	50
Pour point (°C)	–33	–15	–18
Distillation residue (wt%)	<50	–	<1
Higher heating value (MJ/kg)	20–30	35–37	38–40
Elemental composition (wt%)			
C	54–58	–	85
H	5.5–7.0	–	11
N	0–0.2	0	0.3
S	Negligible	0.15–0.5	0.5–3.0
O	35–40	–	1.0

Source: Czernik and Bridgwater (2004); Mohan et al. (2006).

to lower-weight hydrocarbons. An increase of aliphatics, aromatics and olefins and a sharp decrease of asphaltenes and polar groups (highly oxygenated groups) were determined (Pütün et al., 2009).

12.4.1.2 Properties of Bio-oil

Bio-oils are usually dark brown, free-flowing organic liquids that are comprised of highly oxygenated low and high molecular weight compounds. The most distinctive property of bio-oil is its smoky odor. The chemical composition of bio-oil, which arises from the volatilization of biomass constituents, affects its properties. Since bio-oil is produced from lignocellulosic biomass, its elemental composition resembles that of biomass rather than that of petroleum oils. The general properties such as moisture content, pH, specific gravity, elemental composition, higher heating value, viscosity, distillation residue, pour point and flash point of bio-oils and conventional petroleum fuels are compared in Table 12.2. Typical bio-oil properties are very poor in terms of moisture content, pour point, distillation residue, higher heating value and elemental composition compared with commercial fuels. The most important difference in elemental composition is ascribed to the higher oxygen content for bio-oils, which also reduces the higher heating value (Czernik and Bridgwater, 2004; Uzun and Sarioğlu, 2009). However, bio-oils still have higher calorific values (20–30 MJ kg^{-1}) than those of biomass. Due to having no or trace amounts of N and S, bio-oil is cleaner than commercial fuels. To compete with them, its fundamental properties such as water or oxygen content, viscosity and pH should be improved in a refinery.

Bio-oil is a highly viscous and complexly structured organic liquid containing oxygenated compounds such as polycyclic hydrocarbons, phenols, fatty acids, carboxylic compounds and so on. The main reason for the complex structure is the uncontrolled decomposition of lignin that produces high amounts of monocyclic and polycyclic aromatic compounds. Polycyclic aromatic hydrocarbons (PAHs) can be formed during incomplete combustion reactions and are the compounds formed from the combination of two to five aromatic rings. It is well known that PAHs are dangerous for human health due to carcinogenic effects (Bru et al., 2007; Tsai et al., 2007). In general, the organic compounds constituting the bio-oil can be listed as guaiacols (methoxy phenols), cathecol, syringol, vanillin, furancarboxaldehyde, eugenol, isoeugenol, pyrones, indene, acetic acid, formic acid and other carboxylic acids. Also, hydroxyaldehydes, hydroxyketones and phenolic compounds constitute a large portion of the chemical structure of bio-oils. Oligomeric compounds in the bio-oil are mainly formed as a result of lignin decomposition and partly cellulose decomposition. The molecular weight distribution of these oligomeric compounds ranges from 60 to 300 depending on the biomass type, pyrolysis temperature, heating rate, reaction time and other factors (Mohan et al., 2006). Due to the variety in chemical composition, bio-oils show a wide range of boiling temperatures, of between 80 and 280°C (Czernik and Bridgwater, 2004).

The high oxygen (about 35–40%) and water (about 20–25%) contents limit the utilization of bio-oil as a conventional fuel. The oxygen in bio-oil is in the form of water, organic acids, alcohols, aldehydes, ketones and phenols. The content of oxygen affects its stability by changing its properties (pH, viscosity, density, homogeneity etc.) and its behavior between hydrocarbon fuels and biomass pyrolysis oils. It is actually immiscible with liquid hydrocarbons, because of its high polarity and hydrophilic nature (Mohan et al., 2006), which limits its utilization as fuel oil. The aromaticity of bio-oil is related to its hydrogen content. The lower amounts of hydrogen in bio-oil equals higher aromaticity. Due to the negligible amounts of sulfur, vanadium and nickel, bio-oil is an environmentally friendly fuel when compared to petroleum-derived fuels (Tsai et al., 2007).

The pH of bio-oil is relatively low, for example, the pH of the liquid product obtained from wood is about 3. The main reason for low pH value is the presence of organic acids (formic acid and acetic acid) in the composition (Bridgwater and Grassi, 1991). Therefore, bio-oils are corrosive to carbon steel, aluminium and other common construction materials. If other circumstances permit, polyolefins are usually preferred as materials of construction for bio-oil production (Czernik and Bridgwater, 2004).

The viscosity of bio-oils is generally between 35 and 1000 centipoise (cP) at 40°C. The wide range is due to the efficiency of collection of low boiling point components during the pyrolysis process, depending on the feedstock and process conditions. Also, bio-oil ages after its first recovery and this is observed as a viscosity increase. These undesired effects are observed when the oils are stored or handled. This can be explained by both chemical reactions between various compounds present in the oil, leading to the formation of larger molecules, and reaction with oxygen from air (Czernik and Bridgwater, 2004; Mohan et al., 2006).

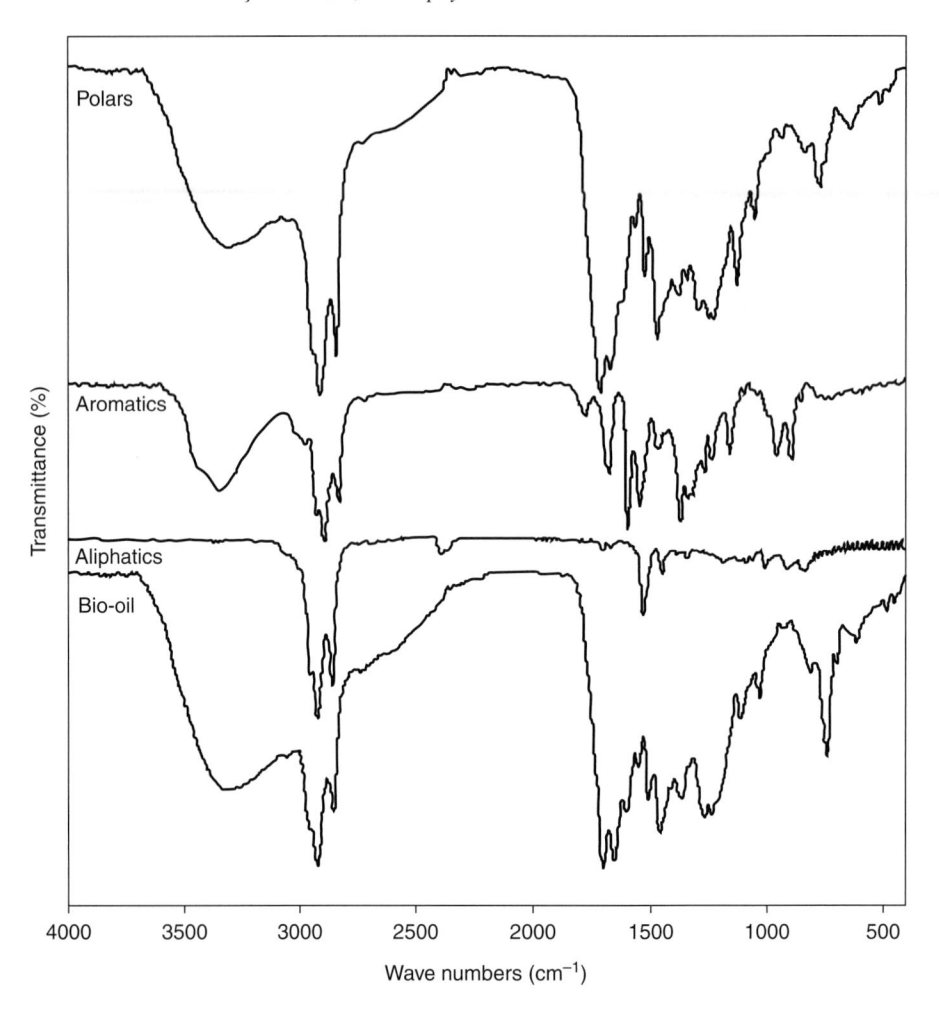

Figure 12.3. FT-IR spectra of bio-oil obtained from tea waste and its sub-fractions. (Reprinted from *Fuel*, 89, B. B. Uzun, E. Apaydın-Varol, F. Ates, N. Ozbay, A. E. Pütün, Synthetic fuel production from tea waste: characterisation of bio-oil and bio-char, 176–184, Copyright (2010), with permission from Elsevier.)

12.4.1.3 Characterization Techniques

Bio-oils contain a very wide range of complex organic chemicals, and hence their detailed characterization depends on mostly instrumental techniques. After the determination of basic properties of bio-oil that are described in the previous section, detailed analysis of bio-oil can be handled with some chromatographic and spectroscopic analyses. Typically, FT-IR, column chromatography, gas chromatography (GC), gas chromatography-mass spectroscopy (GC/MS), nuclear magnetic resonance spectroscopy (NMR), high-performance

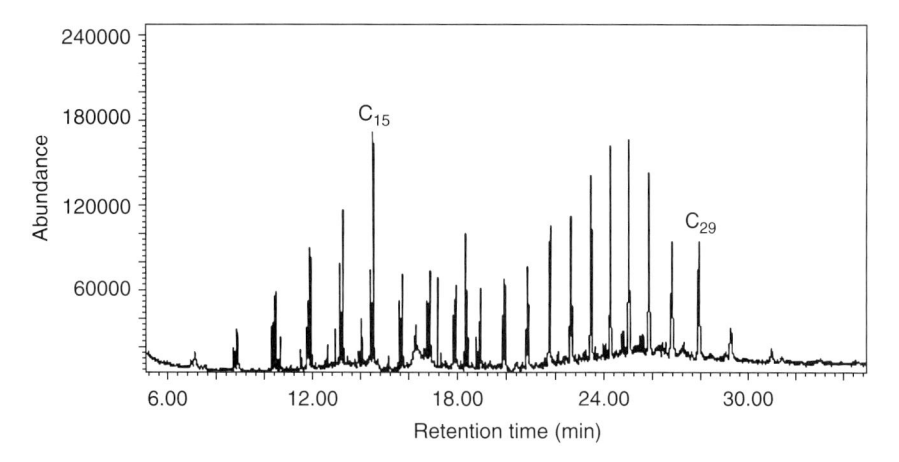

Figure 12.4. GC-MS chromatogram of aliphatic sub-fraction from tea waste bio-oil. (Reprinted from *Fuel*, 89, B. B. Uzun, E. Apaydın-Varol, F. Ates, N. Ozbay, A. E. Pütün, Synthetic fuel production from tea waste: characterisation of bio-oil and bio-char, 176–184, Copyright (2010), with permission from Elsevier.)

liquid chromatography (HPLC) and size-exclusion chromatography (SEC) are the analyses applied for the investigation of the chemical structure of bio-oils.

In a study of the characterization of bio-oil obtained from the fast pyrolysis of tea wastes, FT-IR spectra of the bio-oil and its sub-fractions are given for the functional group determination (Figure 12.3). The sub-fractions are obtained via column chromatography as described in the text. Mainly, it is mentioned in the study that the O–H stretching vibrations between 3200 and 3400 cm^{-1}, indicating the presence of phenols and alcohols does not exist in the aliphatic sub-fraction of bio-oil, and accordingly it is pointed out that the aliphatic sub-fraction does not contain oxygenated compounds like bio-oil. This result is also confirmed by elemental and GC/MS analyses in the study by Uzun et. al. (2010). A GC/MS chromatogram shows that the aliphatic fraction consists of n-alkanes, alkenes and branched hydrocarbons with a majority of the linear chain hydrocarbons distributed in the range of C_{11}–C_{29} (Figure 12.4). It is concluded in the same study that direct utilization of bio-oil as a synthetic fuel without any upgrading may cause many problems since it contains highly oxygenated compounds and low-high molecular weight hydrocarbons (Uzun et al., 2010).

12.4.2 Biochar

The International Biochar Initiative (IBI) describes biochar as: "a solid material obtained from the carbonisation of biomass." Typically, biochar is associated with slow pyrolysis and may be utilized as a solid fuel for energy production, or it may be used for environmental and agricultural purposes as an adsorbent or soil amendment (Sohi et al., 2010). Biochar is a complex material consisting of mainly carbon, inorganics and volatile hydrocarbons. The most important differences between biomass and its related biochar are mainly porosity, surface area, pore structure (micropores, mesopores and macropores) and physicochemical properties such as composition, elemental analysis and ash content. Generally higher contents of ash and solid carbon are observed for the chars than the original samples (Apaydın-Varol and Pütün, 2012).

As described in the previous paragraphs, biochar is an important product of the pyrolysis process. Preferentially, slow pyrolysis is applied if biochar is the target product since the product yield is relatively higher. However, fast pyrolysis or even gasification also derives significant amounts of biochar depending on the feedstock and the pyrolysis conditions. In any case, as a product or byproduct, biochar is a valuable carbonaceous material due to its higher calorific value, surface area, energy density, adsorption capacity and so on.

12.4.2.1 Effect of Various Parameters on Biochar Production

When the production of biochar is under consideration several parameters can affect the yield and quality, such as particle size, moisture content of the feedstock, process/reactor type, heating rate, final temperature and residence time. These effects can be grouped into two major classes: i) effects related to the feedstock and ii) effects related to the process (Table 12.3).

As mentioned earlier, biomass type has a significant effect on pyrolysis product yields and quality. Biomass constituents, namely hemicelluose, cellulose, lignin, ash and extractives, have a significant effect on thermal decomposition. Cellulose and hemicellulose decompose at lower temperatures, giving more volatiles. Since lignin is the constituent that decomposes in a wider temperature range, biomass with a high lignin content gives high biochar yields at moderate pyrolysis temperatures. There are few studies about the effect of inorganics on product yields. However, the inorganics (Si, Ca, K, etc.) in biomass do not volatilize at around 500°C, and hence they are still present in the final biochar, increasing the product yield. Under certain circumstances, it is beneficial to have inorganics in the biochar, for example if it is going to be utilized as a soil conditioner. But in general, the presence of inorganics is a disadvantage during the pyrolysis process since they might accumulate and lead to corrosion. Typically, the lignin content in woody biomass is high and ash content low in contrast to agricultural feedstocks. Therefore, some precautions should be taken if both biomass types are going to be utilized in the same process plant.

Other important factors for biochar production are reactor type, final temperature, residence time, heating rate and pyrolysis atmosphere, details of which were described in Chapter 10. Among these, pyrolysis temperature is the most important. When simple briquette, metal or concrete kilns are used for production, the average temperature that can be reached is around 450°C and it is not easy to control (Brown, 2009). Recently, modern techniques allow production of more controllable furnaces and hence the pyrolysis temperature can be optimized. Laboratory-scale studies showed that the increase in pyrolysis temperature decreases biochar yield, while accelerating the formation of both condensable and non-condensable volatiles. On the other hand, higher carbon content and higher porosity with larger surface areas are obtained at higher temperatures (Kloss et al., 2012; Sun et al., 2014). Therefore, it is important to set the main requirements of the biochar

Table 12.3. *Influence of feedstock specification and process conditions on biochar production*

Parameters		Degree	Expected result
Feedstock related	Hemicellulose+cellulose	Low (<55%)	Fewer volatiles, more biochar
	Lignin	High (>35%)	
	Ash	High (>8%)	
	Moisture content	Low (<10%)	
	Particle size	Low (< 2 mm)	More heat transfer, rapid decomposition, less biochar
Process related	Batch	–	More residence time, more biochar
	Continuous	–	Less residence time, rapid heating, more bio-oil
	Heating rate	Low (< 10°C/min)	Slow heating, more biochar
		High (> 300°C/min)	Rapid heating, more bio-oil
	Pyrolysis/carbonization temperature	Low (<400°C)	More biochar, less C content in biochar
		Moderate (~500°C)	Less biochar, more C content in biochar
		High (>700°C)	Less biochar, more gas products
	Residence time*	Low (0<t<10 min)	Less carbonization and more C content in biochar
		High (>1 h)	More carbonization and more C content in biochar

* Residence time refers to the holding time of biochar at the final pyrolysis temperature.

according to its targeted application before determining the final temperature. The heating rate is also a factor for biochar yield in pyrolytic processes. Although the most influential process parameters are pyrolysis temperature and heating rate, residence time also leads to obvious effects on product yields and characteristics. It is more usual to select a pyrolysis temperature around 500°C at long residence times (many hours to days) to maximize the biochar yield while increasing the surface area and fixed carbon, and changing the volatile matter and ash content (Wang et al., 2013).

For a general review of the effect of various parameters on biochar yield, carbon content and surface area, see Table 12.4. Biomass type is an effective factor for both the yield and final properties. Pyrolysis temperature also has a great influence on the yield either for fast

Table 12.4. List of some examples for biochar production from literature

Biomass type	Reactor type	Biochar production techniques	Biochar production temperature (°C)	Heating rate (°C min^{-1})	Reaction time (min)	Yield (%)	C (wt%)	Surface area m^2 g^{-1}	Reference
Pine cone	Fixed-bed reactor	Slow pyrolysis	550	10	–	29.6	95.16	208	Apaydın-Varol and Pütün, 2012
Soybean cake						25.2	83.95	2.1	
Corn stalk						24.9	94.97	11.8	
Peanut shell						29.7	93.61	211	
Tea waste	Fixed-bed tubular reactor	Fast/slow pyrolysis	400	300	10	43.4	–	2	Uzun et al., 2010
			500	300		35.7		–	
			550	300		30.2		–	
			700	300		21.1		7.5	
			500	5		34.3		–	
			500	500		34.2		–	
			500	700		27.1		–	
Pistachio shell	Fixed-bed tubular reactor	Fast/slow pyrolysis	400	300	–	29.5	–	–	Pütün et al., 2007
			500	300		21.7			
			550	300		20.5			
			700	300		15.4			
			500	5		28.0			
			500	100		24.7			
			500	500		22.0			
			500	700		20.9			
Olive residue	Fixed-bed reactor	Slow pyrolysis	400	7	–	32.4	–	–	Pütün et al., 2005
			500			28.8	69.34		
			550			28.0	–		
			700			27.5	–		
Rice straw	Fixed-bed reactor	Slow pyrolysis	400	5	–	30.5	–	–	Pütün et al., 2004
			500			27.6			
			550			26.1			
			700			23.6			

Oak wood	–	Slow pyrolysis	350			75.9	–	Nguyen and Lehmann, 2009
			600			88.4		
Fruit cuttings	–	Slow pyrolysis	600	10	60	37.5	76.2	431.3–	Agirre et al., 2013
			600		120	37	77.5	474.2	
			600		180	38	72.9		
			750		60	32.5	73.7		
			750		120	31	79.6		
			750		180	33	70.4		
			900		60	29	87.7		
			900		120	29	75.9		
			900		180	29.5	74.4		
Bagasse	Lindberg electric box with retort	Slow pyrolysis	350	–	120	36	75.2	–	Novak et al., 2012
			500			25	85.4		
Peanut hull			400		60	40	74.8	–	
					120	35	81.8		
Pecan shell			350		60	50	64.5	–	
					120	31	91.2		
Pine chip			350		120	42	74.7	–	
			500		120	32	87.2		
Poultry litter			350		60	56	46.1	–	
			700		120	37	44.0		
Barley straw	–	Slow pyrolysis	400	3°C/min	120	31	72.5	–	Sevilla et al., 2011

or slow pyrolysis. For the fast pyrolysis of tea waste and pistachio shell, when the temperature was increased from 400 to 700°C, biochar yields decreased significantly (by about 50%). A decrease was also observed for the same temperature range for slow pyrolysis, however in this case the percentage was not so high. One may also observe from the table that longer residence times resulted in similar or lower biochar yields. In general, the carbon content of biochar is inversely related to biochar yield. Elemental analysis of all biochars showed that they contain a minimum of 70% carbon, which is higher than that of a typical biomass. The carbon contents of biochars slighty increased at higher pyrolysis temperatures. However, no direct correlation is seen between residence time and carbon content.

12.4.2.2 Physicochemical Properties of Biochar

The utilization of char/biochar can be roughly divided into two major groups, the first being energy applications and the second being environmental applications. Generally the solid product of pyrolysis is called char/charcoal when it is used as a solid fuel or as a porous carbon for many purposes, and called biochar if it is going to be utilized in environmental processes such as a soil additive to improve productivity or to provide a potential sink for carbon. However, biochar is best described as a soil conditioner. The use of biochar as a soil conditioner has been proposed as a means to simultaneously mitigate climate change while improving soil fertility. The properties of biochar strongly determine the final management aim. Therefore, it is necessary to know the physical and chemical properties of any biochar prior to amendment. These are mainly determined by the feedstock and the pyrolysis conditions (Antal and Gronli, 2003; Woolf, 2008; Verheijen et al., 2010).

The physicochemical behavior of a biochar is mostly affected by the relative proportion of carbon, volatiles, moisture and mineral matter. Verheijen et. al (2010) stated in their study that for a typical biochar the relative proportions for fixed carbon, volatile matter, moisture and ash are in the ranges of 50–90%, 0–40%, 1–15% and 0.5–5% respectively.

The variety of the composition of biochar influences the heterogeneity and hence surface chemistry. The presence of inorganic elements has a great contribution to the heterogeneity of the surface and reactivity in the soil. This hetorogeneity arises from both the stable and labile components of biochar. The presence of functional groups results in the acidic/basic and hydrophilic/hydrophobic character of the surface. Hydroxyl (-OH), ketone (-OR), aldehyde (-(C=O)H), carboxyl (-(C=O)OH), ester (-(C=O)OR), amino (NH_2) and nitro (NO_2) are the main functional groups that enchance surface chemistry of biochar and interact with organic and inorganic compounds during the utilization of biochar in the soil or in other environmental applications. Pre- or post-treatment of biochar with chemical agents disturbs the structure resulting in a tailored biochar in terms of surface chemistry. This process is called chemical activation and is typically used not only to improve surface chemistry but also to increase porosity (Sohi et al., 2010; Verheijen et al., 2010; Kılıc et al. 2012).

The porosity of a biochar gives information about its pore size distribution, density and surface area. Although the surface morphology is modified during thermal treatment as a result of devolatilization, the vascular structure of the raw material maintains itself and contributes to the formation of macropores (pore diameter > 500 nm). While macropores have been important for many cases in soil applications, micropores (pore diameter < 50 nm) account for the large surface areas. Microporosity in biochar is affected by the type of feedstock and process parameters. At higher pyrolysis temperatures and longer holding times larger areas can be obtained due to the formation of more micropores. Another way to increase microporosity is to apply chemical activation using impregnating chemical compounds such as H_3PO_4, KOH, $ZnCl_2$, K_2CO_3, H_2SO_4 and so on (Apaydın-Varol and Erülken, 2015).

Cation exchange capacity (CEC) and pH are other important properties of biochar that are also related to surface chemistry and heterogeneity. The CEC of the biochar is a measure of the quantity of negatively charged sites on the surface that can retain positively charged ions (cations) such as calcium (Ca^{2+}), magnesium (Mg^{2+}) and potassium (K^+) by electrostatic forces. Cations retained electrostatically are easily exchangeable with cations in the soil solution, thus a soil with a high CEC has a greater capacity to maintain adequate quantities of Ca^{2+}, Mg^{2+} and K^+ than a soil with a low CEC (Ross and Ketterings, 2011). Therefore, it is an important issue to improve the quality of a soil by enhancing its CEC by using biochar with a high CEC as a soil additive. According to Lehmann (2007), CEC variation in biochars ranges from almost zero to around 40 cmol/g.

12.4.2.3 Characterization Techniques

The IBI biochar standards identify three categories (A, B and C) of tests for biochar materials. Category A and Category B are for basic utility properties and toxicant reporting respectively, and required for all biochars. Category C is optional and composed of advanced analysis and soil enhancement indicators.

The required characteristics include the physical properties of particle size and moisture, as well as the chemical properties of elemental composition (H, C and N) as weight percentages, ash proportion, electrical conductivity and pH/liming ability. In addition to basic properties, all biochar materials must meet the soil toxicity assessment thresholds. Toxicants such as metals and polychlorinated biphenyls may arise from the feedstock and toxicants such as PAHs and dioxins may be produced during the thermochemical conversion process.

Furthermore, if biochar is utilized as a soil amendment the last category (C) characteristics should be determined. These advanced analysis characteristics include the volatile matter content, nutrient content and surface area of biochars (IBI Standards, 2013).

Besides these characteristics, there are many instrumental methods used for detailed analysis of biochar such as scanning electron microscopy (SEM), FT-IR, X-ray photoelectron spectroscopy (XPS), energy-dispersive X-ray spectroscopy (EDX), near-edge X-ray absorption fine structure (NEXAFS) spectroscopy and NMR, which have been used to examine the surface morphology and chemistry of biochar. Also, recently many new methods have been developed for the determination of ageing and stability (Cross and Sohi, 2013).

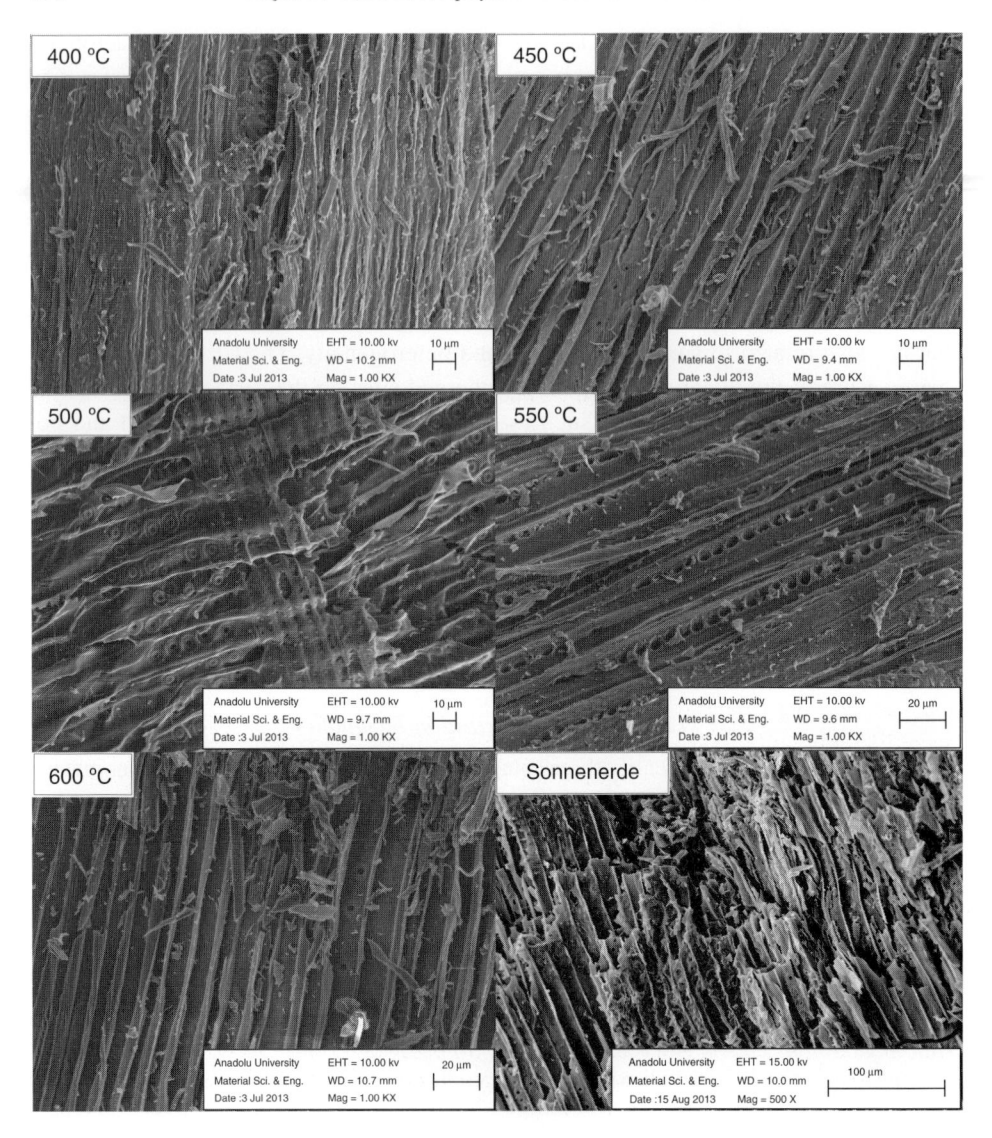

Figure 12.5. SEM images of biochars obtained from spruce wood via slow pyrolysis at different temperatures and the biochar produced at Sonnenerde Company – Austria.

More commonly, surface morphology is observed by SEM for biochars to gain insights about the porous structure before and/or after utilization. Figure 12.5 shows SEM images of biochars obtained from Austrian spruce wood at temperatures of 400, 450, 500, 550 and 600°C via slow pyrolysis under a nitrogen atmosphere using a laboratory-scale batch type pyrolysis reactor, details of which were given in previous studies (Pütün et al., 2005). As the pyrolysis temperature increases, most of the volatiles are removed from biomass, which results in deformation on the surface of the particles. However no major morphological changes are observed except for breakages and new pore openings, and it is detected that the lignocellulosic structure of the biomass is retained during pyrolysis. Figure 12.5 also includes SEM imagery of biochar obtained from a commercial company, Sonnenerde, located in Riedlingsdorf, Austria. The Sonnenerde biochar was also produced from Austrian spruce at 500–600°C using a continuous reactor (Dunst, 2015). It is seen from Figure 12.5 that Sonnenerde biochar has a porous structure with many open canals on the surface arising from the biomass.

12.5 Conclusions

After the oil crises of the 1970s–1980s and strong increases in demand for oil and a limited supply response to rising prices in the mid-2000s, a significant amount of work has been carried out worldwide on the development of new and renewable energy production systems. In this regard, biomass pyrolysis is one of the most promising processes not only in terms of energy supply, but also with regard to the production of valuable chemicals and biochar.

Bio-oil and biochar can be produced from a wide range of organic feedstocks under different pyrolysis conditons, resulting in different product yields and properties. The critical point here is to clearly identify the desired product and apply the most suitable conditions. However, there are still many challenges to overcome before biomass pyrolysis finds large-scale applications for bio-oil and biochar production. The diversity of biomass samples and accordingly bio-oil and bio-char properties emphasizes the need for a case-by-case evaluation of each product prior to its utilization. Further research aiming to fully evaluate the extent and implications of bio-oil as a synthetic fuel and biochar as a soil additive is essential.

References

Agirre, I., Griessacher, T., Rösler, G. and Antrekowitsch J. (2013). Production of charcoal as an alternative reducing agent from agricultural residues using a semi-continuous semi-pilot scale pyrolysis screw reactor. *Fuel Processing Technolgy*, 106, pp. 114–121.

American Statistical Association. (2015). World population likely to surpass 11 billion in 2100: US population projected to grow by 40 percent over next 85 years. *ScienceDaily* [online] Available at: www.sciencedaily.com/releases/2015/08/150810110634.htm [Accessed 10 August 2015]

Amutio, M., Lopez, G., Artetxe, M., Elordi, G., Olazar, M. and Bilbao, J. (2012). Influence of temperature on biomass pyrolysis in a conical spouted bed reactor. *Resources, Conservation and Recycling*, 59, pp. 23–31.

Antal, M. J. and Gronli, M. (2003). The art, science, and technology of charcoal production. *Industrial & Engineering Chemistry Research*, 42, pp. 1619–1640.

Apaydın, E., Pütün, A. E. and Pütün, E. (2003). Pyrolysis of rice straw in a thermogravimetric analyser: how to obtain high yields of char at low temperatures. *Proceedings of the ECOS 2003 Conference*, Copenhagen, 2, pp. 1057–1063.

Apaydın Varol, E. and Pütün, A. E. (2012). Preparation and characterization of pyrolytic chars from different biomass samples. *Journal of Analytical and Applied Pyrolysis*, 98, pp. 29–36.

Apaydin-Varol, E. and Mutlu, Ü. (2013). Biofuels from selected biomass samples via pyrolysis: effect of the interaction between cellulose, hemicellulose and lignin. In: *8th Dubrovnik Conference on Sustainable Development of Energy, Water and Environment Systems-SDEWES 2013 Proceedings cd.* Croatia: Dubrovnik.

Apaydin-Varol, E., Uzun, B. B., Önal, E. and Pütün, A. E. (2014). Synthetic fuel production from cottonseed: fast pyrolysis and a TGA/FT-IR/MS study. *Journal of Analytical and Applied Pyrolysis*, 105, pp. 83–90.

Apaydın-Varol, E. and Erülken, Y. (2015). A study on the porosity development for biomass based carbonaceous materials. *Journal of the Taiwan Institute of Chemical Engineers*, 54, pp. 37–44.

Ates, F. Pütün, A. E. and Pütün, E. (2005). Fixed bed pyrolysis of *Euphorbia rigida* with different catalysts. *Energy Conversion and Management*, 46, pp.421–432.

Bonelli, P. R., Della Rocca, P. A., Cerella, E. G. and Cukierman, A. L. (2001). Effect of pyrolysis temperature on composition, surface properties and thermal degradation rates of Brazil nut shells. *Bioresource Technology*, 76, pp. 15–22.

Brebu, M. and Vasile, C. (2010). Thermal degradation of lignin – a review. *Cellulose Chemistry and Technology*, 44, pp. 353–363.

Bridgwater, A. V. and Grassi, G. (1991). *Biomass Pyrolysis Liquids Upgrading and Utilisation.* London: Elsevier Applied Science.

Bridgwater, A. V. and Meier, D. (1999). An overview of fast pyrolysis of biomass. *Organic Geochemistry*, 30, pp. 1479–1493.

Bridgwater, A. V., Czernik, S. and Piskorz, J. (2001). An overview of fast pyrolysis. *Progress in Thermochemical Biomass Conversion*, 2, pp. 977–997.

Brown, R. (2009). Biochar production technology. In: Lehmann J. and Joseph S. (eds.) *Biochar for Environmental Management: Science and Technology.* Oxford, New York: Earthscan, Routledge, pp. 127–144.

Bru, K., Blin, J., Julbe, A. and Volle, G. (2007). Pyrolysis of metal impregnated biomass: an innovative catalytic way to produce gas fuel. *Journal of Analytical and Applied Pyrolysis*, 78, pp. 291–300.

Carpenter, N. E. (2014). *Chemistry of Sustainable Energy*, Boca Raton, FL: CRC Press.

Cross, A., and Sohi, S. P. (2013). A method for screening the relative long-term stability of biochar. *GCB Bioenergy*, 5, pp. 215–220.

Czernik, S. and Bridgwater A. V. (2004). Overview of applications of biomass fast pyrolysis oil, *Energy & Fuels*, 18, pp. 590–598.

Dorez, G., Ferry, L., Sonnier, R., Taguet, A. and Lopez-Cuesta, J.-M. (2014). Effect of cellulose, hemicellulose and lignin contents on pyrolysis and combustion of natural fibers. *Journal of Analytical and Applied Pyrolysis*, 107 pp. 323–331.

Dunst, G. (2015) *SONNENERDE.* [online] Available at: www.sonnenerde.at/ Austria: Riedlingsdorf. [Accessed 18 August 2015]

Glasser, W. G. (1985). Lignin. In: Overend, R. P., Milne, T. A. and Mudge, L. K. (eds.), *Fundamentals of Thermochemical Biomass Conversion.* London: Elsevier Applied Science, pp. 61–76.

IEA Bioenergy Task 34 (2015). Pyrolysis. [online] Available at: www.pyne.co.uk [Accessed 15 March 2015]

Intergovermental Panel on Climate Change (IPCC) (2015). *Emissions Scenarios.* [online] Available at: www.ipcc.ch/ipccreports/sres/emission/index.php?idp=44#fig28 [Accessed 12 August 2015]

International Biochar Initiative (2013). *IBI Standards Standardized Product Definition and Product Testing Guidelines for Biochar.* [online] Available at: www.biochar-international.org/characterizationstandard [Accessed 10 January 2015]

Islam, M. N., Joardder, M. U. H., Hoque, S. M. N. and Uddin, M. S. (2013). A comparative study on pyrolysis for liquid oil from different biomass solid wastes. *Procedia Engineering*, 56, pp. 643–649.

Islam, M. R., Parveen, M. and Haniu, H. (2010). Properties of sugarcane waste-derived bio-oils obtained by fixed-bed fire-tube heating pyrolysis. *Bioresource Technology*, 101, pp. 4162–4168.

Jahirul, I. and Rasul, M. (2012). Biofuels production through biomass pyrolysis – a technological review. *Energies*, 5, pp. 4952–5001.

Kılıc, M., Apaydın-Varol, E. and Putun, A. E. (2012). Preparation and surface characterization of activated carbons from *Euphorbia rigida* by chemical activation with $ZnCl_2$, K_2CO_3, NaOH and H_3PO_4. *Applied Surface Science*, 261, pp. 247–254.

Klass, D. L. (1998). *Biomass for Renewable Energy, Fuels, and Chemicals*. San Diego, CA: Academic Press.

Kloss, S., Zehetner, F., Dellantonio, A. et al. (2012). Characterization of slow pyrolysis biochars: effects of feedstocks and pyrolysis temperature on biochar properties. *Journal of Environmental Quality*, 41, pp. 990–1000.

Lehmann, J. (2007). A handful of carbon. *Nature*, 447, pp. 143–144.

Lipinsky E. S. (1985). Pretreatment of biomass for thermochemical biomass conversion, In: Overend, R. P., Milne, T. A. and Mudge, L. K. (eds.) *Fundamentals of Thermochemical Biomass Conversion*. London: Elsevier Applied Science, pp. 77–89.

Lv, D., Xu, M., Liu, X., Zhan, Z., Li, Z. and Yao, H. (2010). Effect of cellulose, lignin, alkali and alkaline earth metallic species on biomass pyrolysis and gasification. *Fuel Processing Technology*, 91, pp. 903–909.

McKendry, P. (2002). Energy production from biomass (part 1): overview of biomass. *Bioresource Technology*, 83, pp. 37–46.

Mohan, D., Pittman, C. U. and Steele, P. H. (2006). Pyrolysis of wood/biomass for bio-oil: a critical review. *Energy and Fuels*, 20, pp. 848–889.

Nguyen, B. T. and Lehmann J. (2009). Black carbon decomposition under varying water regimes. *Organic Geochemistry*, 40, pp. 846–853.

Novak, J. M., Cantrell, K. B. and Watts, D. W. (2012). Compositional and thermal evaluation of lignocellulosic and poultry litter chars via high and low temperature pyrolysis: high and low temperature pyrolyzed biochars. *BioEnergy Research*. [online] Available at: www.ars.usda.gov/SP2UserFiles/Place/60820000/Manuscripts/2012/man895.pdf [Accessed 08 March 2015]

Ortega, J. V., Renehan, A. M., Liberatore, M. W. and Herring, A. M. (2011). Physical and chemical characteristics of aging pyrolysis oils produced from hardwood and softwood feedstocks. *Journal of Analytical and Applied Pyrolysis*, 91, pp. 190–198.

Overend, R. P., Milne, T. A., and Mudge, L. (1985). *Fundamentals of Thermochemical Biomass Conversion*. London: Elsevier Applied Science.

Özbay, N., Apaydın-Varol, E., Uzun, B. B. and Pütün, A. E. (2008). Characterization of bio-oil obtained from fruit pulp pyrolysis. *Energy*, 33, pp. 1233–1240.

Pütün, A. E., Apaydın, E. and Pütün, E. (2004). Rice straw as a bio-oil source via pyrolysis and steam pyrolysis. *Energy*, 29, pp. 2171–2180.

Pütün, A. E., Uzun, B. B., Apaydın, E. and Pütün, E. (2005). Bio-oil from olive oil industry wastes: pyrolysis of olive residue under different conditions. *Fuel Processing Technology*, 87 (1), pp. 25–32.

Pütün, A. E., Özbay, N., Apaydın-Varol, E., Uzun, B. B. and Ateş, F. (2007). Rapid and slow pyrolysis of pistachio shell: effect of pyrolysis conditions on the product yields and characterization of the liquid product. *International Journal of Energy Research*, 31, pp. 506–514.

Pütün, E., Uzun, B. B. and Pütün, A. E. (2009). Rapid pyrolysis of olive residue. 2. Effect of catalytic upgrading of pyrolysis vapors in a two-stage fixed-bed reactor. *Energy & Fuels*, 23, pp. 2248–2258.

Ross, D. S. and Ketterings, Q. (2011). Recommended methods for determining soil cation exchange capacity, *Cooperative Bulletin*, 493, pp. 75–86.

Sadaka, S. (2015). *Pyrolysis*. [online] Available at: http://bioweb.sungrant.org/NR/rdonlyres/ 57BCB4D0-1F59-4BC3-A4DD-4B72E9A3DA30/0/Pyrolysis.pdf [Accessed 10 April 2015]

Sanchez-Silva, L., López-González, D., Villasenor, J., Sánchez, P. and Valverde, J. L. (2012). Thermogravimetric-mass spectrometric analysis of lignocellulosic and marine biomass pyrolysis. *Bioresource Technology*, 109, pp. 163–172.

Seebauer, V., Petek, J. and Staudinger, G. (1997). Effects of particle size, heating rate and pressure on measurement of pyrolysis kinetics by thermogravimetric analysis. *Fuel*, 76, pp. 1277–1282.

Sevilla, M., Maciá-Agulló, J. A. and Fuertes, A. B. (2011). Hydrothermal carbonization of biomass as a route for the sequestration of CO_2: chemical and structural properties of the carbonized products. *Biomass & Bioenergy*, 35, pp. 3152–3159.

Shen, D. K. and Gu, S. (2009). The mechanism for thermal decomposition of cellulose and its main products. *Bioresource Technology*, 100, pp. 6496–6504.

Sohi, S. P., Krull, E., Lopez-Capel, E. and Bol, R. (2010). A review of biochar and its use and function in soil. *Advances in Agronomy*, 105, pp. 47–82.

Sorum, L., Gronli, M. G. and Hustad, J. E. (2001). Pyrolysis characteristics and kinetics of municipal solid wastes. *Fuel*, 80, pp. 1217–1227.

Suarez-Garcia, F., Martinez-Alonso, A. and Tascon, J. M. D. (2002). Pyrolysis of apple pulp: effect of operation conditions and a-chemical additives. *Journal of Analytical and Applied Pyrolysis*, 62, pp. 93–109.

Sun, Y. Gao, B., Yao, Y. et al. (2014). Effects of feedstock type, production method, and pyrolysis temperature on biochar and hydrochar properties. *Chemical Engineering Journal*, 240, pp. 574–578.

Theander, O. (1985). Cellulose, hemicellulose and extractives. In: Overend, R. P., Milne, T. A. and Mudge, L. K. (eds.) *Fundementals of Thermochemical Biomass Conversion*. London: Elsevier Applied Science, pp 35–61.

Tsai, W. T., Mi, H. H., Chang, Y. M., Yang, S. Y. and Chang, J. H. (2007). Polycyclic aromatic hydrocarbons (PAHs) in bio-crudes from induction-heating pyrolysis of biomass wastes. *Bioresource Technology*, 98, pp. 1133–1137.

Uzun, B. B. Pütün, A. E. and Pütün, E. (2007). Rapid pyrolysis of olive residue. 1. Effect of heat and mass transfer limitations on product yields and bio-oil compositions. *Energy & Fuels*, 21, pp. 1768–1776.

Uzun, B. B. and Sarioğlu, N. (2009). Rapid and catalytic pyrolysis of corn stalks. *Fuel Processing Technology*, 90, pp. 705–716.

Uzun, B. B., Apaydın-Varol, E., Ateş, F., Özbay, N. and Pütün A. E. (2010). Synthetic fuel production from tea waste: characterisation of bio-oil and bio-char. *Fuel*, 89, pp. 176–184.

Verheijen, F., Jeffery, S., Bastos, A. C., van der Velde, M. and Diafas, I. (2010). Biochar application to soils: a critical scientific review of effects on soil properties, processes and functions. JRC Scientific and Technical Reports, Luxembourg.

Wang, Y., Hu, Y., Zhao, X., Wang, S. and Xing, G. (2013). Comparisons of biochar properties from wood material and crop residues at different temperatures and residence times. *Energy & Fuels*, 27, pp. 5890–5899.

Woolf, D. (2008). *Biochar as a soil amendment: a review of the environmental implications*. [online] Available at www.orgprints.org/13268/1/Biochar_as_a_soil_amendment_-_a_review.pdf [Accessed 2 November 2014]

Zaror, C. A. and Pyle, D. L. (1982). The pyrolysis of biomass: a general review. *Proceedings of the Indian Academy of Science*, 5, pp. 269–285.

13

The Role of Biochar Production in Sustainable Development in Thailand, Lao PDR and Cambodia

MALIWAN HARUTHAITHANASAN, ORRACHA SAE-TUN,
NATTHAPHOL LICHAIKUL, SOKTHA MA,
SITHONG THONGMANIVONG AND
HOUNGPHET CHANTHAVONG

Abstract

The majority of the rural population in Cambodia, Lao PDR and Thailand is forest dependent. Biomass fuel, predominantly fuelwood and charcoal, is the main source of energy for cooking and heating in households, especially in rural areas. Traditional and simple methods to produce charcoal in combination with a range of different feedstocks leads to a wide range of achievable product qualities, production efficiencies and health risks due to emissions. Improvement in charcoal production techniques is therefore increasingly promoted. Utilization of co-products, such as pyroligneous acid, can be beneficial in agricultural and livestock production, for example as insecticides, antimicrobial agents and insect repellents. However, it is necessary to use them in a proper way, therefore more research and ultimately education of farmers and smallholders is needed, along with a robust quality assurance scheme. Moreover, waste from charcoal production such as charcoal dust or chunks of broken charcoal pieces can be used as biochar to improve soil properties. By increasing soil productivity and fertilizer efficiency, the region may contribute to climate change mitigation and sustainable development. This chapter gives an overview of traditional and improved kiln technologies and their characteristics, and potentials for biochar production and byproduct utilization are discussed in detail.

13.1 Forest Resources in Cambodia, Lao PDR and Thailand

Cambodia, Lao PDR (Lao People's Democratic Republic, also known as Laos) and Thailand have diverse tropical forests which comprise evergreen, deciduous, mixed and mangrove forests (the latter does not exist in Lao PDR). Deforestation and forest degradation are serious problems in these countries. The forest area in Cambodia has decreased from 70% in the 1960s to 57% in 2014, an average annual deforestation rate of 0.5% (Forestry Administration, 2010), which represents a loss of 10.36 million ha (Forestry Administration, 2011). FAO (2010) estimated that there are 464 million tons of carbon stock in living forest biomass in Cambodia. Similar to Cambodia, the forest area in Thailand has decreased from 43.21% in 1973 to 31.57% in 2013 (Royal Forest Department, 2013, 2014a). In this context, the forest area in Thailand refers to natural and plantation forests excluding eucalypt plantations (Royal Forest Department, 2014a). In Lao PDR, forest cover in 2010 was approximately 40% of the total land (Department of Forestry, 2012) compared to 41.5% in 2002 (Government of Lao PDR, 2005).

One of the reasons for forest degradation in this region is an increasing demand for wood and non-timber forest products due to high population growth. In Cambodia, forests are obviously a key source of household incomes (Heov et al., 2006): 85% of the rural population relies on forests (UN-REDD, 2011). In Lao PDR, wood-based products (encompassing all timber sector products, such as logs, sawn wood, plywood, veneer, moldings, joinery and furniture, and paper sector products, such as wood chips, pulp and paper) are exported to neighboring countries, that is China, Vietnam and Thailand (Saunders, 2014). Despite a continuing decrease in export value since the 1980s, forest products are still important export commodities.

The three countries aim to increase their forest areas. Cambodia has attempted to increase its forest cover as a consequence of a decreasing trend over the past years. The Royal Government of Cambodia (RGC) has supported conversion of 532,615 ha of non-forest area, such as barren and degraded land, to forest plantations (Forestry Administration, 2011). Consequently, the forest plantation area in Cambodia has increased by 69,000 ha, or 1% of the current forest area.

The government of Lao PDR has promoted plantations to reduce wood extraction from natural forest and to meet the demand for wood. Nowadays, most plantations in Lao PDR supply wood for furniture production. Moreover, plantations of fast-growing species, such as eucalyptus and acacia, provide raw material for wood chips, fuelwood and charcoal production (Southavilay and Castren, 1999). In spite of issuing a number of decrees to reduce illegal logging, the government has not yet accomplished the task due to inadequate enforcement (Saunders, 2014). As part of the effort, it established the Forest Law Enforcement, Governance and Trade Office (FLEGT) in October 2013 with financial and technical support from Germany's Agency for International Cooperation (GIZ).

With concern over adverse environmental impacts and domestic wood demand, Thailand aims to increase forest area to 40% of the national area by reforestation and plantation programs (Royal Forest Department, 2014a), from 31.6% in 2013. With regard to wood demand by industry, Thailand sets a target for plantation at 7.74% of its total land dedicated to supplying the biomass industry. In 2013, industrial plantation comprised 2.88 million ha of rubber tree plantation and 0.48 million ha of eucalypt plantation. Exotic tree species are mainly planted to supply industrial demand, while indigenous tree species are planted for conservation purposes (Royal Forest Department, 2014b).

13.2 Charcoal Production and Consumption in Cambodia, Lao PDR and Thailand

Biomass, often obtained from natural and plantation forests, is the most abundant source of energy (cooking, heating) for rural people in these countries. They mostly rely on fuelwood or charcoal from regional markets.

Cambodia's National Institute of Statistics reported that 97.7% of the population was fuelwood dependent in 1997, decreasing to 75.9% in 2013 (National Institute of Statistics, 1997, 2014). In Phnom Penh, the capital city of Cambodia, around 84% of households relied on fuelwood as the main source of energy for cooking in 1997, and this decreased to 31.2% in 2007 (National Institute of Statistics, 2008). Biomass for the urban center is sourced from forests in nearby provinces and either directly marketed from rural supply areas or through middlemen and/or traders. Unregulated tree cutting has adverse environmental impacts, such as forest degradation and deforestation, and the inhabitants in the supply areas have already experienced some of these, for example longer distances to extract the fuelwood. The overall demand for fuelwood is not expected to decrease in the foreseeable future, although conventional fuels, such as liquefied petroleum gas (LPG) and electricity, are increasingly replacing fuelwood in high-income households (IEA and ERIA, 2013). The majority of people will continue to rely on fuelwood and energy transition is limited by financial constraints.

In Lao PDR, fuelwood is also still the primary source of energy despite several hydropower development programs (Phongoudome, 2014), which were recently introduced. Apart from fuelwood, charcoal is also an important energy source especially for people in urban and suburban areas where fuelwood is scarce. Previously, it was produced for domestic consumption to replace fuelwood. Since around 2005, it has been exported to several countries in Asia. The Ministry of Commerce reported that in 2009, Vietnam was the largest market for charcoal produced in Lao PDR. Vietnam imported 75,000 tons of charcoal in 2009 while Thailand imported nearly 40,000 tons.

However, there is rather limited information on the volume of charcoal produced in Lao PDR for domestic consumption and export. Mekuria et al. (2012) estimated that the net income charcoal producers gained from the sale of charcoal was 57 USD per batch (one batch requires two weeks of production time). According to Barney (2014), the average price of charcoal in Savannakhet, located in Southern Lao PDR, close to the Thai border, was between 167 and 176 USD ton^{-1}, depending on the type of kilns used for charcoal production (see 13.3) and the resulting quality of the charcoal. However, charcoal producers in Lao PDR usually do not consider charcoal production as the primary source of income, since they produce charcoal traditionally after the farming season (Mekuria et al., 2012).

In contrast to Lao PDR and Cambodia, rural households in Thailand predominantly use charcoal for cooking and heating. However, industrial biomass power plants and other industries use biomass in the form of wood chips as an energy source rather than charcoal. The share of charcoal in energy consumption has been increasing. According to the Royal Forest Department (2014b), wood demand by the Thai population is less than 1 m^3 person^{-1} year^{-1}, whereas industrial demand for biomass fuel is equivalent to the wood produced from approximately 56,000 km^2 of forest plantation. The Thailand Department of Alternative Energy Development and Efficiency (2009, 2010, 2011, 2012a, 2012b, 2013) reported a steady increase in imported charcoal for domestic consumption, from 78,000 tons in 2009 to 167,000 tons in 2013. This implies that imported charcoal will play an increasing role in charcoal supply in Thailand.

13.2.1 Feedstock for Charcoal Production

Even though plantations are increasing, the main source of fuelwood and wood for charcoal production in the three countries is still natural forests and residues from wood processing industries and agriculture. Wood from plantations is supplied generally for other industries such as construction and chipping, which have higher demands as compared to charcoal production.

Mixed wood species are used depending on their availability and quality for charcoal production based on local experiences. *Shorea obtusa*, *Dipterocarpus obtusifolius* and *Xylia xylocarpa* are the most preferred species for both fuelwood and charcoal production in the south-west part of Cambodia (San et al., 2012). This area is also one of the main charcoal supply sources for Phnom Penh. In central Cambodia, local species such as *Xylia pierrei* and *Grewia paniculata* are favored for fuelwood (Top et al., 2004). The authors also found that about 70% of fuelwood extracted from forests was living wood while the rest was dead wood. In this region of Cambodia, *Cratoxylum formosum* and *Irvingia malayana* are generally used for charcoal production due to their availability, as well as sawmill residues of mixed hardwood and softwood species.

In Lao PDR, about 70% of wood for charcoal production is sourced from sawmills (residues) and the remaining 30% from managed forests (Barney, 2014). Because of a rapid increase in rubber plantations (Manivong and Cramb, 2008) caused by a government promotion campaign to increase the rubber plantation area from 28,000 ha in 2007 to 250,000 ha in 2010, rubberwood from mature trees is seen as a promising feedstock, not only for furniture but also for charcoal production.

In Thailand, mangrove species were preferred for charcoal production in the past due to their high heating value with low ash content (Table 13.1). However, due to environmental concerns regarding mangrove forest degradation and the associated loss of protective functions and biodiversity, fast-growing wood species such as eucalyptus, acacia and bamboo are playing increasing roles as sources for fuelwood and feedstock for charcoal production. Despite an increase of fast-growing tree plantations, they marginally supply wood for charcoal production in Thailand as residues from wood processing. Charcoal producers mainly use raw materials according to their availability and the properties of charcoal to reduce costs and be able to quickly respond to market demands. Key properties of wood charcoal produced from different tree species are shown in Table 13.1.

Table 13.1. *Properties of traditional wood charcoal from different tree species in Thailand*

Tree species	Tree group	Density (g cm^{-3})	Fixed carbon (%)	Volatile matter (%)	Ash content (%)	Heating value (kcal kg^{-1})
Acacia catechu	EF	0.48	75.2	20.8	4.0	7,240
Acacia auriculiformis	EF	0.41	71.1	24.8	4.9	7,240
Azadirachta indica var. *siamensis*	IL	0.54	88.2	9.41	2.4	7,573
Cassia siamea	IL	0.40	81.5	14.7	3.7	7,524
Casuarina junghuhniana	EF	0.70	83.3	13.8	2.9	7,890
Casuarina equisetifolia	IF	0.45	77.8	18.9	3.3	7,590
Combretum quadrangulare	IL	0.40	79.9	16.2	3.9	6,900
Eucalyptus camaldulensis	EF	0.42	79.8	16.7	3.5	7,350
Leuceana leucocephala	EF	0.44	78.3	18.9	2.7	7,430
Melia azedarach	IL	0.34	75.8	20.5	3.7	7,030
Peltophorum dasyrachis	IL	0.33	75.8	20.5	3.7	7,030
Rhizophora apiculata	IM	0.49	79.9	17.2	2.9	7,500
Spondias pinnata	IL	0.30	73.8	21.6	4.6	7,190

Note: EF = exotic fast-growing tree, IL = indigenous lowland forest tree, IM = indigenous mangrove forest tree. Species in the EF group are commonly grown in forest plantations.
Source: Bunyavejchewin (1989); Kuhakan (2007).

13.3 Charcoal Production Methods in Cambodia, Lao PDR and Thailand

Charcoal production employs a slow carbonization technique through slow pyrolysis (heating rate 5–7°C min^{-1}) producing more solid residuals (char) and less liquid than fast pyrolysis (heating rate over 300°C min^{-1}) (Garcia-Perez et al., 2010). Variations in time and temperature as well as inherent feedstock characteristics affect the properties and yields of the final products (mainly char and pyroligneous acid) (Kumar et al., 1992, 1999; Brewer et al., 2009; Kwon et al., 2014; Missio et al., 2014). Higher temperature during pyrolysis (isothermal) significantly decreases the yield and volatile matter of solid charcoal (Puthson, 1990; Somerville and Jahanshahi, 2015). The structural integrity of charcoal, a very important factor determining the market value, depends on process conditions, but especially on temperature (Kwon et al., 2014). Higher temperatures generally lead to increased brittleness, which is less preferable for customers. Therefore, charcoal producers need to carefully and properly control temperature in the kilns.

The methods for charcoal production in the three countries are generally simple and low cost. Traditional kiln technology is widely used to produce charcoal, but it suffers from low efficiency and negative impacts on human health and the environment. The traditional pit kiln produces charcoal at a process efficiency of approximately 18% (Kuhakan, 2007). Since smoke and dust recovery is not employed, the air pollution of traditional methods reaches high levels due to non-condensable gas emissions, including carbon monoxide (CO), carbon dioxide (CO_2), methane (CH_4) and ethane (C_2H_6), particularly during the drying stage (Garcia-Perez et al., 2010; Woolf et al., 2010). Particulate emissions are another serious issue in regard to human health. Modified kilns have been developed and applied to improve production efficiency, including productivity and product quality, and decrease smoke and particulate emissions due to a higher degree of system closure. Furthermore, product quantity and quality can be improved by more precise temperature control in the modified kilns compared to the traditional kilns.

Charcoal producers in the region use basic indicators, for example exhaust smoke color and temperature at the vent and within the kiln, to determine the current pyrolysis stage. Generally, white smoke indicates dehydration (drying process), pyrolysis starts with black smoke emission and the smoke becomes more transparent as the process continues, and no smoke is produced after carbonization completes (Garcia-Perez et al., 2010). Table 13.2 shows the relationships between smoke color, temperature and pyrolysis stage at different stages of the production process. The table also reflects the subsequent pyrolysis steps of dehydration, carbonization, refinement and cooling in traditional charcoal kilns (Kuhakan, 2007; Panunumpa et al., 2013).

Traditional and modified methods will be discussed in detail separately for each country in the following section. It is important to mention that although each country has developed its own traditional methods, one may find them in other countries as well, especially in regions near the borders. There has been an exchange of methods and knowledge for generations in this entire region, often leading to locally adapted versions of the following kiln types.

13.3.1 Traditional Methods

13.3.1.1 Lao PDR

There are two main charcoal production methods in Lao PDR, the traditional kiln (*Tao Op*) and a kiln which typically uses sawmill residues and larger diameter wood as a feedstock material (*Tao Phi*).

The Tao Op kiln (Figure 13.1) is a small-scale traditional method used for charcoal production for household use. Local people usually build this kiln on their farm and use small-diameter wood from locally available sources to produce charcoal. Depending on the

Table 13.2. *Relationship between smoke, temperature and stage of the pyrolysis process*

Color of smoke	Temperature at vent (°C)	Temperature inside kiln (°C)	Pyrolysis stage
White with light yellow	80–82	320–350	Beginning of dehydration
Brown with gray	82–85	350–380	Beginning of pyroligneous acid collection
Brown with gray	90–100	380–400	Dark color and very viscous pyroligneous acid
Brown with white	100–150	400–430	
Brown with white	150–170	430–450	End of pyroligneous acid collection
Brown with white	170–230	450–500	
Light bluish-purple with white	230–250	500–530	Beginning of carbonization
Bluish-purple with white	250–300	350–570	End of carbonization
Bluish-purple	300–350	570–650	Charcoal refinement and subsequent closure of the kiln

Source: Laemsak and Thornrasin (2005).

expected production quantities, a 1–2 m deep hole is dug and closed with a wooden frame of approximately 2 m height. The frame is then coated with paper or sacks and finally moist pure soil or moist soil mixed with rice chaff or husk. The kiln is filled from an opening at the side with small logs arranged vertically. A fire is ignited to start the pyrolysis process through a separate access trench from the bottom of the kiln. The pyrolysis phase of this traditional charcoal kiln takes about 10–15 days. Readiness of charcoal is generally indicated by smoke observation, as is the case with other traditional kilns in Cambodia and Thailand (see Table 13.2). This traditional Tao Op kiln is also known as a permanent kiln. It can be used for 2–10 years depending on the construction and maintenance (Mekuria et al., 2012), especially of the soil cover. Partial sintering of the clay-rich cover leads to a high stability of the soil dome after its first use. However, the temperature is too low and the material unsuitable for a complete sintering process and therefore the soil cover has to be well maintained by adding moist soil from outside when necessary. In order to prevent erosion of the cover during rainfall, these kilns are often protected by a simple shed.

The Tao Phi kiln uses residues from sawmills, including pieces of whole tree stems, offcuts and plywoods as feedstock material. The diameter of the feedstock material can be larger than in the Tao Op kiln. Wood residuals are piled horizontally on wood pallets (Figure 13.2a) and covered by wet sawdust (Figure 13.2b). The pyrolysis phase for a Tao Phi kiln is about 6–7 days. The quality of charcoal is lower than that from a Tao Op because the Tao Phi kiln operates at a lower temperature controlled only by adding water to the sawdust cover. In addition, the pyrolysis is characterized by a shorter process time.

In central provinces of Lao PDR where sawmills and wood processing factories are nearby, rural villagers build Tao Phi kilns for the commercial production of charcoal. In

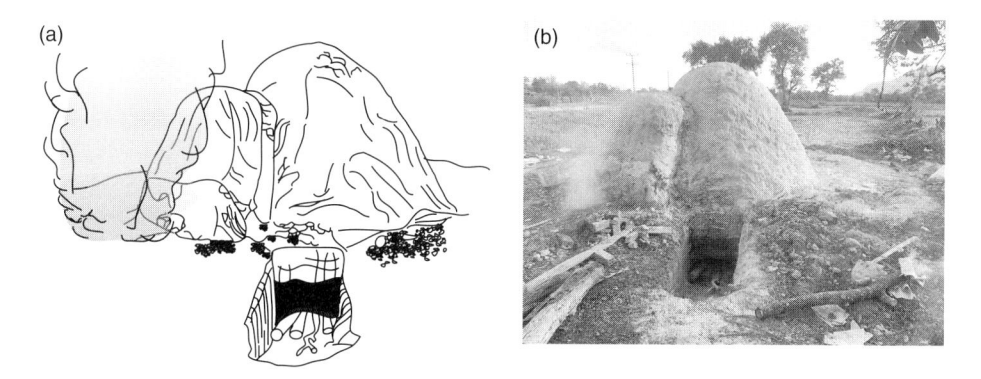

Figure 13.1. Sketch (a) and a photograph (b) of a traditional Tao Op charcoal kiln, commonly used in Lao PDR. The pyrolysis process is started with a fire lit at the access trench at the bottom of the kiln. Photo credit (b): Sithong Thongmanivong.

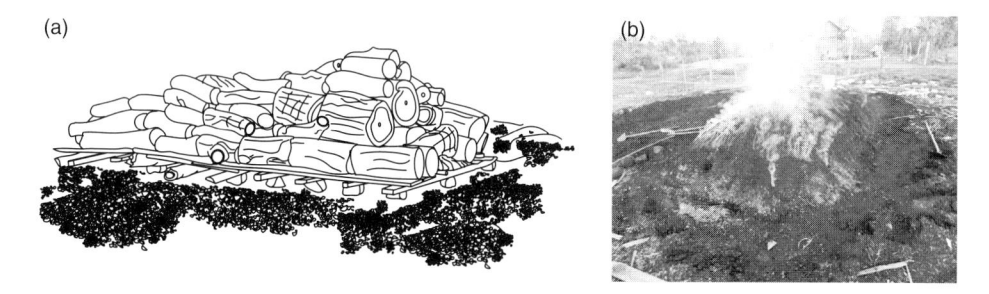

Figure 13.2. (a) Stack of large-diameter wood to be pyrolyzed using a Tao Phi kiln. (b) Tao Phi kiln during pyrolysis. Note that smoke is released from the entire cover, making efficient temperature control impossible. Photo credit (b): Sithong Thongmanivong.

contrast to charcoal from Tao Op kilns, which is mainly used in domestic households and increasingly exported since 2005, the main market for charcoal of this kiln type is the local wood-processing industry to generate process heat. In certain cases, where sawdust is not available in sufficient amounts, soil and sometimes rice husk is used to cover the sawmill residuals instead of sawdust. This slightly adapted version of the Tao Phi kiln is referred to as a "Temerity charcoal production kiln" (Mekuria et al. 2012). The charcoal produced from this method has the lowest quality of all previously introduced methods in Lao PDR, as a consequence of the higher temperature maintained by the sawdust in the original Tao Phi setup (Barney, 2014).

13.3.1.2 Thailand

The ground pit or heap kiln is the traditional charcoal production method in Thailand. Its construction is simple and low-cost; however, only small quantities of charcoal of relatively low quality can be produced. The pit kiln is constructed by firstly digging a 2 m × 3 m

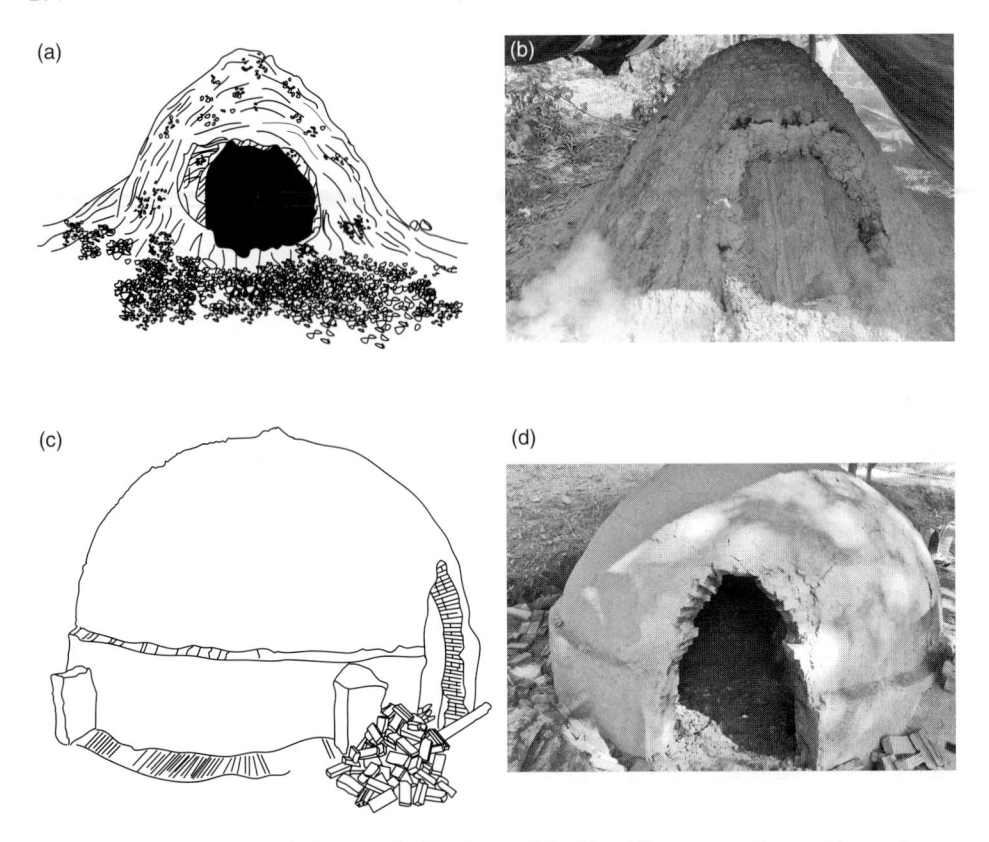

Figure 13.3. Sketch (a) and photograph (b) of a mud beehive kiln, commonly used in rural areas in Thailand. Note the simple shed (b) that is installed to prevent erosion of the mud cover during rainfall events. The brickstone beehive kiln (c, d) improves the longevity of the beehive system. In addition, the capacity per kiln is higher as construction with bricks allows larger volumes. These are commonly found in rural areas operated by professional charcoal producers. Photo credit (b,d): Viktor Bruckman.

and approx. 0.5 m deep pit. Locally available wood is stacked horizontally on a wooden pallet to increase heat convection and therefore improve the pyrolysis process. A fire is started in the pit and subsequently more biomass is added, until the pit is closed with soil, with small holes to allow air circulation. The process time ranges from a few days up to one month depending on the size of the kiln. The yield is low and lies typically between 10 and 15% (wt) of the feedstock material, and the pyrolysis process is uneven, resulting in complete combustion of some parts while a certain fraction might be not even pyrolyzed at the time the kiln is opened. The current pyrolysis stage is estimated from the temperature and characteristics of smoke emitted from the kiln as mentioned above. A major downside of this particular method is that it creates a large amount of emissions, which poses health risks.

In order to overcome the problems associated with low efficiency and quality, the so-called mud and brick beehive kilns (Figure 13.3) were developed, but still represent traditional technology. Both kiln types have holes and a mound-like shape (which looks like a beehive) to improve aeration and ash deposition, resulting in carbonization process improvement (Kuhakan, 2007). However, they differ mainly in the materials used in their construction: mud and bricks respectively. In addition, a pit is excavated in the case of the mud beehive kiln to increase the capacity, while this is not done in the case of the brick kiln. In comparison to the above mentioned Tao Op kiln commonly used in Lao PDR, the fire is in both cases started from an opening at ground level, therefore no separate access pit is necessary. These two kilns produce charcoal that burns with low emissions, and is therefore mainly used for cooking purposes. In recent years, a number of private companies have improved the kilns with the aim to evenly distribute heat inside the kiln during the pyrolysis process. This increases the quality and yield of the charcoal and details about recent developments will be discussed in the next section.

13.3.1.3 Cambodia

The traditional charcoal kilns in Cambodia are comparable to those in Thailand and Lao PDR. The most common type is the beehive kiln for professional producers and the heap kiln for household-scale production. Mangrove charcoal has a long tradition as it provides long-lasting heat. Uniform diameter (20–25 cm) logs of mangrove trunks approximately 1.6 m in length are arranged vertically in the beehive kiln. The vertical arrangement allows even heat distribution during the pyrolysis process. In the next step, the kiln is fully filled with horizontally arranged logs in upper layers. The production process takes about 30 days using the described setting in order to derive high-quality charcoal from mangrove wood species. During the first 10 days of operation, a constant fire will be maintained at the kiln opening, ensuring maximum airflow into the kiln. After this period, the air intake rate will be reduced gradually by sealing the opening of the kiln while reducing the fire intensity at the same time. This pyrolysis phase (air and temperature control) lasts for 10–12 days. During this period, pyroligneous acid can be collected from steam that exits the kiln at the top vent. Finally, the quality of charcoal is improved by completely sealing the kiln for one week.

13.3.2 Modified Methods

Modified kilns aim to enhance charcoal yield and quality in terms of heating value and brittleness and reduce negative impacts on the environment such as harmful emissions during pyrolysis. An additional benefit of using advanced kiln technology is the ability to efficiently collect charcoal co-products such as pyroligneous acid. Various kilns have been developed to suit different conditions for charcoal production. In the entire region under consideration in this chapter, the most effective charcoal production kilns are modified Japanese charcoal kilns. They are the so-called Yoshimura kiln in Cambodia and the Thai-Iwate kiln in Thailand.

13.3.2.1 Cambodia

After three years of trials and intensive field testing between 2008 and 2010, the Yoshimura kiln, a charcoal production kiln initially developed in Japan, was introduced to Cambodia. Charcoal produced with the Yoshimura kiln is of higher quality (15% higher calorific value), can be produced in half of the time and approximately 30% less feedstock material is used to produce the same amount of charcoal as compared to the traditional kilns. The result is a more standardized, reliable product which reduces health risks during production and increases profits for producers. As a consequence of the lower emissions during pyrolysis, the Yoshimura kiln can be installed in close proximity to villages, which is a benefit from the logistics point of view.

13.3.2.2 Lao PDR

In Lao PDR, traditional methods are still widely used but there are recent initiatives aiming at introducing improved kiln technologies. Non-governmental organisations (NGOs) promoting development in Lao PDR, for example the Netherland Development Organisation (SNV), or the Lao Institute for Renewable Energy (LIRE), support local people to improve charcoal production systems. The improved charcoal production kiln comprises a square structure built from cement and bricks. While this increases the longevity of the Tao Op and Tao Phi kilns, the initial costs are higher.

13.3.2.3 Thailand

The "Iwate kiln" was introduced in Thailand in order to improve yields and quality and decrease emissions (Figure 13.4). The design is also based on a Japanese development and this particular kiln is also widely used in China and South Korea. The entire kiln is made of high-quality refractory bricks with cement binding. The major difference compared to all previous kilns introduced here is that this kiln allows an easily controllable operating temperature. The Royal Forest Department has attempted to reduce the construction costs by using non-refractory traditional Thai bricks and mud with fine sand as a binding material. This modified Iwate kiln is called a "Thai-Iwate kiln." The efficiency and charcoal quality does not show a significant difference, however, and the traditional Thai bricks reduce the longevity of this modified kiln.

The moisture content of the feedstock material is significant for this type of kiln. It should be approximately 35% in order to obtain optimal results. A 12 m^3 Thai-Iwate kiln has a typical loading capacity of 9,000 kg feedstock material (dry weight), capable of producing 2,250 kg of charcoal and 600 L of pyroligneous acid per month. The pilot production plant employs two to three workers to operate the kiln. Up to three batches can be processed each month (7 days for charcoal production and 2–3 days for removal of charcoal and refilling of feedstock material). Compared to the traditional kiln, the Thai-Iwate kiln produces charcoal of high uniform quality throughout the entire batch and typical yields are at least 10% higher.

Figure 13.4. The Thai-Iwate kiln is based on Japanese technology and it is widely used in Asia. This particular kiln type allows efficient temperature control. (a) Detailed view of a freshly constructed Thai-Iwate kiln with bamboo sticks supporting the dome during cement curing. (b) The Thai-Iwate kiln can be operated with various dimensions of feedstock material. (c) The front opening is tightly closed prior to pyrolysis with two rows of bricks and a hollow space in between to exactly control the airflow from a vent. Photo credit (a–c): Maliwan Haruthaithanasan. (A black and white version of this figure will appear in some formats. For the colour version, please refer to the plate section.)

13.3.2.4 Drum Kilns

Another recent development in Thailand is drum kilns (Figure 13.5). The drum kilns are made of 200 L crude oil drums, which have been modified to produce charcoal. Varieties include single drum kiln, double drum kiln, Tonga kiln and horizontal drum kiln. These kilns are relatively simple to build and sufficient amounts of charcoal for household consumption can be produced within 1–2 days. However, the collection of pyroligneous acid is impossible except in the horizontal drum kiln setup. Therefore, the latter is widely promoted among rural villages in Thailand due to its high efficiency in terms of production time and capability to collect pyroligneous acid.

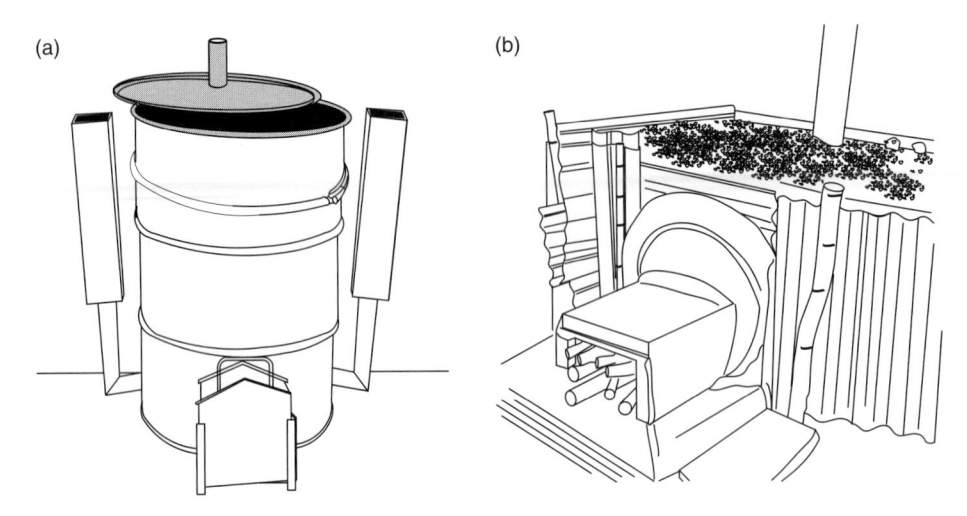

Figure 13.5. Sketch of a vertical single drum kiln (a) and horizontal drum kiln (b) used in rural regions in Thailand.

13.4 Co-products from Charcoal Production

Co-products from charcoal production, which are produced during and at the end of the production process, include pyroligneous acid, ash, charred residues and tar. Pyroligneous acid is a liquid co-product collected by condensing emissions of volatile compounds during the pyrolysis process. Ash is a dust residue from wood or charcoal burning. Charred residues are remains or deposits of broken charcoal which are mostly found at the bottom of the kiln and need to be removed before refilling the kiln with feedstock. Charred residues are currently mostly disposed of as waste. The use of tar from charcoal production is currently neglected in the three countries.

13.4.1 Pyroligneous Acid

As mentioned above, pyroligneous acid is a co-product obtained during charcoal production when the temperature inside the kiln reaches around 300–400°C (Loo et al., 2008; Wang et al., 2009) during the carbonization phase of exothermic decomposition (Burnette, 2010). Traditionally, raw pyroligneous acid and refined pyroligneous acid account for 8% and 5% of dry feedstock weight respectively, but can be increased to 15% by installing a cooling system in the vent (Panunumpa et al., 2013). However, Burnette (2010) reported that one ton of dry wood yielded 0.32 t of pyroligneous acid, which equals a 31% yield.

Pyroligneous acid mainly consists of acetic acid and various organic compounds (Kuhakan, 2007), including phenolic compounds, formaldehyde and methanol. The quality of pyroligneous acid depends on i) wood species and age, ii) temperature and iii) settlement of pyroligneous acid after the condensation (duration ranges from three to six months) (Laemsak and Thornrasin, 2005). An appropriate temperature for pyroligneous acid collection is

Table 13.3. *Characteristics of pyroligneous acid used in agriculture*

Properties	Raw pyroligneous acid	Distilled pyroligneous acid
pH	1.5–3.7	1.5–3.7
Specific gravity	>1.005	>1.001
Color	Yellow, pale brownish red or brownish red	Colorless, pale yellow or pale brownish red
Transparency	Transparent	Transparent
Colloid	No	No
Odor	Smoky odor	Smoky odor
Dissolved tar content	No data	<3%
Ignition residue	No data	<0.2% by weight

Source: Laemsak (2008); Burnette (2010).

80–160°C at the vent or by visual assessment when light yellowish smoke is released. Deposition, filtration and distillation are required to refine pyroligneous acid before use (Laemsak, 2008; Panunumpa et al., 2013). Refinement of pyroligneous acid by activated charcoal and distillation are common technologies for industrial utilization, mostly employed in the pharmaceutical industry. The refinement of raw pyroligneous acid by distillation is not only employed to remove residuals of wood tar but also to prevent negative impacts of other unwanted substances such as oil, as they may block stomata and stick to plant roots resulting in reduced growth or mortality after application. Characteristics of pyroligneous acid commonly used in agriculture are shown in Table 13.3.

After refinement, pyroligneous acid is commonly used as a fertilizer, as pesticide and as a compost catalyst (Rahmat et al., 2014; Mathew and Zakaria, 2015). By dilution with water, a pyroligneous acid solution can be applied as a germicide and insecticide in agricultural fields (Baimark et al., 2008; Theapparat et al., 2015). For livestock production, pyroligneous acid solution is known to control malodor and insects (Takahara et al., 1993, 1994).

A study by Wititsiri (2011) revealed evidence of the effectiveness of pyroligneous acid as an insecticide. In this study, pyroligneous acid from coconut shell charcoal production at a dilution of 1:50 (v/v) had the highest termiticidal activity, killing 82% of termites within 24 hours when applied directly compared to the control (no treatment). Application of 1:10 dilution caused a more than 90% mortality rate in striped mealy bugs. Additionally, wood tar from pyroligneous acid refinement could protect plywood from termite damage.

In Thailand, pyroligneous acid is used widely in agriculture. However, its utilization is generally based on local knowledge and not scientific research (Tiilikkala et al., 2010), which poses certain risks in view of contamination, toxicity and impact on soil microbiology. Research is urgently needed to confirm the proposed positive effects, to understand the mechanisms behind them and to avoid negative consequences. This is underpinned by Thailand's Department of Agriculture, which earlier reported toxicity of pyroligneous acid to fish and a high toxicity to plants if applied excessively (Burnette, 2010).

13.4.2 Ash

Ash is a dust remainder of burnt wood or charcoal. It contains inorganic compounds of biomass. When burning wood at 950°C, 0.5–1.0% (by wt) remains as ash (Laemsak and Thornrasin, 2005). The amount and chemical composition of ash differs among wood species, parts of the trees (Pitman, 2006) and even different provenances or origins of the same tree species.

In agriculture, ash can be applied as a soil amendment to recycle some essential nutrients (e.g. Ca, K and P). Due to its extreme alkalinity, ash is applied to acidic soils to decrease soil acidity (liming effect).

13.5 Charcoal Utilization for Soil Amendment

In the three countries, residuals from charcoal production such as charcoal dust and small charcoal chunks have less or even no market value and are usually disposed of (Figure 13.6a). Depending on the feedstock material, a large share of this material consists of pyrolyzed bark that disintegrates from the stem during the pyrolysis process. In Chapter 4 of this book, the author suggests the use of bark specifically as a feedstock material for pyrolysis, as it contains a higher share of nutrients as compared to the wood fraction. A local charcoal producer using the Thai-Iwate kiln reported that approximately 10% of the total charcoal produced in one batch represents dust and small broken charcoal pieces, and most of the pyrolyzed bark from the feedstock material. This fraction of charcoal has the potential to be considered as biochar for soil amendment. However, there is currently no significant use of this fraction as biochar, apart from a few household-scale applications, as there is a lack of knowledge of using biochar as a soil amendment. Moreover, a proper characterization of this material has to be undertaken to ensure a safe and targeted application. Subsequent analyses are necessary to ensure a stable output quality, especially in view of potential contaminants, regardless of the original biomass' characteristics.

Research on soil improvement by applying charcoal or biochar is underway in the region, particularly in Thailand (Figure 13.6b). Soils in this region are usually heavily weathered and clays degraded, leading to low cation exchange capacities (CEC). Biochar could help to restore CEC and therefore reduce nutrient leaching and increase fertilizer efficiency (Prakongkep et al., 2015; Butphu, et al., 2015; Yooyen et al., 2015). However, to date, there is no wide use of charcoal as a soil amendment in the region, despite a few local initiatives and projects.

Charcoal producers in central Lao PDR occasionally use rice husk to cover the Tao Phi kiln as mentioned in Section 13.3.1, and subsequently use the charred husk as biochar for soil amendment (Mekuria et al., 2012). Rice husk char is mainly applied as biochar for soil amendment in several studies on biochar application in Lao PDR. Biochar application improved soil properties and local economy as a consequence of increased soil fertility (Mekuria et al., 2012, 2013, 2014; Jenkins et al., 2015).

Biochar can be locally produced by a range of simple technologies for charcoal production as introduced above (Section 13.3). The production of biochar also takes advantage of

Figure 13.6. (a) Charcoal residues from a beehive kiln situated near Krabi (Southern Thailand). The feedstock material in the background used here is derived from rubber tree (*Hevea brasiliensis*) plantations. A large share of this material consists of chunks of pyrolyzed bark. (b) The material is occasionally used as a soil amendment after composting with organic waste at household scales. Photo credit (a,b): Viktor Bruckman.

marginal utilization of agricultural wastes to improve soil fertility and productivity (Hugill, 2013; Mekuria et al., 2013; Anonymous, 2013). In north-eastern Thailand, the 200 L horizontal drum kilns are increasingly used to produce charcoal and increasing amounts are used as biochar from tree branches, bamboo and other low-diameter feedstock sources. However, Hugill (2013) reported that rice husk is probably the most abundant feedstock for biochar production in Thailand. In addition, maize cobs and stover, coconut shells and sugar cane bagasse are other potential sources for biochar production.

Sirising (2013) reported that local people in Roi Et province (north-eastern Thailand) were interested in biochar production at household scales for their own crops after observing the success of a pilot model. More than half of the attendees at the biochar training course provided within the framework of this project have adopted biochar application.

13.6 Conclusions and Recommendations

Currently, charcoal in Thailand, Lao PDR and Cambodia is mainly utilized for cooking and heating purposes. The production technologies are based on local knowledge using traditional systems. Biomass for the production of charcoal is mostly extracted from natural or managed forests. While unsustainable and illegal logging activities are the main cause of deforestation and forest degradation in this region, inefficient charcoal production technologies pose health risks for the kiln operators and provide low yields of charcoal. In addition, a higher share of carbon is directly emitted to the atmosphere from inefficient systems. The US National Risk Management Research Laboratory (1999) reported that Thai kilns may significantly contribute to the total GHG emissions of the country. They estimated that Thai kilns emitted 5.0 Mt CO_2-C equivalent annually, which is equal to 7.4% of the total national emissions from fossil fuel burning. The study found that on average, 52% of feedstock carbon was converted to charcoal. The rest is mainly lost in gaseous form during the production process, with the largest share of 20–25% attributed to CO_2 and 9–18% to gaseous products of incomplete combustion (CO, CH_4, HC). The rest is converted to other liquid or solid products, such as ash, aerosol and brands (partially carbonized solids).

To solve these problems, an integrated approach is needed, considering the entire supply chain, starting from a robust biomass energy strategy that is implemented in policy processes and ensures both sustainability and income for rural communities. In addition to that, more efficient kiln technologies currently in the testing phase should be introduced. This requires support as the costs are usually higher and training is necessary in order to ensure an efficient and low-emission pyrolysis process.

Pyroligneous acid is collected as a co-product and used for various purposes. However, inappropriate use of these products may cause long-term soil pollution and degradation. Moreover, charcoal dust or bottom remained charcoal and small broken charcoal chunks are commonly abandoned. There is lack of local knowledge to utilize these residues as a soil amendment. Local farmers may benefit from the soil conditioning properties of biochar and more efficient fertilizer use. However, there is a substantial lack of research proving the benefits, impacts and safety in terms of contaminants from charcoal residues as biochar. The same is true for pyroligneous acid, therefore we suggest further research and development, together with an assessment of the economic impact for local communities and farmers in the long run.

13.6.1 Wood Energy Planning and Policy

Energy policies should promote sustainable biomass energy schemes, including fuelwood and charcoal, to achieve the aim of providing energy to all sectors of society at reasonable prices. Biomass is locally available and a renewable source of energy, and also provides other benefits, such as income for rural communities while supporting climate change mitigation.

Within forest policy, biomass energy provision must be fully integrated into forest management plans and allow participation of local communities to obtain their share of this renewable feedstock. The Department of Forestry of Cambodia implemented this concept, which generates revenue and at the same time local communities will assist in forest management. However, it was noticed that provincially adapted biomass energy policies are necessary to address the local specifics of supply and consumption.

In rural areas, the issue of land tenure must be carefully addressed as most of the fuelwood collectors do not own the land. Accordingly, community forestry projects must include local biomass energy provision, especially in areas with high poverty. In Thailand, 6.5% of the total country area has been converted to plantations to date (Royal Forest Department, 2014b), mostly for providing industrial feedstock for, for example, the pulp and paper industry. It should be assessed whether residual materials can be successfully used to produce biochar as a soil amendment to improve farming productivity. In addition, planting of fast-growing species on small areas by local communities might reduce deforestation in the region. However, policies have to ensure that no natural forest is being cleared for additional plantations.

Further research in the area of sustainable biomass energy supply from forests and plantations is necessary to successfully meet future demands and contribute to sustainable development; biochar as a soil amendment can be a co-product in biomass energy production to sustain productivity in agriculture and plantation. It must be ensured that results of research projects are implemented in both education and training as well as in the relevant policy framework. International and regional collaboration may contribute to successful developments that protect forests from degradation while simultaneously meeting the demands for biomass.

13.6.2 Utilization of Wastes from Charcoal Production

The remainders or wastes such as ash, partially pyrolyzed chunks of wood and charcoal dust and small particles not suitable for selling on the market can be used in other ways aside from disposal. Currently research is focused on the potential use of such materials as soil amendments. It is suggested that this material has beneficial effects, in particular in tropical soils as the soil acidity can be decreased (liming effect of ash), the CEC of the soil can be increased and thus the fertilizer efficiency improved. However, this requires a clear certification and monitoring system that ensures that the material complies with threshold values regarding pollutants. The traditional methods of charcoal production especially likely produce very heterogeneous waste material with varying properties and potential levels of pollutants due to the difficulty controlling pyrolysis conditions. Therefore it will be challenging to ensure that the waste materials are safe to use as soil amendments.

Another proposed way of utilizing charcoal dust and small particles is the production of charcoal pellets prior to use as a soil amendment, as this reduces health risks caused by dust during amendment and increases applicability, for example by spreading devices. Application of pure and untreated biochar can lead to initial negative effects on soil fertility as it has been shown that fresh biochar may compete with plants for nitrogen. There are some research examples from sub-tropic and Western Australia that reported the decrease of NO_3^- and NH_4^+ after biochar application (Bai et al., 2015; Dempster et al., 2012). Moreover, studying the obvious effects of biochar application is time consuming (Hemwong and Cadisch, 2011; Anulaxtipan et al., 2013). As a result, we suggest that long-term monitoring projects be set up in agricultural environments to develop best management practices, including biochar amendment. Data from the Department of Alternative Energy Development and Efficiency (2013) suggest that charcoal trading in these countries was 142,000 tons, while 10% (14,200 tons) was classified as low-value charcoal or charcoal residue. In view of climate change mitigation, it can be concluded that if this amount was used as biochar for soil amendment (and not for combustion), 9,273 tons of additional carbon would have been sequestered in the soil. This calculation is based on a C content of 65.3% (by wt) in the case of hardwood charcoal from slow pyrolysis (Brewer et al., 2009).

References

Anonymous (2013). *Biochar for Soil Amendment Agriculture around Thailand. Dailynews.* [online] Available at: http://m.dailynews.co.th/News.do?contentId=112561 [Accessed 6 February 2015]

Anulaxtipan, Y., Phianphitak, P., Wanichsathian, S. et al. (2013). *Utilization of Biochar to Improve Crop Yields Report*. Thailand.

Bai, S. H., Reverchon, F., Xu, C., et al. (2015). Wood biochar increases nitrogen retention in field settings mainly through abiotic processes. *Soil Biology and Biochemistry*, 90, pp. 232–240.

Baimark, Y., Threeprom, J., Dumrongchai, N., Srisuwan, Y. and Kotsaeng, N. (2008). Utilization of wood vinegars as sustainable coagulating and antifungal agents in the production of natural rubber sheets. *Journal of Environmental Science and Technology*, 1, pp. 157–163.

Barney, K. (2014). *Sparking Regionalisation? Lao Charcoal Commodity Networks in Greater Mekong*. Lao PDR.

Brewer, C. E., Schmidt-Rohr, K., Satrio, J. A. and Brown, R. C. (2009). Characterization of biochar from fast pyrolysis and gasification systems. *Environmental Progress and Sustainable Energy*, 28, pp. 386–396.

Bunyavejchewin, S. (1989). Above-ground net primary productivity, firewood production and charcoal properties of 5 tree species. *Thai Journal of Forestry*, 8, pp. 60–69.

Burnette, R. (2010). An introduction to wood vinegar. *ECHO Asia Notes: A Regional Supplement to ECHO Development Notes*, 7. [online] Available at: https://c.ymcdn.com/sites/members.echocommunity.org/resource/collection/F6FFA3BF-02EF-4FE3-B180-F391C063E31A/Wood_Vinegar.pdf [Accessed 2 February 2015]

Butphu, S., Toomsan, B., Cadisch, G., Rasche, F. and Kaewpradit, W. (2015). *Impact of Biochar Application on Upland Rice Production, N Use Efficiency and Greenhouse Gas Emission in a Rotation System with Sugarcane*. Stuttgart, Germany: Food Security Center.

Dempster, D. N., Gleeson, D. B., Solaiman, Z. M., Jones, D. L. and Murphy, D. V. (2012). Decreased soil microbial biomass and nitrogen mineralisation with Eucalyptus biochar addition to a coarse textured soil. *Plant and Soil*, 354, pp. 311–324.

Department of Alternative Energy Development and Efficiency (2009). *Thailand Alternative Energy Situation 2009*. Bangkok, Thailand: Department of Alternative Energy Development and Efficiency.

Department of Alternative Energy Development and Efficiency (2010). *Thailand Alternative Energy Situation 2010*. Bangkok, Thailand: Department of Alternative Energy Development and Efficiency.

Department of Alternative Energy Development and Efficiency (2011). *Thailand Alternative Energy Situation 2011*. Bangkok, Thailand: Department of Alternative Energy Development and Efficiency.

Department of Alternative Energy Development and Efficiency (2012a). *Energy Balance of Thailand 2012*. Bangkok, Thailand: Department of Alternative Energy Development and Efficiency.

Department of Alternative Energy Development and Efficiency (2012b). *Thailand Alternative Energy Situation 2012*. Bangkok, Thailand: Department of Alternative Energy Development and Efficiency.

Department of Alternative Energy Development and Efficiency (2013). *Thailand Alternative Energy Situation 2013*. Bangkok, Thailand: Department of Alternative Energy Development and Efficiency.

Department of Forestry (2012). *Forest Cover Assessment in 2010*. Vientiane: Lao PDR.

Food and Agriculture Organization of the United Nations (2010). *Global Forest Resource Assessment 2010: Global Tables*. [online] Available at: www.fao.org/forestry/fra/fra2010/en [Accessed 13 January 2015]

Forestry Administration (2011). *Cambodia Forest Cover Change 2006–2010*. Phnom Penh, Cambodia: Forestry Administration.

Forestry Administration (2010). Cambodia Forest Outlook Study. [online] Available at: www.fao.org/docrep/014/am627e/am627e00.pdf [Accessed 13 January 2015]

Garcia-Perez, M., Lewis, T. and Kruger, C. E. (2010). *Methods for Producing Biochar and Advanced Biofuels in Washington State: Part 1: Literature Review of Pyrolysis Reactors; First Project Report*. Pullman, WA: Department of Biological Systems Engineering and the Center for Sustaining Agriculture and Natural Resources, Washington State University.

Government of Lao PDR (2005). *Forestry Strategy to the Year 2020 of Lao PDR*. Vientiane: Lao PDR.

Hemwong, S. and Cadisch, G. (2011). *FSC Brief No. 3: Charcoal Amendments to Improve Soil Fertility and Rice Production in NE Thailand*. Stuttgart, Germany: Food Security Center, Universität Hohenheim.

Heov, K. S., Khlok, B., Hansen, K. and Sloth, C. (2006). The value of forest resources to rural livelihoods in Cambodia. In *Cambodia Development Research Institute (CDRI) Policy Brief 2*. Phnom Penh, Cambodia.

Hugill, B. (2013). *Biochar – An Organic House for Soil Microbes*. ECHO Asia Notes: A Regional Supplement to ECHO Development Notes. [online] Available at: http://c.ymcdn.com/sites/members.echocommunity.org/resource/collection/49B3D109-0DE9-458E-915B-11AAF1A67E20/TN_75_Biochar–An_Organic_House_for_Soil_Microbes.pdf [Accessed 3 February 2015]

IEA and ERIA (2013). *Southeast Asia Energy Outlook: World Energy Outlook Special Report*. Paris, France: IEA Publications.

Jenkins, M., Souvanhnachit, M., Rattanavong, S., et al. (2015). Enhancing productivity and livelihoods among smallholder irrigators through biochar and fertilizer amendments. In: *Centre de cooperation Internationale en Recherche Agronomique pour le Developpement (CIRAD). 3rd Global Science Conference on Climate-Smart Agriculture, Montpellier, France, 16–18 March 2015. Parallel Session L1 Regional Dimensions*. Paris, France: Centre de cooperation Internationale en Recherche Agronomique pour le Developpement (CIRAD), p. 141.

Kuhakan, C. (2007). *Manual of Charcoal Production and Wood Vinegar*. Bangkok, Thailand: Green Media and Products Co.

Kumar, M., Gupta, R. C. and Sharma, T. (1992). Effects of carbonisation conditions on the yield and chemical composition of Acacia and Eucalyptus wood chars. *Biomass and Bioenergy*, 3, pp. 411–417.

Kumar, M., Verma, B. B. and Gupta, R. C. (1999). Mechanical properties of Acacia and Eucalyptus wood chars. *Energy Sources*, 21, pp. 675–685.

Kwon, G., Kim, D., Oh, C., Park, B. and Kang, J. (2014). Tailoring the characteristics of carbonized wood charcoal by using different heating rates. *Journal of the Korean Physical Society*, 64, pp. 1474–1478.

Laemsak, N. (2008). *Utilization of Fast Growing Tree Species for Charcoal and Wood Vinegar Production in Households*. Nakorn Ratchasima, Thailand: Forest Research Center and Udomsap Subdistrict Municipality.

Laemsak, N. and Thornrasin, M. (2005). *The Final Report: The Feasibility Study of Charcoal and Wood Vinegar Production Plant of Biopower Plus Co., Ltd.* Bangkok, Thailand.

Loo, A. Y., Jain, K. and Darah, I. (2008). Antioxidant and radical scavenging activities of the pyroligneous acid from a mangrove plant *Rhizophora apiculata*. *Food Chemistry*, 104, pp. 300–307.

Manivong, V. and Cramb, R. A. (2008). Economics of smallholder rubber expansion in Northern Laos. *Agroforestry Systems*, 77, pp. 113–125.

Mathew, S. and Zakaria, Z. A. (2015). Pyroligneous acid – the smoky acidic liquid from plant biomass. *Applied Microbiology and Biotechnology*, 99, pp. 611–622.

Mekuria, W., Sengtaheuanghoung, O., Hoanh, C. T. and Noble, A. (2012). Economic contribution and the potential use of wood charcoal for soil restoration: a case study of village-based charcoal production in Central Laos. *International Journal of Sustainable Development and World Ecology*, 19, pp. 415–425.

Mekuria, W., Getnet, K., Noble, A., Hoanh, C. T., McCartney, M. and Langan, S. (2013). Economic valuation of organic and clay-based soil amendments in small-scale agriculture in Lao PDR. *Field Crop Research*, 149, pp. 379–389.

Mekuria, W., Noble, A., Hoanh, C. T., McCartney, M., Sengtaheuanghoung, O., Sipaseuth, N., Douangsavanh, S., Langan, S. and Getnet, K. (2014). *The potential role of soil amendments in increasing agricultural productivity and improving the livelihood of smallholders in Lao PDR*. Paper presented at the 15th National Agriculture and Forest Research Institute Anniversary Symposium on Agriculture and Forest Research for Development, Vientaine, Lao PDR, 8–10 April 2014.

Missio, A. L., Mattos, B. D., Gatto, D. A. and de Lima, E. A. (2014). Thermal analysis of charcoal from fast-growing eucalypt wood: influence of raw material moisture content. *Journal of Wood Chemistry and Technology*, 34, pp. 191–201.

National Institute of Statistics (1997). *Socio-Economic Survey of Cambodia 1996*. Phnom Penh, Cambodia: National Institute of Statistics, Ministry of Planning.

National Institute of Statistics (2008). *Statistical Yearbook 2008: General Population Census of Cambodia 2008*. Phnom Penh, Cambodia: National Institute of Statistics, Ministry of Planning.

National Institute of Statistics (2014). *CamInfo 2014*. [online] Available at: http://app.nis.gov.kh/caminfo/libraries/aspx/home.aspx [Accessed 24 February 2015]

National Risk Management Research Laboratory (1999). *Research and Development: Greenhouse Gases from Small-scale Combustion Devices in Developing Countries: Charcoal-making Kilns in Thailand*. Washington, DC: United States Environmental Protection Agency.

Ogawa, T. (2011). *Japan Biochar Association*. [online] Available at www.geocities.jp/yasizato/pioneer.htm [Accessed 2 December 2015]

Panunumpa, N., Tatayanon, S., Kuhakan, C., Sutthiwilairat, L. and Piriyayotha, T. (2013). *Manual of Wood Utilization: Energy and Charcoal*. Bangkok, Thailand: Royal Forest Department.

Payamara, J. (2011). Usage of wood vinegar as new organic substance. *International Journal of ChemTech Research*, 3, pp. 1658–1662.

Phongoudome, C. (2014). Desk study (2000–2014) "Available biomass and their current situation in Lao PDR". In *The Second ACMECS Bioenergy Workshop Biomass for Community Energy: Production and Utilization Technology*. Bangkok, Thailand, pp. 185–196.

Pitman, R. M. (2006). Wood ash use in forestry – a review of the environmental impacts. *Forestry*, 79, pp. 563–588.

Prakongkep, N., Gilkes, R. J. and Wiriyakitnateekul, W. (2015). Forms and solubility of plant nutrient elements in tropical plant waste biochars. *Journal of Plant Nutrition and Soil Science*, 178, pp. 732–740.

Puthson, P. (1990). Effect of heating rates on mass loss and properties of charcoal obtained from pyrolysis of red gum wood. *Thai Journal of Forestry*, 9, pp. 121–128.

Rahmat, B., Pangesti, D., Natawijaya, D. and Sufyadi, D. (2014). Generation of wood-waste vinegar and its effectiveness as a plant growth regulator and pest insect repellent. *BioResources*, 9, pp. 6350–6360.

Royal Forest Department (2013). *Forestry Statistics Data 2013*. Bangkok, Thailand: Royal Forest Department.

Royal Forest Department (2014a). *Executive Summary: Forest Area Assessment Project 2012–2013*. Bangkok, Thailand: Royal Forest Department.

Royal Forest Department (2014b). *Forest Plantation Expansion Plan*. Bangkok, Thailand: Royal Forest Department.

San, V., Spoann, V., Ly, D. and Chheng, N. V. (2012). Fuelwood consumption patterns in Chumriey Mountain, Kampong Chhnang Province, Cambodia. *Energy*, 44, pp. 335–346.

Saunders, J. (2014). *Illegal Logging and Related Trade. The Response in Lao PDR*. London, UK: Chatham House, the Royal Institute of International Affairs.

Sirising, S. (2013). *Development of Learning Process in Applying Biochar for Soil Improvement for Agriculture*. Bangkok, Thailand: Kasetsart University.

Somerville, M. and Jahanshahi, S. (2015). The effect of temperature and compression during pyrolysis on the density of charcoal made from Australian eucalypt wood. *Renewable Energy*, 80, pp. 471–478.

Southavilay, T. and Castren, T. (1999). Timber trade and wood flow-study, Lao PDR. [online] Available at: www.mekonginfo.org/assets/midocs/0002916-environment-timber-trade-and-wood-flow-study-lao-pdr.pdf [Accessed 27 February 2015]

Takahara, Y., Katoh, K., Inaba, R. and Iwata, H. (1993). Study on odor control using wood vinegars. *Japanese Journal of Public Health*, 40, pp. 29–38.

Takahara, Y., Katoh, K., Inaba, R. and Iwata, H. (1994). Study on odor control using wood vinegars (II). Application of wood vinegars to piggery wastes. *Japanese Journal of Public Health*, 41, pp. 147–156.

Theapparat, Y., Chandumpai, A., Leelasuphakul, W. and Leamsak, N. (2015). Pyroligneous acids from carbonisation of wood and bamboo: their components and antifungal activity. *Journal of Tropical Forest Science*, 27, pp. 517–526.

Tiilikkala, K., Fagernäs, L. and Tiilikkala, J. (2010). History and use of wood pyrolysis liquids as biocide and plant protection product. *The Open Agriculture Journal*, 4, pp. 111–118.

Top, N., Mizoue, N., Kai, S. and Nakao, T. (2004). Variation in woodfuel consumption patterns in response to forest availability in Kampong Thom Province, Cambodia. *Biomass and Bioenergy*, 27, pp. 57–68.

UN-REDD (2011). *Cambodia National UN-REDD National Programme Document*. [online] Available at: www.unredd.net/index.php?option=com_docman&task=doc_view&gid=7388&tmpl=component&format=raw&Itemid=53 [Accessed 13 January 2015]

Wang, Z., Cao, J. and Wang, J. (2009). Pyrolytic characteristics of pine wood in a slowly heating and gas sweeping fixed-bed reactor. *Journal of Analytical and Applied Pyrolysis*, 84, pp. 179–184.

Wititsiri, S. (2011). Production of wood vinegars from coconut shells and additional materials for control of termite workers, *Odontotermes* sp. and striped mealy bugs, *Ferrisia virgata*. *Songklanakarin Journal of Science and Technology*, 33, pp. 349–354.

Woolf, D., Amonette, E. J., Alayne Street-Perrott, F., Lehmann, J. and Joseph, S. (2010). Sustainable biochar to mitigate global climate change. *Nature Communication*, 56, pp. 1–8.

Yooyen, J., Wijitkosum, S. and Sriburi, T. (2015). Increasing yield of soybean by adding biochar. *Journal of Environmental Research and Development*, 9, pp. 1066–1074.

Yoshimoto, T. (1994). Toward enhanced and sustainable agricultural productivity in the 2000s: breeding research and biotechnology. In: *Proceedings of The 7th International Congress of the Society for the Advancement of Breeding Researches in Asia and Oceania*, held on 16–20 November 1993, Taipei, Taiwan.

Part IV

Biochar Application as a Soil Amendment

14

Biochar Applications to Agricultural Soils in Temperate Climates – More Than Carbon Sequestration?

GERHARD SOJA, ELENA ANDERS, JANNIS BÜCKER,
SONJA FEICHTMAIR, STEFAN GUNCZY, JASMIN KARER,
BARBARA KITZLER, MICHAELA KLINGLMÜLLER, STEFANIE KLOSS,
MAXIMILIAN LAUER, VOLKER LIEDTKE, FRANZISKA REMPT,
ANDREA WATZINGER, BERNHARD WIMMER,
SOPHIE ZECHMEISTER-BOLTENSTERN AND FRANZ ZEHETNER

Abstract

Biochar as a boon for soil fertility in the tropics still has to show that it is able to provide the same benefits to soils in temperate regions. Here an Austrian study with the objective to analyze the extent of benefits that biochar application offers to agricultural soils in Europe beyond its role as a carbon sequestration strategy is presented. Based on hypothesis testing, several potential benefits of biochar were examined in a series of lab analyses, greenhouse and field experiments. Three hypotheses could be confirmed: biochar can protect groundwater by reducing the nitrate migration in seepage water; biochar can mitigate atmospheric greenhouse gas accumulation by reducing soil N_2O emissions; and biochar can improve soil physical properties by increasing water storage capacity. One hypothesis was only partly confirmed: biochar supports the thriving of soil microorganisms only in specific soil and climate settings. Two hypotheses were refuted: biochar does not generally provide nutrients to plants except when produced from specific feedstocks or by combining it with mineral or organic fertilizers; the cost-effectiveness of biochar application is not given under current production costs if the existing benefits of biochar are not transferable to financial value.

14.1 Introduction

"Black is the new green." This term was coined in 2006 when popularized reports about the astonishing fertility of Amazonian terra preta soils kickstarted global interest in biochar during an era of food production deficits and dwindling soil resources. Based on the pioneering descriptions of nineteenth century explorer Herbert Smith, in the 1990s the soil scientists Wim Sombroek, Johannes Lehmann and Bruno Glaser identified the deliberate addition of charred settlement wastes to the soils as one of the key components for amending soils in the Amazon region (Marris, 2006; Glaser and Birk, 2012). The soil enrichment strategy with charcoal was by no means unique for Amazonia – this nearly forgotten agricultural heritage, frequently displaced by the emergence of mineral fertilizers, was re-discovered also in Japan, China, New Zealand, West Africa and Northern Europe.

The increase of the organic carbon content of the amended Amazonian soils may amount up to additional 150 t ha^{-1} in the top 0.3 m compared to adjacent Ferralsols (Glaser et al., 2002; Thayn et al., 2011). The stability of the additional carbon against physico-chemical or microbial degradation allows for the remarkable persistence of this increase. Mean residence times of biochar in the soil have been assessed in the centennial to millennial time scale by Hammes et al. (2008), Novak et al. (2010) and Major et al. (2010). Although a minor labile fraction of biochar may be mineralized within decades (Hamer et al., 2004; Nguyen et al., 2008), the larger stable fraction is characterized by much longer residence times. This means that carbon sequestration with biochar returns a much higher proportion of carbon permanently to the soil (about 30% of the initial C) than burning or biological decomposition of unmodified, fermented or composted plant residues (<10–20% after 5–10 years; Lehmann et al., 2009). A shorter mean residence time of <100 years may only be expected when the pyrolysis temperature did not exceed 450°C, when cereal straw was the feedstock material and incubation temperature was permanently elevated (>20°C; Fang et al., 2015; Purakayastha et al., 2015). This long-term effect on the removal of carbon dioxide from the atmosphere distinguishes biochar applications as a climate change mitigation measure. Any CO_2 molecule removed from the atmosphere as pyrolyzed organic biomass cannot contribute anymore to the greenhouse effect. Positive priming of native soil organic carbon (increased mineralization) may only occur after addition of biochars pyrolyzed at lower temperatures and in certain clay-poor soils like inceptisols in the first few months after application (Fang et al., 2015). When adding biochars produced at higher temperatures to inceptisols or other soil types, usually negative priming prevails (decreased mineralization of native soil organic carbon). The value of biochar as a permanent carbon-negative soil management and bioenergy generation strategy has been repeatedly confirmed (Matovic, 2011; Galinato et al., 2011; Spokas et al., 2012; Stewart et al., 2013). On that basis, the United Nation Environment Programme has endorsed biochar as one of the options for solving the soil-related problems of the twenty-first century in their education campaigns (UNEP, 2013).

In spite of its potential as a carbon sequestration measure, there is no universal acceptance of biochar application to soil in carbon accounting, yet. Without remuneration for carbon credits, farmers need other incentives for applying biochar to their soils. Derived from the original observations of increased fertility of Amazon dark earths, the idea arose that such improvements could be replicated also in degraded soils of agricultural areas in other parts of the world. However, the soils in many intensively cultivated areas are very different to the acidic, nutrient-poor Oxisols/Ferralsols of the Amazon region. More evidence for the validity of these assumptions would be needed to include biochar in a fertility management strategy for agricultural soils. Similarly, other benefits purported as advantages of biochar application still require confirmation for specific agricultural regions.

This research also needs to be applied to the situation of Austrian agricultural soils. Therefore, a three-year study was performed that had the objective to test the possibility to reap benefits additional to carbon sequestration from the application of biochar to soils typical for Austrian agro-ecosystems and also in a broader sense to soils of temperate climate regions. The broad consensus about the efficacy of biochar deployment for long-term carbon sequestration under various environmental conditions left no doubt that this would also be valid for Austria. However, the difference in soil and climate characteristics required additional investigations of the extent to which other environmental resources could also take advantage of biochar applications. This might create additional incentives for the development of a biochar market beyond the current embryonic stage. So the project team formulated five hypotheses to be tested in the course of this study:

1. The natural resource water (groundwater) will benefit from biochar application because of reduced nitrogen input.
2. The natural resource soil will benefit from biochar application by physical effects and the promotion of soil microorganisms.
3. The natural resource air will benefit from biochar application by reduction of soil greenhouse gas emissions.
4. Crops cultivated on biochar-amended soils will benefit from biochar application by enhanced nutrient availability.
5. Biochar application for taking advantage of biochar benefits will be economically feasible.

The testing of these hypotheses required experiments on the lab scale, a greenhouse pot experiment with a realistic cropping pattern of three crops, and two field experiments over two years. This chapter summarizes the most important results of these studies designed for testing the above hypotheses.

14.2 Materials and Methods

14.2.1 Study Site and Biochar Production

Three soils from the Austrian provinces of Lower Austria and Styria were sampled in summer 2010. The soils were classified as Chernozem (48°19'52.6"N, 15°44'20.5"E; parent material loess; 547 mm mean annual precipitation) and Planosol (48°46'32.9"N, 15°14'28.6 E; parent material granite, 667 mm mean annual precipitation; Kloss et al., 2014a). The third soil, taken in the province of Styria, was classified as Cambisol (47°13'46.0"N, 15°50'40.6"E; parent material pellitic Tertiary sediments; 883 mm mean annual precipitation; Kloss et al., 2014a).

The soils used for the experiment were taken from the top 30 cm. After transport, the soils were air-dried and homogenized. The soils were not sieved, but large aggregates were shredded to approximately 2–3 cm and rocks > 3 cm were removed before generating soil-biochar mixtures (Kloss et al., 2014a).

Biochar (BC) was produced from three different feedstocks: vineyard pruning (*Vitis vinifera* L.), wheat straw (*Triticum aestivum* L.) and mixed woodchips. BCs from wheat straw (WS) and woodchips (WC; maximum dimensions 2 cm × 2 cm × 2 cm) were pyrolyzed in a rotary furnace at a highest treatment temperature (HTT) of 525°C (residence time 60 min). Vineyard prunings (VP) were pyrolyzed at an HTT of 400°C (VP400) and 525°C (VP525) (residence time eight and six hours, respectively) in a stainless tube furnace (for further details see Kloss et al., 2012). After cooling in the furnace under an argon atmosphere, WS- and VP-derived biochars (but not WC-derived biochar) were ground to < 2 mm (Kloss et al., 2014a).

14.2.2 Soil and Biochar Characterization

The basic soil and BC characterizations are summarized in Table 14.1 (Kloss et al., 2014a). The soil types differed in soil texture, pH, carbonate content and CEC. The lowest pH and CEC were found for the sandy Planosol. The highest pH was found in the Chernozem due to the high carbonate content. Standard methods for the determination of pH, EC, ash content, C and N of the BCs were used as described in Kloss et al. (2012). CEC was determined with sequential water-$BaCl_2$ extraction: 40 mL of water was added to 2 g BC and allowed to stand overnight. The next day, the suspension was shaken for 1 h and centrifuged. Twenty mL of the supernatant was drawn with a Pasteur pipette, filtered and the water-extractable cations were measured in the filtrate. The remaining suspension was treated with 20 mL of 0.2 mol L^{-1} $BaCl_2$, shaken for 2 h and filtered. In the filtrate, Ca, Mg, K, Na, Al, Fe and Mn were measured using atomic absorption spectroscopy (AAS). The results were corrected for the water-extracted amounts that remained in the extract and CEC was calculated as the sum of the above-mentioned cations. Specific surface area (SSA) was determined with the N_2 adsorption method according to Brunauer et al. (1938). The individual BCs differed in most investigated parameters, and pyrolysis temperature significantly affected BC properties (Kloss et al., 2014a).

14.2.3 Experimental Set-up in the Greenhouse

The pot experiment (soil columns also used as microlysimeters) was installed and maintained under greenhouse conditions. Before filling the pots, the soil and biochar were mixed in a closed cement mixer at BC concentrations of 1% and 3% (which corresponded to 30 and 90 t fresh matter (fm) ha^{-1} or 24 and 72 t dry matter (dm) ha^{-1} at an incorporation depth of 20 cm in the field). The resulting bulk densities of the treatments within the microlysimeter experiment ranged from 1.35 g cm^{-3} (Planosol control) to 1.11 g cm^{-3} (Cambisol, 3% w/w, WC). Each column had a siphon-like outlet with a fixed hosepipe

Table 14.1. *Basic soil and biochar characterizations (n = 3)*

Soil		Texture	pH (CaCl$_2$)	EC (dS m^{-1})	CEC (mmol$_c$ kg^{-1})	C: N ratio	Carbonate (w.-%)
Planosol		sandy loam	5.4 ± 0.0 a	0.04 ± 0.01 a	75.1 ± 0.4 a	14.9 ± 1.6 b	0.0 ± 0.0 a
Cambisol		clay loam	6.6 ± 0.1 b	0.10 ± 0.00 b	209.4 ± 2.2 b	13.8 ± 0.8 ab	0.0 ± 0.0 a
Chernozem		silt loam	7.4 ± 0.1 c	0.17 ± 0.01 c	208.6 ± 3.6 b	11.9 ± 0.7 a	15.8 ± 0.1 b

Biochar	Pyrolysis temperature	Ash content (%)	pH (CaCl$_2$)	EC (dS m^{-1})	CEC (mmolc kg^{-1})	C: N ratio	BET -N2 SSA (m^{2}g^{-1})
Straw	525°C	28.10	9.7 ± 0.0 c	5.18 ± 0.06 d	148.5 ± 0.8 d	63.8 ± 1.6 c	12.26 ± 1 c
Woodchips	525°C	15.20	8.9 ± 0.1 b	1.58 ± 0.02 c	93.0 ± 1.9 b	58.2 ± 0.7 b	26.41 ± 1 d
Vineyard pruning	400°C	4.30	8.3 ± 0.0 a	1.48 ± 0.03 b	123.5 ± 1.3 c	52.0 ± 2.4 a	1.69 ± 0 a
Vineyard pruning	525°C	7.70	8.8 ± 0.1 b	1.08 ± 0.03 a	78.8 ± 1.4 a	58.0 ± 1.9 b	4.85 ± 0 b

Note: Analyses were carried out as described in Kloss et al. (2012). Different letters indicate significant difference within one column (P < 5%; Tukey's test). EC: electrical conductivity; CEC: cation exchange capacity; BET-N$_2$ SSA: Brunauer-Emmett-Teller specific surface area (N$_2$ adsorption).
Source: Kloss et al. (2014a).

connected to the seepage water collector and a 3 cm sand layer at the bottom with a mesh sieve underneath. The pots were 40 cm high and had a diameter of 23.5 cm (resulting volume: 17.3 L). The pots were arranged in a randomized block design. Altogether, 18 treatments with five replicates as well as controls were installed (Kloss et al., 2014a).

The pots were sown with mustard (*Sinapis alba* L. cv. Serval) as first crop, followed by barley (*Hordeum vulgare* L. cv. Xanadu) and red clover (*Trifolium pratense* L. cv. Reichersberger Neu). The standard fertilization rate for mustard was 40 kg N ha^{-1} using a commercial combination fertilizer (N: P: K: S = 15: 7: 12: 3; Linzer Star), which was increased to 100 kg N ha^{-1} for barley. To investigate the effect of varying N-fertilization rates, an additional set of pots with Planosol was treated with four varying N-fertilizer rates ranging from 0 (N0) to 80 (N80) kg N ha^{-1} for mustard and 0 (N0) to 200 (N200) kg N ha^{-1} for barley, both with and without BC (WC). Red clover as third crop received no fertilizer. The water content of the pots was surveyed by equipping one replicate of each treatment with TDR (time-domain reflector) probes and ECHO probes, respectively. Irrigation was regularly carried out according to the measured water content using artificial rain water (3 mg Ca L^{-1}; 50% CaCl$_2$, 50%

CaSO$_4$). The experiment was run under semi-controlled temperature conditions simulating a day–night temperature profile according to average field conditions (Kloss et al., 2014a).

14.2.4 Soil and Plant Analyses

After preparation of the soil-BC mixtures and before filling of the pots, a basic soil sampling was carried out. In addition, the soils in the pots were sampled after barley harvest approximately seven months after the start of the pot experiment using a small core drill (four cores unified to one composite sample per pot). The control treatment refers to soil without BC but with standard fertilization (Kloss et al., 2014a).

At each harvest, the above-ground biomass was cut off manually 1 cm above soil surface, dried at 80°C to constant weight and weighed for dry mass (DM) determination. Barley ears were threshed to allow for nutrient analyses separated according to grains and straw. Plant N concentration was determined using a CHNS-O elemental analyzer. Subsamples of mustard, barley straw and clover were ground using a ball mill and analyzed for elemental composition after full acid digestion (HNO$_3$: HClO$_4$ = 20 + 4 mL), measured by inductively coupled plasma optical emission spectrometry (ICP-OES) (Kloss et al., 2014b).

For CEC measurements, 5 g of soil was mixed with 100 mL 0.1 M BaCl$_2$ solution and settled overnight. After shaking for 2 h and subsequent filtration, exchangeable cations (Ca, Mg, K, Na, Al, Fe, Mn) were measured by AAS; CEC was calculated as the sum of the exchangeable cations (in mmol$_c$ kg^{-1}). Total C and N were determined by dry combustion (Tabatabai and Bremner, 1991) and measured with an elemental analyzer. Subsequently, organic carbon (C$_{org}$) was calculated as the difference between total C and carbonate C, which had been determined gas volumetrically (Burt, 2004). Phosphorus (P$_{CAL}$) and potassium (K$_{CAL}$) were extracted with calcium-acetate-lactate (CAL). In detail, 5 g of soil was mixed with 100 mL of the 1:5 diluted stock solution (77 g (CH$_3$COO)$_2$Ca, 39.5 g C$_6$H$_{10}$CaO$_6$ and 89.5 mL 100% CH$_3$COOH, filled up to 1 L with distilled water) and shaken for 2 h. The solution was filtered. P$_{CAL}$ was measured with a photometer at 710 nm using the molybdate method 1; K$_{CAL}$ was measured using AAS (Kloss et al., 2014a).

14.2.5 Leachate Sampling Campaigns and Chemical Analyses

The leachates were collected at intervals of four (mustard) and eight weeks (barley), respectively, and analyzed for their volume, pH value, EC, nitrate (NO$_3^-$) using ion chromatography, potassium (K$^+$) and dissolved phosphorus (P$_{DISS}$) with an ICP-MS, and DOC with a TOC/DOC analyzer.

14.2.6 Experimental Set-up in the Field

At the study sites with Chernozem and Cambisol, two field experiments were installed in 2011. Each circular net plot (used for harvest analysis, soil and plant sampling) with a diameter of 3.5 m was positioned in the center of a circular gross plot 6.5 m in diameter. The minimum distance between the outer borders of net plots was 6.5 m; the minimum distance between the outer borders of gross plots was 0.5 m. There were four different treatments with four replicates (n=4), arranged in a Latin square. Nutrients (N, P, K) were supplied according to standard agricultural practices in the respective region. The treatments consisted of three different BC application rates (0, 24 and 72 t dm ha^{-1}) with identical mineral N fertilization and one additional treatment without N supplement but with a BC application rate of 72 t ha^{-1}. Further methodological details are given in Karer et al. (2013).

Samples of the soil and the soil–biochar mixtures were packed into metal cylinders of a size of 100 cm³ to a bulk density (BD) representing field conditions, in order to determine water holding capacity (WHC) and water retention characteristics (pF curve). For determination of the WHC, the soil cores were fully saturated and placed on a moist sand bed until the excess water had drained by gravity. The soil cores were weighed after equilibrium was reached. The water retention characteristics of the samples were determined using the pressure chamber method. The soil cores in the metal cylinders were saturated and drained at three pressure steps (6, 30 and 1500 kPa, equivalent to pF 1.8, 2.5 and 4.2) until equilibrium was reached (Klute, 1986). The water content of the samples was measured after each pressure step and after drying the samples at 105°C for 24 h at the end of the procedure. Plant available water (PAW) was calculated as the difference between pF 1.8 and 4.2, in percentage.

14.2.7 Analysis of Phospholipid Fatty Acids (PLFAs) from Soil Microorganisms

Microorganisms were investigated using phospholipid fatty acids (PLFAs) analyses. PLFAs were extracted from soil samples according to the procedure of Bligh and Dyer (1959) as described by Frostegård et al. (1991). We used 2±0.2 g soil for each extraction. Details on the extraction and measuring method used are described in Anders et al. (2013) and Watzinger et al. (2014).

14.2.8 Greenhouse Gas Flux Measurements

Gas samples were taken from the greenhouse experiment described above at several stages of plant development. To allow for the initial flush of respiration after rewetting (Xiang et al., 2008), each sampling took place 2–3 days after an irrigation event, with the first measurement three days after the start of the experiment (= day 0) and the last at day 211. The closed chamber technique was used for measuring the gas exchange between soil and atmosphere. The greenhouse gas fluxes were measured by collecting gas samples from a closed chamber at regular intervals (0, 5, 10 and 30 minutes) after closing and determining the changes with time in CH$_4$, CO$_2$ and N$_2$O concentrations of the closed chamber (Rolston,

1986). Chambers were manually installed and air samples of 15 ml were extracted with a gas tight syringe and injected into 10 ml air-tight evacuated glass vials, which were proofed with silicon sealed rubber lids and aluminium crowns. Until analysis, vials were stored in a refrigerator at 4°C and kept for 14 days at maximum. Analogous methods were applied for studying greenhouse gas emissions from the field experiment with the Chernozem.

Gas concentrations were analyzed by gas chromatography (GC):CH_4 and CO_2 concentrations were determined with a flame ionization detector (FID) and Helium as carrier gas; N_2O concentrations were detected using a ^{63}Ni-electron- capture detector (ECD) and N_2 in ECD quality as carrier gas. The flux rates of CH_4 (µg C m^{-2} h^{-1}), CO_2 (mg C m^{-2} h^{-1}) and N_2O (µg N m^{-2} h^{-1}) were determined using a linear regression of the four data points per measurement.

14.3 Results and Discussion

14.3.1 Biochar Benefits for the Natural Resource Groundwater

The migration of nutrients beyond the plant rooting zone may pose a hazard to groundwater quality and prevents the root uptake of elements vital for plant growth. The biochar effects on nutrient transport in the soil were studied by analyzing leachate concentrations of nitrate, phosphorus and potassium in the microlysimeter experiment. Nitrate retention is of special importance for groundwater protection to avoid violations of water quality thresholds according to Chapter 4.1 of the Water Framework Directive (European Commission, 2000). Phosphorus and potassium concentrations are of less concern for groundwater quality but their release or retention is an indicator of biochar sorption or nutrient supply characteristics.

The results showed that during a period of six months with five seepage water events the soils differed significantly in their nutrient release behavior. Biochar caused the most pronounced reduction of total nitrate loss in the Planosol (by up to 81%) whereas in the Chernozem and Cambisol the reductions were <50%. The soils also differed in their dependence on the biochar concentration: whereas in the Planosol the motto "the more, the better" was valid, in the Chernozem an application rate of 1% biochar withheld nitrate similarly to 3% biochar and in the Cambisol only 3% biochar produced a small reduction (Figure 14.1). Biochar clearly contributed to the release of potassium in all soils, but total dissolved phosphorus was more influenced by the soil type than by the biochar concentration. Only the Planosol exhibited a slightly enhanced phosphorus release in the 3% biochar treatment. The amounts of eluted dissolved organic carbon were only marginally decreased at higher biochar concentrations but the differences did not reach significance thresholds.

The observed reductions of nitrate concentrations in the seepage water clearly indicate the potential of biochar to prevent the undesired migration of nitrogen beyond the rooting zone. Similar results were obtained in the field experiments of Ventura et al. (2013), who found that the nitrate reduction effects were evident even a year after biochar application. Yao et al. (2012) remind that different biochar qualities may have varying sorption

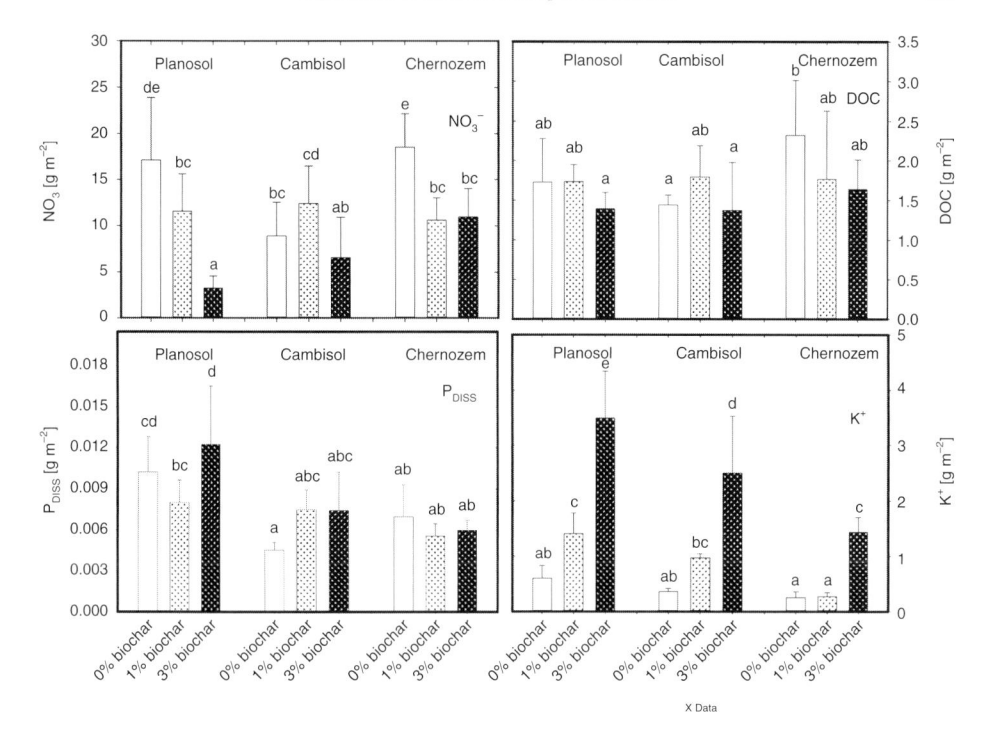

Figure 14.1. Cumulated loads (after five leachate collections) of nitrate (NO$_3^-$), dissolved phosphorus (P$_{DISS}$), potassium (K$^+$) and dissolved organic carbon (DOC) in leachates of three different soils and increasing biochar application rate (1 and 3% w/w); feedstock = woodchips; columns show means ± sd; different letters show significant differences (Duncan's multiple range test; p<0.05).
Source: Bücker (2012).

characteristics; a wood-based biochar was best also in their column experiments to withhold nutrients. The effectiveness of biochar to retain nitrogen from different organic and mineral fertilizer sources was confirmed for soil columns without plants by Angst et al. (2013). If the soil columns were planted, crops like maize could take advantage of the reduced nitrogen losses and use the higher nutrient availability as a growth stimulant (Zheng et al., 2013).

Both the results of our study and recent investigations of other groups confirm the hypothesis that biochar provides benefits for groundwater protection by reducing nitrate migration beyond the rooting zone.

14.3.2 Biochar Benefits for the Natural Resource Soil

Besides the well-documented biochar-induced liming effects (Hass et al., 2012; Slavich et al., 2013; Wang et al., 2014), two other aspects are of special interest: the effects on soil water balance and the effects on soil microorganisms. We hypothesized that biochar would

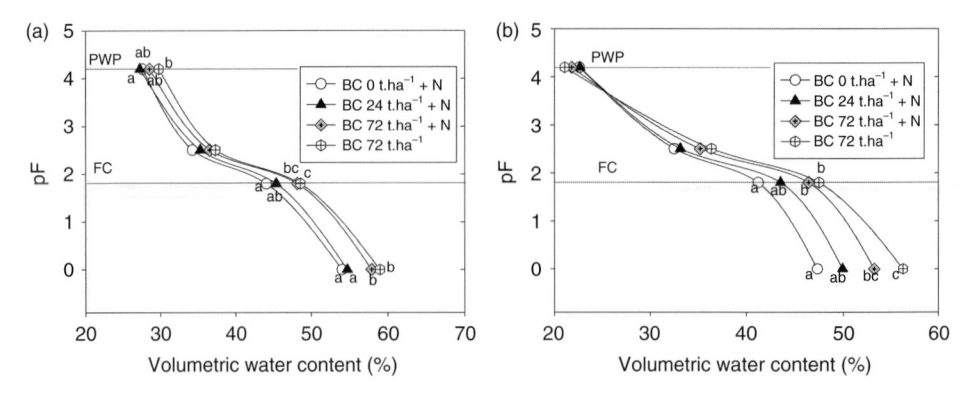

Figure 14.2. pF curves of the Cambisol (A) and the Chernozem (B) (n=4). Volumetric water contents were determined at pF 0, 1.8, 2.5 and 4.2. Different letters indicate significant differences at p = 0.05 (Duncan's multiple range test). PWP: Permanent wilting point; FC: Field capacity; BC: biochar; N: nitrogen fertilization according to common agricultural practice.
Source: Karer et al. (2013). With permission.

support both, thereby improving the potential of soil to supply plants with water and the function of soil as a microbial habitat. In our investigations we analyzed soils from the field experiments with and without biochar.

In the Cambisol, biochar application caused significant increases in water retention. The pF curve (Figure 14.2) shows that both 72 t ha^{-1} biochar treatments had significantly higher volumetric water contents at saturation. Similarly, at pF 1.8, these two treatments had the highest volumetric water contents, while at pF 4.2, the differences were less pronounced. Plant available water significantly increased after 72 t ha^{-1} biochar amendment, with and without N fertilization (Figure 14.2). In the Chernozem, biochar effected similar trends for water retention as in the Cambisol. The pF curve showed, similar to the Cambisol, that both biochar 72 t ha^{-1} treated plots had significantly higher volumetric water contents at saturation and pF 1.8. Plant available water increased through BC application, albeit not statistically significantly (Figure 14.2; Karer et al., 2013).

The positive effects on soil water retention in our study were consistent with studies from Liu et al. (2012) and Kammann et al. (2011). Moreover, increased water retention in biochar-treated plots was also reported by Petter et al. (2012). Similarly, Cornelissen et al. (2013) observed increased plant-available water in their field experiment. Based on these studies and our own results, we assume that the hypothesis concerning improved soil water retention as a biochar benefit is confirmed.

The effects on soil microorganisms were studied by PLFA analyses in the greenhouse column experiment with different biochars at different supplementary nutrient supply levels. At the same nitrogen levels, the biochar treatments caused shifts in microbial community compositions (Figure 14.3), but did not significantly increase or decrease the microbial biomass. Whereas at the beginning of experiment the differences in microbial community composition between biochar-amended and non-amended soils were similarly distinct

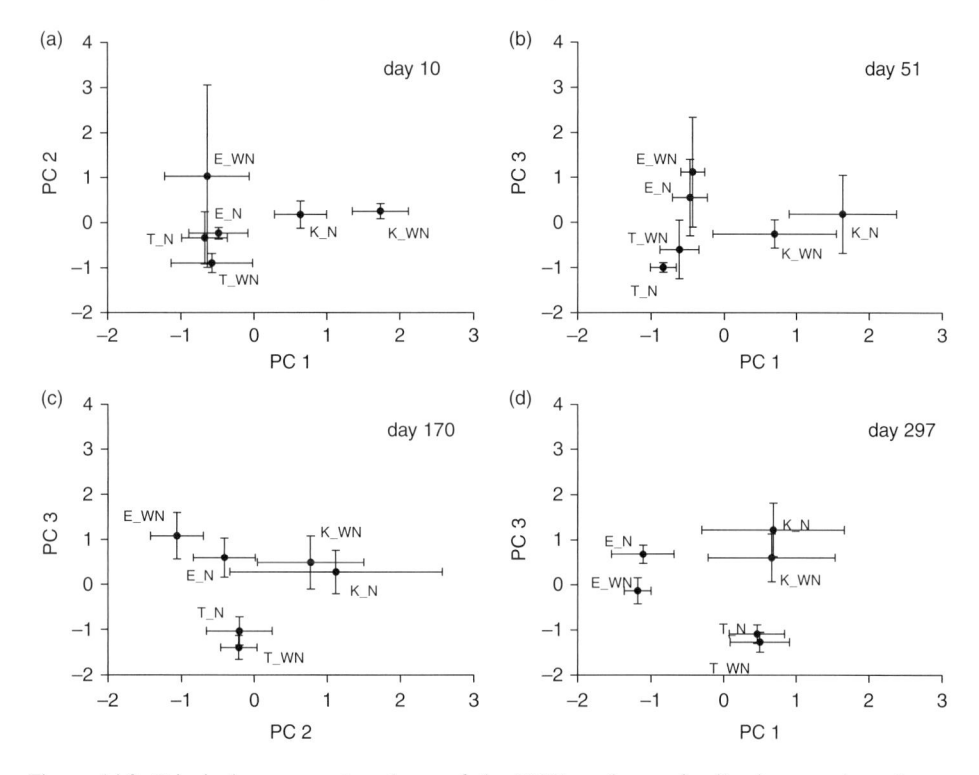

Figure 14.3. Principal component analyses of the PLFA-analyses of soil microorganisms from a greenhouse pot experiment with three different soils with and without biochar. Nitrogen levels were kept constant in all treatments. Days indicate the time of sampling after installation of the experiment. E: Planosol; K: Cambisol; T: Chernozem; WN: woodchip-based biochar; N: no biochar added.
Source: Anders et al., 2013. With permission.

as between different soils, over the course of one year the differences caused by biochar decreased, leaving the differences due to soil as the main discrimination parameter (Figure 14.3). Only the treatment with vineyard pruning biochar pyrolyzed at 400°C showed a significant increase of microbial biomass, presumably because non-volatized low molecular weight hydrocarbons could be utilized by microorganisms (Spokas et al., 2011). The shift in the microbial community correlated with chemical soil properties (e.g. C/N, pH value) and plant growth after biochar application. The most influential factors on microbial PLFAs were, in the order of diminishing importance: soil type, fertilization regime and biochar amendment. We concluded that the shifts in the microbial community structure were an indirect rather than direct effect and depended on soil conditions and nutrient status (Anders et al., 2013).

In an incubation experiment with the Planosol, microbial biomass was elevated by wheat husk biochar. The increase in PLFAs was mainly attributed to Gram-negative bacteria and actinomycetes; fungi and Gram-positive bacteria were less affected. In the case of willow biochar addition, microorganisms hardly responded. It appears that pH change as a consequence of biochar incorporation was the most dominant factor driving changes in microbial community structure. Thus, sandy acidic soils might profit from biochar addition, but application to loamy and calcareous soils is unlikely to foster the build-up of soil microorganisms (Watzinger et al., 2014).

These results show that a positive effect of biochar on soil microbiota cannot be assumed automatically. The effects of biochar feedstock, pyrolysis conditions and the original soil characteristics determine if microorganisms can take advantage of newly available nutrients and of biochar pores as habitats. Under certain conditions, even negative effects on soil microorganisms cannot be excluded (Dempster et al., 2012). In conclusion, results about benefits of biochar for soil organisms are too ambiguous to accept a clear-cut confirmation of our hypothesis. However, if organic fertilizers are applied in parallel with or as a supplement to pure biochar, the combined effects of these additives for soil life might be much more positive.

14.3.3 Biochar Benefits for the Natural Resource Air

The interactions of biochar with soil nitrogen compounds and the enzymatic processes involved in their transformations play an important role in the availability of nitrogen species for plant roots, for microorganisms, for leaching or gaseous losses. If nitrogen is released to the atmosphere as nitrous oxide (N_2O), the high greenhouse gas potential of N_2O is a matter of concern. Therefore, the intervention of biochar in nitrogen metabolism processes in the soil, as could be observed for example concerning the retention of nitrates, has raised hopes that increasing immobilization of individual nitrogen species and a deceleration of the turnover processes would decrease the N_2O emissions (Kammann et al., 2012; Ameloot et al., 2013).

Our monitoring measurements under field conditions have shown that increasing biochar application rates decreased the N_2O emissions significantly. This was most apparent in the early phase of our experiments and after mineral fertilizer additions. Over the course of one year, a reduction of N_2O losses to the atmosphere by about one-third to one-half could be verified (Figure 14.4). A similar extent of N_2O reduction was also confirmed in the studies of Alho et al. (2012). Saarnio et al. (2013) stated that the reductions of N_2O emissions were more pronounced during dry soil conditions. As a large proportion of Austrian agricultural areas are dominated by drought-prone climate conditions, this supports the regional importance of climate change mitigation efficacy of biochar. An additional specification of Austrian agriculture is the importance of animal husbandry on meadows and pastures in alpine regions. The biochar-mediated reduction of gaseous nitrogen losses from animal excreta on pastures as described by Clough et al. (2010) and Taghizadeh-Toosi et al. (2011) would provide additional mitigation options for grassland agriculture. Besides, an incorporation of biochar into animal feed has the potential to reduce ruminant methane emissions (Hansen et al., 2012) and would provide a delivery system for biochar to pasture soils (Calvelo Pereira et al., 2014).

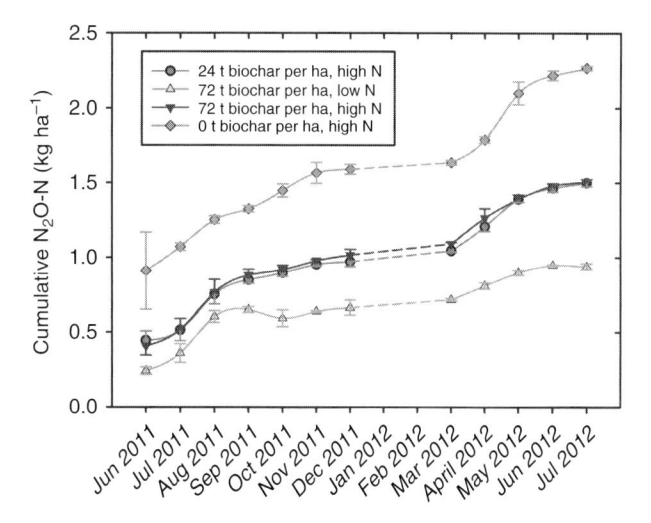

Figure 14.4. Cumulative nitrous oxide fluxes from differently biochar-amended plots (see graph legend) of a field experiment on a Chernozem in the period July 2011–July 2012.
Source: Kitzler et al., in preparation. (A black and white version of this figure will appear in some formats. For the colour version, please refer to the plate section.)

In conclusion, the hypothesis that biochar contributes to the protection of the natural resource air could be confirmed. Although the main mechanisms of biochar as a climate change mitigation strategy will rely on the long-term carbon sequestration potential, the reduction of nitrous oxide emissions will support the mitigation efficacy.

14.3.4 Biochar Benefits for Crop Nutrient Supply

The bioavailability of plant nutrients in biochar and the liming effect are important determinants of the soil fertility effects of biochar. The fertility improvement potential of biochar supplemented with organic fertilizers is undisputed, not only because this combination resembles more the original additives that have led to the anthropogenic dark earths known as terra preta (Glaser and Birk, 2012). However, the nutrient supplying potential of pure biochar is a matter of discussion. On the one hand, the use of nitrogen-rich feedstocks for pyrolysis is known to enable the transfer of bioavailable nitrogen directly to crops (Chan et al., 2008; Dai et al., 2013). On the other hand, nutrient-poor feedstocks like wood chips can only provide a significant amount of nitrogen if the application rate is very high, the pyrolysis temperature has been low to moderate, thereby decreasing the gaseous N losses, or if the low N content of wood is supplemented by adding N-rich feedstock like manure to the input material. Because of the heterocyclic nature of the N compounds in biochar, the N bioavailability is usually low. The influence of different feedstocks on biochar characteristics like cation exchange capacity and nutrient availability has also been stressed by Kloss et al. (2012, 2014a). To test the availability of nitrogen and other plant nutrients from

(a) Mustard (above-ground dm)

Treatment	Group
8 g/m² N without BC	a
4 g/m² N without BC	a
2 g/m² N without BC	b
0 g/m² N without BC	b
8 g/m² N + 3% BC	b
4 g/m² N + 3% BC	c
2 g/m² N + 3% BC	d
0 g/m² N + 3% BC	e

Relative dry matter yield (%)

(b) Barley (above-ground dm)

Treatment	Group
20 g/m² N without BC	a
10 g/m² N without BC	c
5 g/m² N without BC	e
0 g/m² N without BC	h
20 g/m² N + 3% BC	b
10 g/m² N + 3% BC	d
5 g/m² N + 3% BC	f
0 g/m² N + 3% BC	g

Relative dry matter yield (%)

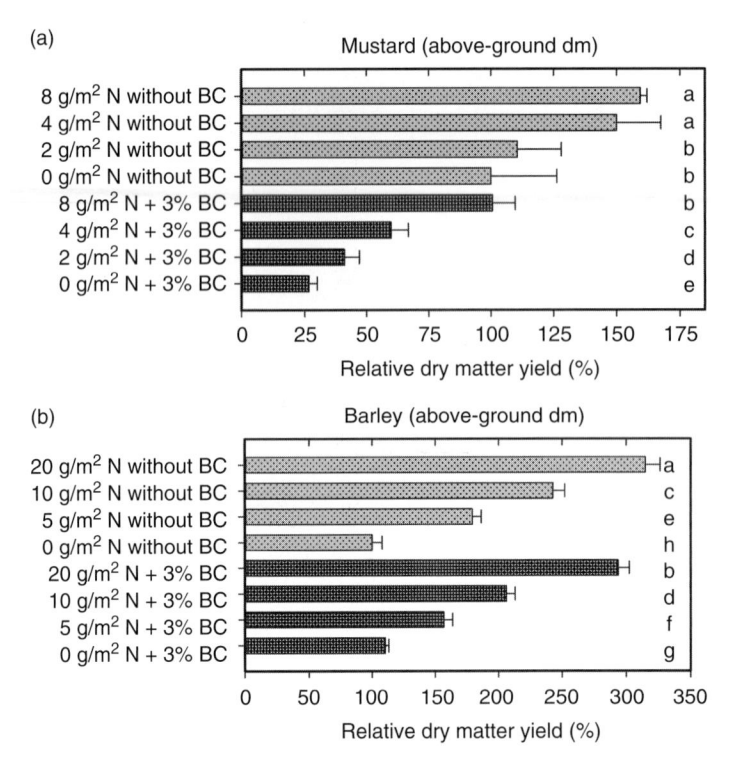

Figure 14.5. Relative dry matter yields of a cropping pattern with mustard as first crop (a) and barley as second crop (b) in a soil column experiment. Biochar (BC) and nitrogen (N) additions are indicated as axis labels. Biochar addition rate to soil is on a dry weight basis. One hundred per cent of relative yields correspond to the yields of unamended controls (no biochar, no additional nitrogen). Columns labeled with different letters are significantly different according to Tukey's test (p = 0.05).
Source: Kloss et al. (2014b).

a wood-based biochar, we studied the soil fertility effects in a greenhouse pot experiment and in two field experiments.

In the pot experiment, both mustard and barley yields significantly increased with increasing N fertilization for both BC-treated and untreated pots. Without added N, the first crop (mustard) showed significantly lower yields when treated with BC, but in the second crop (barley) this effect disappeared (Kloss et al., 2014a). Yield of mustard treated with 8 g N m^{-2} and 3% BC was merely in the range of mustard grown on the unfertilized soil without BC (Figure 14.5). As pH increased for all BC treatments, this may also have affected nutrient availability (Marschner and Rengel, 2012). Micronutrient availability decreased and may have caused soil pH-induced micronutrient deficiencies (Bolan et al., 2003). However, barley yields on BC-treated pots were less depressed compared to mustard (Kloss et al., 2014a). The competition for nitrogen between plant roots, soil microorganisms and biochar surfaces may have been more severe shortly after biochar application in the experiment with mustard, whereas for barley as subsequent crop the adverse effects of pure biochar gradually disappeared.

Agricultural soils in a temperate climate

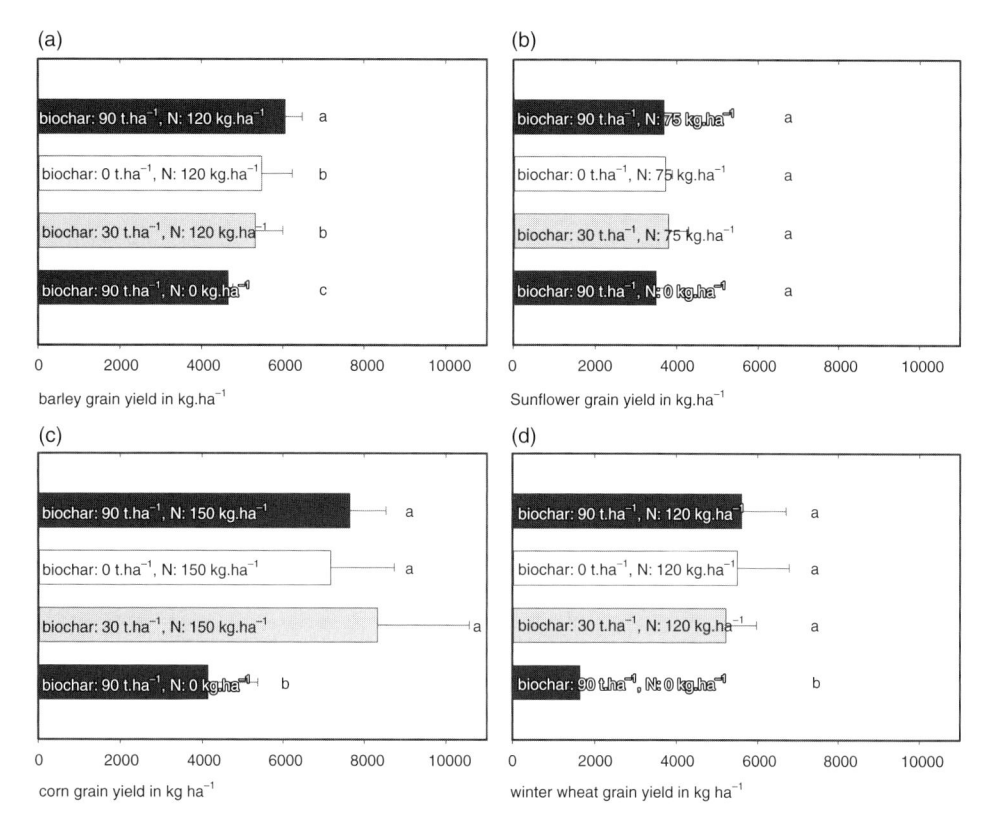

(a)

	barley grain yield in kg.ha^{-1}
biochar: 90 t.ha^{-1}, N: 120 kg.ha^{-1}	a
biochar: 0 t.ha^{-1}, N: 120 kg.ha^{-1}	b
biochar: 30 t.ha^{-1}, N: 120 kg.ha^{-1}	b
biochar: 90 t.ha^{-1}, N: 0 kg.ha^{-1}	c

(b)

	Sunflower grain yield in kg.ha^{-1}
biochar: 90 t.ha^{-1}, N: 75 kg.ha^{-1}	a
biochar: 0 t.ha^{-1}, N: 75 kg.ha^{-1}	a
biochar: 30 t.ha^{-1}, N: 75 kg.ha^{-1}	a
biochar: 90 t.ha^{-1}, N: 0 kg.ha^{-1}	a

(c)

	corn grain yield in kg ha^{-1}
biochar: 90 t.ha^{-1}, N: 150 kg.ha^{-1}	a
biochar: 0 t.ha^{-1}, N: 150 kg.ha^{-1}	a
biochar: 30 t.ha^{-1}, N: 150 kg.ha^{-1}	a
biochar: 90 t.ha^{-1}, N: 0 kg.ha^{-1}	b

(d)

	winter wheat grain yield in kg ha^{-1}
biochar: 90 t.ha^{-1}, N: 120 kg.ha^{-1}	a
biochar: 0 t.ha^{-1}, N: 120 kg.ha^{-1}	a
biochar: 30 t.ha^{-1}, N: 120 kg.ha^{-1}	a
biochar: 90 t.ha^{-1}, N: 0 kg.ha^{-1}	b

Figure 14.6. Crop yields in two field experiments with different biochar and nitrogen application rates. In the graphs, biochar rates are given as fm including 20% moisture, corresponding to 24 and 72 t biochar dm ha^{-1}. N was added as mineral fertilizer (NH$_4$NO$_3$). Barley (a) and sunflower (b) were cultivated on a Chernozem; maize (c) and winter wheat (d) on a Cambisol. Columns labeled with different letters are significantly different according to Tukey's test (p = 0.05).
Source: Karer et al. (2013).

Also the field experiment showed the dominant effect of sufficient nitrogen supply. If supplementary N application was absent, all crops showed significant yield decreases after the highest biochar application rate (Figure 14.6). On the Cambisol, maize total above-ground biomass yield decreased by 37%, and winter wheat decreased by 71%. However, varying BC amendments (0, 24 and 72 t dm ha^{-1}) with identical mineral N supply did not cause significant differences in the grain yields of maize and wheat. In the Chernozem, missing N after 72 t ha^{-1} BC application without N supplement caused a lower decrease than in the Cambisol: barley and sunflower grain yields were reduced by 23% and 15%, respectively. With sufficient N supply, barley grain yield took advantage of the high biochar application rate (72 t dm ha^{-1}) with a 10% yield increase. This was probably due to a severe precipitation deficit during the growth period of barley (50% of the long-term mean) and an improved water holding capacity in the biochar-amended plots (+21%; Karer et al., 2013).

Biochar addition requires consideration of its effects on nitrogen availability and potential supplementary nitrogen sources. Due to the high C/N ratio of biochar, nitrogen immobilization and/or decreased availability for plant roots is to be expected. An increase of soil N immobilization after BC addition was also observed by Lehmann et al. (2003), Bruun et al. (2012), Zheng et al. (2013) and Tammeorg et al. (2014). Additionally, the possibility of decreased plant uptake of cationic trace elements caused by soil pH increases and by the sorption capacity of biochar has to considered (Kloss et al., 2015). However, this effect depends on the specific biochar feedstock and soil conditions. In the case of anionic trace elements, mobilization and increased plant uptake may be observed (Kloss et al., 2014b).

These results show that a positive effect of biochar on the fertility of soils that are neither acidic nor impoverished in nutrients cannot be guaranteed. The biochar feedstock is the main determinant if the additive becomes a source or sink for nutrients, additionally to the influence of the local soil characteristics and the effect of pyrolysis temperature. Though some types of pure straw-based biochars are able to deliver relevant amounts of nutrients (Kloss et al., 2012), usually the supplementary fertilizer component in a mixture like compost-biochar will be the dominant nutrient vector. Exceptions might be the intentional selection of nutrient-rich feedstocks for pyrolysis, such as animal wastes or sewage sludge that could be mixed with appropriate proportions of carbon-rich vegetation-based materials to produce biochars with the desired properties for both carbon sequestration and plant nutrition (Cascarosa et al., 2013; Mendez et al., 2013; Vassilev et al., 2013).

14.3.5 Economic Feasibility of Biochar Application

From a business economic viewpoint, currently high costs of biochar are not balanced by the small to moderate (and frequently even missing) increases in crop yields and thus agricultural revenues. The costs for supplementary organic (e.g. compost) or mineral fertilizers have to be added to the biochar costs but these supplementary additives have their own costs and benefits, which were not considered in the frame of our study. Improved water retention due to biochar, however, might justify biochar as an adaptation measure to global warming, especially when considering also overall economic aspects besides business economic aspects.

When not assuming crop failures but increased soil fertility, even an inclusion of avoided social (=societal) costs by sequestering carbon and thereby helping to mitigate climate change does not economically justify the application of biochar at the high rates used in our field experiments. The price of biochar would need to decrease by at least 40% to achieve a break-even from the overall economic viewpoint (if optimistic assumptions about the social value of sequestered carbon are applied; at pessimistic assumptions the price for biochar would need to decrease by 80% or more in order to break even). Society would have to be willing to pay around €138 per ton of sequestered CO_2 for costs of biochar of €300 / t to be justified. This price of biochar, however, only applies for low-quality charcoal grit; biochar of certified quality produced in modern commercial pyrolysis units currently costs in the range of €400–600 / t, depending on the feedstock price. This calculation also considers the energy input as a specific electricity consumption of 250 kWh$_{electricity}$ per ton of biochar.

Concluding, it can be stated that pure biochar cannot be applied economically at current costs and prices in average soils of temperate climate regions that are neither extremely degraded, contaminated nor acidic. Also when including external benefits of carbon sequestration and avoided N$_2$O emissions, cost of biochar still dominates. However, biochar deployment in dry climate conditions might be considered as an adaptation measure to global warming, especially from an overall economic point of view. The situation might be more favorable if synergistic effects of a parallel biochar and compost application are considered (Schulz et al., 2013). The interactions of biochar with supplementary additives still need further economic evaluations for monetizing these effects. Further upscaling of the biochar production units, including an economic and sellable co-production of methanol (via conversion of syngas) or electricity, could be the next steps to improve the economic situation of biochar production (Harsono et al., 2013; Shabangu et al., 2014). Another important aspect of biochar utilization is its use as animal feed in small proportions to prevent diseases from fodder contaminants, for example fungal metabolites or organic pollutants (Huwig et al., 2001; Jouany, 2007; McHenry, 2010; Fujita et al., 2012, Chu et al., 2013). This use has no fixed financial value yet but undoubtedly contributes to savings in veterinary spends. Administration as a feed component may be the first step in a multi-cascadal use of biochar that continues as an odor and NH$_3$ sorbent in the bedding and in manure to end up as sequestered carbon in agricultural soils.

14.4 Conclusions

The experimental design of this study, which included an extensive microlysimeter experiment under controlled greenhouse conditions and two field experiments, allowed the authors to investigate several research questions in parallel and to test the hypotheses formulated in the introduction to this chapter.

An important result that will have an impact on future applications of biochar is the ambiguous role of non-supplemented biochar for soil fertility. Nutrients are only released in the short term from straw biochar but not from wood-based biochars. However, concerning other physico-chemical effects on the soil, straw biochar offers fewer benefits than other feedstocks. The combination of biochar with either mineral or organic fertilizers such as compost is indispensable to avoid yield losses. Therefore, the nutrient content of biochar plays a minor role for crop nutrition; rather, there is likely an additional demand for nutrients to counteract the competition between biochar sorption sites and microbiological nutrient demand. A timely consideration of this nutrient demand creates additional costs for the farmer that lower the economic efficiency of biochar production and application. The biochar-mediated decrease in greenhouse gas emissions, notably N$_2$O, is an advantage that is not yet rewarded financially because of the lack of financial instruments that consider soil in the frame of GHG accounting. Only regional campaigns that design locally adapted certificate trade systems have the possibility to account for soil GHG emissions, but these are not yet widespread. If it is possible to put the concept of a knowledge-based global bioeconomy and smart specialization in the regional context, this progress achieved at a local level could serve as an example for larger scale economies. The problematic economic situation of pure biochar application to fertile agricultural soils currently prevents a large-scale implementation of this strategy as soil amendment. Although the quality of certified biochar nowadays is satisfying, and the amendments can be applied without risk to any soils if modern biochar production techniques have been applied and the pollutant threshold values of the European Biochar Certificate (or its IBI (International Biochar Initiative) equivalent) are not exceeded, these achievements are not enough. At the moment, the price of biochar cannot be translated into corresponding yield increases by the farmers. In the long run, however, the additional benefits of biochar – such as reduced GHG emissions, reduced nitrate leaching to groundwater, pollutant immobilization, long-term carbon sequestration in soil and increased soil water holding capacity – may translate into farmers' income. Currently this is not the case, partly because these benefits are not yet reproducibly quantified under different environments, and partly because no financial remuneration systems exist. In future agricultural management strategies (e.g. the Austrian Agri-environmental Programme) such effects could be related to benefit points that are related to financial support. Such considerations could change the economic assessment of biochar considerably, but the basic quantitative functions necessary to connect ecological with financial effects will need a broader basis than a single study can offer. In addition, for specific local situations where soil contaminations prevent an agricultural use, the sorption potential of biochar may provide an excellent remediation tool. In such cases, the remediation costs may be lower than alternative remediation techniques and the increase of land value for a successful remediation might easily compensate for the biochar costs. If the scientific literature on individual aspects of biochar were combined, for example by meta-analytical techniques, quantitative assessments could be derived that are needed for such strategic considerations (Jeffery et al., 2011; Crane-Droesch et al., 2013).

Our field experiments and the investigation of water retention of biochar-amended soils have shown that during drought periods field crops may take advantage of prolonged water availability. This result appeared although the biochar was a "general purpose" wood-based biochar not specifically designed to hold more water. Biochar amendment resulted in a higher stability of soil structure, a lower bulk density and thus a higher soil pore volume throughout the vegetation period that could have contributed to the improved water retention characteristics, too. If the feedstock selection considered the porosity architecture of the original plant material and if the pyrolysis conditions supported a high specific surface area formation, biochar with improved water storage capacity could result. Also, suitable post-treatment modifications might further "design" desired biochar characteristics and open a field for new product developments.

The results of our case study have provided new information not only for research but also for companies interested in commercializing biochar to a larger extent. In the first line these might be producers of composts who can take advantage of creating new compost varieties by the addition of biochar. But also the production of biochar itself with technical-scale reactors can be a business opportunity if scientific results are considered and the economic framework changes. Whereas under current conditions the economically feasible production of biochar is not yet possible, at a larger production scale the exploitation and marketing of energy-rich co-products or electricity might change the situation. Besides, the establishment of a financial value for the non-yield related biochar benefits to animal health, groundwater, air, soil remediation and physico-chemical soil characteristics would also add to the profitability of a biochar strategy.

The legal situation is one of the most important framework conditions that will determine the future importance of biochar as a soil amendment. In Austria, large-scale application to agricultural soils is bound to special licences. But examples both in Switzerland and in California have shown that for certain production and application scenarios a framework can be created to legally allow for a regulated biochar use in agricultural soils, even for acceptance in organic agriculture. Only with a reliable European regulation framework may biochar become a real business opportunity and hold up to its promises.

Finally, the test results of the hypotheses put forward at the beginning of our case study require a sophisticated judgment about the future prospects of biochar in agriculture:

- Benefits for the natural resource groundwater because of reduced nitrogen input – confirmed
- Benefits for the natural resource soil (physical effects, water holding capacity) – confirmed
- Benefits for the natural resource soil (soil microorganisms) – only partly confirmed
- Benefits for the natural resource air by reducing soil GHG emissions – confirmed for N_2O
- Benefits for crops by enhanced nutrient availability – refuted
- Cost-effectiveness of biochar application under current production costs and benefit rewards – refuted

This analysis shows that there is significant potential to supplement the carbon sequestration ability of biochar with other benefits, but these still need an allocated monetized value and some of them are not as effective as expected for all soil conditions.

Acknowledgements

The authors gratefully acknowledge the financing of this study by FFG (project nr. 825438) and the Austrian KLI-EN funds (3rd call "Neue Energien 2020"). Technical support by Christian Mayer, Patrick Kobe and Walter Frank was highly valued. We are grateful to the landowners Rudolf Hofmann and Johann König for permitting the installation and maintenance of the field experiments on their premises.

References

Alho, C. F. B. V., Cardoso, A. D., Alves, B. J. R. and Novotny, E. H. (2012). Biochar and soil nitrous oxide emissions. *Pesquisa Agropecuaria Brasileira*, 47, pp. 722–725.

Ameloot, N., Neve, S., Jegajeevagan, K., Yildiz, G., Buchan, D., Funkuin, Y. N., Prins, W., Bouckaert, L. and Sleutel, S. (2013). Short-term CO2 and N2O emissions and microbial properties of biochar amended sandy loam soils. *Soil Biology & Biochemistry*, 57, pp. 401–410.

Anders, E., Watzinger, A., Rempt, F., Kitzler, B., Wimmer, B., Zehetner, F., Stahr, K., Zechmeister-Boltenstern, S. and Soja, G. (2013). Biochar affects the structure rather than the total biomass of microbial communities in temperate soils. *Agricultural and Food Science*, 22, pp. 404–423.

Angst, T. E., Patterson, C. J., Reay, D. S., Anderson, P., Peshkur, T. A. and Sohi, S. P. (2013). Biochar diminishes nitrous oxide and nitrate leaching from diverse nutrient sources. *Journal of Environmental Quality*, 42, pp. 672–682.

Bolan, N., Adriano, D. and Curtin, D. (2003). Soil acidification and liming interactions with nutrient and heavy metal transformation and bioavailability, *Advances in Agronomy*, 78, pp. 216–272.

Brunauer, S., Emmett, P. H. and Teller, E. (1938). Adsorption of gases in multimolecular layers. *Journal of the American Chemical Society*, 60, pp. 309–331.

Bruun, E. W., Ambus, P., Egsgaard, H. and Hauggaard-Nielsen, H. (2012). Effects of slow and fast pyrolysis biochar on soil C and N turnover dynamics. *Soil Biology & Biochemistry*, 46, pp. 73–79.

Bücker, J. (2012). *Effects of biochar on leachate characteristics and crop production of mustard (Sinapis alba) and barley (Hordeum vulgare) in a micro-lysimeter experiment on three agricultural soils in Austria*. Diploma thesis, BTU Cottbus/Germany.

Burt, R. (2004). *Soil Survey Laboratory Methods Manual*. Soil survey investigations report, 42. Washington, DC: USDA-NRCS.

Calvelo Pereira, R., Muetzel, S., Camps Arbestain, M., Bishop, P. Hina, K. and Hedley, M. (2014). Assessment of the influence of biochar on rumen and silage fermentation: a laboratory-scale experiment. *Animal Feed Science and Technology*, 196, pp. 22–31.

Cascarosa, E., Boldrin, A. and Astrup, T. (2013). Pyrolysis and gasification of meat-and-bone-meal: energy balance and GHG accounting. *Waste Management*, 33, pp. 2501–2508.

Chan, K. Y., Van Zwieten, L., Meszaros, I., Downie, A. and Joseph, S. (2008). Using poultry litter biochars as soil amendments. *Australian Journal of Soil Research*, 46, pp. 437–444.

Chu, G. M., Jung, C. K., Kim, H. Y., Ha, J. H., Kim, J. H., Jung, M. S., Lee, S. J., Song, Y., Ibrahim, R. I. H., Cho, J. H., Lee, S. S. and Song, Y. M. (2013). Effects of bamboo charcoal and bamboo vinegar as antibiotic alternatives on growth performance, immune responses and fecal microflora population in fattening pigs. *Animal Science Journal*, 84, pp. 113–120.

Clough, T. J., Bertram, J. E., Ray, J. L., Condron, L. M., O'Callaghan, M., Sherlock, R. R. and Wells, N. S. (2010). Unweathered wood biochar impact on nitrous oxide emissions from a bovine-urine-amended pasture soil. *Soil Science Society of America Journal*, 74, pp. 852–860.

Cornelissen, G., Martinsen, V., Shitumbanuma, V., Alling, V., Breedveld, G. D., Rutherford, D. W., Sparrevik, M., Hale, S. E., Obia, A. and Mulder, J. (2013). Biochar effect on maize yield and soil characteristics in five conservation farming sites in Zambia. *Agronomy*, 3, pp. 256–274.

Crane-Droesch, A., Abiven, S., Jeffery, S. and Torn, M. S. (2013). Heterogeneous global crop yield response to biochar: a meta-regression analysis. *Environmental Research Letters*, 8, open access nr. 044049 (8 pp.).

Dai, Z. M., Meng, J., Muhammad, N., Liu, X. M., Wang, H. Z., He, Y., Brookes, P. C. and Xu, J. M. (2013). The potential feasibility for soil improvement, based on the properties of biochars pyrolyzed from different feedstocks. *Journal of Soils and Sediments*, 13, pp. 989–1000.

Dempster, D. N., Gleeson, D. B., Solaiman, Z. M., Jones, D. L. and Murphy, D. V. (2012). Decreased soil microbial biomass and nitrogen mineralization with Eucalyptus biochar addition to a coarse textured soil. *Plant and Soil*, 354, pp. 311–324.

European Commission (2000). Directive 2000/60/EC of the European Parliament and of the Council of 23 October 2000 establishing a framework for Community action in the field of water policy.

Fang, Y., Singh, B. and Singh, B. P. (2015). Effect of temperature on biochar priming effects and its stability in soils. *Soil Biology & Biochemistry*, 80, pp. 136–145.

Frostegård, Å., Tunlid, A. and Bååth, E. (1991). Microbial biomass measured as total lipid phosphate in soils of different organic content. *Journal of Microbiological Methods*, 14, pp. 151–163.

Fujita, H., Honda, K., Iwakiri, R., Guruge, K. S., Yamanaka, N. and Tanimura, N. (2012). Suppressive effect of polychlorinated dibenzo-p-dioxins, polychlorinated dibenzofurans and dioxin-like polychlorinated biphenyls transfer from feed to eggs of laying hens by activated carbon as feed additive. *Chemosphere*, 88, pp. 820–827.

Galinato, S. P., Yoder, J. K. and Granatstein, D. (2011). The economic value of biochar in crop production and carbon sequestration, *Energy Policy*, 39, pp. 6344–6350.

Glaser, B., Lehmann, J. and Zech, W. (2002). Ameliorating physical and chemical properties of highly weathered soils in the tropics with charcoal – a review. *Biology Fertility Soils*, 35, pp. 219–230.

Glaser, B. and Birk, J. J. (2012). State of the scientific knowledge on properties and genesis of anthropogenic dark earths in Central Amazonia (terra preta de Indio). *Geochimica et Cosmochimica Acta*, 82, pp. 39–51.

Hamer, U., Marschner, B., Brodowski, S. and Amelung, W. (2004). Interactive priming of black carbon and glucose mineralization. *Organic Geochemistry*, 35, pp. 823–830.

Hammes, K., Torn, M. S., Lapenas, A. G. and Schmidt, M. W. I. (2008). Centennial black carbon turnover observed in a Russian steppe soil. *Biogeosciences*, 5, pp. 1339–1350.

Hansen, H. H., Storm, I. M. L. D. and Sell, A. M. (2012). Effect of biochar on in vitro rumen methane production. *Acta Agriculturae Scandinavica, Section A – Animal Science*, 62, pp. 305–309.

Harsono, S. S., Grundman, P., Lau, L. H., Hansen, A., Salleh, M. A. M., Meyer-Aurich, A., Idris, A. and Ghazi, T. I. M. (2013). Energy balances, greenhouse gas emissions and economics of biochar production from palm oil empty fruit bunches. *Resources Conservation and Recycling*, 77, pp. 108–115.

Hass, A., Gonzalez, J. M., Lima, I. M., Godwin, H. W., Halvorson, J. J. and Boyer, D. G. (2012). Chicken manure biochar as liming and nutrient source for acid Appalachian soil. *Journal of Environmental Quality*, 41, pp. 1096–1106.

Huwig, A., Freimund, S., Kappeli, O. and Dutler, H. (2001). Mycotoxin detoxication of animal feed by different adsorbents. *Toxicology Letters*, 122, pp. 179–188.

Jeffery, S., Verheijen, F., van der Velde, M. and Bastos, A. (2011). A quantitative review of the effects of biochar application to soils on crop productivity using meta-analysis. *Agriculture Ecosystems Environment*, 144, pp. 175–187.

Jouany, J. P. (2007). Methods for preventing, decontaminating and minimizing the toxicity of mycotoxins in feeds. *Animal Feed Science and Technology*, 137, pp. 342–362.

Kammann, C. I., Linsel, S., Goessling, J. W. and Koyro, H. W. (2011). Influence of biochar on drought tolerance of *Chenopodium quinoa* Willd and on soil-plant relations. *Plant and Soil*, 345, pp. 195–210.

Kammann, C., Ratering, S., Eckhard, C. and Muller, C. (2012). Biochar and hydrochar effects on greenhouse gas (carbon dioxide, nitrous oxide, and methane) fluxes from soils. *Journal of Environmental Quality*, 41, pp. 1052–1066.

Karer, J., Wimmer, B., Zehetner, F., Kloss, S. and Soja, G. (2013). Biochar application to temperate soils: effects on nutrient uptake and crop yield under field conditions. *Agricultural and Food Science*, 22, pp. 390–403.

Klinglmüller, M. (2013). *Effects of biochar on greenhouse gas fluxes from agricultural soils and resulting greenhouse gas abatement costs – an Austrian case study*. Masters thesis, University for Natural Resources and Life Sciences, Vienna, Austria.

Kloss, S., Zehetner, F., Dellantonio, A., Hamid, R., Ottner, F., Liedtke, V., Schwanninger, M., Gerzabek, M. H. and Soja, G. (2012). Characterization of slow pyrolysis biochars: effects of feedstocks and pyrolysis temperature on biochar properties. *Journal of Environmental Quality*, 41, pp. 990–1000.

Kloss, S., Zehetner, F., Wimmer, B., Buecker, J., Rempt, F. and Soja, G. (2014a). Biochar application to temperate soils: effects on soil fertility and crop growth under greenhouse conditions. *Journal of Plant Nutrition and Soil Science*, 177, pp. 3–15.

Kloss, S., Zehetner, F., Oburger, E., Buecker, J., Kitzler, B., Wenzel, W. W., Wimmer, B. and Soja, G. (2014b). Trace element concentrations in leachates and mustard plant tissue (*Sinapis alba* L.) after biochar application to temperate soils. *Science of the Total Environment*, 481, pp. 498–508.

Kloss, S., Zehetner, F., Buecker, J., Oburger, E., Wenzel, W. W., Enders, A., Lehmann, J. and Soja, G. (2015). Trace element biogeochemistry in the soil-water-plant system of a temperate agricultural soil amended with different biochars. *Environmental Science and Pollution Research*, 22, pp. 4513–4526.

Klute, A. (1986). Water retention: laboratory methods. In: Klute, A. (ed.) *Methods of Soil Analysis, Part 1, Physical and Mineralogical Methods*. Agronomy Monograph 9, 2nd Edition. Madison, WI: American Society of Agronomy, Soil Science of America, pp. 635–662.

Lehmann, J., Czimczik, C., Laird, D. and Sohi, S. (2009). Stability of biochar in the soil. In: Lehmann, J. and Joseph, S. (eds.) *Biochar for Environmental Management*. London: Earthscan, pp. 183–205.

Lehmann, J., Kern, D. C., German, L. A., McCann, J., Martins, G. C. and Moreira, A. (2003). Soil fertility and production potential. In: Lehmann, J., Kern, D. C., Glaser, B. and Woods, W. (eds.) *Amazonian Dark Earths: Origin, Properties, Management*. Dordrecht, The Netherlands: Kluwer Academic Publishers, pp. 105–124.

Liu, J., Schulz, H., Brandl, S., Miehtke, H., Huwe, B. and Glaser, B. (2012). Short-term effect of biochar and compost on soil fertility and water status of a Dystric Cambisol in NE Germany under field conditions. *Journal of Plant Nutrition and Soil Science*, 175, pp. 698–707.

Major, J., Lehmann, J., Rondon, M. and Goodale, C. (2010). Fate of soil-applied black carbon: downward migration, leaching and soil respiration. *Global Change Biology*, 16, pp. 1366–1379.

Marris, E. (2006). Black is the new green, *Nature*, 442, pp. 624–626.

Marschner, P. and Rengel, Z. (2012). Nutrient availability in soils. In: Marschner, P. (ed.) *Marschner's Mineral Nutrition of Higher Plants*. 3rd Edition. Amsterdam: Elsevier, pp. 315–330.

Matovic, D. (2011). Biochar as a viable carbon sequestration option: global and Canadian perspective. *Energy*, 36, pp. 2011–2016.

McHenry, M. P. (2010). Carbon-based stock feed additives: a research methodology that explores ecologically delivered C biosequestration, alongside live weights, feed use efficiency, soil nutrient retention, and perennial fodder plantations. *Journal of the Science of Food and Agriculture*, 90, pp. 183–187.

Mendez, A., Terradillos, M. and Gasco, G. (2013). Physicochemical and agronomic properties of biochar from sewage sludge pyrolysed at different temperatures. *Journal of Analytical and Applied Pyrolysis*, 102, pp. 124–130.

Nguyen, B. T., Lehmann, J., Kinyangi, J., Smernik, R., Riha, S. J. and Engelhard, M. H. (2008). Long-term black carbon dynamics in cultivated soil. *Biogeochemistry*, 89, pp. 295–308.

Novak, J. M., Busscher, W. J., Watts, D. W., Laird, D. A., Ahmedna, M. A. and Niandou, M. A. S. (2010). Short-term CO(2) mineralization after additions of biochar and switchgrass to a Typic Kandiudult. *Geoderma*, 154, pp. 281–288.

Petter, F. A., Madari, B. E., da Silva, M. A. S., Carneiro, M. A. C., Carvalho, M. T. de M., Marimon Jr., B. H. and Pacheco, L. P. (2012). Soil fertility and upland rice yield after biochar application in the Cerrado. *Pesquisa Agropecuária Brasileira*, 47, pp. 699–706.

Purakayastha, T. J., Kumari, S. and Pathak, H. (2015). Characterisation, stability, and microbial effects of four biochars produced from crop residues. *Geoderma*, 239–240, pp. 293–303.

Rolston, D. E. (1986). Gas flux. In: Klute, A. (ed.) *Methods of Soil Analysis. Part 1*. Madison, WI: Soil Science Society of America and American Society of Agronomy, pp. 1103–1119.

Saarnio, S., Heimonen, K. and Kettunen, R. (2013). Biochar addition indirectly affects N2O emissions via soil moisture and plant N uptake. *Soil Biology and Biochemistry*, 58, pp. 99–106.

Schulz, H., Dunst, G. and Glaser, B. (2013). Positive effects of composted biochar on plant growth and soil fertility. *Agronomy for Sustainable Development*, 33, pp. 817–827.

Shabangu, S., Woolf, D., Fisher, E. M., Angenent, L. T. and Lehmann, J. (2014). Techno-economic assessment of biomass slow pyrolysis into different biochar and methanol concepts. *Fuel*, 117, pp. 742–748.

Slavich, P. G., Sinclair, K., Morris, S. G., Kimber, S. W. L., Downie, A. and Van Zwieten, L. (2013). Contrasting effects of manure and green waste biochars on the properties of an acidic ferralsol and productivity of a subtropical pasture. *Plant and Soil*, 366, pp. 213–227.

Spokas, K. A., Novak, J. M., Stewart, C. E., Cantrell, K. B., Uchimiya, M., DuSaire, M. G. and Ro, K. S. (2011). Qualitative analysis of volatile organic compounds on biochar. *Chemosphere*, 85, pp. 869–882.

Spokas, K. A., Cantrell, K. B., Novak, J. M., Archer, D. W., Ippolito, J. A., Collins, H. P., Boateng, A. A., Lima, I. M., Lamb, M. C., McAloon, A. J., Lentz, R. D. and Nichols, K. A. (2012). Biochar: a synthesis of its agronomic impact beyond carbon sequestration. *Journal of Environmental Quality*, 41, pp. 973–989.

Stewart, C. E., Zheng, J. Y., Botte, J. and Cotrufo, M. F. (2013). Co-generated fast pyrolysis biochar mitigates green-house gas emissions and increases carbon sequestration in temperate soils. *Global Change Biology Bioenergy*, 5, pp. 153–164.

Tabatabai, M. A. and Bremner, J. M. (1991). Automated instruments for determination of total carbon, nitrogen, and sulfur in soils by combustion techniques. In: Smith, K. A. (ed.). *Soil Analysis*. New York: Marcel Dekker, pp. 261–286.

Taghizadeh-Toosi, A., Clough, T. J., Condron, L. M., Sherlock, R. R., Anderson, C. R. and Craigie, R. A. (2011). Biochar incorporation into pasture soil suppresses in situ nitrous oxide emissions from ruminant urine patches. *Journal of Environmental Quality*, 40, pp. 468–476.

Tammeorg, P., Simojoki, A., Mäkelä, P., Stoddard, F. L., Alakukku, L. and Helenius, J. (2014). Biochar application to a fertile sandy clay loam in boreal conditions: effects on soil properties and yield formation of wheat, turnip rape and faba bean. *Plant and Soil*, 374, pp. 89–107.

Thayn, J. B., Price, K. P. and Woods, W. I. (2011). Locating Amazonian Dark Earths (ADE) using vegetation vigour as a surrogate for soil type. *International Journal of Remote Sensing*, 32, pp. 6713–6729.

UNEP, United Nations Environment Programme (2013). Black soil, black gold. *TUNZA*, 9, pp. 14–15.

Vassilev, N., Martos, E., Mendes, G., Martos, V. and Vassileva, M. (2013). Biochar of animal origin: a sustainable solution to the global problem of high-grade rock phosphate scarcity? *Journal of the Science of Food and Agriculture*, 93, pp. 1799–1804.

Ventura, M., Sorrenti, G., Panzacchi, P., George, E. and Tonon, G. (2013). Biochar reduces short-term nitrate leaching from a horizon in an apple orchard. *Journal of Environmental Quality*, 42, pp. 76–82.

Wang, L., Butterly, C. R., Wang, Y., Herath, H. M. S. K., Xi, Y. G. and Xiao, X. J. (2014). Effect of crop residue biochar on soil acidity amelioration in strongly acidic tea garden soils. *Soil Use and Management*, 30, pp. 119–128.

Watzinger, A., Feichtmair, S., Kitzler, B., Zehetner, F., Kloss, S., Wimmer, B., Zechmeister-Boltenstern, S. and Soja, G. (2014). Soil microbial communities responded to biochar application in temperate soils and slowly metabolized 13C labelled biochar as revealed by 13C PLFA analyses – results from a short term incubation and pot experiment. *European Journal of Soil Science*, 65, pp. 40–51.

Xiang, S.-R., Doyle, A., Holden, P. A. and Schimel, J. P. (2008). Drying and rewetting effects on C and N mineralization and microbial activity in surface and subsurface California grassland soils. *Soil Biology and Biochemistry*, 40, pp. 2281–2289.

Yao, Y., Gao, B., Zhang, M., Inyang, M. and Zimmerman, A. R. (2012). Effect of biochar amendment on sorption and leaching of nitrate, ammonium, and phosphate in a sandy soil. *Chemosphere*, 89, pp. 1467–1471.

Zheng, H., Wang, Z. Y., Deng, X., Herbert, S. and Xing, B. S. (2013). Impacts of adding biochar on nitrogen retention and bioavailability in agricultural soil. *Geoderma*, 206, pp. 32–39.

15

Opportunities and Uses of Biochar on Forest Sites in North America

DEBORAH S. PAGE-DUMROESE, MARK D. COLEMAN AND
SEAN C. THOMAS

> **Abstract**
>
> Biochar may be useful for restoring or revitalizing degraded forest soils and help with carbon sequestration, nutrient leaching losses, and reducing greenhouse gas emissions. However, biochar is not currently widely used on forested lands across North America. This chapter provides an overview of several biochar experiments conducted in North America and discusses the feasibility of using in-woods mobile pyrolysis systems to convert excess forest biomass into biochar. Biochar may be applied to forest sites in order to positively influence soil properties (nutrient leaching, water holding capacity), but its biggest benefit may be in facilitating reforestation of degraded or contaminated sites, and in sequestering carbon in soils. The majority of data on biochar applications on forest sites focus on seedling responses and short-term impacts on nutrients, soil physical properties and microbial changes. Long-term field research is necessary to determine water use, carbon sequestration, nutrient use, and greenhouse gas emissions, and the subsequent alteration of forest growth and stand dynamics.

15.1 Introduction

Many North American forests face management challenges related to wildfire, insect and disease outbreaks, and invasive species, resulting in part from overstocked or stressed stands. These sources of forest stress are already being exacerbated by global climate change (Dale et al., 2001). For example, changes in the pattern, distribution, and severity of fire may result in large-scale impacts on species diversity and regeneration (Stocks et al., 1998). Further, commercial forestry in many regions faces challenges related to decreased commodity values and increasing operational expenses, such that the cost of biomass removal often exceeds its value, despite increasing interest in forest biomass utilization (Rummer et al., 2003). Large quantities of forest residues – including tops, limbs, cull sections, and unmerchantable round wood – are potentially available for use in the production of energy, fuels, and biochar. These byproducts of forest operations could also be used to offset the use of fossil fuels and reduce greenhouse gas (GHG) emissions (Jones et al., 2010). In the USA, there are approximately 303 million hectares of forestlands which could yield approximately 320 million dry tons annually of forest residues for bioenergy production (US Department of Energy, 2011).

Currently, forest restoration or rehabilitation treatments involve forest thinning and regeneration harvests that can produce 40–60 million dry metric tons of woody biomass per year (Buford and Neary, 2010). Reducing wildfire hazard by fuel reduction can be costly (Desrochers et al., 1993; Zamora-Cristales et al., 2014), but in-woods processing to create chips (Jones et al., 2010), slash forwarding to recover previously discarded material (Harrill and Han, 2010), or mobile pyrolysis (i.e. thermochemical conversion of wood; Anderson et al., 2013) may all be used to decrease costs. The use of in-woods fast pyrolysis is also one method to potentially produce a viable byproduct, biochar from "waste" wood left on log landings or in slash piles (Dymond et al., 2010; Coleman et al., 2010). In addition sawmills and other wood product facilities produce large quantities of woody biomass in the form of chips, sawdust, bark, and wood shavings that could be used to create biochar at centralized bioenergy facilities.

Biochar is defined as "a solid material obtained from thermochemical conversion of biomass in an oxygen-limited environment" (IBI, 2012), and can be analogous to charcoal naturally found in fire-prone ecosystems (DeLuca and Aplet, 2008). Biochar has been tested as a soil amendment in many agricultural systems (Lehmann and Joseph, 2009; Liu et al., 2013); however, there has been considerably less work on biochar in forest systems, and in particular few published field trials (Thomas and Gale, 2015). In addition to a long residence time that results in C sequestration, biochar can improve soil properties by enhancing cation exchange capacity (CEC), increasing water holding capacity, increasing soil pH as a liming agent, and reducing soil bulk density and physical resistance to water and gas flow within the soil matrix (Mukherjee and Lal, 2013). All of these properties are thought to play a role in enhancing plant growth in biochar-amended soils (Atkinson et al., 2010).

Production of biochar, coupled with new national and international policies that promote large-scale biomass utilization (Abbas et al., 2011), could potentially lead to changes in how forest soils and stands are sustainably managed (Homagain et al., 2014). Bioenergy coupled with biochar as a co-product is a promising alternative for green energy (Homagain et al., 2014). Removal of forest residues can improve stand health and reduce the risk of wildfire (IEA, 2002), but residues also may serve as essential habitat for wood decay fungi and other organisms (Siitonen, 2001), provide cover for wildlife, reduce soil erosion, and play an important role in soil nutrient dynamics and hydrology (Lattimore et al., 2009). Therefore, how much biomass is left or removed should take into account multiple management objectives and should be determined on a site-specific basis (Wood and Layzell, 2003; Lamers et al., 2013).

Although biochar application in forest ecosystems may be logistically more challenging than in agricultural systems, forest sites are prime candidates for soil improvement from biochar additions (Page-Dumroese et al., 2010; Coleman et al., 2010; Jarvis et al., 2014). Biochar has the potential to reduce fire risks by removing highly flammable excess woody residues from forest sites, and improve soil water and nutrient retention, and to enhance vegetation growth through improved soil physical or chemical properties. In addition, since charcoal is a major component of the fire-adapted ecosystems as a result of wildfires or prescribed burns (Certini, 2005), application of biochar is expected to mimic many of the soil properties associated with wildfire-generated charcoal (Harvey et al., 1979; Deluca and Aplet, 2008; Matovic, 2011) and thus better emulate natural disturbance processes (Thomas, 2013).

In this chapter we review current progress in biochar as applied to managed forest ecosystems in North America. We specifically address the properties of biochar generated from forest residues and wood "waste" material, management scenarios and objectives in which biochar is most likely to play a role, and the effects of biochar additions on forest soil properties and tree growth. Field studies on biochar effects in forests are few, and we present novel data from field trials conducted in the western USA. We conclude with a discussion of barriers to applied use of biochar in the North American context, and of related research priorities.

15.2 Biochar Production and General Properties

Biochar can be produced in any number of ways, including traditional kilns and earth mounds and engineered systems for slow pyrolysis, fast pyrolysis, flash pyrolysis, gasification, and microwave pyrolysis (Brown, 2009; Garcia-Perez et al., 2011). See Chapter 10 for a detailed overview on different biochar production technologies. Fast-pyrolysis biochar (involving rapid heating rates to peak temperatures) has been more readily available for field and lab testing and will be the focus of the following discussions. In addition to variation in pyrolysis methods, many different feedstocks can be used, such as mill residues (sawdust, bark, wood chips), slash, and thinning residues. All production methods and feedstocks will result in differences in biochar physical and chemical properties; likewise, the same method at a different temperature or residence time will yield biochar with differing properties. For example, biochar produced between 400–600°C generally has the least amount of hydrophobicity and highest water holding capacity, while those created under higher temperatures have much stronger hydrophobic tendencies (Kinney et al., 2012; Page-Dumroese et al., 2015).

Black carbon encompasses a spectrum of carbonaceous materials, including char, high-carbon ash, coke, and soot, a subset of which can be considered biochar (Spokas et al., 2012). Biochar itself varies greatly, and even biochar created from woody residues can be inconsistent in terms of chemical properties, with tree species being particularly important in determining char chemistry, pH, and electrical conductivity (EC). Table 15.1 lists the chemical composition of several biochar samples produced from the same equipment (Abri Tech Incorporated, Namur, QC) operated by Biochar Products in Halfway, OR, USA, with similar residence times (5–7 min) and temperature ranges (388–450°C). In particular, the

Table 15.1. Selected chemical properties, pH, and EC of biochar created from woody feedstocks in the western USA

Tree species or species mix	Chemical element									pH	EC
	N	C	Ca	Mg	K	P	S	Fe	Zn		
	– wt% –					– µg g^{-1} –					
Mixed conifer	0.26	89	6700	990	3900	490	120	3900	33	8.1	103
Fire salvage	0.34	94	8700	1400	4600	730	200	9700	94	7.4	258
Beetle-killed salvage	0.18	86	5100	930	3400	280	120	13000	86	8.1	90
Quercus garryana Douglas ex Hook	0.62	87	35000	2300	8600	880	250	13000	65	7.9	180
Cytisus scoparious (L.) Link	1.10	94	8000	3100	12000	1300	270	6000	91	7.5	235
Thuja plicata Donn ex D.Don	0.31	92	9800	1300	4300	960	170	10000	65	5.4	789
Pinus edulis Englem. and *Juniperus communis* L.	0.50	76	5500	350	1200	200	<75	380	8	6.5	330
Arbutus menziesii Pursh.	0.21	85	4500	630	1600	240	96	8500	35	4.5	789
Mean	0.44	88	10413	1375	4950	635	175	8060	59	6.9	347
Coefficient of variation	69	7	97	66	73	63	39	55	53	19	82

Note: Fast pyrolysis was conducted on each feedstock using the same reactor, feed rate, residence time, and temperature range. Mixed conifer consisted of 70% *Pseudotsuga menziesii* Mirb. Franco, 20% *Tsuga heterophylla* (Raf.) Sarg., and 10% *Abies concolor* (Gord. & Glend.) Lindl. ex Hilebr. Fire salvage consisted of 60% *Pseudotsuga menziesii*, 30% *Tsuga heterophylla*, and 10% *Abies concolor*. Material was salvaged three years after fire. Beetle-killed salvage material consists of 60% *Pinus contorta* Douglas ex Loudon and 40% *Pseudotsuga menziesii*.

wide range of pH, EC, and macro- and micro-nutrients indicates that care should be taken to understand how soil properties might be altered after application of a given biochar. Information about supply chains for biomass production, feedstock logistics, conversion, distribution logistics, and end uses are described in Chapter 2.

15.3 Field Applications

Large-scale, centralized biomass and biochar facilities require large quantities (potentially thousands of tons) of feedstock biomass each year and a transportation infrastructure to move biomass from a harvest unit to the facility and transport biochar to an application site. There are examples of such large-scale facilities in North America in situations where there is both feedstock availability and good access to markets for biochar. In many cases such large-scale facilities are not logistically or economically feasible; however, advanced thermochemical technologies currently being developed are targeted to small-scale demand and processing (Fransham and Badger, 2006; Biochar Solutions Incorporated, 2011; Anderson et al., 2013). Using smaller scale, in-woods (or near woods) biochar processing is one alternative for creating biochar from "waste" wood using residues that would normally be left on-site (lop and scatter) or burned in slash piles. Both the economic feasibility and carbon benefits of these systems are enhanced by reducing transportation of low-value woody biomass. If excess forest residues are pyrolyzed, rather than burned in slash piles, large quantities of the byproduct biochar would result (Mohan et al., 2006).

Generating biochar from waste wood has additional advantages; soil damage is minimized when slash pile burning is avoided or reduced (Page-Dumroese et al., 2010) and there are fewer particulates and GHG emissions from pyrolysis as compared to slash burning (Anderson et al., 2013). Distributed, small-scale facilities would be able to make biochar from local sources and have the potential to allow individuals to match biochar properties to particular sites. Matching biochar may be particularly useful for remediation of specific soil chemical or physical properties (Novak et al., 2009a). In addition to in-woods pyrolysis systems, other in-woods portable equipment for feedstock preparation, such as dryers, chippers, grinders, and pellet mills, would potentially provide the means for moving slash within a local harvest unit and processing it into biochar that can be applied on-site or sold as a commercial product.

Unlike agricultural soils where biochar can be added and tilled into the soil profile, application of biochar on forest sites is more difficult since trees, stumps, and downed wood hinder movement across a harvest unit. However, in managed forests log landings, skid trails, abandoned roads, or abandoned mine land soils all require some form of restoration. Biochar added to the surface or mixed into the mineral soil during restoration activities (e.g. decompaction or invasive species removal) can help increase water retention, reduce leaching, or improve bulk density (Ippolito et al., 2012) and can be applied with existing forest harvest equipment. However, biochar applications should not disturb the surface organic horizons (Page-Dumroese et al., 2010). Ease of biochar application

Figure 15.1. Pellets made from biochar and logging slash (photo by Deborah S. Page-Dumroese).

will depend on the equipment used to make the char, where material size varies from several centimeters to sub-millimeters. Fine-textured biochar could potentially be applied to forest sites using modified agricultural machinery similar to that used in forest liming, as has been widely practiced in high-value hardwood stands in eastern North America (Long et al., 1997). Formal evaluations of use of spreaders for wood ash have indicated challenges in efficiency and uniformity (Wilhoit and Ling, 1996). If biochar is pelletized on site, a log forwarder-pulled pellet spreader (see Figure 2.8) could potentially be used on skid trails and move throughout relatively open harvested stands. Pellets, such as shown in Figure 15.1, can be produced using fresh slash as a binder (Dr K. Englund personal communication, 2015). Moreover, the spreader has the capability to be used on slopes ($\leq 35\%$) with spread width and quantity adjusted based on need or terrain. Care will have to be taken with the spreader so that soil conditions (i.e. high moisture content, low bulk density) do not result in excess compaction.

Another important use for biochar in a forestry context is in mine tailings restoration. Abandoned hardrock mines dot much of North America, and in western USA forested landscapes they are extraordinarily common. In many places, signs of their existence are simply holes in the ground or cliff wall; in other places, there are square kilometers of unproductive, exposed tailing features. Environmental concerns with the latter scenario include soil instability, sediment transport into nearby streams, limited revegetation, and natural succession processes that are extremely slow, or occurring with undesirable species. In cases of acid-generating metal-leaching tailings, there are additional critical concerns involving soil and stream acidification and mobilization of toxic metals. Biochar amendments have the potential to reduce leaching and bioavailability of heavy metals such as copper, zinc, lead, and cadmium (Beesley and Marmiroli, 2011; Beesley et al., 2014; Bakshi et al., 2014), mainly as a result of char sorption characteristics and biochar effects on soil pH. Furthermore, in addition to retaining heavy metals, biochar may also be useful in adsorbing mineral salts near urban areas (de-icing) or on mine spoils. In areas where road salt is routinely applied, biochar could mitigate salt-induced stresses. In a greenhouse experiment, biochar applied at 50 t ha^{-1} alleviated salt-induced mortality in two herbaceous plant species (*Abutilon theophrasti* Medik. and *Prunella vulgaris* L.) (Thomas et al., 2013). Changes to plant growth and survival were attributed to salt sorption on the biochar rather than increased plant growth.

15.4 Biochar Effects on Forest Soil Properties

15.4.1 Physical Properties

Biochar is highly porous and its application to forest soil can improve a range of soil physical properties, including soil porosity, pore-size distribution, bulk density, moisture holding capacity, infiltration, and hydraulic conductivity (Atkinson et al., 2010; van Zweiten et al., 2012). Of particular importance to forestry operations are the beneficial effects related to reduced soil bulk density on skid trails or log landings. In many areas, road removal on National Forests in the USA is being used to restore ecosystem processes. Often roads are ripped to decompact the soil surface and this is typically done with a bulldozer pulling a plow over the roadbed or a grappler lifting the roadbed. Once the road surface has been decompacted, soil amendments can be either surface applied or mixed in. Removing old or unused roads presents an opportunity to use biochar to add organic matter, help maintain a lower bulk density by forming micro-aggregates (Verheijen et al., 2009), and help establish vegetation (Adams et al., 2013). In addition, mulching with biochar or other organic amendments may prevent the soil surface from sealing, which might increase sedimentation and runoff (Luce, 1997; Bradley, 1997).

Direct empirical data from field trials in forests are limited. Data from a road decommissioning project in central Montana show that after two years, biochar did not improve soil bulk density or soil moisture to a much greater extent than just ripping (Table 15.2), which is similar to other findings (e.g. Switalski et al., 2004) for soil physical properties. Although positive effects on soil hydrological properties have been found in agricultural systems, even at a rate of 47 Mg ha^{-1} in an apple orchard, biochar did not alter soil porosity or water holding capacity (Hardie et al., 2014).

Table 15.2. *Average ground cover and moisture content, bulk density, and organic matter content two years after road restoration and biochar additions in central Montana*

Treatment	Soil surface cover				Soil moisture content in August	Soil organic matter	Soil bulk density Mg m^{-3}
	Bare ground	Forbs	Grass	Organic horizon			
	–% cover –				wt%		
25 Mg/ha biochar	65 (4)	10 (1)	10 (1)	4 (0.7)	29.4 (5.4)	4.8 (0.2)	0.9 (0.1)
10 Mg/ha biochar	62 (3)	7 (1)	8 (1)	3 (0.5)	11.9 (3.8)	3.5 (0.2)	0.6 (0.2)
2 Mg/ha biochar	69 (4)	9 (1)	9 (1)	9 (0.6)	17.6 (1.0)	4.3 (0.2)	1.1 (0.1)
2 Mg/ha biochar pellets	68 (4)	3 (0.5)	8 (1)	9 (.8)	20.9 (1.5)	3.7 (0.2)	0.9 (0.1)
Ripping only	68 (3)	13 (2)	1 (0.2)	14 (1)	19.2 (1.2)	3.9 (0.3)	1.1 (0.1)
14 Mg/ha wood straw	44 (2)*	5 (1)	7 (1)	1 (0.2)	13.4 (4.4)	2.7 (0.1)*	0.9 (0.2)
Untreated road	13 (3)*	6 (1)	81 (3)*	0	5.7 (0.5)*	2.9 (0.1)*	1.5 (0.3)*

Note: Biochar was created using mobile fast pyrolysis at ~400°C. Feedstock was beetle-killed lodgepole (*Pinus contorta* Douglas ex Loudan). Asterisks indicate significant differences from the other treatments at $p < 0.05$.

15.4.2 Chemical Properties

Nutrient transformations are dependent on the type and quality of biochar when it is added to the soil. During pyrolysis, heating causes some nutrients to volatilize, especially at the surface of the biochar, while other nutrients become concentrated (DeLuca et al., 2009). Nitrogen is usually lost from the char during high-temperature pyrolysis (Tyron, 1948). High-temperature (800°C) biochar produced from wood waste feedstocks generally shows higher pH, EC, and extractable NO_3^- relative to low-temperature (350°C) biochar; however, biochar density, extractable PO_4^-, and NH_4^+ are generally lower in high-temperature biochars (Gundale and DeLuca, 2006). Biochar produced from wood waste material is generally high in soluble potassium, and to a variable extent in phosphorus and calcium. In a Northern hardwood forest Sackett et al. (2014) found an initial increase in soil-available potassium following biochar additions, followed later by increases in soil-available calcium and magnesium.

15.4.3 Biological Properties

Recent research suggests that biochar commonly initially stimulates microbial communities, with this effect diminishing over time (Kuzyakov et al., 2009) as labile C is metabolized (Smith et al., 2010). Soil enzyme activity, similar to soil chemical and physical property changes, is related to biochar quality and soil type (Bailey et al., 2011). In a comparison of a forest soil (Andisol) and agricultural soil (Mollisol), enzymes responsible for

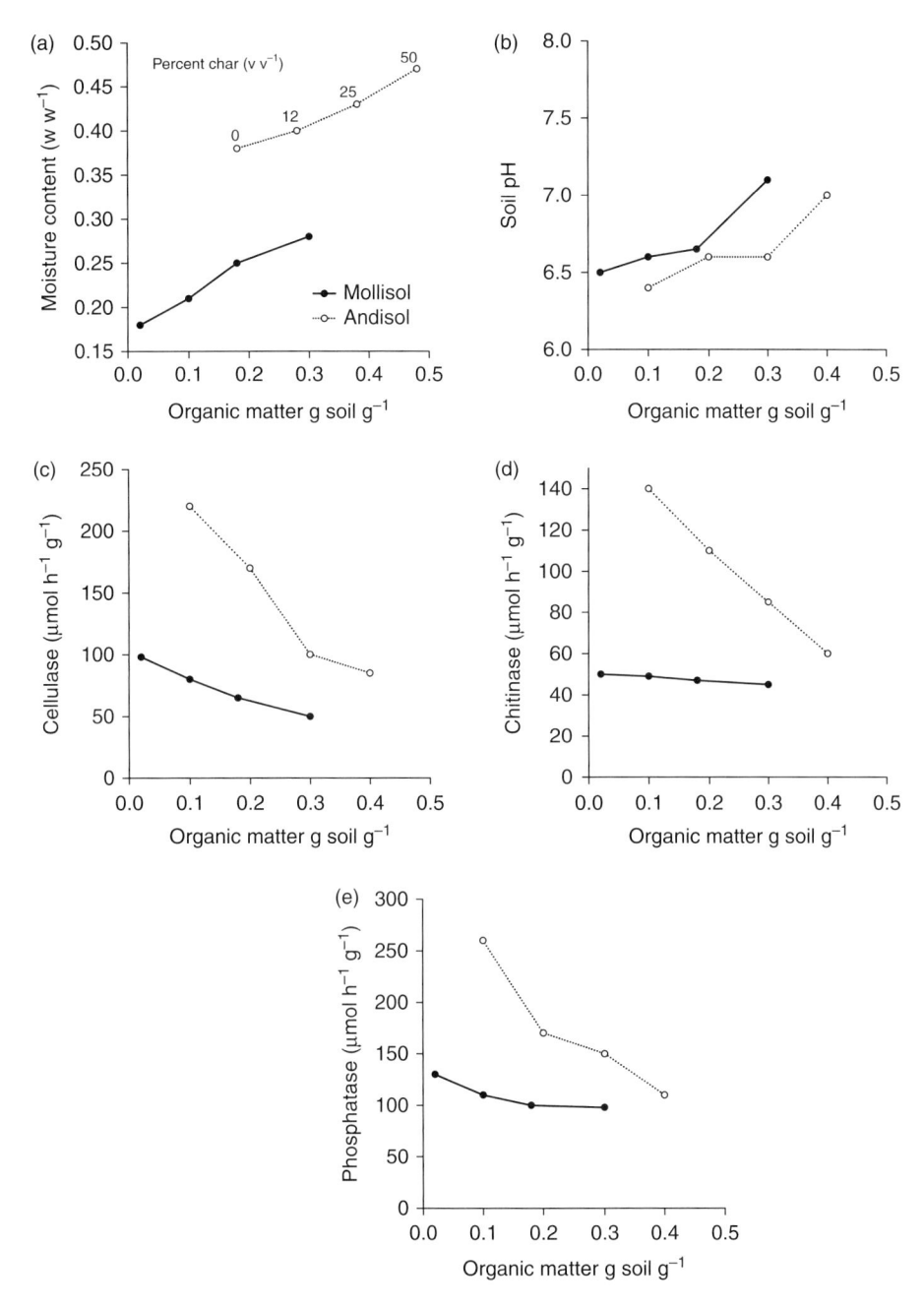

Figure 15.2. Changes in (A) soil moisture, (B) soil pH, (C) cellulase, (D), chitinase, and (E) phosphatase in biochar-amended Andisol and Mollisol soil types after laboratory incubation. CQuest Biochar was used for amendment (Dynamotive, Vancouver, BC, Canada) and was produced using fast pyrolysis of hardwood residue (McElligott, 2011). This biochar had a total surface area of 1.6 m^2 g^{-1}, 16–23% organic volatile compounds, and with 100% particle size distribution <2 mm in size, 95% of the particles <1 mm, and 60% of the particles <0.5 mm. Physical and chemical analyses at the University of Idaho indicated a bulk density of 0.33 Mg m^{-3}, a pH of 6.8, a CEC of 30 cmol$^{(+)}$kg^{-1}, 62% total C, and 0.18% total N.

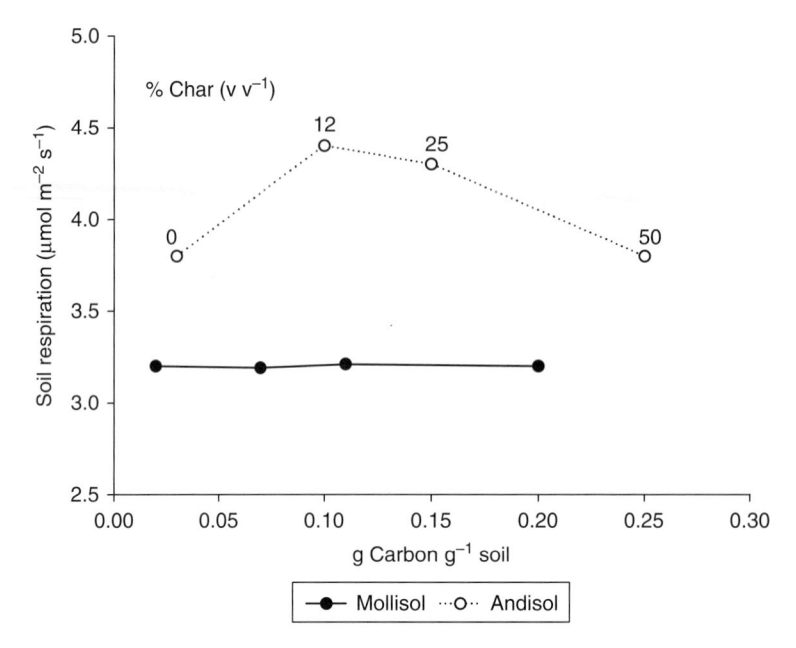

Figure 15.3. Biochar amendment changes to soil respiration in an Andisol and Mollisol after laboratory incubation. CQuest Biochar was used for amendment (see Figure 15.2).

decomposition processes decreased with increased biochar additions (Figure 15.2), but soil respiration was unaffected (Figure 15.3) indicating that organic matter is likely not lost as biochar is added to the soil.

Soil microbial composition is also likely to change in response to biochar additions to forest soils. Biochar has sometimes been portrayed as being particularly beneficial to fungi (Ishii and Kadoya, 1994; Warnock et al., 2007); however, recent studies indicate that biochar additions result in increased soil bacterial populations and increased bacterial:fungal ratio in a variety of systems (Chen et al., 2013; Farrell et al., 2013; Gomez et al., 2014). In a Northern hardwood forest soil only minor effects on soil microbial community structure were found with low rates of biochar addition (5 t ha^{-1}) with a small but significant increase in bacterial: fungal ratio (Noyce et al., 2015). Laboratory soil incubations in the same system showed a pronounced shifts in the soil microbial community at higher biochar addition rates (10 and 20 t ha^{-1}), with an increase in the bacterial:fungal ratio and a transient increase in Gram-negative bacteria (Perry et al., 2015).

15.4.4 Greenhouse Gas Flux

Biochar is thought to be an important potential tool for mitigating increasing atmospheric levels of CO_2, firstly by sequestering carbon, and secondarily by increasing net primary productivity and reducing GHG emissions from the soil or plant materials. Studies of both soil CO_2 and methane flux (Rondon et al., 2005; Spokas et al., 2010; Stewart et al., 2013) have given conflicting data on the value of adding biochar. Biochar is generally expected to result in at least a transient increase in soil CO_2 efflux (sometimes termed "priming") as a result of microbial responses to labile carbon and nutrients (Ameloot et al., 2013). Some studies have also found increased soil C mineralization in response to char additions (Wardle et al., 2008). However, recent studies suggest highly variable responses, including "negative priming" effects in which biochar additions reduce soil respiration (Zimmerman et al., 2011; Jones et al., 2011; Ameloot et al., 2013). In agricultural systems biochar is expected to reduce soil methane emissions by enhancing soil porosity and oxygen levels, and indeed complete suppression of methane emissions from field plots in the tropics has been observed (Rondon et al., 2005). As noted previously, many of the responses associated with biochar added to soils will be dependent on the original feedstock for biochar and the soil, as well as the pyrolysis conditions (e.g. temperature).

The limited data available on soil GHG flux responses to biochar amendments in forest systems likewise appear variable. Lab incubation studies with forest soils have found increases in soil respiration in the short term, but positive "priming" effects are commonly transient (Steinbeiss et al., 2009; Zimmerman et al., 2011), or show complicated dynamics (Mitchell et al., 2015). Responses are also highly dependent on soil type. For example, soil respiration from the forested Andisol and the agricultural Mollisol were different, but there was no response to the addition of biochar (Figure 15.3). In a 12-month laboratory incubation of temperate hardwood forest soils, Sackett et al. (2015) found higher microbial respiration in soils treated with biochar from maple feedstocks than in soils treated with spruce feedstock biochar. Spokas and Reicosky (2009) noted that after testing 16 different biochar samples on agricultural, forest, and landfill soils changes in GHGs were dependent on both soil and biochar types. Field responses may also show strong deviations from laboratory incubations since half or more of total soil CO_2 efflux is attributable to root respiration. Sackett et al. (2015) found no detectable effect of biochar additions on soil CO_2 efflux in a field trial, in spite of significant effects in laboratory incubations.

Forest soils, particularly those of upland temperate forests, are a globally significant sink for methane (Price et al., 2003); however, there is substantial heterogeneity in soil methane flux patterns in forest ecosystems, linked to local variation in hydrology (Dalal and Allen, 2008; Wang et al., 2013). Methane uptake by forest soils is thought to be strongly substrate-limited (Bradford et al., 2001; Dalal and Allen, 2008; Wang et al., 2013), suggesting the importance of soil porosity and aeration. We are aware of only one field study that has tested biochar effects on soil methane uptake (Sackett et al., 2015); although this study did not find a significant effect, the biochar addition rate used was low (5 t ha^{-1}), and at the time of measurements biochar was not fully incorporated in the mineral soil.

15.4.5 Growth Responses

There has been a rapid increase in studies examining plant growth responses to biochar additions: recent meta-analyses that now incorporate hundreds of independent experiments suggest that agricultural crops show average increases in the range of 10–25% (Biederman and Harpole, 2013; Liu et al., 2013). A recent meta-analysis restricted to tree response studies found an average 41% increase in biomass (Thomas and Gale, 2015). However, it should be emphasized that both agricultural and forestry studies show high variability, with individual studies showing positive, negative, or no significant change in vegetative growth (Spokas et al., 2012). This variability arises due to inherent differences in the soil, fertilizer application, the nature of the biochar, and differences in responses among plant species. Biochar additions to infertile soil can improve cation exchange capacity (Cheng et al., 2006; Lee et al., 2010), but no or minimal changes in cation exchange have also been observed (Novak et al., 2009b). Further, there are complex relationships between biochar and the soil matrix, leading to altered pH, soil nutrient availability, and microbial communities (Major et al., 2010). In addition, vegetation responses may be delayed initially, followed by yield increases in subsequent years (Gaskin et al., 2010; Major et al., 2010). Delayed responses could be due to "aging" of the biochar (e.g. oxidation) (Spokas et al., 2012), or sorption of volatile organic compounds (Spokas et al., 2010; Gale et al., 2016). Biochar sorptive properties can mitigate impacts on plants by reducing exposure to the stress agent (Spooks et al. 2012; Thomas et al. 2013, 2015). In addition, aging or weathering of biochar often results in alteration of the surface chemistry (Azargohar and Dalai, 2006; Nuithitikul et al., 2010), and in out-gassing substances such as ethylene (Fulton et al., 2013), but there is commonly little documentation regarding handling, storage, or post-treatment of biochars.

Thomas and Gale (2015) published a review of tree responses to biochar mostly involved with laboratory and greenhouse trials (e.g. 14 of 17 studies included in meta-analysis). In the Inland Northwest USA, there are several ongoing biochar field trials examining tree growth responses to biochar (McElligott, 2011). Short-term (1–2 years) changes in diameter increment on two sites (Inceptisol and Andisol soils) were not significantly impacted by biochar additions (Figure 15.4).

The Andisol is a fine-textured, highly productive soil (Page-Dumroese et al., 2015) and here tree growth was not affected by biochar amendment, but could be improved by leaving the residual slash in place. This result is similar on the coarser-textured Inceptisol, but higher biochar application rates had a greater tree response. Again, tree growth in the biochar plots was not significantly different from the residual slash retention plots. On this relatively infertile soil type (Inceptisol), biochar with fertilization also did not offer additional growth gains. Longer-term (five years) results from a coarse-textured Andisol in south-central Oregon also indicate that biochar application at 25 Mg ha^{-1} was similar to retaining forest residues (Figure 15.5) (McElligott, 2011). However, lower levels of biochar application were not as effective as slash retention for increasing growth, but did increase height growth slightly over the control trees. Although slash provides similar growth gains as biochar application, slash has a short residence time on the soil surface, whereas biochar

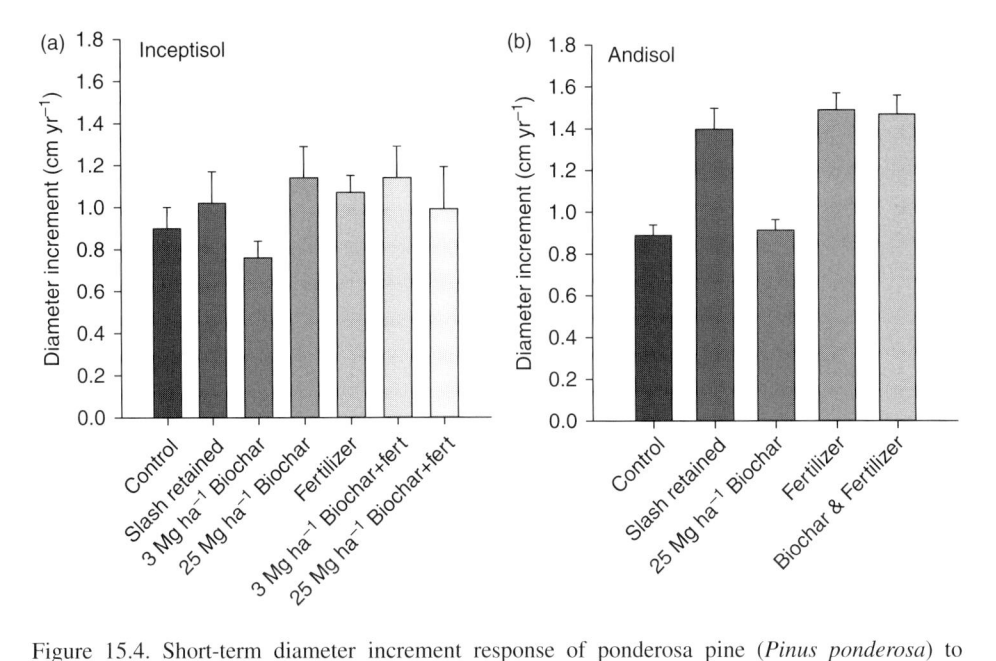

Figure 15.4. Short-term diameter increment response of ponderosa pine (*Pinus ponderosa*) to biochar, fertilizer, or slash retention on (A) Inceptisol or (B) Andisol soils in the Inland Northwest, USA. The Andisol location used CQuest Biochar for amendment (see Figure 15.2). The Inceptisol used char from Biochar Solutions, Inc. and was produced using fast pyrolysis on a mixture of western conifer residues. The biochar had a total surface area of 12 m² g^{-1}, with particles sizes ranging from 6.5–0.2 mm; 80% of the particles were <2 mm in size (Anderson et al., 2013). Physical and chemical characteristics tests were conducted at the Rocky Mountain Research Station and indicated a bulk density of 0.13 Mg m^{-3}, a pH of 8.7, 76% total C, and 0.45% total N. Means plotted +1 SE.

provides long-term soil C once it migrates into the mineral soil (Lehmann, 2007). For all forest sites, biochar was applied to the surface (on top of the existing forest floor) to limit soil disturbance and maintain nutrient cycling, and this may explain the lack of pronounced tree growth responses. To alter the mineral soil, biochar must first be transported through the forest floor to provide benefits of soil water retention and subsequent tree growth. This study is described in McElligott (2011), but we have collected height growth data in subsequent years (Figure 15.5). There have not been large gains in productivity, but neither has tree growth been significantly reduced. In addition, at the application rate of 25 Mg ha^{-1} with approximately 80% C, 15 Mg C ha^{-1} was sequestered with no deleterious effects.

15.4.6 Invasive Species

Biochar has the potential to improve soil quality and thereby increase desirable species restoration by the addition of organic C. Biochar additions may also result in greater microbial uptake and immobilization of N (Perry et al., 2004). On a tallgrass prairie site in Minnesota, soil C additions resulted in a 54% reduction of weed biomass and a seven-fold

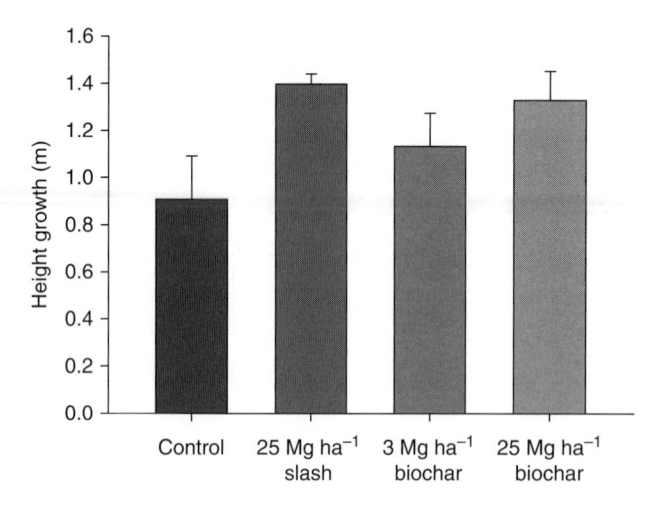

Figure 15.5. Five-year height growth response of ponderosa pine to biochar or slash retention on the Umpqua National Forest, Oregon, USA. CQuest Biochar was created by fast pyrolysis from a fixed plant using mixed hardwood feedstock (see Figure 15.2). Means plotted +1 SE.

increase in native prairie species biomass, which was attributed to a large reduction in soil N (Blumenthal et al., 2003). Other authors have noted similar results with C additions reducing weed growth and/or greater growth of desired species (Blumenthal et al., 2003; Perry et al., 2004; Grygiel et al., 2010). However, other studies have reported no effect on invasive or desired species after soil C additions (Corbin and D'Antonio, 2004; Mangold and Sheley, 2008), or found that C additions reduced growth of desired species (Averett et al., 2004). Rapid establishment of vegetation is important for ripped roads and skid trails or after harvest operations. Vegetation growth is one of the first signs of ecosystem recovery (Wright and Blaser, 1981). Unused roads are typically nutrient poor and commonly dominated by invasive species (Switalksi et al., 2004). In a central Montana road decommissioning project, the road surface vegetation was dominated by invasive grasses. However, after ripping, forbs and native grass species were beginning to revegetate both the ripped only and the biochar plots after two years (Table 15.2). While this study did not show definitive increases in desirable species in response to biochar, biochar additions did not impede revegetation efforts.

15.5 General Prospectus and Critical Research Needs

The potential benefits of adding biochar to agricultural sites have received considerable recent attention (e.g. Spokas et al., 2012), but few studies to date have examined analogous approaches in the forestry sector. There is a clear need for long-term field trials examining a range of biochars, soils, and forest types. A repeated theme in the present review is that responses observed in short-term lab or greenhouse studies do not necessarily translate into comparable responses in the field. It is certainly the case that careful planning to match biochar with site properties can result in C sequestration and improved soil conditions such as organic matter content, porosity, and water hold capacity. No deleterious impacts of biochar additions on forest vegetation have been found to date, though effects on a broader range of forest organisms, such as soil invertebrates, have received almost no attention. Site access and transport considerations are certain to be of critical importance in all practical applications of biochar to managed forests. Highly impacted areas such as skid trails and log landings will likely be a priority for applications due to both potential benefits for site remediation, and ease of access. Pelletizing biochar improves the ease with which it can be applied and reduces dust and particulates in the air. In addition, pellets made with fresh slash return many nutrients inherent in the biomass back to the site, thereby reducing the risk of nutrient depletion.

Acknowledgements

Partial funding was provided by grants from The Agriculture and Food Research Initiative, Biomass Research and Development Initiative, Competitive Grant no. 2010–05325 from the USDA National Institute of Food and Agriculture, the Department of Energy, Biomass and Research and Development Initiative, Competitive Grant no. DE-EE0006297, and the Natural Sciences and Engineering Research Council of Canada.

References

Abbas, D., Current, D., Phillips, M., Rossman, R., Hoganson, H. and Brooks, K. N. (2011). Guidelines for harvesting forest biomass for energy: a synthesis of environmental considerations. *Biomass and Bioenergy*, 35, pp. 4538–4546.

Adams, M. M., Benjamin, T. J., Emery, N. C., Brouder, S. J. and Gibson, K. D. (2013). The effect of biochar on native and invasive prairie plant species. *Invasive Plant Science and Management*, 6, pp. 197–207.

Ameloot, N., Graber, E. R., Verheijen, F. G. A. and De Neve, S. (2013). Interactions between biochar stability and soil organisms: review and research needs. *European Journal of Soil Science*, 64, pp. 379–390.

Anderson, N., Jones, J. G., Page-Dumroese, D., et al. (2013). A comparison of producer gas, biochar, and activated carbon from two distributed scale thermochemical conversion systems used to process forest biomass. *Energies*, 6, pp. 164–183.

Atkinson, C. J., Fitzgerald, J. D. and Hipps, N. A. (2010). Potential mechanisms for achieving agricultural benefits from biochar application to temperate soils: a review. *Plant and Soil*, 337, pp. 1–18.

Averett, J. M., Klops, R. A., Nave, L. E., Frey, S. D. and Curtis, P. S. (2004). Effects of soil carbon amendment on nitrogen availability and plant growth in an experimental tallgrass prairie restoration. *Restoration Ecology*, 12, pp. 568–574.

Azargohar, R. and Dalai, A. K. (2006). Biochar as a precursor of activated carbon. *Applied Biochemistry and Biotechnology*, 1131, pp. 762–773.

Bailey, V. L., Fansler, S. J., Smith, J. L. and Bolton, H., Jr. (2011). Reconciling apparent variability in effects of biochar amendment on soil enzyme activities by assay optimization. *Soil Biology and Biochemistry*, 43, pp. 296–301.

Bakshi, S., He, Z. L. and Harris, W. G. (2014). Biochar amendment affects leaching potential of copper and nutrient release behavior in contaminated sandy soils. *Journal of Environmental Quality*, 43, pp. 1894–1902.

Beesley, L. and Marmiroli, M., 2011. The immobilisation and retention of soluble arsenic, cadmium and zinc by biochar. *Environmental Pollution*, 159, pp. 474–480.

Beesley, L., Inneh, O. S., Norton, G. J., Moreno-Jimenez, E., Pardo, T., Clemente, R. and Dawson, J. J. (2014). Assessing the influence of compost and biochar amendments on the mobility and toxicity of metals and arsenic in a naturally contaminated mine soil. *Environmental Pollution*, 186, pp. 195–202.

Biederman, L. A. and Harpole, S. (2013). Biochar and its effects on plant productivity and nutrient cycling: a meta-analysis. *GCB Bioenergy*, 5, pp. 202–214.

Biochar Solutions Incorporated (BSI) (2011). *BSI Biochar Base Unit: Technical Specifications, Version 3.0.* Carbondale, CO: Biochar Solutions, Incorporated.

Blumenthal, D. M., Jordan, N. R. and Russelle, M. P. (2003). Soil carbon addition controls weeds and facilitates prairie restoration. *Ecological Applications*, 13, pp. 605–615.

Bradford, M. A., Ineson, P., Wookey, P. A. and Lappin-Scott, H. M. (2001). Role of CH4 oxidation, production and transport in forest soil CH4 flux. *Soil Biology & Biochemistry*, 33, pp. 1625–1631.

Bradley. K. (1997) Evaluation of two techniques for the utilization of logging residues: organic mulch for abandoned road revegetation and accelerated decomposition in small chipped piles. Theses, Dissertations, Professional Papers. University of Montana. Paper 2264.

Brown, R. (2009). Biochar production technology. In: Lehmann, J. and Joseph, S. (eds.) *Biochar for Environmental Management: Science and Technology.* London: Earthscan, pp. 127–146.

Buford, M. A. and Neary, D. G. (2010). *Sustainable biofuels from forests: meeting the challenge. The Ecological Society of America.* [online] Available at: http://esa.org/biofuelsreports [Accessed 5 January 2015]

Certini, G. (2005). Effects of fire on properties of forest soils: a review. *Oecologia*, 143, pp. 1–10.

Chen, J., Liu, X., Zheng, J., Zhang, B., Lu, H., Chi, Z. and Yu, X. (2013). Biochar soil amendment increased bacterial but decreased fungal gene abundance with shifts in community structure in a slightly acid rice paddy from Southwest China. *Applied Soil Ecology*, 71, pp. 33–44.

Cheng, C. H., Lehmann, J., Thies, J. E., Burton, S. D. and Engelhard, M. H. (2006). Oxidation of black carbon by biotic and abiotic processes. *Organic Geochemistry*, 37, pp. 1477–1488.

Chintala, R., Mollinedo, J., Schumacher, T. E., Malo, D. D. and Julson, J. L. (2014). Effect of biochar on chemical properties of acidic soils. *Archives of Agronomy and Soil Science*, 60, pp. 393–404.

Coleman, M., Page-Dumroese, D., Archuleta, J., et al. (2010). Can portable pyrolysis units make biomass utilization affordable while using biochar to enhance soil productivity and sequester carbon? In: Jain, T. B., Graham, R. T. and Sandquist, J. (technical eds.) *Proceedings of the National Silviculture Workshop on Integrated Management of Carbon Sequestration and Biomass Utilization Opportunities in a Changing Climate.* ID: RMRS-P-61. Fort Collins, CO: USDA Forerst Service Rocky Mountain Research Station, pp. 159–168.

Corbin, J. D. and D'Antonio, C. M. (2004). Can carbon addition increase competitiveness of native grasses? A case study from California. *Restoration Ecology*, 12, pp. 36–43.

Dale, V. H., Joyce, L. A., McNulty, S, Neilson, R. P., et al. (2001). Climate change and forest disturbances. *Bioscience*, 51, pp. 723–733.

Dalal, R. C. and Allen, D. E. (2008). Greenhouse gas fluxes from natural ecosystems. *Australian Journal of Botany*, 56, pp. 369–407.

DeLuca, T. H. and Aplet, G. H. (2008). Charcoal and carbon storage in forest soils of the Rocky Mountain West. *Frontiers in Ecology and Environment*, 6, pp. 18–24.

DeLuca, T. H., MacKenzie, M. D. and Gundale, M. J. (2009). Biochar effects on soil nutrient transformations. In: Lehmann, J. and Joseph, S. (eds.) *Biochar for Environmental Management: Science and Technology*. London: Earthscan. pp. 421–454.

Desrochers, L., Puttock, G. and Ryans, M. (1993). The economics of chipping logging residues at roadside: a study of three systems. *Biomass & Bioenergy*, 5, pp. 401–411.

Dymond, C. C., Titus, B. D., Stinson, G. and Kurz, W. A. (2010). Future quantities and spatial distribution of harvesting residue and dead wood from natural disturbances in Canada. *Forest Ecology and Management*, 260, pp. 181–192.

Farrell, M., Kuhn, T. K., Macdonald, L. M., Madden, T. M., et al. (2013). Microbial utilization of biochar-derived carbon. *Science of the Total Environment*, 465, pp. 288–297.

Fellet, G., Marchiol, L., Delle Vedove, G. and Peressotti, A. (2011). Application of biochar on mine tailing: effects and perspectives for land reclamation. *Chemosphere*, 83, pp. 1262–1267.

Fransham, P. and Badger, P. (2006). Use of mobile fast pyrolysis plants to densify biomass and reduce biomass handling cost – a preliminary assessment. *Biomass and Bioenergy*, 30, pp. 321–325.

Fulton, W., Gray, M., Prahl, F. and Kleber, M. (2013). A simple technique to eliminate ethylene emissions from biochar amendment in agriculture. *Agronomy for Sustainable Development*, 33, pp. 469–474.

Gale, N. V., Sackett, T. and Thomas, S. C. (2016). Thermal treatment and leaching biochar alleviates plant growth inhibition from mobile organic compounds. PeerJ. doi: 10.7717/peerj.2385.

Garcia-Perez, M., Lewis, T. and Kruger, C. E. (2011). Methods for producing biochar and advanced biofuels in Washington state. Part 1: Literature review of pyrolysis reactors. Ecology publication Number 11-07-017 Washington State Department of Ecology, Spokane, WA. p. 137. [online] Available at: www.ecy.wa.gov/pubs/1107017.pdf [Accessed 1 March 2015]

Gaskin, J. W., Speir, R. A., Harris, K., et al. (2010). Effect of peanut hull and pine chip biochar on soil nutrients, corn nutrient status, and yield. *Agronomy Journal*, 102, pp. 623–633.

Gomez, J. D., Denef, K., Stewart, C. E., Zheng, J. and Cotrufo, M. F. (2014). Biochar addition rate influences soil microbial abundance and activity in temperate soils. *European Journal of Soil Science*, 65, pp. 28–39.

Grygiel, C. E., Jorland, J. E. and Biondini, M. E. (2010). Can carbon and phosphorus amendments increase native forbs in a restoration process? A case study in northern tallgrass prairie (USA). *Restoration Ecology*, 20, pp. 122–130.

Gundale, M. J. and DeLuca, T. H. (2006). Temperature and source material influence ecological attributes of ponderosa pine and Douglas-fir charcoal. *Forest Ecology and Management*, 231, pp. 86–93.

Gundale, M. and DeLuca, T. (2007). Charcoal effects on soil solution chemistry and growth of *Koeleria macrantha* in the ponderosa pine/Douglas-fir ecosystem. *Biology and Fertility of Soils*, 43, pp. 303–311.

Hardie, M., Clothier, B., Bound, S., Oliver, G. and Close, D. (2014). Does biochar influence soil physical properties and soil water availability? *Plant and Soil,* 376, pp. 347–361.

Harrill, H. and Han, H.-S. (2010). Application of hook-lift trucks in centralized logging slash grinding operations. *Biofuels*, 1, pp. 399–408.

Harvey, A. E., Larsen, M. J. and Jurgensen, M. F. (1979). Comparative distribution of ectomycorrhizae in soils of three western Montana forest habitat types. *Forest Science*, 25, pp. 350–358.

Homagain, K., Shahi, C., Luckai, N. and Sharma, M. (2014). Biochar-based bioenergy and its environmental impact in Northwestern Ontario Canada: a review. *Journal of Forestry Research*, 25, pp. 737–748.

IBI (International Biochar Initiative) (2012). Standardized product definition and product testing guidelines for biochar that is used in soil. [online] Available at: www.biochar-international.org/sites/default/files/guidelines_for_specifications_of_biochars_for_use_in_soils_January-2012-draft.pdf [Accessed 24 June 2016]

IEA (International Energy Agency) (2002). Sustainable production of woody biomass for energy. Position Paper. [online] Available at: www.ieabioenergy.com/library/157_PositionPaper-SustainableProductionofWoodyBiomassforEnergy.pdf [Accessed 24 June 2016]

IEA (International Energy Agency) (2009). *From 1st to 2nd- generation biofuel technologies.* [online] Available at: www.iea.org/textbase/publications/index.asp. [Accessed 5 January 2015]

Ippolito, J. A., Laird, D. A. and Busscher, W. J. (2012). Environmental benefits of biochar. *Journal of Enviornmental Quality*, 41, pp. 967–972.

Ishii, T. and Kadoya, K. (1994) Effects of charcoal as a soil conditioner on citrus growth and vesicular-arbuscular mycorrhizal development. *Journal of the Japanese Society of Horticultural Science*, 63, pp. 529–535.

Jarvis, J. M., Page-Dumroese, D. S., Anderson, N. M., Corilo, Y. and Rodgers, R. P. (2014). Characterization of fast pyrolysis products generated from several western USA woody species. *Energy & Fuels*, 28, pp. 6438–6446.

Jauss, V., Johnson, M., Krull, E., Daub, M. and Lehmann, J. (2015). Pyrogenic carbon controls across a soil catena in the Pacific Northwest. *Catena*, 124, pp. 53–59.

Jones, D. L., Murphy, D. V., Khalid, M., Ahmad, W., Edwards-Jones, G. and DeLuca, T. H. (2011). Short-term biochar-induced increase in soil CO_2 release is both biotically and abiotically mediated. *Soil Biology & Biochemistry*, 43, pp. 1723–1731.

Jones, G., Loeffler, D., Calkin, D. and Chung, W. (2010). Forest treatment residues for thermal energy compared with disposal by onsite burning: emissions and energy return. *Biomass & Bioenergy*, 34, pp. 737–746.

Kinney, T. J., Masiello, C. A., Dugan, B., et al. (2012). Hydrologic properties of biochars produced at different temperatures. *Biomass & Bioenergy*, 41, pp. 34–43.

Kuzyakov, Y., Subbotina, I., Chen, H. Q., Bogomolova, I. and Xu, X. L. (2009). Black carbon decomposition and incorporation into soil microbial biomass estimated by C-14 labeling. *Soil Biology & Biochemistry*, 41, pp. 210–219.

Lamers, P., Thiffault, E., Paré, D. and Junginger, M. (2013). Feedstock specific environmental risk levels related to biomass extraction for energy from boreal and temperate forests. *Biomass & Bioenergy*, 55, pp. 212–226.

Lattimore, B., Smith, C. T., Titus, B. D, Stupak, I. and Egnell, G. (2009). Environmental factors in woodfuel production: opportunities, risks, and criteria and indicators for sustainable practices. *Biomass & Bioenergy*, 33, pp. 1321–1342.

Lee, J. W., Kidder, M., Evans, B. R., et al. (2010). Characterization of biochars produced from cornstovers for soil amendment. *Environmental Science and Technology*, 44, pp. 7970–7974.

Lehmann, J. (2007). A handful of carbon. *Nature*, 446, pp. 143–144.
Lehmann, J. and Joseph, S. (eds.) (2009). *Biochar for Environmental Management: Science and Technology*. London: Earthscan.
Liu, X., Zhang, A., Ji, C., et al. (2013). Biochar's effect on crop productivity and the dependence on experimental conditions – a meta-analysis of literature data. *Plant and Soil*, 373, pp. 583–94.
Long, R. P., Horsley, S. B. and Lilja, P. R. (1997). Impact of forest liming on growth and crown vigor of sugar maple and associated hardwoods. *Canadian Journal Forest Research*, 27, pp. 1560–1573.
Luce, C. H. (1997). Effectiveness of road ripping. *Restoration Ecology*, 5, pp. 265–270.
Major, J., Lehmann, J., Rondon, M. and Goodale, C. (2010). Fate of soil-applied black carbon: downward migration, leaching and soil respiration. *Global Change Biology*, 16, pp. 1366–1379.
Mangold, J. M. and Sheley, R. L. (2008). Controlling performance of bluebunch wheatgrass and spotted knapweed using nitrogen and sucrose amendments. *Western North American Naturalist*, 68, pp. 129–137.
Matovic, D. (2011). Biochar as a viable carbon sequestration option: global and Canadian perspective. *Energy*, 36, pp. 2011–2016.
McElligott, K. (2011). *Biochar Amendments to Forest Soils: Effects on Soil Properties and Tree Growth*. Moscow, ID: Master of Science Thesis, University of Idaho.
Mitchell, P. J., Simpson, A. J., Soong, R. and Simpson, M. J. (2015). Shifts in microbial community and water-extractable organic matter composition with biochar amendment in a temperate forest soil. *Soil Biology and Biochemistry*, 81, pp. 244–254.
Mohan, D., Pittman, C.U., Jr. and Steele, P. H. (2006). Pyrolysis of wood/biomass for bio-oil: a critical review. *Energy & Fuels*, 20, pp. 848–889.
Mukherjee, A. and Lal, R. (2013). Biochar impacts on soil physical properties and greenhouse gas emissions. *Agronomy*, 3, pp. 313–339.
Novak, J. M., Lima, I., Xing, B., et al. (2009a). Characterization of designer biochar produced at different temperatures and their effects on a loamy sand. *Annals of Environmental Science*, 3, pp. 195–206.
Novak, J. M., Busscher, W. J., Laird, D. L., Ahmedna, M., Watts, D. W. and Niandou, M. A. S. (2009b). Impact of biochar amendment on fertility of a southeastern coastal plain soil. *Soil Science*, 174, pp. 105–112.
Noyce, G., Basiliko, N., Fulthorpe, R., Sackett, T. E. and Thomas, S. C. (2015) Soil microbial responses over 2 years following biochar addition to north temperate forest. *Biology and Fertility of Soils*, 51, pp. 649–659.
Nuithitikul, K., Srikhun, S. and Hirunpraditkoon, S. (2010). Influences of pyrolysis condition and acid treatment on properties of durian peel-based activated carbon. *Bioresource Technology*, 101, pp. 426–429.
Page-Dumroese, D. S., Jurgensen, M. and Terry, T. (2010). Maintaining soil productivity during forest biomass-to-energy thinning harvests in the western United States. *Western Journal of Applied Forestry*, 25, pp. 5–11.
Page-Dumroese, D. S., Robichaud, P. R., Brown, R. E. and Tirocke, J. M. (2015). Water repellency of two forest soils after biochar addition. *Transactions of the American Society of Agricultural and Biological Engineers*, 58, pp. 335–342.
Perry, L. G., Galatowitsch, S. M. and Rosen, C. J. (2004). Competitive control of invasive vegetation: a native wetland sedge suppresses *Phalaris arundinacea* in carbon-enriched soil. *Journal of Applied Ecology*, 41, pp. 151–162.

Price, S. J., Sherlock, R. R., Kelliher, F. M., McSeveny, T. M., Tate, K. R. and Condron, L. M. (2003). Pristine New Zealand forest soil is a strong methane sink. *Global Change Biology*, 10, pp. 16–26.

Rondon, M. A., Ramirez, J. and Lehmann, J. (2005). Charcoal additions reduce net emissions of greenhouse gases to the atmosphere. In: *Proceedings of the 3rd USDA Symposium on Greenhouse Gases and Carbon Sequestration in Agriculture and Forestry, March 21–24*. Baltimore, MD: University of Delaware Press.

Rummer, B., Prestemon, J., May, D., et al. (2003). *A Strategic Assessment of Forest Biomass and Fuel Reduction Treatments in Western States*. Washington, DC: US Department of Agriculture. Forest Service, Research and Development.

Sackett, T. E., Basiliko, N., Noyce, G. L., et al. (2014). Soil and greenhouse gas responses to biochar additions in a temperate hardwood forest. *GCB Bioenergy*, 7, pp. 1062–1074.

Shan, S. (2013). *Soil Soluble Nitrogen Pools in Northern Idaho Temperate Forest and Biochar Influence*. Moscow, ID: Master of Science Thesis. University of Idaho.

Siitonen, J. (2001). Forest management, coarse woody debris and saproxylic organisms: Fennoscandian boreal forests as an example. *Ecological Bulletins*, 49, pp. 11–41.

Smith, J. L., Collins, H. P. and Bailey, V. L. (2010). The effect of young biochar on soil respiration. *Soil Biology & Biochemistry*, 42, pp. 2245–2247.

Spokas, K. A. and Reicosky, D. C. (2009). Impacts of sixteen different biochars on soil greenhouse gas production. *Annals of Environmental Science*, 3, pp. 179–193.

Spokas, K. A., Baker, J. M. and Reicosky, D. C. (2010). Ethylene: potential key for biochar amendment impacts. *Plant and Soil*, 333, pp. 443–452.

Spokas, K. A., Cantrell, K. B., Novak, J. M., et al. (2012). Biochar: a synthesis of its agronomic impact beyond carbon sequestration. *Journal of Environmental Quality*, 41, pp. 1–23.

Steinbeiss, S., Gleixner, G. and Antonietti, M. (2009). Effect of biochar amendment on soil carbon balance and soil microbial activity. *Soil Biology & Biochemistry*, 41, pp. 1301–1310.

Steiner, C., Teixeira, W. G., Lehmann, J., et al. (2007). Long term effects of manure, charcoal, and mineral fertilization on crop production and fertility on a highly weathered Central Amazonian upland soil. *Plant and Soil*, 291, pp. 275–290.

Stewart, C. E., Zheng, J., Botte, J. and Cotrufo, M. F. (2013). Co-generated fast pyrolysis biochar mitigates greenhouse gas emissions and increases carbon sequestration in temperate soils. *GCB Bioenergy*, 5, pp. 153–164.

Stocks, B. J., Fosberg, M. A., Lynham, T. J., et al. (1998). Climate change and forest fire potential in Russian and Canadian boreal forests. *Climate Change*, 38, pp. 1–13.

Switalski, T. A., Bissonette, J. A., DeLuca, T. H., Luce, C. H. and Madej, M. A. (2004). Benefits and impacts of road removal. *Frontiers in Ecology and Environment*, 2, pp. 21–28.

Thomas, S. C. (2013). Biochar and its potential in Canadian forestry. *Silviculture Magazine*, January 2013, pp. 4–6.

Thomas, S. C., Frye, S., Gale, N., et al. (2013). Biochar mitigates negative effects of salt additions on two herbaceous plant species. *Journal of Environmental Management*, 129, pp. 62–68.

Thomas, S. C. and Gale, N. (2015). Biochar and forest restoration: a review and meta-analysis of tree responses. *New Forests*, 46, pp. 931–946.

Tyron, E. H. (1948). Effect of charcoal on certain physical, chemical, and biological properties of forest soils. *Ecological Monographs*, 18, pp. 81–226.

Uchimiya, M., Lima, I. M., Klasson, K. T. and Wartelle, L. H. (2010). Contaminant immobilization and nutrient release by biochar soil amendment: roles of natural organic matter. *Chemosphere*, 80, pp. 935–940.

US Department of Energy (2011). U.S. Billion-ton update: biomass supply for a bioenergy and bioproducts industry. R.D. Perlack and B.J. Stokes (leads). ORNL/TM-2011/334. Oak Ridge National Laboratory, Oak Ridge, TN. 337.

Uslu, A. Faaij, A., and Bergman, P. C. A. (2008). Pretreatment technologies and their effect on international bioenergy supply chain logistics: techno-economic evaluation of torrefaction, fast pyrolysis and pelletization. *Energy*, 33, pp. 1206–1223.

Van Zwieten, L. V., Singh, B. P. and Cox, J. (2012). Biochar effects on soil properties. In: Cox, J. (ed.) *Biochar in Horticulture: Prospects for the Use of Biochar in Australian Horticulture.* New South Wales, Australia: Horticulture Australia, NSW Department of Primary Industries, Chapter 4.

Verheijen, F., Jeffery, S., Bastons, A. C., van der Velde, M. and Diafas, I. (2009). Biochar application to soils. A critical scientific review of effects on soil properties, processes and functions. Office for the Official Publication of the European Communities. Luxembourg.

Wang, J. M., Murphy, J. G., Geddes, J. A., Winsborough, C. L., Basiliko, N. and Thomas, S. C. (2013). Methane fluxes measured by eddy covariance and static chamber techniques at a temperate forest in central Ontario, Canada. *Biogeosciences*, 10, pp. 4371–4382.

Wardle, D. A., Nilsson, M. C. and Zackrisson, O. (2008). Fire-derived charcoal causes loss of forest humus. *Science*, 320, pp. 629.

Warnock, D. D., Lehmann, J., Kuyper, T. W. and Rillig, M. C. (2007). Mycorrhizal responses to biochar in soil – concepts and mechanisms. *Plant and Soil*, 300, pp. 9–20.

Wilhoit, J. H. and Ling, Q. (1996). Spreader performance evaluation for forest land application of wood and fly ash. *Journal of Environmental Quality*, 25, pp. 945–950.

Wood, S. A. and Layzell, D. B. (2003). *A Canadian Biomass Inventory: Feedstocks for a Bio-based Economy.* Canada: BIOCAP Canada Foundation Report.

Wright, D. L. and Blaser, R. E. (1981). Establishment of vegetation on graded road cuts as influenced by topsoiling and tillage. *Soil Science Society of America Journal*, 45, pp: 419–422.

Zamora-Cristales, R., Sessions, J., Boston, K. and Murphy, G. (2014). Economic optimization of forest biomass processing and transport in the Pacific Northwest USA. *Forest Science*, 60, pp. 1–15.

Zimmerman, A. R. (2010). Abiotic and microbial oxidation of laboratory-produced black carbon (biochar). *Environmental Science and Technology,* 44, pp. 1295–1301.

Zimmerman, A. R., Gao, B. and Ahn, M.-Y. (2011) Positive and negative carbon mineralization priming effects among a variety of biochar-amended soils. *Soil Biology and Biochemistry*, 43, pp. 1169–1179.

16

The Role of Mycorrhizae and Biochar in Plant Growth and Soil Quality

İBRAHIM ORTAŞ

Abstract

Arbuscular mycorrhizal fungi (AMF) are key organisms of the soil–plant system, and contribute to the uptake of nutrients and water. AMF also play a role in the aggregation and structural stability of soil. Biochar may provide a suitable habitat for mycorrhizal growth, which is also favorable in view of productive soils. The impact of a combination of biochar and mycorrhizae on plant growth was assessed in case studies. One of the main findings was that biochar application alone at 40 Mg ha-1 did not influence growth of citrus seedlings. However, biochar and mycorrhizae species inoculation significantly increased shoot and root dry weight. In general, the contribution of mycorrhizae seemed to be higher than the biochar contribution alone and a combination performed best. This might be a result of the mycorrhizae inoculum activating the biochar when added into the rhizosphere, which is a rich medium for other beneficial soil organisms. Although the relationship between biochar and mycorrhizal colonization is not yet entirely clear, the effect of physical protection of hyphae from fungal grazers might be facilitated as a consequence of the pore structure of biochar. The results also provide evidence for competition between biochar and plant nutrient uptake.

16.1 Introduction

Since the onset of the industrial revolution around 1750, the concentration of carbon dioxide (CO_2) has increased by 42.9% from 280 ppm to 400 ppm in 2015, (NOAA, 2015) and is presently increasing at a rate of 2.25 ppm per year. Despite improving crop production, the Green Revolution of the 1960s has been a major contributor to the increase in atmospheric CO_2 via intensification of agricultural production and use of chemical fertilizers and pesticides, together with increased practices of tillage and irrigation. However, after achieving a significant rise in production, at a certain point of time, further increasing agricultural input does not lead to an increase in yields. Loss of soil organic matter (SOM) and the degradation of soil structure can have a detrimental effect on soil fertility and crop productivity (Singh et al., 2009). On the other hand, agricultural soils, being depleted of a large amount of soil organic carbon (SOC) due to intensive cultivation, have significant potential to sequester atmospheric CO_2 (Lal and Kimble, 1997). There is strong interest in stabilizing the atmospheric abundance of CO_2 in order to reduce its effect on global warming.

There are three strategies for reducing CO_2 emissions to mitigate climate change (Schrag, 2007), namely:

- reducing global energy use,
- developing low or no-carbon fuel, and
- sequestering CO_2 from point sources or from the atmosphere through natural (photosynthesis) and engineering techniques.

Through the photosynthesis process, plants absorb CO_2 from the atmosphere, transform it into plant carbon (C), and sequester it in either above- or below-ground biomass and/or as soil C. The basic pathways of the C balance are atmospheric CO_2 uptake via photosynthesis and CO_2 release via decomposition. The net gain of C in the soil is a function of the balance between input (i.e. net primary productivity plus any external input) and loss (i.e. decomposition, erosion, leaching). Since organic material is usually subject to decomposition where C is released into the atmosphere as CO_2 due to heterotrophic respiration, new approaches are needed to retain SOC. Thus, a promising solution is to convert feedstock rich in C into biochar by using pyrolysis (see Chapters 10 and 12) to stabilize C in the soil over the long term. In addition, biochar application in soils may offer interesting benefits for the mitigation of climate change through decreasing greenhouse gas (GHG) emissions, such as N_2O (Jeffery et al., 2015). Further, subsoil biochar application may have a notable contribution to climate change mitigation (Lorenz and Lal, 2014) as pyrogenic C would be less prone to decomposition and erosional processes. Lehmann (2007) reported that the effectiveness of using biochar as an approach to mitigate climate change rests on its relative recalcitrance to microbial decay, which may have a decomposition inhibiting effect on inherent SOC as well.

SOM is an important component of soil fertility, productivity and quality because of its crucial role in the chemical, physical and biological properties of soil (Ortas, 2006).

While biochar may also impact soil fertility (Whitman et al., 2010), mycorrhizae can also sequester considerable amounts of C, especially via its turnover. Therefore, it is necessary to review the synergistic effects of biochar and mycorrhizae on soil quality and plant health.

The aim of this chapter, therefore, is to elaborate on the effects of biochar–mycorrhizae interaction on crop growth, nutrient uptake, mycorrhizal development and soil quality.

16.2 Mycorrhizae, Soil and Biochar Relationships

The mycelia network of mycorrhizae and/or other fungi and bacteria can have a binding action in the soil and improve soil structure (Tisdall et al., 1997; Rillig et al., 2010;

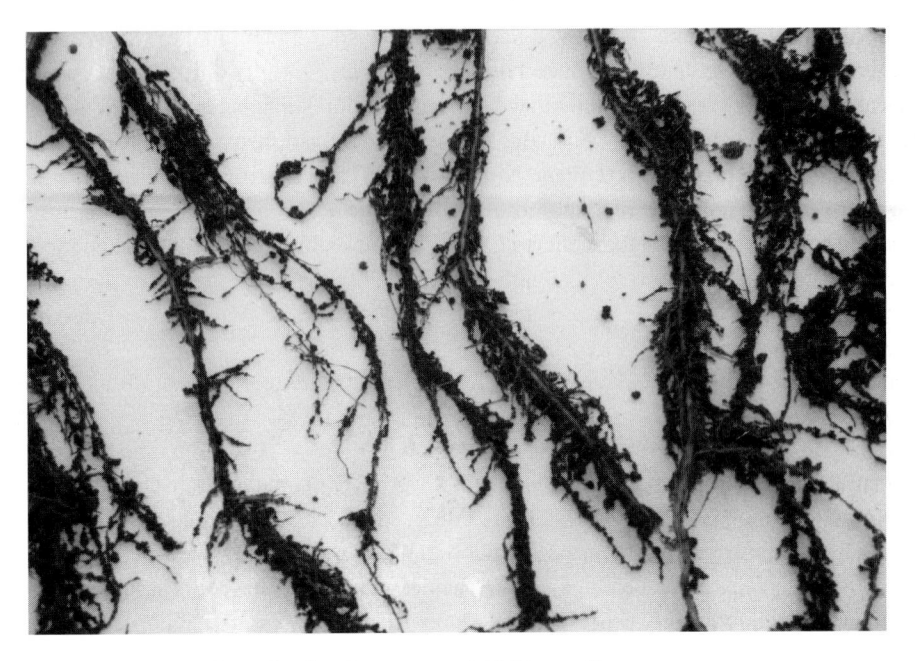

Figure 16.1. Aggregate formation by plant roots and AMF mycelium.

Caravaca et al., 2006). Arbuscular mycorrhizal fungi (AMF) comprise the largest symbiotic relationship between plants and fungi, and make significant contributions to the physical, chemical and biological aspects of soil quality via hyphae extending into the rhizosphere, thereby improving the absorption of nutrients, especially P and micronutrients (Karandashov and Bucher, 2005; Smith and Read, 2008; Ortaş, 2003).

The potential role of AMF in C sequestration in the soil includes several direct and indirect mechanisms (Ortaş, 2006; Ortaş et al., 2013):

- increasing direct C sink (increased transfer of photosynthate, large hyphal and spore biomass turnover, exudates),
- transforming biomass into different forms of SOC that are relatively resistant to degradation,
- protecting SOC physically through improved aggregation,
- increasing total biomass and residual material as a source for SOC, and
- contributing to additional C sequestration (amplifying effect) by combined application of mycorrhizae and biochar.

Soil aggregates play an important role in sequestering and stabilizing SOC (Sundermeier et al., 2011). Daynes et al. (2013) reported that in the presence of adequate organic matter, plant roots are key contributors to the development of soil structure, which is further stabilized by AMF. An extensive soil-based fungal mycelium located around roots creates a favorable soil structure via aggregation processes (Figure 16.1). Soil biota, especially mycorrhizal fungi, significantly influence soil fertility and the habitat for microbial communities. Soil quality is also improved by increasing the total microbial biomass and species abundance. Soil microbes are known to produce polymers that also serve as aggregate binding agents (Jastrow et al., 1998). However, studies on applied mycorrhizae are still rare in the field of soil quality research.

AMF symbiosis has been proposed as one of the most effective mechanisms of increasing heavy metal and water stress tolerance due to their effect on rhizospheric physical, chemical and biological activities (Smith and Read, 2008).

The contribution of mycorrhizal inoculation to the SOC content and aggregate formation depends on root and hyphae productivity, turnover rates and exudation. For instance, Nadian et al. (2009) demonstrated that the root length of colonized berseem clover (*Trifolium alexandrinum* L.) was improved by 20% with increase in diameter of soil aggregate. In addition plant roots influence soil structure in diverse ways, such as through root exudation, which includes polysaccharides that help to bind micro-aggregates and clay particles, and also through root pressure and penetration, which help to increase the proportion of stable aggregates through root entanglement (Six et al., 2004).

AMF further enhance the aggregate stability of soil due to the production of a protein known as glomalin, which is a brown to red-brown colored glycoprotein and a major component of SOM with a strong cementing effect (Rillig et al., 2003; Rillig, 2004). It is currently determined by the extraction method proposed by Nichols and Wright (2005), and it has interesting properties and positive effects on aggregate stability, water infiltration, air permeability, root development, microbial activity, resistance to surface sealing and erosion, altogether leading to enhanced productivity (Wright and Upadhyaya, 1998).

Biochar may also be stabilized by the exudates from mycorrhizal mycelia and glomalin. In turn, biochar may significantly affect glomalin production as a consequence of habitat suitability for mycorrhizal hyphae development. However, Ameloot et al. (2013) highlighted that this hypothesis is still pending confirmation. The role of glomalin in the ecosystem is still not clear and should be studied thoroughly in long-term and diverse management systems (Treseder and Turner, 2007).

More C is generally sequestered in the soil through root and mycorrhizal hyphae turnover than through above-ground biomass input (litter) in many ecosystems. The preferential preservation of root and mycelial C in soils has been attributed to:

- higher chemical recalcitrance of root C (Rasse et al., 2005),
- lower decomposition in deeper soil layers due to environmental conditions, which are detrimental to the decomposition of plant tissues (Gill et al., 2002),
- more physico-chemical protection of root exudates and decomposition byproducts through interactions with minerals (formation of organo-mineral compounds) (Rillig et al., 2010; Jones and Edwards, 1998), and
- higher physical protection from microbial decomposers via aggregation by roots and hyphae (Jastrow et al., 1998; Golchin et al., 1994; Bearden, 2001).

16.3 The Interaction Between Biochar and Mycorrhizae

Biochar has a strong effect on nutrient availability, which can affect root mycorrhizal colonization (Ishii and Kadoya, 1994). Both mycorrhizae and biochar improve plant growth and may enhance soil quality and soil health (Lehmann et al., 2011). However, the effects of biochar on plant growth and mycorrhizal fungi are often contradictory as a consequence of the biochar production methodology employed, such as heating rates, reactor residence time and feedstock material used. LeCroy et al. (2013) reported, based on observations on a *Sorghum bicolor* experiment, that the addition of biochar, mycorrhizae and N promoted mycorrhizal root colonization. Likewise, several other studies have reported a positive relationship between biochar addition and fungal colonization (Rillig et al., 2010; Solaiman et al., 2010; Vanek and Lehmann, 2015). The results of Blackwell et al. (2010) showed that during dry seasons and in low-P soils, the addition of biochar improved crop nutrient availability and increased AMF colonization. Medina and Azcon (2010) suggested that soil inoculated with mycorrhizal fungi may be used as a successful bio-strategy to improve plant performance in P-deficient soils under Mediterranean conditions. A potential process behind the increased colonization was suggested by Elmer and Pignatello (2011), who discovered that biochar increased mycorrhizal colonization via the sorption of allelochemicals that usually suppress colonization.

An example from Japan suggests the stimulating effect of charcoal on indigenous AMF under field conditions. Soybean (*Glycine max* L.) grown in a humus-rich Andosol was significantly more infected with AMF when biochar was applied (Saito, 1990).

Biochar influences the production of AMF extraradical hyphae, which facilitate plant nutrient uptake (Warnock et al., 2007). The degree of mycorrhizal association and its relationship with the soil structure depends on the root morphology (Miller and Jastrow, 1990) and abundance of mycorrhizal hyphae. However, AMF contributions to the physical development of soil depend on several factors and have a significant effect on ecosystem services. The ecosystem services provided by AMF are based on modification of the root morphology and development of a complex ramified mycelial network in the soil.

However, not all studies indicated a positive relationship between biochar amendment and mycorrhizal colonization. Biochar can have relatively high contents of selected elements in the ash compartment, such as P and K, which are not volatilized during the pyrolysis process. Elevated concentrations of P may lead to decreased mycorrhizal hyphae development (Grant et al., 2005). In addition, biochar may have a high alkalinity (pH > 9), which also exerts a negative effect on hyphae colonization. Biochar can also reduce mycorrhizal colonization by decreasing nutrient availability or by creating unfavorable nutrient ratios in the soil (Zuccarini, 2010). Most studies have suggested that biochar has the potential to improve ecosystem services and sequester atmospheric C, which can help to mitigate climate change (Lehmann et al., 2006; Sohi et al., 2010; Lehmann, 2007). The increase of C in soil must be more than what is added through biochar for it to cause sequestration of atmospheric CO_2. The combination of biochar and mycorrhizae is a promising strategy to contribute to C sequestration in soils (Warnock et al., 2007). Synergistic effects between

16.4 Case Studies

16.4.1 Effect of Biochar and Different Mycorrhizae Species on Citrus *Growth*

In order to determine the effect of the application of mycorrhizae and biochar on the growth of sour orange (*Citrus aurantium* L.) seedlings, a pot experiment was conducted using a clay loam soil with inoculation of four mycorrhizal species: *Glomus mosseae*, *G. etunicatum*, *G. clarium*, *G. intraradices* and a cocktail of those four inoculums (equal amounts). Increasing the temperature from 450°C to 500°C led to an increase of the *Eucalyptus* biochar nutrient element concentration (carbon, nitrogen, phosphorus, potassium, micronutrients). The aim was to identify the most suitable mycorrhizal species for a synergistic interaction with biochar. The experiment was conducted under greenhouse conditions for eight months. A consortia of these species was treated with a relatively large amount of 40 Mg ha^{-1} biochar produced from *Phragmites* feedstock. At the transplanting stage, each seedling received 1,000 spores, which were placed 50 mm below the seedling roots. Control pots with and without biochar received the same amount of mycorrhizae-free inoculum.

At harvest, plant height, shoot weight and root dry weight were measured. Compared with the control treatment, the application of biochar alone did not increase sour orange plant height, fresh and dry shoot weight or root dry weight.

The results showed that mycorrhizae-inoculated sour orange plants exhibited significantly higher plant height and fresh and dry weight than the non-inoculated control and biochar alone application (Table 16.1). Biochar and *G. intraradices* inoculated seedlings produced 2.61 g dry weight per pot, which is significantly higher than all other treatments. It appears that mycorrhizal inoculation significantly affects citrus plant growth rather than biochar application alone.

No significant differences between the control and biochar addition pots were observed for shoot weight. The shoot-to-root ratio was calculated and mycorrhizae-inoculated plants had a significantly lower shoot-to-root ratio. The overall results presented in Table 16.1 show that biochar application alone did not make a significant difference compared with the control application. This simple experiment confirms that pure biochar does not always improve plant growth, but it may serve as a catalyst for mycorrhizae-induced growth performance increase.

16.4.2 Biochar Dose and Mycorrhizae Inoculation Effects on Citrus *Seedling Growth*

In another pot experiment, the effect of different levels of biochar application on sour orange (*C. aurantium*) seedling development was tested. The aim of the work was to determine the optimum biochar doses in combination with mycorrhizal inoculation for citrus seedling

Table 16.1. Effect of 40 Mg ha^{-1} biochar application with several mycorrhizal species on citrus seedlings (sour orange (Citrus aurantium L.)) shoot dry weight after eight months in greenhouse conditions

Treatments	Plant height (cm)	Shoot fresh weight (g/pot)	Shoot dry weight (g/pot)	Root dry weight (g/pot)	Root: shoot ratio
Control	28.70±1.83c	2.98±1.09b	0.78±0.41b	0.87±0.39b	1.11
Biochar	29.13±4.63c	3.00±0.61b	0.81±0.45b	0.88±0.26b	1.10
biochar+G. clarium	29.73±5.64c	3.64±0.44b	0.81±0.33b	0.88±0.66b	1.08
biochar+G. etunicatum	40.33±6.24a	7.33±1.44a	1.35±0.30b	1.59±0.42ab	1.18
biochar+G. mosseae	37.57±8.32a-c	6.11±3.49ab	1.75±0.87ab	1.26±0.15ab	0.72
biochar+G. intraradices*	39.43±3.40ab	6.95±2.33a	2.61±0.82a	1.99±1.00a	0.76
biochar+cocktail inoculation	30.50±3.36bc	3.15±0.79b	0.82±0.14b	1.02±0.26ab	1.23
Grand mean	**33.63**	**4.74**	**1.27**	**1.21**	**1.03**

Note: The mean of three replicates and standard error is presented. Columns with different letters denote significant differences ($P < 0.05$) according to Duncan's multiple-range test.

* renamed to *Rhizophagus irregularis* after recent molecular analysis.

Figure 16.2. Growth response to mycorrhizal inoculation and different biochar application rates (10 and 20 Mg ha^{-1}) of sour orange (*Citrus aurantium*) seedlings. (A black and white version of this figure will appear in some formats. For the colour version, please refer to the plate section.)

growth over eight months. Under greenhouse conditions, *G. mosseae* mycorrhizae inoculated and non-inoculated sour orange seedlings were treated with 10 Mg ha^{-1} and 20 Mg ha^{-1} biochar derived from *Eucalyptus* sp. (Figure 16.2) under low fertile clay loam soil. At harvest, the plant responses to mycorrhizae and biochar were determined. It was found that mycorrhizae-inoculated plants produced more fresh shoot and root dry weight than non-inoculated ones.

The treatment containing biochar and mycorrhizae significantly increased the shoot dry weight. In non-inoculated treatments, the control, 10 Mg ha^{-1} and 20 Mg ha^{-1} biochar addition produced 0.97, 0.99, 1.88 g shoot dry weight (SDW) per pot respectively (Table 16.2). In the mycorrhizae-inoculated sour orange control, 10 Mg ha^{-1} and 20 Mg ha^{-1} biochar application produced 1.63, 1.92 and 3.28 g SDW per pot respectively. In non-inoculated plants the root dry weight (RDW) also increased with increasing biochar application levels. Usually, poorer soil conditions without mycorrhizae inoculation lead to increased root growth of the plant. Our results suggest that biochar competes for nutrient uptake, which may directly and/or indirectly be related to the effect of biochar on soil fertility.

16.4.3 Effect of Different Biochar and Mycorrhiza Species on Sorghum Growth

Eucalyptus- and *Phragmites*-derived biochars were used with three mycorrhizae species (*G. mosseae, G. clarium* and *G. intraradices*) on *Sorghum* host plants to study the effects of biochar and mycorrhizae inoculation on plant growth. *Eucalyptus* and *Phragmites* biomass

Table 16.2. Effect of 10 and 20 Mg ha⁻¹ biochar application with and without mycorrhizal inoculation on citrus seedlings' (sour orange) shoot dry weight after eight months from seedling stages under greenhouse conditions

Mycorrhiza treatments	Biochar treatments	Plant height (cm)	Shoot fresh weight (g/pot)	Shoot dry weight (g/pot)	Root dry weight (g/pot)	Root: Shoot Ratio
No mycorrhiza	Control	20.60±7.07a	1.89±0.30a	0.97±0.00b	0.84±0.42a	0.86±0.43
	10 Mg ha⁻¹	19.10±5.94a	2.04±0.18a	0.99±0.13b	0.93±0.24a	0.48±0.04
	20 Mg ha⁻¹	23.75±2.19a	3.39±1.55ab	1.88±0.78	1.39±1.00a	0.96±0.37
	Mean	**21.15**	**2.10**	**1.28**	**1.05**	**0.77**
Mycorrhiza	Control	25.30±7.92a	2.67±0.96ab	1.63±0.16b	0.78±0.01a	0.75±0.14
	10 Mg ha⁻¹	24.70±6.79a	3.03±0.80ab	1.92±0.14b	1.44±0.18a	0.69±0.25
	20 Mg ha⁻¹	28.50±1.70a	4.32±0.77a	3.28±0.78a	1.09±0.02a	0.38±0.20
	Mean	**26.17**	**3.34**	**2.28**	**1.10**	**0.60**

Note: The mean of three replicates and standard error is presented. Columns with different letters denote significant differences ($P < 0.05$) according to Duncan's multiple range test.

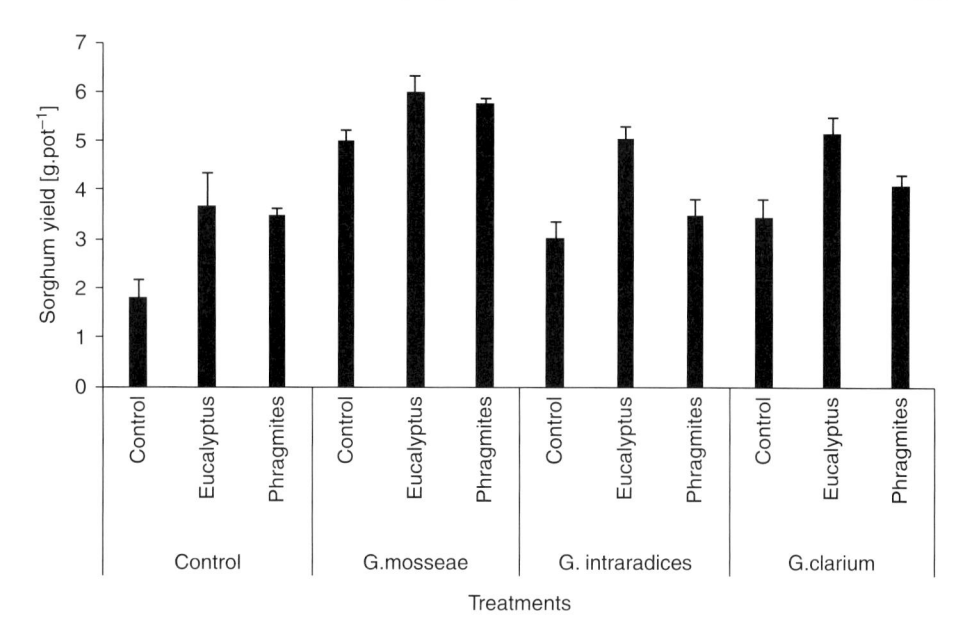

Figure 16.3. Effect of biochar feedstock and mycorrhizae species on Sorghum plant growth after 65 days under greenhouse conditions.

was pyrolyzed at a temperature between 450 and 500°C for 30 minutes. Before amending with soil, the biochar was milled and then sieved through a 2 mm sieve. Experiments were conducted using a clay loam soil containing low P and Zn concentrations. *Sorghum bicolor* L. plants were grown for 65 days under greenhouse conditions and it was found that biochar significantly increased the *Sorghum* shoot dry weight. Figure 16.3 shows that *Eucalyptus* feedstock-derived biochar produced more dry matter than *Phragmites* biochar. It is assumed that this is related to varying characteristics of biochar in terms of nutrient contents. In general, the contribution of mycorrhizae seems to be higher than the biochar contribution. In particular, *G. mosseae*-inoculated plants produced more shoot dry weight compared with other mycorrhizae species. However, compared with control treatments, biochar application significantly increased *Sorghum* growth. The combined application of mycorrhizae and biochar increased *Sorghum* plant growth further, which is a very important finding for future applications.

16.5 Mechanisms Between Biochar, Mycorrhizae and Other Soil Organisms

Since AMF and biochar interactions have the potential to improve crop performance, there is an increasing interest in understanding the underlying mechanisms. Biochar is enriched in C relative to the feedstock material, and thus it may support mycorrhizal development. Biochar addition may change the soil nutrient availability, depending on pretreatment such as mixing with humus or manure. The resulting shift in nutrient availability certainly impacts mycorrhizal development (LeCroy et al., 2013). This was also confirmed by Warnock et al. (2010), who indicated that AMF functionality can be improved by the addition of a soil additive such as biochar.

Warnock et al. (2007) suggested four possible mechanisms for how biochar could influence mycorrhizal abundance and/or function: (a) alteration of the physico-chemical properties of soil; (b) indirect effects on mycorrhizae via effects on other soil microbes; (c) plant–fungus signaling interference and detoxification of allelochemicals in the biochar; and (d) physical protection from fungal grazers. In addition, biochar has a large pore size, which may possibly be a niche for mycorrhizae spores to grow in the soil–biochar medium. Ameloot et al. (2013) indicated that biochar is a suitable physical habitat for soil microorganisms. Hammer et al. (2014) worked with a P-33 radiotracer and found that AMF hyphae access microsites within biochar, that are too small for most plant roots to enter (<10 μm), and hyphae can hence utilize nutrients, for example P, from biochar. Biochar pores originate from the inherent structure of fresh biomass and from the development of cracks during pyrolysis and may also protect hyphae against fungal grazers. It seems that biochar can be a protective microhabitat for many other microorganisms too, such as N-fixing bacteria, for instance. Abundant microbial populations may further stimulate mycorrhizal performance. Figure 16.4 shows char of *Picea abies*, full of life despite being buried in the soil for 110 years. The material was accumulated at a forest site due to the formerly common practice of slash burning for forest site preparation after harvesting. The sample was obtained from the mineral soil (~10 cm soil depth). Note that the biomass microstructure is still intact after more than a century of weathering under acidic soil conditions (pH = 3–4). Bordered pits, characteristic for *Picea abies,* can still be identified in the lower section of the image. Line structures indicate abundant existence of likely basidiomycete hyphae. The successful water retention of biochar can be confirmed by the occurrence of testaste amoebae (TA), as they require moist conditions which obviously were maintained even during periods of summer droughts. It was also suggested that TA may have an important influence on soil nutrient and also silica cycles, which are both in turn important in regard to GHG mitigation (Wilkinson, 2008).

However, it was reported that biochar with a high P concentration and produced at a low temperature may be toxic to mycorrhizae (Warnock et al., 2010). In addition, sorption of organic plant compounds and allelochemical sorbing materials may regulate mycorrhizal colonization. Since biochar has indirect effects on extensive root systems, biochar may have contributed to the observed increased yield related to nutrient mobilization and uptake.

16.6. Conclusions

Biochar addition increases the activity of root and mycorrhizal systems, leading to a higher specific surface area and therefore better accessibility of nutrients and water. It may also provide a habitat for mycorrhizal growth and serve as a refuge for mycorrhizal hyphae, providing protection from fungal grazers. It is also important to manage mycorrhizal fungi, which can be a powerful tool to fix more CO_2 from the atmosphere in the soil via plant litter and roots. The effect of multiple treatments of biochar and AMF inoculation on mycorrhizal spore production is unclear. Although the relationship between biochar and mycorrhizae

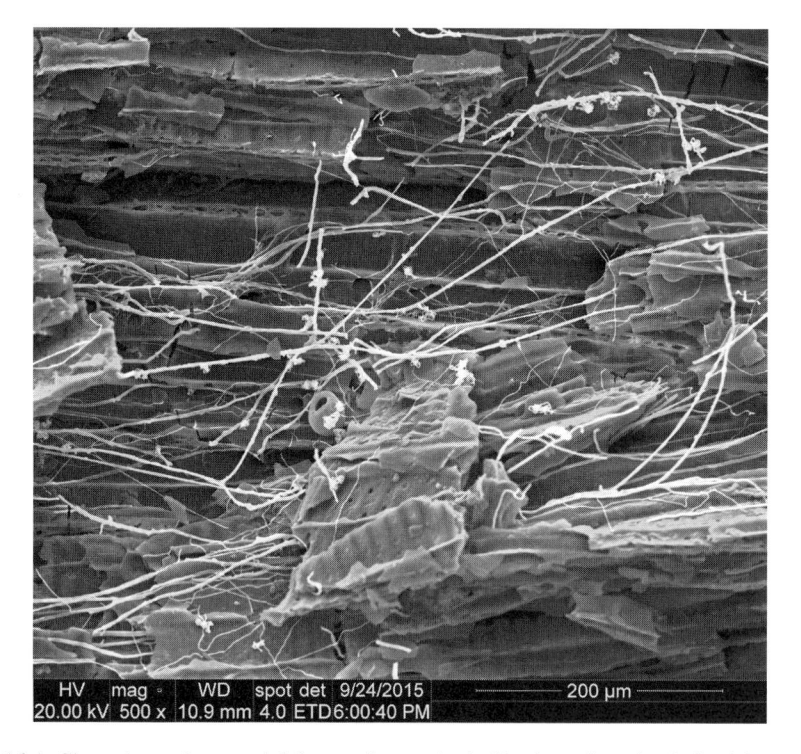

Figure 16.4. Charred woody material from a forest site in Northern Austria, derived from ~10 cm cm soil depth. The material was identified as 110 years old *Picea abies* charcoal from slash burning. Intensive colonization with mycorrhiza and the frequent occurrence of testaste amoebae (round object in the center) suggests an active role in maintaining soil functions. (A black and white version of this figure will appear in some formats. For the colour version, please refer to the plate section.)

has not been thoroughly elucidated, biochar may contribute to the development of mycorrhizae, and biochar increases mycorrhizal colonization. On the other hand, mycorrhizae may activate biochar to contribute to soil development and nutrient uptake. In three case studies it has been found that:

1. Biochar application alone even at 40 Mg ha^{-1} did not influence growth of citrus (sour orange) seedlings. However, biochar plus mycorrhizae species inoculation significantly increased shoot and root dry weight.
2. Biochar application with and without mycorrhizal inoculation at 20 Mg ha^{-1} significantly influenced *Sorghum* shoot yields (fresh and dry weight).
3. Biochar derived from *Eucalyptus* feedstock increased *Sorghum* dry weight. The interaction between mycorrhizae and biochar further improved yields of *Sorghum*. Biochar with *G. mosseae* inoculation yielded the highest dry weight compared with biochar + *G. intraradices* and biochar + *G. clarium* applications.

The overall results suggest that biochar application alone does not ultimately cause a significant growth response. However, biochar with mycorrhizal inoculation significantly increased plant growth, while the contribution of mycorrhizae seems to be higher than that of biochar.

It can be concluded that a combined application of biochar and mycorrhizae spores may lead to synergistic effects that increase plant growth, which may in turn lead to more efficient nutrient and water use and therefore less consumed energy and associated GHG emissions. Further field experiments are required to explore the hidden potential of the underlying synergistic effects.

Acknowledgements

This study was supported by the Scientific and Technological Research Council of Turkey (Project No: TÜBİTAK-TOVAG-112O785).

References

Ameloot, N., Graber, E. R., Verheijen, F. G. A. and De Neve, S. (2013). Interactions between biochar stability and soil organisms: review and research needs. *European Journal of Soil Science*, 64, pp. 379–390.

Bearden, B. N. (2001). Influence of arbuscular mycorrhizal fungi on soil structure and soil water characteristics of vertisols. *Plant and Soil*, 229, pp. 245–258.

Blackwell, P., Krull, E., Butler, G., Herbert, A. and Solaiman, Z. (2010). Effect of banded biochar on dryland wheat production and fertiliser use in south-western Australia: an agronomic and economic perspective. *Australian Journal of Soil Research*, 48, pp. 531–545.

Caravaca, F., Alguacil, M. M., Azcon, R. and Roldan, A. (2006). Formation of stable aggregates in rhizosphere soil of *Juniperus oxycedrus*: Effect of AM fungi and organic amendments. *Applied Soil Ecology*, 33, pp. 30–38.

Daynes, C. N., Field, D. J., Saleeba, J. A., Cole, M. A. and McGee, P. A. (2013). Development and stabilisation of soil structure via interactions between organic matter, arbuscular mycorrhizal fungi and plant roots. *Soil Biology and Biochemistry*, 57, pp. 683–694.

Elmer, W. H. and Pignatello, J. J. (2011). Effect of biochar amendments on mycorrhizal associations and Fusarium Crown and Root Rot of asparagus in replant soils. *Plant Disease*, 95, pp. 960–966.

Gill, R. A., Kelly, R. H., Parton, W. J., Day, K. A., Jackson, R. B., Morgan, J. A., Scurlock, J. M. O., Tieszen, L. L., Castle, J. V., Ojima, D. S. and Zhang, X. S. (2002). Using simple environmental variables to estimate below-ground productivity in grasslands. *Global Ecology and Biogeography*, 11, pp. 79–86.

Golchin, A., Oades, J. M., Skjemstad, J. O. and Clarke, P. (1994). Soil-structure and carbon cycling. *Australian Journal of Soil Research*, 32, pp. 1043–1068.

Grant, C., Bittman, S., Montreal, M., Plenchette, C. and Morel, C. (2005). Soil and fertilizer phosphorus: effects on plant P supply and mycorrhizal development. *Canadian Journal of Plant Science*, 85, pp. 3–14.

Hammer, E. C., Balogh-Brunstad, Z., Jakobsen, I., Olsson, P. A., Stipp, S. L. S. and Rillig, M. C. (2014). A mycorrhizal fungus grows on biochar and captures phosphorus from its surfaces. *Soil Biology and Biochemistry*, 77, pp. 252–260.

Ishii, T. and Kadoya, K. (1994). Effects of charcoal as a soil conditioner on citrus growth and vesicular-arbuscular mycorrhizal development. *Journal of the Japanese Society for Horticultural Science*, 63, pp. 529–535.

Jastrow, J. D., Miller, R. M. and Lussenhop, J. (1998). Contributions of interacting biological mechanisms to soil aggregate stabilization in restored prairie. *Soil Biology and Biochemistry*, 30, pp 905–916.

Jeffery, S., Bezemer, T. M., Cornelissen, G., Kuyper, T. W., Lehmann, J., Mommer, L., Sohi, S. P., van de Voorde, T. F. J., Wardle, D. A. and van Groenigen, J. W. (2015). The way forward in biochar research: targeting trade-offs between the potential wins. *Global Change Biology Bioenergy*, 7, pp. 1–13.

Jones, D. L. and Edwards, A. C. (1998). Influence of sorption on the biological utilization of two simple carbon substrates. *Soil Biology and Biochemistry*, 30, pp. 1895–1902.

Karandashov, V. and Bucher, M. (2005). Symbiotic phosphate transport in arbuscular mycorrhizas. *Trends in Plant Science*, 10, pp. 22–29.

Lal, R. and Kimble, J. M. (1997). Conservation tillage for carbon sequestration. *Nutrient Cycling in Agroecosystems*, 49, pp. 243–253.

LeCroy, C., Masiello, C. A., Rudgers, J. A., Hockaday, W. C. and Silberg, J. J. (2013). Nitrogen, biochar, and mycorrhizae: alteration of the symbiosis and oxidation of the char surface. *Soil Biology and Biochemistry*, 58, pp. 248–254.

Lehmann, J. (2007). A handful of carbon. *Nature*, 447, pp. 143–144.

Lehmann, J., Gaunt, J. and Rondon, M. (2006). Biochar sequestration in terrestrial ecosystems – a review. *Mitigation and Adaptation Strategies for Global Change*, 11, pp. 403–427.

Lehmann, J., Rillig, M. C., Thies, J., Masiello, C. A., Hockaday, W. C. and Crowley, D. (2011). Biochar effects on soil biota – a review. *Soil Biology and Biochemistry*, 43, pp. 1812–1836.

Lorenz, K. and Lal, R. (2014). Biochar application to soil for climate change mitigation by soil organic carbon sequestration. *Journal of Plant Nutrition and Soil Science*, 177, pp. 651–670.

Medina, A. and Azcon, R. (2010). Effectiveness of the application of arbuscular mycorrhiza fungi and organic amendments to improve soil quality and plant performance under stress conditions. *Journal of Soil Science and Plant Nutrition*, 10, pp. 354–372.

Miller, R. M., and Jastrow, J. D. (1990). Hierarchy of root and mycorrhizal fungal interactions with soil aggregation. *Soil Biology and Biochemistry*, 22, pp. 579–584.

Nadian, H., Hashemi, M. and Herbert, S. J. (2009). Soil aggregate size and mycorrhizal colonization effect on root growth and phosphorus accumulation by Berseem clover. *Communications in Soil Science and Plant Analysis*, 40, pp. 2413–2425.

Nichols, K. A. and Wright, S. F. (2005). Comparison of glomalin and humic acid in eight native US soils. *Soil Science*, 170, pp. 985–997.

NOAA (2015). Global carbon dioxide concentrations surpass 400 parts per million for the first month since measurements began. [online] Available at: http://research.noaa.gov/News/NewsArchive/LatestNews/TabId/684/ArtMID/1768/ArticleID/11153/Greenhouse-gas-benchmark-reached-.aspx [Accessed 18 January 2016]

Ortas, I. (2003). Effect of selected mycorrhizal inoculation on phosphorus sustainability in sterile and non-sterile soils in the Harran Plain in South Anatolia. *Journal of Plant Nutrition*, 26, pp. 1–17.

Ortas, I. (2006). Soil biological degradation. In: Lal, R. (ed.) *Encyclopedia of Soil Science*. New York: Marcel Dekker, pp. 264–267.

Ortas, I., Akpinar, C. and Lal, R. (2013). Long-term impacts of organic and inorganic fertilizers on carbon sequestration in aggregates of an Entisol in Mediterranean Turkey. *Soil Science*, 178, pp. 12–23.

Rasse, D. P., Rumpel, C. and Dignac, M. F. (2005). Is soil carbon mostly root carbon? Mechanisms for a specific stabilisation. *Plant and Soil*, 269, pp. 341–356.

Rillig, M. C. (2004). Arbuscular mycorrhizae, glomalin, and soil aggregation. *Canadian Journal of Soil Science*, 84, pp. 355–363.

Rillig, M. C., Mardatin, N. F., Leifheit, E. F. and Antunes, P. M. (2010). Mycelium of arbuscular mycorrhizal fungi increases soil water repellency and is sufficient to maintain water-stable soil aggregates. *Soil Biology and Biochemistry*, 42, pp. 1189–1191.

Rillig, M. C., Ramsey, P. W., Morris, S. and Paul, E. A. (2003). Glomalin, an arbuscular-mycorrhizal fungal soil protein, responds to land-use change. *Plant and Soil*, 253, pp. 293–299.

Saito, M. (1990). Charcoal as a microhabitat for VA mycorrhizal fungi, and its practical implication. *Agriculture Ecosystems and Environment*, 29, pp. 341–344.

Schrag, D. P. (2007). About climate change – reply. *Elements*, 3, pp. 375–375.

Singh, S., Mishra, R., Singh, A., Ghoshal, N. and Singh, K. P. (2009). Soil physicochemical properties in a grassland and agroecosystem receiving varying organic inputs. *Soil Science Society of America Journal*, 73, pp. 1530–1538.

Six, J., Bossuyt, H., Degryze, S. and Denef, K. (2004). A history of research on the link between (micro) aggregates, soil biota, and soil organic matter dynamics. *Soil and Tillage Research*, 79, pp. 7–31.

Smith, S. E. and Read, D. J. (2008). *Mycorrhizal Symbiosis*. 3rd Edition. San Diego, CA: Academic Press.

Sohi, S. P., Krull, E., Lopez-Capel, E. and Bol, R. (2010). A review of biochar and its use and function in soil. In: Sparks, D. L. (ed.) *Advances in Agronomy*, 105. San Diego, CA: Elsevier Academic Press, pp. 47–82.

Solaiman, Z. M., Blackwell, P., Abbott, L. K. and Storer, P. (2010). Direct and residual effect of biochar application on mycorrhizal root colonisation, growth and nutrition of wheat. *Australian Journal of Soil Research*, 48, pp. 546–554.

Sundermeier, A. P., Islam, K. R., Raut, Y., Reeder, R. C. and Dick, W. A. (2011). Continuous no-till impacts on soil biophysical carbon sequestration. *Soil Science Society of America Journal*, 75, pp. 1779–1788.

Tisdall, J. M., Smith, S. E. and Rengasamy, P. (1997). Aggregation of soil by fungal hyphae. *Australian Journal of Soil Research*, 35, pp. 55–60.

Treseder, K. K. and Turner, K. M. (2007). Glomalin in ecosystems. *Soil Science Society of America Journal*, 71, pp.1257–1266.

Vanek, S. J. and Lehmann, J. (2015). Phosphorus availability to beans via interactions between mycorrhizas and biochar. *Plant and Soil*, 395, pp. 105–123.

Warnock, D. D., Lehmann, J., Kuyper, T. W. and Rillig, M. C. (2007). Mycorrhizal responses to biochar in soil – concepts and mechanisms. *Plant and Soil*, 300, pp. 9–20.

Warnock, D. D., Mummey, D. L., McBride, B., Major, J., Lehmann, J. and Rillig, M. C. (2010). Influences of non-herbaceous biochar on arbuscular mycorrhizal fungal abundances in roots and soils: results from growth-chamber and field experiments. *Applied Soil Ecology*, 46, pp. 450–456.

Whitman, T., Scholz, S. M. and Lehmann, J. (2010). Biochar projects for mitigating climate change: an investigation of critical methodology issues for carbon accounting. *Carbon Management*, 1, pp. 89–107.

Wilkinson, D. M. (2008) Testate amoebae and nutrient cycling: peering into the black box of soil ecology. *Trends in Ecology & Evolution*, 23, 596–599.

Wright, S. F. and Upadhyaya, A. (1998). A survey of soils for aggregate stability and glomalin, a glycoprotein produced by hyphae of arbuscular mycorrhizal fungi. *Plant and Soil*, 198, pp. 97–107.

Zuccarini, P. (2010). Biological and technological strategies against soil and water salinization rhizosphere. *Journal of Plant Nutrition*, 33, pp. 1287–1300.

17

The Use of Stable Isotopes in Understanding the Impact of Biochar on the Nitrogen Cycle

REBECCA HOOD-NOWOTNY

> **Abstract**
>
> The practice of applying biochar to soil could increase crop production and sequester carbon, whilst tightening leaky nitrogen cycles. Biochar has been shown to improve soil properties and even reduce greenhouse gas emissions, however the underlying mechanisms that lead to yield increases and GHG mitigation still elude us. Recent and ongoing studies have demonstrated that detailed analysis of the inherent biogeochemical processes using stable isotope techniques can unravel the complex soil–plant interactions and begin to tease out the multifaceted impacts of biochar on soil processes and plant growth. Here we present a range of nitrogen isotope techniques that could be, or have been, used to understand the changes in dominant processes in the nitrogen cycle following biochar addition.

17.1 Introduction

Biochar discovered by the ancient Amazonians has been heralded as one of the possible saviours of the planet (Lehmann, 2007). Biochar production yields bio-fuel and biochar, a continuum of black carbon (BC) compounds (Schmidt and Noack, 2000) with a macromolecular structure dominated by poly-aromatic hydrocarbons, which when added to soil has been shown to improve soil properties (Jeffery et al., 2015) and even reduce greenhouse gas emissions (Van Zwieten et al., 2015; Glaser et al., 2001; Steiner et al., 2008; Pietikainen et al., 2000), in addition to sequestering significant amounts of carbon for millennia (Singh and Cowie, 2014). Given the great hype and potential of biochar there are still many issues to be systematically and scientifically investigated. Here we provide a review of the utility of stable isotopes in studying the impact of biochar on the soil nitrogen cycle.

17.2 Biochar: the Product and the Process

Biochar is formed by partial combustion or pyrolysis of plant-derived biomass or waste stream products, yielding a continuum of BC-rich compounds. Restricting the oxygen supply prevents complete combustion, resulting in carbon-rich biochar material with a macromolecular structure dominated by poly-aromatic hydrocarbons, with oxygen and hydrogen functional groups (Schmidt and Noack, 2000). We know from paleoecological or archaeological research that charcoal is highly resistant to decomposition due to its condensed aromatic structures. It is biochar's rare aromatic-macromolecular structure that renders it more recalcitrant to microbial decomposition than uncharred organic matter; basically the soil microbial biomass does not have the enzymatic wherewithal readily available (Schimel and Weintraub, 2003) to degrade these novel products rapidly, making biochar a potential long-term carbon sink (Lehmann and Joseph, 2015; Baldock and Skjemstad, 2000). Evidence suggests that biochar has a mean residence time (MRT) in the range of hundreds of millions of years depending on whether a single pool, two pool or infinite pool model is used for the calculation (Lehmann et al., 2015). To put this in perspective the majority of uncharred agricultural organic matter entering the soil is mineralized within months, as shown in numerous mass loss and litterbag experiments (Christensen and Jonasson, 1999), with a small portion, stabilized through interactions with mineral surfaces, resulting in MRTs in the region of 50–200 years (Trumbore, 2000; Six and Jastrow, 2002)

17.3 Biochar as a Soil Amendment

Biochar has been shown to improve structure and fertility of soil, increasing biomass production (Lehmann, 2007; Steiner et al., 2007). It has also been shown to increase cation exchange capacity (Cheng et al., 2006). Increasingly research suggests that presence of biochar may reduce emissions of two potent greenhouse gases: nitrous oxide and methane (Van Zwieten et al., 2015; Cayuela et al., 2013 and references therein).

The benefits of adding biochar to tropical soils are clear from the success of the fertile Terra Pretas and an emerging body of evidence (Glaser et al., 2001; Jeffery et al., 2015). However biochar could also play a major role in reducing environmentally damaging nutrient losses from temperate agricultural and forestry systems, whilst increasing production (Laird and Rogovska, 2015). Biochar is an acceptable organic (bio) soil amendment, and has been approved in Japan (Fomsgaard, 2006). It may play a significant role in the Austrian organic (bio) farming systems, which currently comprise 20% of Austrian agriculture (Arbenz et al., 2015). It should be stressed that further research and legislation is still required to realize this carbon sequestration strategy.

17.4 The Nitrogen Cycle

Globally the nitrogen cycle is one of the most important biological cycles, from the perspective of crop production, GHG emissions and ecosystem services (Tilman et al., 2002;

Galloway et al., 2008). Nitrogen is a limiting factor to production in most agricultural and natural ecosystems, and is essential to the survival of all life forms; however nitrogen application is not without associated problems when applied at rates excessive to demand, leading to the release of reactive nitrogen species (Galloway et al., 2008; Steffen et al., 2015). The applications of nitrogen fertilizer and subsequent increases in crop production have supported continued economic growth, meeting global food demands over the last century (Fanzo et al., 2011).

The massive anthropogenic input of inorganic nitrogen has led to a number of environmental problems, including groundwater pollution, the release of potent GHGs such as nitrous oxide (N_2O), ammonia volatilization and coastal and terrestrial eutrophication, all of which have an impact on people's health, biodiversity and ecosystem function on a regional and global basis (Sutton et al., 2012; Steffen et al., 2015).

Ammonia (NH_3), N_2O and nitric oxide (NO) are released into the atmosphere as a result of agricultural activities. Nitrate (NO_3) leaching into aquatic systems is also a danger if soil management is inappropriate. N_2O is a direct GHG with a global warming potential 300 times that of carbon dioxide (CO_2). Agriculture is the main source of N_2O emissions and is responsible for some 65% of total anthropogenic N_2O emissions (Brennan et al., 2012; Hu et al., 2015). Collectively these pollutant nitrogen species have been termed reactive nitrogen. Biochar has been shown to be effective in mitigating the release of a number of reactive nitrogen species (Xiang et al., 2014). For example biochar addition has been shown to reduce N_2O emissions by an average of 50% across a range of soils and ecosystems (Van Zwieten et al., 2015).

There are a number of ecosystem services that biochar directly influences, including soil organic matter dynamics and fertility, regulation of carbon flux and consequent climate control, and soil water storage.

The evidence of the benefits of biochar addition is mounting (Jeffery et al., 2015), however more detailed information on the mechanisms and processes which drive these changes is required to inspire the confidence of soil custodians and to allow us to make predictions and evidence-based recommendations. One of the most important and dominant soil processes is the soil nitrogen cycle, and providing information on how biochar affects the nitrogen cycle would allow us to extrapolate and predict its long-term effects with greater confidence (Clough et al., 2013; Clough and Condron, 2010).

Soil processes are notoriously complex and interdependent, therefore, holistic approaches to studying numerous N pools and the gross N fluxes between them simultaneously is required. Stable isotope techniques facilitate element tracing with minimal perturbation to the natural system. Using stable isotopes it is possible to study the fate of applied fertilizer nitrogen directly and follow it into the various soil/plant pools, and to trace nitrogen losses from the system directly. The versatility of the isotope also means it is possible to study the natural supply of inorganic nitrogen from organic matter.

The main biogeochemical processes in the nitrogen cycle are shown in Figure 17.1. As the term cycle implies, all nitrogen that enters the system will ultimately leave it, closing the cycle. Natural soil N inputs are biological nitrogen fixation, atmospheric lightning,

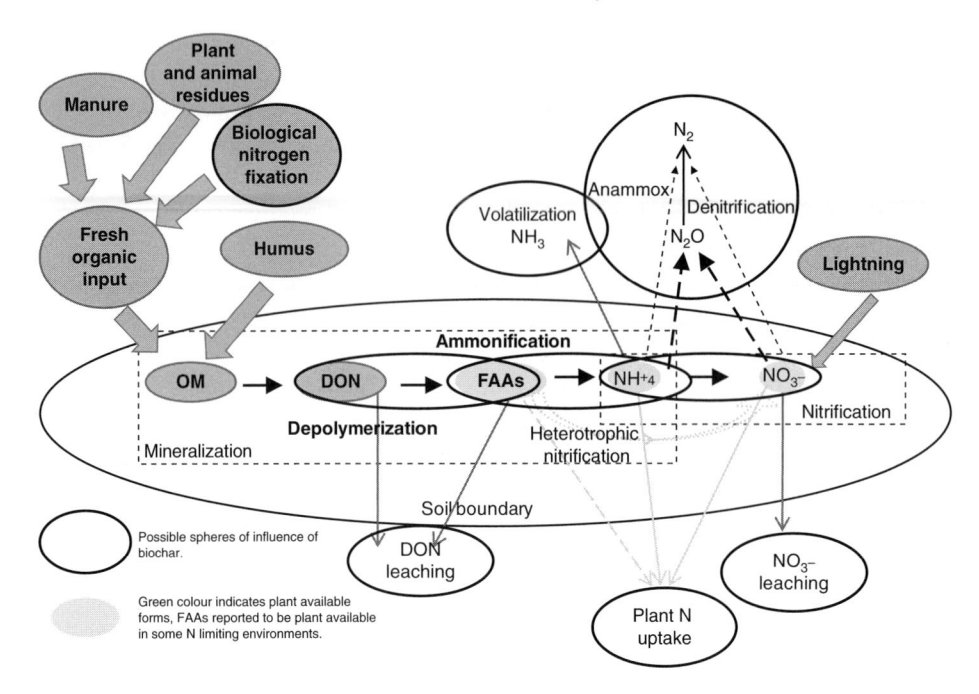

Figure 17.1. The nitrogen cycle and possible spheres of influence of biochar. OM = organic matter, FAAs = free amino acids, DON = dissolved organic nitrogen. Black arrows indicate microbially mediated processes. (A black and white version of this figure will appear in some formats. For the colour version, please refer to the plate section.)

or subsequently after cycling through the food web animal excretion or organism death and decomposition. The organic forms of nitrogen are broken down into inorganic forms through mineralization processes, where they can be taken up by plants and recycled back into the soil system. Because there is a large background of plant unavailable organic nitrogen in soils, inorganic ^{15}N-labelled stable isotopes have proved to be useful tracers to track the turnover and movement of nitrogen in soils. Simultaneous differential isotope labelling (i.e. producing chemically 'identical' material with different isotopic signatures) allows a number of concurrent processes to be studied and their interactions to be assessed.

17.5 Introduction to Stable Isotopes

An isotope of an element has the same atomic number (protons) but a different number of neutrons and thus a different atomic weight; therefore stable isotopes react chemically identically to the more common isotope, because they occupy the same place in the periodic table. However different rates of reaction at an enzymatic level and kinetics can result in slight variations in isotopic composition in nature. It is these natural signatures that can be used in ecological studies to trace food web structure, migration patterns, feeding preferences and so on. Studies of natural variations in naturally occurring isotopic signatures are commonly referred to as natural abundance studies. Studies in which artificially enriched isotopes are added to the system are known as enrichment studies.

Stable isotopes occur naturally in the environment. They are safe and non-radioactive and therefore do not decay, thus they are ideal natural tracers. For example, approximately 1% of global carbon is in the form of the stable isotope ^{13}C, the rest ^{12}C; about 0.4% of global nitrogen is ^{15}N, with the rest being ^{14}N.

Stable isotope analysis is traditionally undertaken using isotope ratio mass spectrometry (IRMS), separating ionized isotopes based on their mass to charge ratio. There are a range of other successful spectroscopic techniques, established and emerging, to measure stable isotopes, many of which are laser based.

Stable isotopes are particularly useful in environmental science, as they provide a means to follow pathways with minimal disturbance or impact to the natural system, allowing holistic unperturbed systems to be studied. It is possible purchase a range of substances that are highly labelled with stable isotopes, such as ^{15}N-labelled fertilizers. These labelled substances can be used to trace their fate and longevity in the soil-plant-atmosphere continuum.

17.6 Measuring Soil N Transformation Processes in the Soil

As stressed above ultimately all organic matter (OM) entering the soil is degraded from organic to inorganic forms; that is, it is mineralized by the soil microbial biomass (SMB) (Figure 17.1). Macrofauna or physical processes break down OM into particulate organic forms. The ensuing enzymatic depolymerization of the particulate organic nitrogen transforms the complex polymeric-N into bio-available monomeric dissolved organic nitrogen (DON) forms such as free amino acids (FAAs), amino sugars and so on (Jackson et al., 2008; Schimel and Bennett, 2004). In natural systems DON may also be a significant source for plant nutrition. However in agricultural systems it is thought the DON is further assimilated or immobilized by the SMB, where depending on the soil N status, N is eventually released as ammonium (ammonification) or rarely as nitrate (heterotrophic nitrification). The processes involved include microbial mineralization-immobilization, micro/meso faunal grazing – the 'microbial loop' (Coleman, 1994) – and microbial cell death and damage due to desiccation or viral lysis (Birch, 1958; Bowatte et al., 2010). Resultant excess available ammonium is either taken up by plants, lost to the atmosphere via volatilization as ammonia, adsorbed, or further assimilated by different guilds of SMB, for growth or energy, through the conversion of ammonium into nitrate (nitrification). The resultant inorganic nitrate can be readily taken up by plants, leached into the groundwater or denitrified to nitrous oxide or di-nitrogen gas. The main processes in the nitrogen cycle on which biochar has shown to have some effects are ammonification, nitrification, volatilization, denitrification, nitrate leaching, DON leaching and plant N uptake. Using stable isotopes it is possible to determine the impact of biochar on these processes with a greater degree of resolution and reliability than non-isotopic methods.

17.6.1 Isotope Dilution

Measuring the rates of N transformation processes is confounded by the simultaneous occurrence of mineralization (inorganic N release from organic matter), nitrification (conversion of ammonium to nitrate) and immobilization (biological assimilation of available N by the microbial biomass), in addition to nitrate leaching and gaseous N losses (Barraclough, 1991). This is particularly the case when studying the effect of something like biochar addition to soil, as biochar addition will impact nitrogen turnover processes in different ways. Fortunately it is possible to study each transformation process independently using ^{15}N isotope dilution techniques (Barraclough, 1991; Watkins and Barraclough, 1996; Hood et al., 2003). Measures of independent transformation fluxes are conventionally referred to as gross fluxes, for example gross nitrification, as opposed to net fluxes, which are the sum of all the above concurrent N processes. The principle of the isotope dilution technique is that the target pool is labelled with ^{15}N and then the dilution of the ^{15}N and change in the pool size are monitored (Barraclough, 1991). From this the amount of unlabelled N coming into the pool (gross flux) can be calculated (Figure 17.2, for example ammonification). The assumptions inherent in this technique have been outlined by Hart et al. (1994) and are: 1) all rate processes over the experimental period can be described by zero order kinetics; 2) microorganisms do not discriminate between ^{14}N and ^{15}N (possibly not completely true but at enrichments used in these type of experiments true enough); 3) there is uniform mixing of added label with the target N pool; and 4) labelled N immobilized over the experimental period is not remobilized. In addition care should be taken to avoid initial immobilization of added ^{15}N, due to fixation by clays or microbial action, and to ensure that equilibrium conditions have been reached prior to initial sampling (Hood et al., 2003). Modelling approaches have attempted to overcome some of the difficulties in achieving or assuming zero order kinetics but at the expense of increased experimental and interpretive complexity (Inselsbacher et al., 2013; Nelissen et al., 2012).

17.6.2 Measuring Ammonification and Nitrification

Experimental measurement of gross transformation rates is achieved by uniformly labelling a series of replicate (n≈10) intact soil cores or soil aliquots simultaneously with identical quantities of labelled ammonium or nitrate (depending on target process, ammonification or nitrification, to be studied), usually no more than 10% of the initial pool size (concentration) is added as label, with sufficient enrichment to detect after incubation. In practice initial enrichments in the region of 2–5 atom% are sufficient. The label is usually applied using either a soil injection device or to the surface and thoroughly mixed (Monaghan, 1995). After a period allowing for equilibration of the label, usually 4–24 hours, the first set of cores (n≈5) is extracted using either KCl or K_2SO_4 and the concentration and isotope signature of the inorganic N pool are determined by standard chemical methods (Hood-Nowotny et al., 2010; Hood et al., 2003). A second set of cores is extracted, usually 24–48 hours after the initial set of cores has been sampled, and again the N isotope enrichment

The Impact of Biochar on the Nitrogen Cycle 357

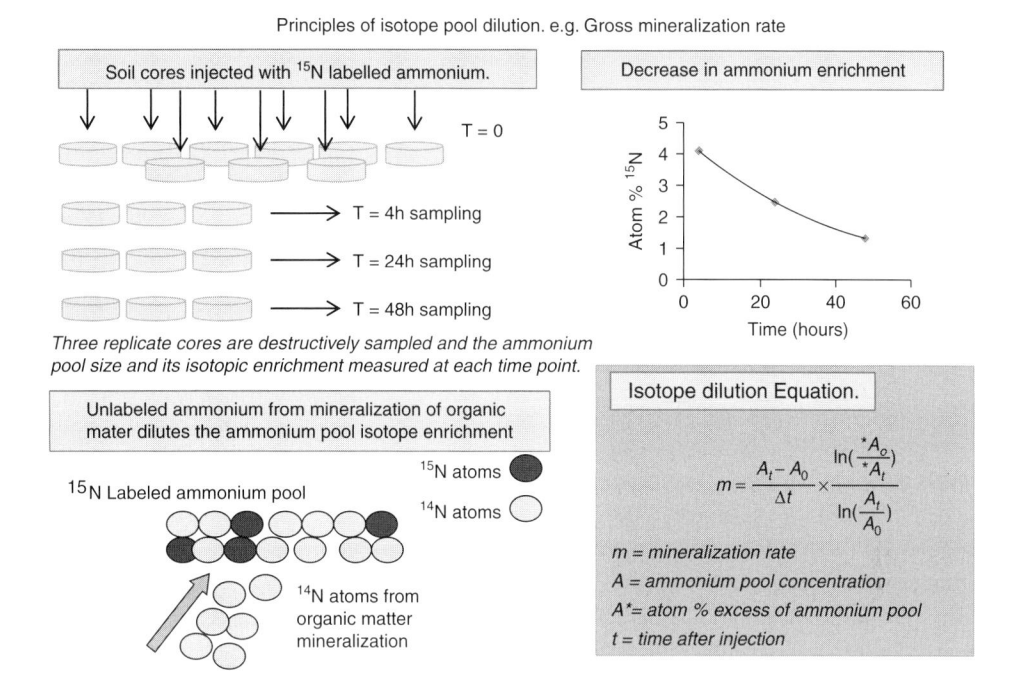

Figure 17.2. The principle of the isotope dilution technique is that the target pool (in this case the ammonium pool) is labelled with ^{15}N ammonium and then the dilution of the ^{15}N in the ammonium pool and the change in the pool concentration are monitored (Barraclough, 1991). The enrichment of the pool decreases as unlabelled N comes into the pool from unlabelled organic matter mineralization; from the dilution of the label we can calculate gross mineralization.

and N concentration are determined. The dilution of the labelled N pool is quantified, so that the dilution of the labelled pool by unlabelled N provides an estimate of the gross N delivery rate to the labelled pool. Process rates are determined from the mean of the two measurements using the isotope dilution equation in Figure 17.2. Using the differential labelling of the ammonium and nitrate pool, it is possible to determine both gross mineralization and gross nitrification independent of other conflicting soil processes. This approach can be extended and forms the basis of the modelling approach (Inselsbacher et al., 2013).

Obviously to determine the impacts of biochar on these fluxes in soil requires at least two treatments; with and without biochar. It would then be necessary to determine the gross rates in both treatments and statistically prove whether they are different or not. This can prove difficult due to high variability in rate process measurement, therefore careful experimental design is required, and usually also a high degree of replication.

One of the original underlying assumptions of the isotope dilution technique for measuring gross nitrogen mineralization rates in soils is that all organic to inorganic N transformations pass through the ammonium pool window prior to uptake by microbial biomass, classically termed the mineralization-immobilization turnover (MIT) route. This was pivotally disproved by Barraclough (1997), who showed unequivocally that amino acids can be taken up directly by the microbial biomass (the direct route) and that in most soils, both processes occur concurrently. This paradigm shift in conceptualization of the nitrogen cycle (Schimel and Bennett, 2004) has also led to an awareness of the role of DON in plant nutrition in natural ecosystems. To what extent the new paradigm is relevant to N-replete agricultural systems, where biochar is likely to be added to soils, is still to be tested, but the new concept allows an overarching understanding of soil processes irrespective of whether they are natural or agricultural.

17.6.3 Measuring Depolymerization

Using isotope dilution techniques it is also possible to determine depolymerization rates (Wanek et al., 2010). Rates are calculated by measuring the concentrations and isotope enrichments of amino acid pools, and measuring their decline over time. Similarly it is also possible to determine the rate of protein turnover in soils (Prommer et al., 2014). There is evidence to suggest that there is increased oxidation of labile organic matter on biochar surface. Be they plant derived or enzyme derived these properties will have a major impact on the organic N turnover mechanisms in soils, particularly in relation to organic N cycling. Whether these properties then translate into differences in ecosystem function, such as whether there are differences in the amino acid turnover rates, is unclear (Dempster et al., 2012). Prommer et al. (2014) showed that biochar addition had strong negative effects on soil organic N cycling rates in temperate soils as it reduced protein production and concurrently decreased free amino acid production. This observation is important given the predominant role that protein depolymerization plays in organic N recycling. Furthermore enhanced gross nitrification rates were measured on soils amended with biochar and the biochar had no significant effects on gross mineralization. Future isotope studies will play a vital role in understanding the fate and location of organic N, to study the impact of biochar on soils' nitrogen turnover processes.

17.6.4 Measuring Losses

There are three major pathways of N loss from the soil: nitrate leaching, denitrification and ammonia volatilization. Biochar has been shown to have significant impacts on all three of these processes and in recent years there has been a wealth of studies using stable isotopes of nitrogen, clearly demonstrating their utility (Taghizadeh-Toosi et al., 2011a; Cayuela et al., 2013; Harter et al., 2014) and how they can be used to gain a more complete understanding of the mechanisms underlying the effects caused by biochar.

One of the simplest ways to determine the pathways of fertilizer N is to label a specific N within a labelled isotope pool to carry out a mass balance of the ^{15}N applied. However, care must be taken to sample every available N pool and minimize N losses during the sample preparation procedures, so that all possible N can be accounted for. Mass balance studies are notoriously difficult and are therefore scarce in the literature.

Measuring di-nitrogen losses as a result of denitrification appears to be one of the biggest challenges; however methods have been developed that appear to overcome some of these (Butterbach-Bahl et al., 2002). Another approach is to label the pool in question and trace the losses from the pool and ultimate fate of the isotopes. This approach has been nicely demonstrated using an isotope of nitrogen to show that ammonia-N adsorbed by biochar is stable in ambient air, but readily bioavailable when placed in the soil (Taghizadeh-Toosi et al., 2011b).

17.7 Nitrate Leaching

Efficient use of soil N requires minimal losses by leaching. Not only is nitrate leaching a loss of a valuable resource – it is also environmentally detrimental. Nitrate can contaminate groundwater and causes accelerated soil acidification. The WHO safety limit for nitrate in drinking water is 11.3 mg NO_3^- –N L^{-1}, the US Environmental Protection Agency's (USEPA's) maximum contaminant level (MCL) is 10 mg N L^{-1}. Nitrogen is usually leached in the form of nitrate, as ammonium is more strongly bound to the cation exchange sites. In soils with net negative charge, nitrate moves freely with the water. A number of factors determine the rate of leaching: 1) the quantity of water passing down across the surface of interest; 2) the concentration of nitrate in the water; 3) the water-holding capacity of the soil volume in question; and 4) uptake by actively growing vegetation (when present). Rainfall intensity and surface soil conditions determine the rate at which the water enters the soil. The soil structure and the size and shape of the soil pores determine the rate and distance that the water moves through the soil and thus the rate of nitrate leaching. In heavier textured soils water moves through cracks and macropores (bypass flow), and the rate of movement through the soil aggregates may be very different than through the pores. It has been shown that biochar addition can reduce nitrate leaching by up to 70% under field conditions (Ventura et al., 2012), 25% in coarse textured soils (Dempster et al., 2012), and can lead to nitrate leaching reductions in sugar cane systems (Chen et al., 2010). Güereña et al. (2012) showed how stable isotopes could have solved some of the questions raised as to the mechanisms involved in the reduction in leaching. They showed that biochar significantly reduced nitrate leaching when fertilized with recommended rates of inorganic N fertilizer and they attributed the differences not to physical adsorption but to biological retention in the soil microbial biomass. Using labelled nitrate allows tracing of the particular source of nitrate applied to be considered. This could be particularly important in biochar research where the incorporation of biochar into the soil could lead to intrinsic increases or decreases in soil nitrification rates, thus making it difficult to source specific losses.

17.8 Reduction in Greenhouse Gas Emissions with Biochar Amendment

Numerous studies have now shown that biochar addition to soils can reduce globally significant GHG emissions, especially of N_2O, one of the most important GHGs in agriculture (Bruun et al., 2012; Cheng et al., 2012; Cayuela et al., 2014; Van Zwieten et al., 2015). However, there also a number that show no or negative effects of biochar on nitrous oxide emissions (Clough et al., 2013; Xiang et al., 2014). Recently meta-analysis showed significant reductions in N_2O emissions, on average 54%, when soils were amended with biochar (Cayuela et al., 2014).

There are a number of explanations for this phenomenon, from reduction in nitrification rate, changes in pH, impacts on fungal to bacterial ratios, inhibition by ethylene and the electron shuttle effect (Baggs and Philipott, 2010; Laughlin et al., 2008; Cayuela et al., 2013; Spokas et al., 2009). Interestingly there were also a number of studies which showed that the effect of biochar on nitrous oxide emissions was dependent on the soil type, suggesting that reactions to biochar are soil context dependent (Malghani et al., 2013). Recently the importance of the molar H:Corg ratio of biochar in determining mitigation of N_2O has been stressed, suggesting that it links to the degree of polymerization and aromaticity of biochar (Cayuela et al., 2015).

Numerous pathways of nitrous oxide emission are postulated (Butterbach-Bahl et al., 2013). The principal pathways are ammonium oxidation (primary step in nitrification), nitrifier denitrification and co-denitrification, and dissimilatory nitrate or nitrite reduction to ammonium (DNRA). Using labelled isotopes it is possible to determine the dominant pathways of nitrous oxide production. For example the contribution of nitrification and denitrification to the flux of N_2O can be studied by differentially ^{15}N-labelling the nitrate and ammonium in soils (the ^{15}N gas flux method; Stevens et al., 1997). By periodically measuring and comparing the enrichments of the N_2O, NH_{4+} and NO_3- pools, the relative importance of the processes can be quantified. The estimations of nitrous oxide contribution from the various pools are based on calculations which assume that the ^{15}N atom fractions of the nitrification and denitrification pools remain uniform throughout the incubation (Arah, 1997). Using this ^{15}N gas flux method combined with gross rate measurements it is possible to determine the parameters which drive the emissions of the important GHG N_2O and to evaluate the specific impact of biochar on the N loss processes (Bateman and Baggs, 2005; Stevens et al., 1997). It has been shown that nitrification and denitrification N_2O fluxes are mediated by different microbial guilds in the soil (Hu et al., 2015). Fungi are capable of nitrification and denitrification and are often the dominant group in the microbial biomass (Laughlin et al., 2008). Application of biochar has been demonstrated to alter the composition of the soil microbial biomass, in terms of fungal to bacterial ratios (Pietikainen et al., 2000; Jones et al., 2011). It has been hypothesized that it is the change in microbial biomass community structure that leads to the observed reductions in GHG loss (Han et al., 2013). This community-driven functional response to biochar has also been observed in temperate soils, where measured increases in ammonium-oxidizing bacteria and archaea were associated with increases in nitrification and decreases in nitrous oxide emissions (Prommer et al., 2014). Using isotope techniques, Cayuela et al. (2013) showed that biochar consistently reduced N_2O emissions in agricultural soils. They demonstrated that biochars promote the last step of denitrification and proposed that biochar is able to facilitate the transfer of electrons to denitrifying microorganisms in soil, acting as an electron shuttle, and that this together with its liming effect and high surface area would promote the reduction of N_2O to N_2. More recently isotope experiments have shown that the dominant pathway of nitrous oxide emission in soils has a big impact on whether biochar soil addition leads to a reduction or enhancement of emissions (Sánchez-García, 2014).

It is also possible to use triplet tracer experiments (TTEs) combined with inverse abundance approaches to reduce errors associated with the analytical models, and knowing the isotopic composition of each pool and the isotope ratio of the nitrous oxide it is possible to trace the pathways of denitrification, immobilization, DNRA and anaerobic ammonium oxidation (anammox), and to separate out the source pools of the various processes (Spott and Stange, 2007).

Natural abundance approaches based on isotopomers could possibly also be used to understand N_2O emission pathways. Isotopomer site preference (SP) of nitrous oxide is based on the different intra-molecular distributions of ^{15}N, whether central, next to the oxygen atom (α) or terminal (β) in a linear N_2O molecule, and is often expressed as $SP = \delta^{15}N_\alpha - \delta^{15}N_\beta$. Site preference has been shown to be a good tool for distinguishing the pathways of N_2O emissions from classical denitrification compared to nitrification-associated N_2O emissions, in culture (Sutka and Ostrom, 2006).

Obviously there is a wealth of research underway in this area. Studies using stable isotopes will be necessary to elucidate the mechanisms involved in the reduction in nitrous oxide emissions due to biochar addition. As Cayuela et al. (2015) pointed out there is an urgent need to identify the mechanisms underpinning N_2O mitigation.

17.9 Free-living N_2 Fixation

Although *nifH* genes involved in biological nitrogen fixation (BNF) have been shown to be up-regulated in biochar-amended soils, to date there have been few quantitative measurements of the impact of biochar on free-living nitrogen fixation (Ducey et al., 2013; Harter et al., 2014). However, there are fairly simple methods available to get at these data. The estimated contribution of free-living N-fixing prokaryotes to the N input of soils ranges from 0 to 60 kg ha^{-1} year^{-1} (Cleveland et al., 1999). Physical protection in pore structures and improved bacterial attachment are thought to lead to higher bacterial growth rates in biochar-amended soils (Pietikainen et al., 2000). It is possible that pore structures could also lead to improved micro-habitats for free-living N_2-fixing prokaryotes. Using a direct method of measuring N_2 fixation, by incubation of soil samples in a synthetic air atmosphere of highly labelled N_2 gas, it is possible to determine N fixation directly without any compounding artefacts (Fürnkranz et al., 2008; McNeill et al., 1994). The soil is then analysed and the increase in ^{15}N abundance in the soil measured; if there is significant BNF there is a significant increase over time in the enrichment of the soil. This method is the most functionally precise measurement as it measures the amount of nitrogen fixation for a specific period under specific conditions.

17.10 Plant N Uptake

In agriculture we are ultimately interested in how much nitrogen plants take up as this has significant impacts on both crop productivity and quality. In fertilized systems mineralization of SOM provides around 50% of the crop nitrogen (Hood et al., 2003), and although inorganic fertilizer can result in high yields they are inherently inefficient, with rarely more than 40% of the nitrogen that is applied in the form of fertilizer ending up in the plant, and the remaining nitrogen immobilized in the soil, lost as described above or denitrified.

Meta-analysis has shown that biochar addition to soils generally increases productivity (Biederman and Harpole, 2013; Jeffery et al., 2015). However there are few studies that show that this is directly attributable to an increase in nitrogen uptake or availability (Jones et al., 2011; Li et al., 2012; Vaccari et al., 2011), although many have shown that relatively large additions of biochar carbon have a slight or no significant impact on crop growth, in contrast to soils that receive equivalent quantities of unpyrolysed carbon, where yield depression due to immobilization of nitrogen can depress yield by up to one-third (Ganry et al., 1978; Recous et al., 1998). Tracing the fate of stable isotope-labelled nitrogen in biochar amended soils will enable us to tease out some of the important impacts of biochar-addition on crop growth. This in turn will allow us to determine whether improvements are due to nutritional gains, gains in water usage, or both.

17.11 Conclusions

It is clear that stable isotopes will play an important role in discovering the true mechanisms behind the yield increases observed in biochar-amended systems. Understanding the role of biochar in these processes is the key to being able to manage the systems to their full potential and to make informed decisions about possible outcomes and consequences of biochar addition, which in turn will allow for more confident adoption of biochar application practices around the world.

Acknowledgement

This work was funded by FWF Project Does carbon negative biochar tighten the nitrogen cycle? V-138.

References

Arah, J. R. M. (1997). Apportioning nitrous oxide fluxes between nitrification and denitrification using gas-phase mass spectrometry. *Soil Biology and Biochemistry*, 29, pp. 1295–1299.

Arbenz, M., Willer, H., Lernoud, J., Huber, B. and Amarjit, S. (2015). The World of Organic Agriculture – Statistics and Emerging Trends (Session at the BioFach Congress 2015, Nuremberg, Germany, 11–15.2.2015).

Baggs, E. M. and Philippot, L. (2010). Microbial terrestrial pathways to nitrous oxide. In: Smith, K. (ed.) *Nitrous Oxide and Climate Change*. Oxford: Earthscan, pp. 4–35.

Baldock, J. and Skjemstad, J. (2000). Role of the soil matrix and minerals in protecting natural organic materials against biological attack. *Organic Geochemistry*, 31, pp. 697–710.

Barraclough, D. (1991). The use of mean pool abundances to interpret 15N tracer experiments. *Plant and Soil*, 131, pp. 89–96.

Barraclough, D. (1997). The direct route or MIT route for nitrogen immobilization: a 15N 'mirror image' study with leucine and glycine, *Soil Biology and Biochemistry*, 29, pp. 101–108.

Bateman, E. J. and Baggs, E. M. (2005). Contributions of nitrification and denitrification to N_2O emissions from soils at different water-filled pore space. *Biology and Fertility of Soils*, 41, pp. 379–388.

Biederman, L. A. and Harpole, W. S. (2013). Biochar and its effects on plant productivity and nutrient cycling: a meta-analysis. *GCB Bioenergy*, 5, pp. 202–214.

Birch, H. F. (1958). The effect of soil drying on humus decomposition and nitrogen availability. *Plant and Soil*, 10, pp. 9–31.

Bowatte, S., Newton, P. C. D., Takahashi, R. et al. (2010). High frequency of virus-infected bacterial cells in a sheep grazed pasture soil in New Zealand. *Soil Biology and Biochemistry*, 42, pp. 708–712.

Brennan, M. D., Cheong, R. and Levchenko, A. (2012). Systems biology. How information theory handles cell signaling and uncertainty. *Science*, 338, pp. 334–335.

Bruun, E. W., Petersen, C., Strobel, B. W. et al. (2012). Nitrogen and carbon leaching in repacked sandy soil with added fine particulate biochar. *Soil Science Society of America Journal*, 76, pp. 1142–1148.

Butterbach-Bahl, K., Baggs, E. M., Dannemann, M. et al. (2013). Nitrous oxide emissions from soils: how well do we understand the processes and their controls? *Philosophical Transactions of the Royal Society of London. Series B, Biological Sciences*, 368, No.20130122.

Butterbach-Bahl, K., Willibald, G. and Papen, H. (2002). Soil core method for direct simultaneous determination of N2 and N2O emissions from forest soils. *Plant and Soil*, 240, pp. 105–116.

Cayuela, M. L., Sánchez-Monedero, M. A., Roig, A. et al. (2013). Biochar and denitrification in soils: when, how much and why does biochar reduce N₂O emissions? *Scientific Reports*, 3, No.1732.

Cayuela, M. L., van Zwieten, L., Singh, B. P. et al. (2014). Biochar's role in mitigating soil nitrous oxide emissions: a review and meta-analysis. *Agriculture, Ecosystems & Environment*, 191, pp. 5–16.

Cayuela, M. L., Jeffery, S. and van Zwieten, L. (2015). The molar H: Corg ratio of biochar is a key factor in mitigating N_2O emissions from soil. *Agriculture, Ecosystems & Environment*, 202, pp. 135–138.

Chen, Y., Shinogi, Y. and Taira, M. (2010). Influence of biochar use on sugarcane growth, soil parameters, and groundwater quality. *Australian Journal of Soil Research*, 48, pp. 526–530.

Cheng, C., Lehmann, J. and Thies, J., 2006. Oxidation of black carbon by biotic and abiotic processes. *Organic Geochemistry*, 37, pp. 1477–1488.

Cheng, Y., Cai, Z.-C., Chang, S. X. et al. (2012). Wheat straw and its biochar have contrasting effects on inorganic N retention and N_2O production in a cultivated Black Chernozem. *Biology and Fertility of Soils*, 48, pp. 941–946.

Christensen, T. and Jonasson, S. (1999). On the potential CO_2 release from tundra soils in a changing climate. *Applied Soil Ecology*, 11, pp. 127–134.

Cleveland, C. C., Townsend, A. R., Schimel, D. S. et al. (1999). Global patterns of terrestrial biological nitrogen (N_2) fixation in natural ecosystems. *Global Biogeochemical Cycles*, 13, pp. 623–645.

Clough, T., Condron, L. M., Kammann, C. et al. (2013). A review of biochar and soil nitrogen dynamics. *Agronomy*, 3, pp. 275–293.

Clough, T. J. and Condron, L. M. (2010). Biochar and the nitrogen cycle: introduction. *Journal of Environment Quality*, 39, pp. 1218–1223.

Coleman, D. C. (1994). The microbial loop concept as used in terrestrial soil ecology studies. *Microbial Ecology*, 28, pp. 245–250.

Dempster, D. N., Jones, D. L. and Murphy, D. V. (2012). Clay and biochar amendments decreased inorganic but not dissolved organic nitrogen leaching in soil. *Soil Research*, 50, pp. 216–221.

Ducey, T. F., Ippolito, J. A., Cantrell, K. B. et al. (2013). Addition of activated switchgrass biochar to an aridic subsoil increases microbial nitrogen cycling gene abundances. *Applied Soil Ecology*, 65, pp. 65–72.

Fanzo, J., Remans, R. and Sanchez, P. (2011). The role of chemistry in addressing hunger and food security. In: Garcia-Martinez, J. and Serrano-Torregrosa, E. (eds.) *The Chemical Element: Chemistry's Contribution to our Global Future*. London: Wiley, pp. 71–97.

Fomsgaard, S. (2006). *Organic Agriculture Movement at a Crossroad – a Comparative Study of Denmark and Japan*. Working Paper OASE. Aalborg, Denmark: Aalborg University, Department of Economics Politics and Public Administratration.

Fürnkranz, M., Wanek, W., Richter, A. et al. (2008). Nitrogen fixation by phyllosphere bacteria associated with higher plants and their colonizing epiphytes of a tropical lowland rainforest of Costa Rica. *The ISME Journal*, 2, pp. 561–570.

Galloway, J. N., Townsend, A. R., Erisman, J. W. et al. (2008). Transformation of the nitrogen cycle: recent trends, questions, and potential solutions. *Science*, 320, pp. 889–892.

Ganry, F., Guiraud, G. and Dommergues, Y. (1978). Effect of straw incorporation on the yield and nitrogen balance in the sandy soil-pearl millet cropping system of Senegal. *Plant and Soil*, 50, pp. 647–662.

Glaser, B., Haumaier, L., Guggenberger, G. et al. (2001). The 'Terra Preta' phenomenon: a model for sustainable agriculture in the humid tropics. *Naturwissenschaften*, 88, pp. 37–41.

Güereña, D. Lehmann, J., Hanley, K. et al. (2012). Nitrogen dynamics following field application of biochar in a temperate North American maize-based production system. *Plant and Soil*, 365, pp. 239–254.

Han, G. M., Meng, J., Zhang, W. M. et al. (2013). Effect of biochar on microorganisms quantity and soil physicochemical property in rhizosphere of spinach (*Spinacia oleracea* L.). *Applied Mechanics and Materials*, 295–298, pp. 210–219.

Hart, S. C., Stark, J. M., Davidson, E. A. et al. (1994). Methods of soil analysis: part 2 – microbiological and biochemical properties. In: Bottomley, P. S., Angle, J. S. and Weaver, R. W. (eds.) *Methods of Soil Analysis: Part 2 – Microbiological and Biochemical Properties*. SSSA Book Series. Washington, DC: Soil Science Society of America, pp. 985–1018.

Harter, J., Krause, H.-M., Schuettler, S. et al. (2014). Linking N_2O emissions from biochar-amended soil to the structure and function of the N-cycling microbial community. *The ISME Journal*, 8, pp. 660–674.

Hood, R., Bautista, E. and Heiling, M. (2003). Gross mineralization and plant N uptake from animal manures under non-N limiting conditions, measured using ^{15}N isotope dilution techniques. *Phytochemistry Reviews*, 2, pp. 113–119.

Hood-Nowotny, R., Umana, N. H.-N., Inselbacher, E. et al. (2010). Alternative methods for measuring inorganic, organic, and total dissolved nitrogen in soil. *Soil Science Society of America Journal*, 74, pp. 1018–1027.

Hu, H.-W., Chen, D. and He, J.-Z. (2015). Microbial regulation of terrestrial nitrous oxide formation: understanding the biological pathways for prediction of emission rates. *FEMS Microbiology Reviews*, 39, pp. 729–749.

Inselsbacher, E., Wanek, W., Strauss, J. et al. (2013). A novel ^{15}N tracer model reveals: plant nitrate uptake governs nitrogen transformation rates in agricultural soils. *Soil Biology and Biochemistry*, 57, pp. 301–310.

Jackson, L. E., Burger, M. and Cavagnaro, T. R. (2008). Roots, nitrogen transformations, and ecosystem services. *Annual Review of Plant Biology*, 59, pp. 341–363.

Jeffery, S., Abalos, D., Spokas, K. A. et al. (2015). Biochar effects on crop yield. In: Lehmann, J. and Joseph, S. (eds.) *Biochar for Environmental Management: Science, Technology and Implementation*. London: Earthscan, Routledge, pp. 301–325.

Jones, D. L., Rosuk, J., Edwards-Jones, G. et al. (2011). Biochar-mediated changes in soil quality and plant growth in a three year field trial. *Soil Biology and Biochemistry*, 45, pp. 113–124.

Laird, D. and Rogovska, N. (2015). Biochar effects on nutrient leaching. In: Lehmann, J. and Joseph, S. (eds.) *Biochar for Environmental Management: Science, Technology and Implementation*. London: Earthscan, Routledge, pp. 521–542.

Laughlin, R. J., Stevens, R. J. and Müller, C. (2008). Evidence that fungi can oxidize NH_4^+ to NO_3^- in a grassland soil. *European Journal of Soil Science*, 59, pp. 285–291.

Lehmann, J. (2007). A handful of carbon. *Nature*, 447, pp. 10–11.

Lehmann, J. and Joseph, S. (eds.) (2015). *Biochar for Environmental Management: Science, Technology and Implementation*, London: Earthscan, Routledge.

Li, M., Zhen, A. H., Ning, L. et al. (2012). Effects of biochar application on wheat growth and nitrogen balance. *Xinjiang Agricultural Sciences*, 49, pp. 589–594.

Malghani, S., Gleixner, G. and Trumbore, S. (2013). Chars produced by slow pyrolysis and hydrothermal carbonization vary in carbon sequestration potential and greenhouse gases emissions. *Soil Biology and Biochemistry*, 62, pp. 137–146.

McNeill, A., Hood, R. and Wood, M. (1994). Direct measurement of nitrogen fixation by *Trifolium repens* L. and *Alnus glutinosa* L. using $^{15}N_2$. *Journal of Experimental Botany*, 45, pp.749–755.

Monaghan, R. (1995). Errors in estimates of gross rates of nitrogen mineralization due to non-uniform distributions of ^{15}N label, *Soil Biology and Biochemistry*, 27, pp. 855–859.

Nelissen, V., Rüttig, T., Huygens, D. et al. (2012). Maize biochars accelerate short-term soil nitrogen dynamics in a loamy sand soil. *Soil Biology and Biochemistry*, 55, pp. 20–27.

Pietikainen, J., Kiikkila, O. and Fritze, H. (2000). Charcoal as a habitat for microbes and its effect on the microbial community of the underlying humus. *Oikos*, 89, pp. 231–242.

Prommer, J., Wanek, W., Hofhansl, F. et al. (2014). Biochar decelerates soil organic nitrogen cycling but stimulates soil nitrification in a temperate arable field trial. *PloS One*, 9, No.e86388.

Recous, S., Aita, C. and Mary, B. (1998). In situ changes in gross N transformations in bare soil after addition of straw. *Soil Biology and Biochemistry*, 31, pp. 119–133.

Sánchez-García, M. (2014). Biochar increases soil N$_2$O emissions produced by nitrification-mediated pathways. *Frontiers in Environmental Science*, 2, pp. 1–10.

Schimel, J. P. and Bennett, J. (2004). Nitrogen mineralization: challenges of a changing paradigm. *Ecology*, 85, pp. 591–602.

Schimel, J. P. and Weintraub, M. N. (2003). The implications of exoenzyme activity on microbial carbon and nitrogen limitation in soil: a theoretical model. *Soil Biology and Biochemistry*, 35, pp. 549–563.

Schmidt, M. W. I. and Noack, A. G. (2000). Black carbon in soils and sediments: analysis, distribution, implications, and current challenges. *Global Biogeochemical Cycles*, 14, pp. 777–793.

Singh, B. P. and Cowie, A. L. (2014). Long-term influence of biochar on native organic carbon mineralisation in a low-carbon clayey soil. *Scientific Reports*, 4, No.3687.

Six, J. and Jastrow, J. (2002). Organic matter turnover. In: *Encyclopedia of Soil Science*. New York: Marcel Dekker, pp. 936–942.

Spokas, K. A., Koskinen, W. C., Baker, J. M. and Reicosky, D. C. (2009). Impacts of wood-chip biochar additions on greenhouse gas production and sorption/degradation of two herbicides in a Minnesota soil. *Chemosphere*, 77, pp. 574–581.

Spott, O. and Stange, C. F. (2007). A new mathematical approach for calculating the contribution of anammox, denitrification and atmosphere to an N$_2$ mixture based on a ^{15}N tracer technique. *Rapid Communications in Mass Spectrometry*, 2, pp. 2398–2406.

Steffen, W., Richardson, K., Rockström, J. et al. (2015). Planetary boundaries: guiding human development on a changing planet. *Science*, 347, doi: 10.1126/science.1259855.

Steiner, C., Teixeira, W. G., Lehmann, J. et al. (2007). Long term effects of manure, charcoal and mineral fertilization on crop production and fertility on a highly weathered Central Amazonian upland soil. *Plant and Soil*, 291, pp. 275–290.

Steiner, C., Glaser, B., Teixeira, W. G. et al. (2008). Nitrogen retention and plant uptake on a highly weathered central Amazonian Ferralsol amended with compost and charcoal. *Journal of Plant Nutrition and Soil Science*, 171, pp. 893–899.

Stevens, R. J., Laughlin, R. J., Burns, L. C. et al. (1997). Measuring the contributions of nitrification and denitrification to the flux of nitrous oxide from soil. *Soil Biology and Biochemistry*, 29, pp. 139–151.

Sutka, R. and Ostrom, N. (2006). Distinguishing nitrous oxide production from nitrification and denitrification on the basis of isotopomer abundances. *Applied and Environmental Microbiology*, 72, pp. 638–644.

Sutton, M. A., Reis, S., Billen, G. et al. (2012). Preface: 'Nitrogen and global change'. *Biogeosciences*, 9, pp. 1691–1693.

Taghizadeh-Toosi, A., Clough, T. J., Sherlock, R. R. et al. (2011a). A wood based low-temperature biochar captures NH$_3$-N generated from ruminant urine-N, retaining its bioavailability. *Plant and Soil*, 353, pp. 73–84.

Taghizadeh-Toosi, A., Clough, T. J., Sherlock, R. R. et al. (2011b). Biochar adsorbed ammonia is bioavailable. *Plant and Soil*, 350, pp. 57–69.

Tilman, D., Cassman, K. G., Matson, P. A. et al. (2002). Agricultural sustainability and intensive production practices. *Nature*, 418, pp. 671–677.

Trumbore, S. (2000). Age of soil organic matter and soil respiration: radiocarbon constraints on belowground C dynamics. *Ecological Applications*, 10, pp. 399–411.

Vaccari, F., Baronti, S., Lugato, E. et al. (2011). Biochar as a strategy to sequester carbon and increase yield in durum wheat. *European Journal of Agronomy*, 34, pp. 231–238.

Van Zwieten, L., Kammann, C., Cayuela, M. L. et al. (2015). Biochar effects on nitrous oxide and methane emissions from soil. In: Lehmann, J. and Joseph, S. (eds.) *Biochar for Environmental Management: Science, Technology and Implementation.* London: Earthscan, Routledge, pp. 489–520.

Ventura, M. Sorrenti, G., Panzacchi, P. et al. (2012). Biochar reduces short-term nitrate leaching from a horizon in an apple orchard. *Journal of Environmental Quality*, 42, pp. 76–82.

Wanek, W., Mooshammer, M., Blöchl, A. et al. (2010). Determination of gross rates of amino acid production and immobilization in decomposing leaf litter by a novel ^{15}N isotope pool dilution technique. *Soil Biology and Biochemistry*, 42, pp. 1293–1302.

Watkins, N. and Barraclough, D. (1996). Gross rates of N mineralization associated with the decomposition of plant residues. *Soil Biology and Biochemistry*, 28, pp. 169–175.

Xiang, J., Liu, D., Yuan, J. et al. (2014). Effects of biochar on nitrous oxide and nitric oxide emissions from paddy field during the wheat growth season. *Journal of Cleaner Production*, 104, pp. 52–58.

18

Biochar Amendment Experiments in Thailand: Practical Examples

THAVIVONGSE SRIBURI AND SAOWANEE WIJITKOSUM

Abstract

This chapter briefly summarizes the use of biochar to increase the productivity of crop yields and to improve the soil properties in Thailand. The data presented in this chapter are based on research experiments in various types of problematic soils such as infertile sandy clay soil and clay loam. The characteristics of the biochar obtained from the controlled temperature biochar retort for slow pyrolysis are similar to those of laboratory-scale produced biochar. The retort is cost-efficient and can be built easily using locally available materials and in addition can use locally available biomass as feedstock.

The study investigated the effects on crop yield of incorporating soil with biochar in the experimental area. Both soil samples and biochar samples were collected before and after cultivation. The results showed that biochar amendment improved the soil properties in terms of organic matter, nutrients and cation exchange capacity. The yield and growth of crops increased significantly when the soil was treated with biochar. In addition, the soil properties, yield and growth of crops increased even more when the soil had been incorporated with both biochar and organic fertilizer. The combination offers a significant improvement of soil and crop yield.

18.1 Introduction

Biochar is a fine-textured and highly porous, industrially manufactured substance similar to natural charcoal. It is produced by heating biomass in a limited supply of oxygen in a process known as pyrolysis (Xu et al., 2011); see Chapter 10 for production details. It is rich in carbon, highly stable and has a large internal surface area with a huge amount of anions on its surface (Lehmann and Rondon, 2006; Sohi et al., 2010; Kookana et al., 2011). These attributes allow biochar to adsorb and retain water and other primary nutrients essential for plant growth. This makes biochar suitable for soil amelioration (Kookana et al, 2011). Indeed, biochar has been used in this way for over a thousand years (Lehmann and Rondon, 2006). Currently, there is interest in the potential of carbon sequestration by biochar to mitigate climate change and increase crop yields. However, its value in both these areas can vary widely, depending on pyrolysis conditions.

Thailand is a predominantly agricultural country and one of the major food global sources. However, aggressive agriculture has led to negative impacts such as soil degradation and other environmental consequences, such as groundwater pollution from excessive use of agrochemicals. Moreover, low soil fertility contributes to poverty and indebtedness among rural people.

Biochar created from locally available biomass has been proposed as an innovative alternative for soil amelioration and to increase crop yields over the long term. Recent research has focused on developing a simple, cost-efficient and practical production method that does not negatively impact the environment.

18.2 Biochar Retort

A 'Controlled Temperature Biochar Retort for Slow Pyrolysis' (patented) has been developed for production of standard-quality biochar that can be used to improve both crop yields and soil quality. The retort is cost-efficient and can be built easily using locally available materials. It can be operated on regional biomass resources as a feedstock.

The 200 L retort was developed from a type of pyrolysis retort that had been successful in commercial use. The design was tested in several studies (Sriburi, 2011a; Sriburi, 2011b; Sriburi, 2013) to ensure its effectiveness in producing biochar of the required quality and characteristics, as well as to ensure its suitability for agricultural use. The retort was designed to allow control over pyrolysis temperature (Figure 18.1).

The pyrolysis process heats organic materials in limited oxygen conditions inside the retort. Several experiments were conducted to determine the most effective means to control air intake to allow air circulation in the space between the biochar retort (Figure 18.2a, no. 1) and the outer furnace (Figure 18.1, no. 2). Temperature in the retort was controlled by air inflow into the kiln (Figure 18.1, no. 5), which regulates the combustion in the space between the retort and the furnace. Air intake holes were designed by trial and error with the purpose of maintaining the desired temperature that triggers the release of syngas and volatiles into the combustion chamber. The number of holes was gradually increased from zero to 12 (2.5 cm diameter per hole) in order to find the optimal airflow rate and combustion temperature.

18.2.1 The Construction of the Biochar Retort

Construction materials were inexpensive and easy to acquire locally; the total cost per kiln was approximately 50 USD. The materials are listed below and shown in Figure 18.2:

1) Standard 200 L cylindrical steel drum with lockable lid and clamp (Figure 18.2b, no.2 and 3) (outer height 85 cm; inner height 83 cm; diameter 58 cm)

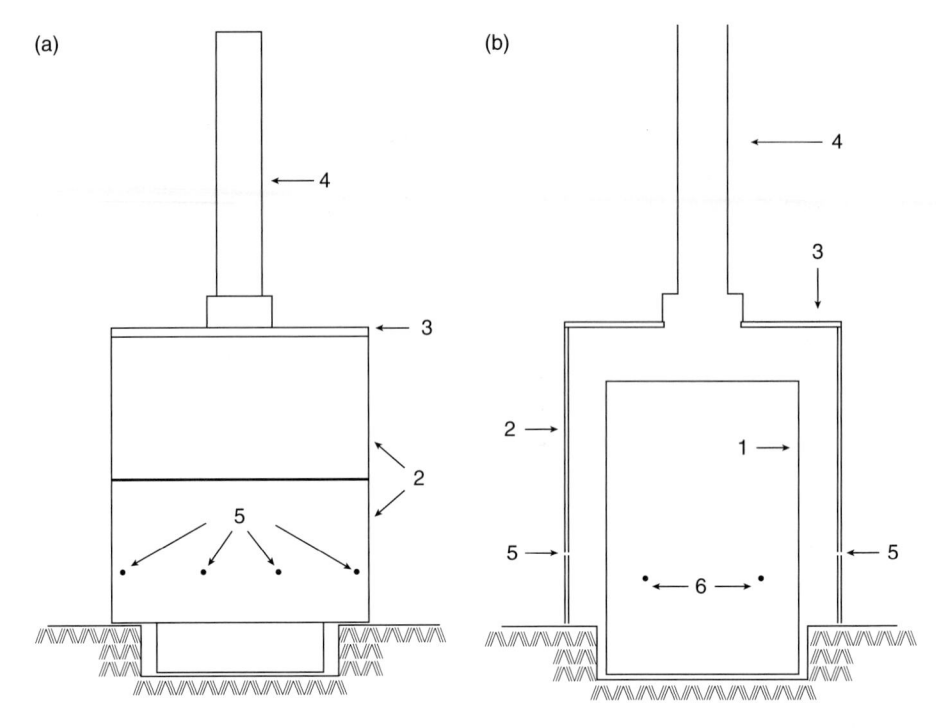

Figure 18.1. Exterior (a) and interior (b) sketch of a simple, low-cost and effective biochar kiln produced with locally available materials used for our experiments. 1 = biochar retort (200 L steel barrel); 2 = outer furnace (two concrete rings); 3 = concrete lid of outer furnace; 4 = chimney (concrete pipe); 5 = holes in the outer concrete furnace to control temperature of the pyrolysis reaction; 6 = syngas outlet holes in the biochar retort (steel barrel).

2) Two concrete pipes (Figure 18.2a, no.5) (height 40 cm; outer diameter 100 cm; inner diameter 92 cm)
3) A concrete pipe lid (Figure 18.2a, no.6) (diameter 100 cm)
4) A long concrete pipe (Figure 18.2a, no.7) (height 100 cm; diameter 17 cm, inner diameter 14 cm)
5) A hand drill and drill bits for drilling 2.5 cm wide holes in the steel tank (Figure 18.2b, no.4) and concrete pipes (Figure 18.2a, no.8).

The retort was designed for slow pyrolysis at 450–600°C according to the Food and Agriculture Organization guidelines (FAO, 2009). In order to control the temperature, temperature data were collected from inside the biochar retort and at the end of the chimney at different times throughout the heating process (Sriburi, 2011a).

During development, a number of tests were conducted to determine the optimal temperature, number of air intake and exhaust holes and the ratio of biomass to fuel required to produce high-quality biochar. In each experiment, the temperature was measured both in the retort and in the kiln. The study indicated that the outer furnace

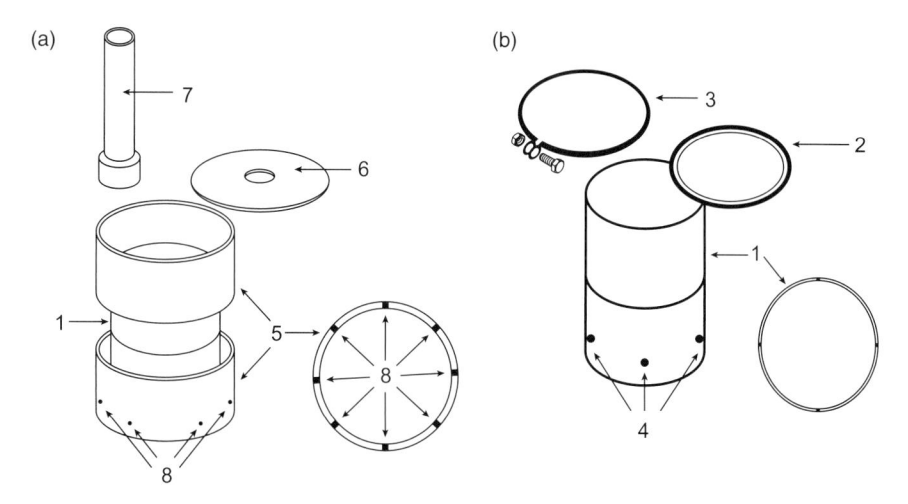

Figure 18.2. Locally available materials and construction of the experimental biochar kiln showing details of the furnace (a) and biochar retort (b). 1 = steel barrel (200 L); 2 = steel lid; 3 = barrel bind (to be closed using a nut bolt); 4 = syngas outlet holes; 5 = concrete rings; 6 = concrete lid with a hole for the chimney; 7 = chimney (concrete); 8 = air inlet holes with a diameter of 2.5 cm.

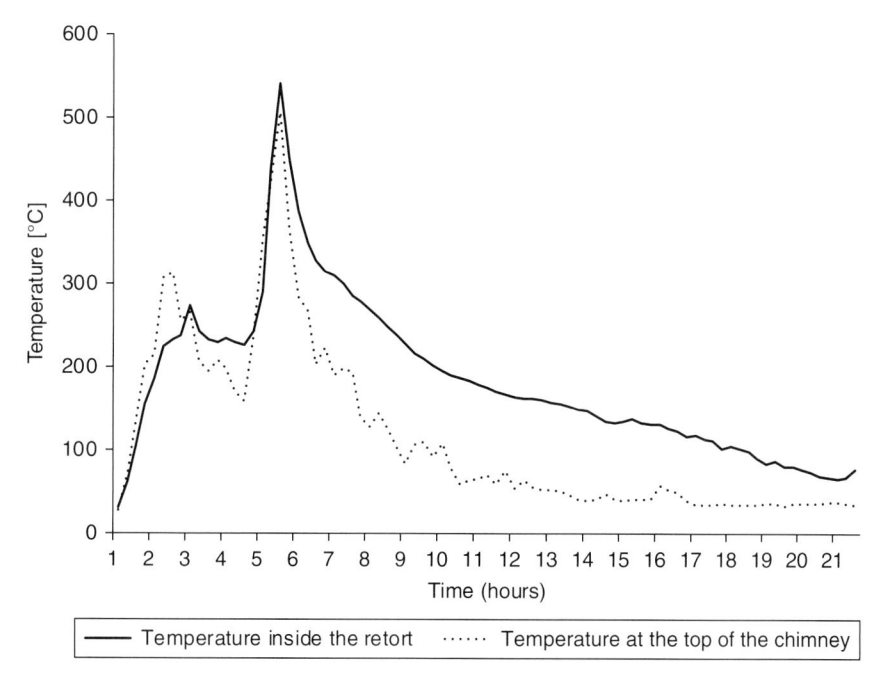

Figure 18.3. Relationship between heating temperatures and time (hours) for the biochar retort with eight air intake holes. The pyrolysis process took a total of 21 hours to complete. The temperature was gradually increased until it remained stable at 540°C during the fifth hour of the process.

with eight drilled holes and the biochar retort (steel tank) with four drilled holes with a diameter of 2.5 cm each showed a ratio of materials to turn into biochar (biomass) per fuel of 1: 0.60 (40 kg: 25 kg). The total duration of the pyrolysis process was 21 hours as shown in Figure 18.3.

The diameter of the holes must be exactly proportioned as an incorrect hole diameter could lead to excessively high temperatures, which may adversely affect the quality of the biochar produced. In addition, toxins such as dioxins can be emitted in such conditions (Garcia-Perez, 2008).

18.2.2 Biochar Production Process

The production process entails a number of critical steps that determine the quality of the biochar. These are the amount and distribution of the biomass fuel, as well as the temperature control techniques. The processes are explained below and they are shown in Figure 18.4.

1) Biomass needs to be arranged vertically in the 200 L retort. The diameter of the feedstock sticks should be approximately 2.5 cm and they should be placed at the bottom until the retort is packed. Sticks larger than 3 cm in diameter should be placed on top.
2) After closing the 200 L steel tank tightly, fuel needs to be arranged horizontally in the space between the retort and the furnace. The fuel should be small and of similar size. The fuel is ignited from the top of the furnace and the flames subsequently propagate towards the bottom. The lid of the furnace should be closed and the chimney installed in the next step.
3) Biomass in the retort will be heated at a temperature of about 450–600°C with limited oxygen.
4) After the pyrolysis process is completed, the retort is left to cool for about three hours (temperature below 100°C). Opening the retort at high temperature or immediately after pyrolysis introduces oxygen that could cause the biochar to combust.

18.3 Materials and Characteristics

18.3.1 Feedstock Materials

18.3.1.1 Crop Residues

The agricultural process typically generates large amounts of crop residue. Disposing of such wastes is counted as part of the production cost for farmers. Biochar can be produced from crop residues such as corncobs, coconut leaves, palm clusters and prunings. In this study, corncobs were used. Corn is widely grown in every region of Thailand and large volumes of corncobs are disposed of as agricultural waste. The corncobs must be sun-dried for at least one day before use in order to reduce the moisture content.

(a) (b)
(c) (d)

Figure 18.4. Biochar production process: (a) = arranging biomass vertically in the retort for producing biochar; (b) = fuelwood is being placed horizontally in the space between the retort and the furnace; (c) = biomass is being heated to a temperature of about 450–600°C in the retort, after ignition and closing of the furnace; (d) = the retort can be opened after a cooling down phase of about three hours. (A black and white version of this figure will appear in some formats. For the colour version, please refer to the plate section.)

18.3.1.2 Hardwood

Hardwood typically contains two types of cells: wood fibers and vessel elements (Henricson, 2004). The fibers are considerably shorter than softwood tracheids, with an average length of less than 1 mm. Vessel elements are found only in hardwoods, and are not found in softwoods. The vessel elements form tubes along the stems by joining end-to-end, appearing as pores in cross-sections. Hardwoods are plentiful in the western part of the country; *Streblus ilicifolius* is a commonly found hardwood that is locally used as feedstock for biochar production.

18.3.1.3 Softwood

Softwood refers to structurally non-porous wood such as conifers (Chadthasing, 2008). Softwood consists of tracheids and ray cells, and is rich in lignin that consists of 90–95% spindle-shaped cells known as tracheids. The tracheids form fibers typically $25–50 \times 10^{-6}$ m wide (Browning, 1963; Fengel and Wegener, 1984; Retulainen et al., 1998; Henricson, 2004). Softwoods used as biomass include *Albizia myriophylla* Benth, *Combretum punctatum* Blume, *Leucaena leucocephala* (Lamk.) de Wit and *Samanea saman* (Jacq.) Merr., which are commonly found in the area. Pruning is necessary for growing softwoods; the prunings can be used to produce biochar. Some softwoods are vines, locally considered as weeds.

18.3.2 Biochar characteristics

In this study, biochar produced from five different types of wood (*Streblus ilicifolius* (Vida) Corner), *Albizia myriophylla* Benth, *Combretum punctatum* Blume, *Leucaena leucocephala* (Lamk.) de Wit and *Samanea saman* (Jacq.) Merr., and *Zea mays* Linn. (corncobs) was analyzed to identify and compare characteristics (Table 18.1).

The biochar made from *Streblus ilicifolius* had the highest surface area and porosity volume. Biochar from corncobs was ranked second by surface area and porosity volume. The data also revealed that biochar from *Albizia myriophylla* Benth had the highest average pore diameter. The lowest pore diameter was found in *Streblus ilicifolius* biochar. The amount of carbon, hydrogen and nitrogen in biochar produced from the six different biomasses changed after the pyrolysis process; the results showed that *Samanea saman* and *Leucaena leucocephala* had similar amounts of carbon, which were higher than that of the other feedstock materials.

Biochar can be used either alone or in combination with biochar from several sources. In a subsequent experiment, biochar from five feedstock materials was used to produce blended biochars in order to determine the properties of using mixed materials. In some areas, it might be difficult for the locals to find a single source to produce biochar. Mixing locally available materials together may be beneficial for soil amelioration and increasing yields. Biochar used

Table 18.1. *Surface area, porosity volume, diameter and elemental analysis of biochar from different types of biomass*

Feedstock biomass	Multipoint BET (m²/g)	Total pore volume (cc/g)	Average pore diameter (Å)	%C	%H	%N
Samanea saman (Jacq.) Merr.	45.62	0.0372	32.61	84.78	2.43	0.64
Leucaena leucocephala (Lamk.) de Wit.	42.08	0.0374	35.53	84.61	2.47	1.03
Streblus ilicifolius (Vidal) Corner.	137.91	0.0977	28.34	78.88	1.38	0.78
Combretum punctatum Blume.	2.92	0.0104	143.1	74.56	3.23	1.20
Albizia myriophylla Benth.	3.67	0.0150	163.1	66.64	3.77	1.00
Zea mays Linn.	56.35	0.0405	28.72	81.35	2.42	1.00

Note: BET = Brunauer-Emmett-Teller surface area.

Table 18.2. *Characteristics and chemical properties of mixed biochar obtained from the controlled temperature biochar retort for slow pyrolysis and from laboratory-produced biochar*

Characteristics and chemical properties	Units	Controlled temperature biochar retort	Laboratory-produced biochar
pH		8.50±0.02	9.10±0.03
Electrical conductivity	dS/m	0.49±0.15	2.41±0.08
Organic matter	%w/w	20.25±0.54	35.83±2.12
Cation exchange capacity	cmol/kg	26.97±0.56	27.98±0.15
%C		70.51±1.24	68.73±1.35
%H		2.92±0.06	2.85±0.04
%N		1.21±1.24	1.25±0.05
Specific surface area	m^2/g	21.02±1.74	31.34±1.00
Total pore volume	cc/g	0.0255±0.01	0.0312±0.02
Total nitrogen	%	0.68±0.03	1.10±0.02
Available phosphorus	%	0.25±0.04	0.15±0.02
Exchangeable potassium	%	0.77±0.03	0.83±0.03

Source: Kunlayasiri (2014).

in this study was produced from the combination of five different sources, that is *Streblus ilicifolius*, *Albizia myriophylla* Benth, *Combretum punctatum* Blume, *Leucaena leucocephala* and *Samanea saman*. The mixture consisted of equal amounts of each type of biochar (by weight) prior to being used for soil amendment.

A comparison of the characteristics of the mixed biochar produced from the controlled temperature biochar retort for slow pyrolysis with biochar from laboratory-produced biochar is shown in Table 18.2. The results indicated no statistically significant differences at 95% (T-test, $p > 0.05$), indicating that biochar obtained from the controlled temperature biochar retort produced biochar similar in quality to laboratory-produced biochar.

18.4 Application of Biochar for Increasing Yield and Amelioration of Soil Properties

18.4.1 Indigenous Upland Rice (Oryza sativa L.)

Millions of farmers depend on paddy cultivation on dry hillsides, especially those in Southeast Asia, South Asia and Africa (International Rice Research Institute, 1986; Cox, 2009). Upland rice (*Oryza sativa* L.) is grown in rain-fed areas with naturally well-drained soils. Generally, the upland environment is often perceived as sloping drought-prone areas with erosion problems. The physical and chemical conditions of the soil are poor and without surface water accumulation (International Rice Research Institute, 1984; Huke, 1982).

Thailand has a large genetic diversity of indigenous upland rice, consisting of both photosensitive and photo-insensitive species. Nasarn rice is one of the indigenous upland rice cultivars. It has very strong roots and can reach an average of 130–150 cm in height. Nasarn rice, a photosensitive species, was selected for study since it is drought-tolerant with harvesting after 130–150 days of cultivation (Julsrigival and Tiyawalee, 1984; Watanesk, 2004). Because of its photosensitiveness, it only flowers when the daylight periods are shorter than night periods. In Thailand, photosensitive rice flowers in months with day length periods of approximately 11 hours and 40 minutes or shorter. The study on increasing the yield of indigenous upland rice by using biochar-incorporated soils was carried out with the mixed biochar at an experimental plot in the Pa-deng Biochar Research Center, Petchaburi Province, in the southern part of Thailand. Because the soil characteristic in the experiment site was sandy-clay with a neutral pH of 6–7, any type of wood biochar could be used. The study investigated the effect on crop yields in the area by incorporating biochar in the soil. The randomized complete block design (RCBD) (Major, 2009) was employed, each with three replications. ANOVA (analysis of variance) and Duncan's New Multiple Range Test (DMRT) were used for statistical analysis. The study was comprised of four treatments: T1 was soil with organic fertilizer (10 t/ha), T2 was soil with biochar (10 t/ha), T3 was soil with organic fertilizer (5 t/ha) and biochar (5 t/ha), and T0 was an untreated soil (control). Moreover, the organic fertilizer was applied twice; two weeks before planting at 6.25 t/ha and before the booting stage at 3.75 t/ha. The organic fertilizer contained N (0.89%), P (0.40%), K (0.30%), OM (8.61%) and organic carbon (4.99%); EC was 0.35 dS/m and pH 7.3.

Yield components were evaluated. Data were collected during the six growth stages of the rice from tillering through to harvest. The analysis of biochar and its impact on crop growth was done by studying the amount of panicles of paddy per area, number of seeds per panicle, percentage of good seeds, weight of 1,000 seeds and number of seeds per area (Table 18.3).

The results above showed that the biochar incorporated in plot (T2) could statistically significantly increase the amount of yield and plant growth higher than the treatment without biochar incorporation (T0). The treatments incorporated with biochar yielded more than the treatment incorporated solely with organic fertilizer (T1). However, the results did not show statistically significant differences between treatments. The T3 treatment gave the highest yields. For this treatment, the improvements were shown in various factors: number of seeds per panicle, 1,000-seed weight and percentage of good seeds. Regarding these three factors, the T3 treatment showed a statistically significant difference when compared to the other treatments. However, the panicles per area showed no statistical difference in any of the treatments. Evaluating the effect of biochar on rice yield by calculating the number of seeds per area revealed that Nasarn rice grown in the T3 treatment produced the highest yields when compared to the other treatments at a 95% confidence level.

Table 18.3. *Yields of Nasarn rice*

Treatments	Composition of Nasarn rice yield				
	Panicles per area	Seeds per panicle	1,000-seed weight (g)	Percentage of good seeds (%)	Seeds per area (kg/ha)
T0	25.33[a]	131.50[c]	24.25[b]	88.00[b]	3,661.25
T1	27.67[a]	138.80[b]	25.51[ab]	89.13[bc]	4,012.88
T2	27.33[a]	140.77[b]	25.81[ab]	90.00[b]	4,379.56
T3	28.67[a]	221.80[a]	28.48[a]	94.60[a]	4,556.81

Note: T0 = control, T1 = soil with organic fertilizer (10 t/ha), T2 = soil with biochar (10 t/ha), T3 = soil with organic fertilizer (5 t/ha) and biochar (5 t/ha).

The increases in rice yields are a direct effect of the primary nutrients (N, P and K), which are major chemical components in organic fertilizer and biochar. The biochar's surface area contained a large amount of anions which allowed it to absorb more primary nutrients from organic fertilizers. Moreover, biochar is highly porous, which allows it to retain nutrients within it for long-term use. At the early growth stage, rice captures nitrogen for development in the vegetative phase. Later, the absorbed nitrogen will be used for grain formation (Osotsapar, 2000; Thasanasongchan, 1984; De Datta, 1981; Yoshida, 1981). Phosphorus could help increase reproductive growth: faster blooming flowers and seed set of rice. Phosphorus also captures potassium, balances nitrogen in the roots and improves the quality of seeds (Thasanasongchan, 1984; Yoshida, 1981). Potassium has a great effect on the amount of flowers and the amount of good seeds per ear of paddy. The rice will use potassium to create flowers, strengthen the pollen (Von Uexkull, 1976) and to create carbohydrates that will ensure complete grains that are larger in size (Thasanasongchan, 1984; De Datta, 1981).

18.4.1.1 The Impact of Biochar as a Soil Amendment on the Soil Properties of Rice Plantation

A pilot project on soil amelioration of infertile sandy clay soil was conducted using upland rice grown over a 120-day period. The physical and chemical properties of biochar were measured both before and after the experiments by analyzing key parameters such as pH, organic matter (OM), electrical conductivity (EC), total C, total N, available K, exchangeable P (Soil Survey Staff, 1996), and cation exchange capacity (CEC) (Bremner and Mulvaney, 1982). Soil samples were collected before and after the experiment at a depth of 15 cm below the soil surface. The samples were dried, crushed and sieved through a 2 mm mesh and the parameters of interest were analyzed. The soil sample was analyzed for CEC (Soil Survey Staff, 1996), pH, EC, available K and exchangeable P (Gavlak et al., 1994). OM was tested with Walkley and Black's method, total C was tested with a

Table 18.4. The soil properties in different treatments of upland rice planting

Treatment	pH pre	pH post	EC (dS/m) pre	EC (dS/m) post	OM (%) pre	OM (%) post	CEC (cmol/kg) pre	CEC (cmol/kg) post
T0	6.65±0.15c	ⁿ6.67±0.12b	0.19±0.02c	ⁿ0.15±0.04b	0.69±0.15c	ᶞ0.60±0.06c	11.61±0.67c	ⁿ9.19±0.15b
T1	6.67±0.14bc	ⁿ6.60±0.17b	0.19±0.34bc	ᶜᵉⁿ 0.27±0.08ab	0.97±0.02b	ⁿ 1.55±0.13b	12.62±0.08b	ᶜᵉ17.25±0.65a
T2	6.73±0.25ab	ⁿ 7.47±0.14a	0.20±0.21ab	ᶜᵉ 0.31±0.08a	1.02±0.03b	ᶜᵉⁿ1.74±0.15ab	12.94±0.21b	ᶜᵉ17.70±0.36a
T3	6.77±0.72a	ⁿ 7.52±0.15a	0.22±0.10a	ᶜᵉ 0.33±0.09a	1.08±0.03a	ᶜᵉ 1.81±0.13a	13.47±0.35a	ᶜᵉ17.88±0.72a

Treatment	Total C (%) pre	Total C (%) post	Total N (%) pre	Total N (%) post	Available P (mg/kg) pre	Available P (mg/kg) post	Exchangeable K (mg/kg) pre	Exchangeable K (mg/kg) post
T0	0.06±0.002c	ᶲ0.01±0.11d	0.11±0.05c	ⁿ 0.04±0.01b	10.86±0.08d	ᶲ5.34±0.22d	87.12±0.16d	ᶝ78.28±0.15c
T1	0.67±0.05c	ᶞ0.18±0.02c	0.13±0.27bc	ᶜᵉⁿ0.16±0.05ab	11.40±0.09c	ᶝ14.87±0.32c	90.53±0.41c	ⁿ98.07±5.52b
T2	0.14±0.16b	ⁿ 0.32±0.01b	0.16±0.02b	ᶜᵉ 0.23±0.10a	12.89±0.12b	ⁿ18.17±0.83b	95.32±0.18b	ᶜᵉ103.19±1.92a
T3	0.19±0.10a	ᶜᵉ0.38±0.01a	0.19±0.01a	ᶜᵉ0.25±0.11a	15.58±0.48a	ᶜᵉ21.26±0.07a	100.89±0.37a	ᶜᵉ107.25±0.15a

CHNOS analyzer, while total N was assessed with Kjeldahl's method (Soil Survey Staff, 1996). The ANOVA and DMRT were used for statistical analysis. The effects of biochar on the quality of soil used to cultivate upland rice were evaluated by comparing the pre- and post-experimental soil properties. The findings are summarized in Table 18.4.

The results of this study show us that after rice plantation, the control plot (T0) showed decreased levels of EC, OM, CEC, total C and N, available P and exchangeable K. On the other hand, the other treatments showed various increases in soil parameters. The soil in T2 is fertile and the total N, exchangeable K, pH, EC and CEC have increased statistically significantly.

The comparative analysis parameter results between the four treatments after the rice harvest indicate statistically significant differences in pH values with each treatment. The T3 treatment has the highest value in all parameter results and the T0 treatment has the lowest value in all parameter results. For the T2 and the T3 treatments, there are no significant differences with the exception of EC, total N and exchangeable K. The CEC for T1, T2 and T3 treatment is not statistically significantly different.

Analyzing the biochar properties found that after the rice plantation, the pH, EC, OM, CEC, total C and total N had decreased in both the T2 and T3 treatments. On the other hand, the values of available P and exchangeable K had increased both in T2 and T3. All the parameters of the biochar in T3 are statistically significantly different from T2 with regards to pH, OM, total C, available P and exchangeable K. (Table 18.5).

The results indicated that biochar increases pH, EC, CEC, OM, total C, total N, available P and exchangeable K of the soil. The increase of EC is due to the capacity of the biochar to retain nutrients in soil directly through the negative charge that develops on its surface, and the biochar's anion can buffer acidity in the soil, as does organic matter in general (Novak et al., 2009). Biochar can increase soil pH mainly due to the ash content. Acidic soil has cationic H$^+$ dissolved in the soil solution. When adding biochar to acidic soil, the biochar causes an ion exchange that reduces the pH of the soil.

Table 18.5. *The biochar properties in different treatments of upland rice planting*

Parameters	Experiments		
	pre	post	
		T2	T3
pH	8.50±0.02	6.73±0.25[ab]	6.77±0.14[bc]
EC (dS/m)	0.49±0.15	0.31±0.08[a]	0.33±0.09[a]
OM (%)	20.25±0.54	1.74±0.15[ab]	1.81±0.13[a]
CEC (cmol/kg)	26.97±0.56	17.70±0.36[a]	17.88±0.72[a]
Total C (%)	70.51±1.24	0.32 ± 0.01[b]	0.38 ± 0.01[a]
Total N (%)	0.68±0.03	0.23 ± 0.10[a]	0.25 ± 0.11[a]
Available P (mg/kg)	0.25±0.04	18.17±0.83[b]	21.26±0.07[a]
Exchangeable K (mg/kg)	0.77±0.03	103.19±1.92[ab]	107.25±0.15[a]

Adding biochar to rice plantations can cause an increase of OM in the soil. As a result, the soil becomes more porous and soil fertility increases (Sohi et al., 2010; Zheng et al., 2010). In addition, the low bulk density of soil increases soil porosity and soil aeration, and may have a positive effect on microbial respiration (Steiner, 2009; Laird et al., 2010). The highly porous structure of biochar helps retain water and primary nutrients, contributing to increased soil fertility, lower soil bulk density (Steiner, 2009) and increased soil aeration (Islam et al., 1999). The large internal surface area of biochar also provides a conductive habitat for soil microorganisms, and allows the slow release of primary nutrients (Wijitkosum and Jiwnok, 2014). The surface of the biochar particles oxidizes and interacts with soil constituents, resulting in functional groups and a higher surface negative charge (Liang et al., 2006), which ultimately leads to an increase in CEC.

Adding biochar with a high CEC content to the soil increases its capacity to exchange cations, which improves nutrient retention (Masulili et al., 2010; Zheng et al., 2010; Steiner, 2009). Soil nutrient levels would be expected to decline over the growing season due to crop uptake. In the T3 treatment, the organic fertilizer, containing the primary macronutrients (N, P, K), was mixed with the biochar and the nutrients from the fertilizer were then adsorbed and retained in the biochar (Sukartono et al., 2011). The comparison between the treatments revealed that the T3 treatment had a higher amount of N, P and K than the T2 treatment. Moreover, the organic fertilizer contained nutrients in the form of ions. Once absorbed, the biochar's ability to absorb cations would increase due to negative charges arising from the carboxyl groups of organic matter (Masulili et al., 2010). Thus, incorporation of biochar and organic fertilizer shows the highest increase in CEC.

The results indicated that adding biochar could ameliorate soil chemical properties and increase yield. Adding biochar with organic fertilizer could improve the chemical properties of the soil and increase the rice yield at a higher rate than only adding biochar. This is a result of the additional input of primary nutrients and organic matter contained in the organic fertilizer (Novak et al., 2009; Limpothong et al., 2011; Petter et al., 2012; Sriburi, 2013; Yooyen, 2014; Wijitkosum and Kallayasiri, 2015). The results were statistically significant at a 95% confidence level. Therefore, biochar incorporation in the soil is recommended as a cost-effective way of improving soil quality and boosting crop yield, whilst also utilizing crop residues that are otherwise treated as waste material.

18.4.2 Soybean (Glycine max (L.) Merrill)

The effect of biochar on crop yield was also tested using soybean, an important and widely grown food crop in Thailand. The experiment was conducted in four 2 × 5 m plots, each treatment with three replicates. Levees were constructed to separate the plots, preventing horizontal movement of water and nutrients between adjacent plots. The four treatments were: i) untreated soil (BC0), ii) soil incorporated with biochar 1 kg/m^2 (BC1), iii) soil incorporated with biochar 2 kg/m^2 (BC2) and iv) soil incorporated with biochar 3 kg/m^2 (BC3).

The biochar used in this study was prepared by pyrolysis of *Blachia siamensis* Gagnep at 500–600°C in a low-cost, locally designed retort (Figure 18.1) for approximately 24 hours. Following pyrolysis, the biochar was ground and sieved through a 2 mm sieve. Chemically, the biochar comprised 78.88% C, 1.38% H and 0.78% N. Physically, the inner matrix surface area was 137.91 m^2/g, the average pore diameter was 28.34 Å and the water adsorption capacity was 2.51 times higher than its weight (Sriburi, 2011b; Yooyen, 2014). Mixing biochar with the soil was done by spreading the biochar on the soil surface and manually incorporating it using a rake. Cow manure containing N (0.55%), P (0.4%) and K (0.35%) was added to all treatments at 1 kg per m^2 and incorporated using a rake.

The soybean seeds were inoculated with rhizobia at 60 g of soybean per 1 g of rhizobia. The seeds were sown immediately after inoculation. Six seeds were planted into each hole, spaced at a density of 30 × 30 cm. The soil was watered before sowing and once a day after sowing using a sprinkler set. For each plot, single soybean plants were selected randomly for sampling during five growth stages from beginning pods to harvesting. Data collected included the number of pods per plant, the weight of fresh pods per plant, the numbers of seeds per plot, the size of seeds and the length of the roots. ANOVA and DMRT were used for statistical analysis (Table 18.6).

The results revealed that biomass and yields of the soybean were higher in biochar incorporated treatments. The results showed statistically significant differences (p < 0.05) for all factors with the exception of the number of seeds per pod and the seed size. The results indicate that BC3 treatment significantly increased biomass in the different parts of the soybean plant, including pod and seed biomass, when compared with the control. The results corresponded with other previous research indicating that adding biochar in agricultural areas could increase the biomass of pods and the biomass of seeds in comparison with the areas without biochar incorporation (Asai et al., 2009; Chan et al., 2007; Steiner et al., 2007; Suppadit et al., 2012; Vaccari et al., 2011).

Adding biochar into the soybean plots (BC2 and BC3) significantly increased the number of pods per plant and the weight of the seeds. The application of biochar to the soybean plot

Table 18.6. *Yield components of soybean*

Treatments	Number of pods per plant Pod/plant	Number of seeds per pod Seed/pod	Seed size Grams/100 seeds	Weight of seeds per plant Grams/plant	Weight of seeds per area kg/ha
BC0	33.3±2.9[a]	2.1±0.2[a]	21.9±1.0[a]	13.6±1.8[a]	3,922.50±82.9[a]
BC1	35.9±0.8[ab]	2.0±0.6[a]	21.3±1.4[a]	14.2±1.0[a]	4,078.75±44.2[a]
BC2	39.8±1.5[bc]	2.1±0.8[a]	22.1±0.9[a]	17.4±1.0[b]	5,021.88±47.2[b]
BC3	44.1±3.9[c]	2.1±0.3[a]	20.0±0.1[a]	18.6±1.8[b]	5,367.5082.1[b]

Note: BC0 = control, BC1 = soil with biochar (1 kg/m^2), BC2 = soil with biochar (2 kg/m^2), BC3 = soil with biochar (3 kg/m^2).

had a statistically significant effect on both the stem height and number of pods in all cropping cycles. The study clearly demonstrates that applying a sufficient quantity of biochar to soil has a statistically significant effect on the plants' height and number of pods. The results of this study are consistent with other studies regarding the impact of applying biochar to plant crops; biochar treatments were found to significantly or potentially increase the plant growth in all the cropping cycles in comparison to no biochar treatment crops (Suppadit et al., 2012; Carter et al., 2013). The influence of biochar in increasing plant height and the number of pods may be influenced by the biochars' characteristics that lead to increasing soybean growth. A pivotal property of biochar is adsorption of nitrate (NO_3^-) and ammonia (NH_4^+), which can help to increase soil nitrogen stocks by increasing nitrogen retention in soil (Knowles et al., 2011; Dempster et al., 2012). Other important properties of biochar include a decrease of nitrogen losses from nitrous oxide emission during denitrification, which occurs in the absence of oxygen or in flooded soil. The level of nitrogen in the soil plays an important role in plant growth due to its essential nutrients, which are a major factor in leaf and stem growth. The soybean plots with biochar application (BC2 and BC3) showed a significant increase in plant height and node number.

Furthermore, the application of biochar leads to a statistically significant increase in the root length of soybeans when compared to the control plot. At the beginning of the pod stage, the longest root length was found in the BC2 plot, which is significantly different from the other biochar treatments. At the maturity stage (at harvest), the longest root length was found in the BC3 plot, which is significantly different from other biochar treatments (Table 18.7). Soybeans have a tap root system and they generally produce nodules at the base of the primary and lateral roots, normally formed close to where the rhizobia (*Rhizobium japonicum*) inhabit. The bacteria capture nitrogen from the air. Nitrogen is crucial for plant growth and increasing yields. The development of the roots and their growth process leads to an increase in bacterial activity. This will influence the roots' capacity to sequester nutrients. Natural nitrification can occur by symbiotic bacteria or free-living organisms, which are considered as major sources of nitrogen in plant tissue.

Table 18.7. *The root length of soybeans in different treatments*

Treatments	Length of soybean roots (cm)				
	Beginning pod	Beginning seed	Full seed	Beginning maturity	Harvesting
BC0	17.3±0.7[a]	14.5±1.4[a]	22.4±1.9[a]	18.4±5.9[a]	14.1±1.9[a]
BC1	21.7±1.0[b]	15.1±2.3[a]	21.5±1.5[a]	17.8±1.7[a]	17.4±1.5[a]
BC2	25.0±2.2[c]	19.5±1.4[b]	26.2±2.2[b]	20.9±3.5[a]	20.0±3.4[b]
BC3	23.6±1.1[b]	20.7±1.9[b]	27.3±1.4[b]	23.0±1.3[a]	22.1±1.7[c]

Note: BC0 = control, BC1 = soil with biochar (1 kg/m^2), BC2 = soil with biochar (2 kg/m^2), BC3 = soil with biochar (3 kg/m^2).

Furthermore, the symbiotic relationship of mycorrhizae and plant roots plays an important role in the absorption of phosphate (Russell, 1977; Norman, 1961). Root growth is also influenced by the surrounding environment. A well-developed root system will support stem growth and plant biomass production, as well as plant adaptation to environmental stress (Turner et al., 1985; Gollan et al., 1985). In addition, application of biochar may help increase the amount of primary nutrients and water absorption, reducing soil bulk density, promoting the growth of soil microbes, and adjusting soil conditions to be more suitable for plant root growth (Cavigelli and Robertson, 2001; Chan et al., 2007; Knowles et al., 2011; Lehmann and Rondon, 2006; Liang et al., 2006; Warnock et al., 2007; Yamato et al., 2006; Limpothong et al., 2011; Sriburi, 2011a).

During the vegetative phase of soybean growth, leaves synthesize carbohydrates, which are stored in leaves, roots and the stem base in order to be later translocated to pods and seeds. Biomass is stored as protein, oil and starch. The BC2 and BC3 treatments had higher amounts of foliage than the control plot. The results were statistically significant ($p < 0.05$). The number of stem nodes correlates with the stem height; soybeans in the BC3 treatment were taller and had more nodes than in the control plot. The number of nodes also reflected the number of pods, and the number of pods was linked to seed numbers. Soybean samples taken from the BC2 and BC3 treatments, which had a higher number of pods, also had a significantly higher number of pods per plant and seed weight than the control plot.

18.4.2.1 The Impact of Biochar as a Soil Amendment on the Soil Properties of Soybean

The soil in this research is clay loam with a low fertility and slightly alkaline conditions. These are per se not suitable for planting soybeans. The objective of this research experiment is to improve the quality of the soil by adding biochar. In this study, three soil samples were collected from each plot before and after the experiment. The results of the analysis of soil properties are shown in Table 18.8.

The results indicate that after planting soybeans, the control plot (BC0) had the lowest amounts of total N, OM, available P and exchangeable K. The pH of the soil did not change significantly. The amount of OM has increased tendentially with all treatments, but it is not statistically significant.

CEC in all treatments showed no statistically significant difference, which indicated that adding biochar cannot increase it. The clay loam has similar CEC ability to that of biochar, so adding biochar does not change the ability of the soil to exchange nutrient elements. However, adding biochar can significantly increase CEC when the biochar is added to a sandy soil or any soil with a low cation exchange capacity (Masulili et al., 2010; Shenbagavalli and Mahimairaja, 2012).

The quantity of nitrogen in all treatments shows a statistically significant increase when compared to the soil before planting; however the BC3 treatment demonstrated the largest effect among the results of total N when compared to the other treatments. We conclude that adding biochar can increase nitrogen stocks in soil. There are two reasons behind this

Table 18.8. *The soil properties in different treatments of soybean planting*

Parameters	Pre-experiment	Treatments Post-experiment			
		BC0	BC1	BC2	BC3
pH	7.8±0.10	7.8±0.06[a]	7.8±0.06[a]	7.8±0.06[a]	7.8±0.06[a]
Total N (mg/kg)	500±0.0	1026.07±58.65[a]	1196.20±73.47[ab]	1190.87±94.53[ab]	1386.80±167.28[b]
%OM	1.14±0.12	1.16±0.06[a]	1.24±0.10[a]	1.19±0.09[a]	1.18±0.12[a]
Available P (mg/kg)	27.0±8.19	30.67±9.02[a]	34.00±6.80[a]	47.00±9.54[a]	43.33±10.07[a]
Exchangeable K (mg/kg)	167.0±17.60	145.00±20.00[a]	190.00±18.03[b]	171.67±15.28[ab]	180.00±10.00[b]
CEC (cmol/kg)	20.12±1.10	18.19±1.01[a]	17.90±0.93[a]	19.84±0.52[ab]	19.44±0.64[ab]

Note: BC0 = control, BC1 = soil with biochar (1 kg/m^2), BC2= soil with biochar (2 kg/m^2), BC3= soil with biochar (3 kg/m^2).

increase: firstly, the biochar's properties enable it to increase total N in the soil by allowing it to absorb nitrate (NO_3^-) and ammonium (NH_4^+) and prevent the soil from becoming eroded by water. This property retains nitrate and ammonium within the soil (Dempster et al., 2012; Knowles et al., 2011). Secondly, the biochar's high internal porosity increases the flow of oxygen in the soil and prevents it from losing nitrogen due to emission of nitrous oxide from the denitrification process as mentioned above. The denitrification process could incur once the soil lacks oxygen or becomes flooded (Yanai et al., 2007).

Soils post-cultivation of soybeans contain a higher amount of available phosphorus and exchangeable potassium than the pre-cultivation soils. Thus, the two parameters do not show a statistically significant difference in any of the treatments. However, the amounts of available phosphorus and exchangeable potassium found in biochar-incorporated treatments were higher than in the non-biochar treatment. The research results showed that adding biochar could increase the amount of available phosphorus and exchangeable potassium in the soil. The increase of exchangeable potassium came from wood ashes that were released from the biochar (Major et al., 2010). The potassium was retained by clay and accumulated in the soil. Furthermore, biochar can also retain the available phosphorus in the soil and prevent it from being leached. The chemical bonds between calcium and phosphorus may encourage the capture of phosphorus and allow the soil to retain a higher amount of phosphorus (Novak et al., 2009; Xu et al., 2011).

The soybean study results showed that adding biochar into the soil has the potential to increase soil fertility. The Land Development Department has assessed the soil fertility according to the Soil Classification System and the results showed that post-cultivation soils from the two treatments (BC2 and BC3) had moderate fertility levels (Sagwansupyakorn, 2009).

Adding biochar to the clay loam reduced soil density and allowed for better root development, which resulted in longer soybean roots. It can be concluded that adding biochar into soybean agricultural areas could increase soil fertility resulting in taller crops, longer roots and more soybean nodes.

18.5 Summary

This study used biochar to ameliorate the soil and increase the yield of rice and soybeans. The study was run as a pilot project in a carefully selected area in the southern part of Thailand. The pyrolysis reactors used were made using locally available materials, with the intent that local farmers could easily replicate the experiment at low costs. The biochar obtained from the homemade retort was not different in terms of quality from those produced in a laboratory. Several biomass types were converted into biochar and the biochar's characteristics were largely based on the type of biomass materials used. In this study, the biochar was obtained from locally available branches and agricultural waste. This will allow an increase in the recycling capacity of farmers, while at the same time maximizing the use of their agricultural waste.

It is worth mentioning that the relation between the biochar and the soil needs to be taken into account when ameliorating the soil. Otherwise, the amelioration could have adverse effects on yield. The correct ratios depend on the characteristics and properties of the soil.

The study showed that biochar can be used for soil amelioration and it significantly increases rice and soybean yields. Mixing biochar into the soil with fertilizers allows more fertile soils and higher yields both in terms of quality and quantity. The improvements lead to an overall increase of the value of the products on the market. Moreover, lower production costs from using less agrofertilizers will also improve the quality of life of the farmers and promote environmentally safe farming.

Acknowledgements

Research funding was provided by the Higher Education Research Promotion and National Research University Project of Thailand. Partial support by the Integrated Innovation Academic Center (IIAC), Chulalongkorn University Centenary Academic Development Project and the Ratchadaphiseksomphot Endowment Fund of Chulalongkorn University.

References

Asai, H., Samson, B. K., Stephan, H. M., Songyikhangsuthor, K., Homma, K., Kiyono, Y., Inoue, Y., Shiraiwa, T. and Horie, T. (2009). Biochar amendment techniques for upland rice production in Northern Laos 1. Soil physical properties, leaf SPAD and grain yield. *Field Crops Research*, 111, pp. 81–84.

Bremner, J. M. and Mulvaney, C. S. (1982). Nitrogen – total. In: Page, L. A. (ed.) *Methods of Soil Analysis. Part 2. Chemical and Microbiological Properties*. Madison, WI: American Society of Agronomy, pp. 595–624.

Browning, B. L. (1963). *The Chemistry of Wood*. New York: Interscience Publishers, pp. 18–55.

Carter, S., Shackley, S., Sohi, S., Suy, T. B. and Haefele, S. (2013). The impact of biochar application on soil properties and plant growth of pot grown lettuce (*Lactuca sativa*) and cabbage (*Brassica chinensis*). *Agronomy*, 3, pp. 404–418.

Cavigelli, M. A. and Robertson, G. P. (2001). Role of denitrifier diversity in rates of nitrous oxide consumption in a terrestrial ecosystem. *Soil Biology and Biochemistry*, 33, pp. 297–310.

Chadthasing, B. (2008). *Hard wood of Thailand.* [online] Available at: www.baannatura.com/th/mat/content/detail/136.html [Accessed 10 July 2014]

Chan, K. Y., Van Zwieten, L., Meszaros, I., Downie, A. and Joseph, S. (2007). Agronomic values of greenwaste biochar as a soil amendment. *Australian Journal of Soil Research*, 45, pp. 629–634.

Cox, T. P. (2009). *Perennial upland rice takes root.* [online] Available at: www.new-ag.info/en/developments/devItem.php?a=798 [Accessed 10 July 2014]

DeDatta, S. K. (1981). *Principles and Practices of Rice Production.* New York: John Wiley.

Dempster, D. N., Gleeson, D. B., Solaiman, Z. M., Jones, D. L. and Murphy, D. V. (2012). Biochar addition to soil changed microbial biomass carbon and net inorganic nitrogen mineralized. In: *World Congress of Soil Science, Soil Solution for a Changing World. Brisbane, Australia, 1–6 August 2010.* Brisbane: International Union of Soil Sciences.

FAO (2009). *The research progress of biomass pyrolysis processes.* [online] Available at: www.fao.org/docrep/t4470e/t4470e0a.htm [Accessed 10 July 2014]

Fengel, D. and Wegener, G. (1984). *Wood: Chemistry; Ultrastructure; Reactions.* Berlin, New York: Walter de Gruyter.

Garcia-Perez, M. (2008). *The Formation of Polyaromatic Hydrocarbons and Dioxins during Pyrolysis: A Review of the Literature with Descriptions of Biomass Composition, Fast Pyrolysis Technologies and Thermochemical Reactions.* Washington, DC: Washington State University, p. 63.

Gavlak, R. G., Horneck, D. A. and Miller, R. O. (1994). *Plant, Soil, and Water Reference Methods for the Western Region.* 3rd Edition. Western Regional Extension Publication, 125. Fairbanks: University of Alaska Cooperation Extension Service.

Gollan, T., Turner, N. C. and Schulze, E. D. (1985). The responses of stomata and leaf gas exchange to vapour pressure deficits and soil water content. III. In the Scierophyllous species Nerium Oleander. *Oecologia* (Berlin), 65, pp. 356–362.

Henricson, K. (2004). Wood structure and fibres. In: *An Introduction to Chemical Pulping Technology.* Educational course material, Lappeenranta University of Technology, August 2014, pp. 1–17.

Huke, R. E. (1982). *Rice Area by Type of Culture: South, Southeast and East Asia.* Philippines, Los Baños: International Rice Research Institute.

International Rice Research Institute (1984). Upland rice in Asia. An overview of upland rice research. *Bouake, Ivory Coast, Upland Rice Workshop.* Los Baños, Philippines: International Rice Research Institute, pp. 45–68.

International Rice Research Institute (1986). *Upland Rice: A Global Perspective.* Los Baños, Philippines: International Rice Research Institute.

Islam, M. R., Islam, M. S., Jahiruddin, M. and Hoque, M. S. (1999). Effects of sulphur, zinc and boron on yield, yield components and nutrient uptake of wheat. *Pakistan Journal of Scientific and Industrial Research*, 42, pp. 137–140.

Julsrigival, S. and Tiyawalee, D. (1984). Varietal improvement program for upland rice at Chiang Mai University. In: *Proceedings, The Highland Rice Development Workshop.* Northern Region Agricultural Development Centre, Chiang Mai, Thailand, 13–14 September 1984.

Kunlayasiri, W. (2014). Effect of biochar on rice product and sandy clay quality. Case study of Pa-deng sub-district, Kaeng Krachan District, Petchaburi Province. Thesis, MSc Interdisciplinary Program, Graduate School, Chulalongkorn University.

Knowles, O. A., Robinson, B. H., Contangelo, A. and Clucas, L. (2011). Biochar for the mitigation of nitrate leaching from soil amended with biosolids. *Science of the Total Environment*, 409, pp. 3206–3210.

Kookana, R. S., Sarmah, A. K., Van Zwieten, L., Krull, E. and Singh, B. (2011). Biochar application to soil: agronomic and environmental benefits and unintended consequences. *Advances in Agronomy*, 112, pp. 103–143.

Laird, D., Fleming, P., Davis, D. D., Horton. R., Wang, B. and Karlen, D. L. (2010). Impact of biochar amendments on the quality of a typical Midwestern agricultural soil. *Geoderma*, 158, pp. 443–449.

Lehmann, J. and Rondon, M. (2006). Bio-char soil management on highly weathered soils in the humid tropics. In: Uphoff, N. (ed.) *Biological Approaches to Sustainable Soil Systems*. Boca Raton, FL: CRC Press, pp. 517–530.

Liang, B., Lehmann, J., Solomon, D., Kinyangi, J., Grossman, J., O'Neill, B., Skjemstad, J. O., Thies, J., Luizão, F. J., Petersen, J. and Neves, E. G. (2006). Black carbon increases cation exchange capacity in soils. *Soil Science Society of America Journal*, 70, pp. 1719–1730.

Limpothong, W., Susing, S. and Disathaporn, C. (2011). *Study Type and Rate of Biochar Incorporated with Chemical Fertilizer to Increase of Rice Pathumthani 1 in Sandy Soil*. Nongkhai: Nongkhai Local Land Development Station, Land and Development Regional Office.

Major, J. (2009). A Guide to Conducting Biochar Trials. International Biochar Initiative (IBI), version 1.3.

Major, J., Rondon, M., Molina, D., Riha, S. J. and Lehmann, J. (2010). Maize yield and nutrition 4 years after biochar application to a Colombian savanna oxisol. *Plant and Soil*, 333, pp. 117–128.

Masulili, A., Utomo, W. H. and Syechfani, M. (2010). Rice husk biochar for rice based cropping system in acid soil: the characteristics of rice husk biochar and its influence on the properties of acid sulfate soils and rice growth in West Kalimantan, Indonesia. *Journal of Agricultural Science*, 2, pp. 39–47.

Norman, A. G. (1961). Microbial products affecting root development. In: *Transactions of the 7th International Congress of the Soil Science Society*, 2, pp. 531–536.

Novak, J. M., Lima, I. M., Xing, B., Gaskin, J. W., Steiner, C., Das, K. C., Ahmedna, M., Rehrah, D., Watts, D. W., Busscher, W. J. and Schomberg, H. (2009). Characterization of designer biochar produced at different temperatures and their effects on a loamy sand. *Annals of Environmental Science*, 3, pp. 195–206.

Osotsapar, Y. (2000). *Mineral Plant Nutrition*. Bangkok: Department of Soil Science, Faculty of Agriculture, Kasetsart University.

Petter, F. A., Beata E. M., Soler Da Silva, M. A. et al. (2012). Soil fertility and upland rice yield after biochar application in the Cerrado. *Pesquisa Agropecuária Brasileira*, 47, pp. 699–706.

Retulainen, E. Niskanen, K. and Nilsen, N. (1998). Fibres and bonds. In: Niskanen, K. (ed.) *Paper Physics*. Helsinki: Fapet Oy, pp. 54–87.

Russell, R. S. (1977). *Plant Root Systems: Their Function and Interaction with Soil*. London: McGraw-Hill Book Company.

Sagwansupyakorn, C. (2009). *The Utilization of Soil Analysis Data for Soil Improvement. In a presentation documents course of Soil and Water Conservation in the Development*

Zone Land. [online]. Available at: http://e-library.ldd.go.th/Web_KM/Data/re_7.pdf [Accessed 14 April 2012]

Shenbagavalli, S. and Mahimairaja, S. (2012). Charaterization and effect of biochar on nitrogen and carbon dynamics in soil. *International Journal of Advanced Biological Research*, 2, pp. 249–255.

Soil Survey Staff (1996). *Soil Survey Laboratory Method Manual.*Washington D.C.: United States Department of Agriculture, Natural Resources Conservation Service.

Sohi, S., Krull, E., Lopez-Capel, E. and Bol, R. (2010). A review of biochar and its use and function in soil. *Advances in Agronomy,* 105, pp. 47–82.

Sriburi, T. (2011a). Biochar researches for soil amendment at Pa-deng Biochar Research Center (PdBRC (CC294I)). In: *2011 International Symposium on Biochar for Climate Change Mitigation & Soil and Environmental Management. Kangwon, South Korea, 8–9 December 2011.* Kangwon: Biochar Research Center.

Sriburi, T. (2011b). Testing properties of biochar from wood residues before using as a soil amendment. In: *Proceedings, The Conference on Natural Resource Management and Quality of Life Improvement under the Royal Initiation.* Petchburi, Thailand: Huay Sai Royal Development Study Center.

Sriburi, T. (2013). *Evaluating the Life Cycle of Greenhouse Gas Emissions and Retention from a Project for Sustainable Development, Huay Sai Royal Development Study Center, Petchburi.* Bangkok: Final Report of National Research University, Office of Higher Education Commission.

Steiner, C., Teixeira, W. G., Lehmann, J., Nehls, T., Macedo, J. L. V., Blum, W. E. H. and Zech, W. (2007). Long term effects of manure, charcoal and mineral fertilization on crop production and fertility on a highly weathered central Amazonian upland soil. *Plant and Soil*, 291, pp. 275–290.

Steiner, C. (2009). Soil charcoal amendments maintain soil fertility and establish a carbon sink – research and prospects. In: Liu, T. X. (ed.) *Soil Ecology Research Developments*. New York: Nova Science, pp. 1–4.

Sukartono, Utomo, W. H., Kusuma, Z. and Nugroho, W. H. (2011). Soil fertility status, nutrient uptake, and maize (*Zea mays* L.) yield following biochar application on sandy soils of Lombok, Indonesia. *Journal of Tropical Agriculture*, 49, pp. 47–52.

Suppadit, T., Phumkokrak, N. and Pongsuk, P. (2012). The effect of using quail litter biochar on soybean (*Glycine max* [L.] production. *Chilean Journal of Agricultural Research*, 72, pp. 244–251.

Thasanasongchan, A. (1984). *The Story of Rice*. 2nd Edition. Bangkok: Department of Agronomy, Faculty of Agriculture, Kasetsart University.

Turner, S. M. and Liss, P. S. (1985). Measurement of various sulphur gases in a coastal marine environment. *Journal of Atmospheric Chemistry*, 2, pp. 223–232.

Vaccari, F. P., Baronti, S., Lugato, E., Genesio, L., Castaldi, S., Fornasier, F. and Miglietta, F. (2011). Biochar as a strategy to sequester carbon and increase yield in durum wheat. *European Journal of Agronomy*, 34, pp. 231–238.

von Uexkull, H. R. (1976). *Aspects of Fertilizer Use in Modern, High-Yield Rice Culture.* IPI-Bulletin No. 3. Worblaufen-Bern, Schwitzerland: International Potash Institute.

Warnock, D. D., Lehmann, J., Kuyper, T. W. and Rillig, M. C. (2007). Mycorrhizal responses to biochar in soil – concepts and mechanisms. *Plant and Soil*, 300, pp. 9–20.

Watanesk, O. (2004). *Characterization and Evaluation of Local Lowland Rice Accessions in Central West and East Regions.* Bangkok: Rice Economy Research Group, Rice Research Center.

Wijitkosum, S. and Jiwnok, P. (2014). Biochar for soil amelioration and increasing crops yield. *Environmental Journal*, 18, pp. 30–40.

Wijitkosum, S. and Kallayasiri, W. (2015). The use of biochar to increase productivity of indigenous upland rice (*Oryza sativa* L.) and improve soil properties. *Research Journal of Pharmaceutical, Biological and Chemical Sciences*, 6, pp. 1326–1336.

Xu, R., Ferrante, L., Hall, K., Briens, C. and Berruti, F. (2011). Thermal self-sustainability of biochar production by pyrolysis. *Journal of Analytical and Applied Pyrolysis*, 95, pp. 55–66.

Yamato, M., Okimori, Y., Wibowo, I. F., Anshori, S. and Ogawa, M. (2006). Effects of the application of charred bark of *Acacia mangium* on the yield of maize, cowpea and peanut, and soil chemical properties in South Sumatra, Indonesia. *Soil Science and Plant Nutrition*, 52, pp. 489–495.

Yanai, Y., Toyota, K. and Okazaki, M. (2007). Effects of charcoal addition on N2O emissions from soil resulting from rewetting air-dried soil in short-term laboratory experiments. *Soil Science and Plant Nutrition*, 53, pp. 181–188.

Yoshida, S. (1981). *Fundamentals of Rice Crop Science*. Laguna, Philippines: The International Rice Research Institute.

Yooyen, J. (2014). Use of Biochar in Soybean Fields for Increasing Yield and Carbon Sequestration. MSc Thesis. Department of Environmental Science, Graduate School, Chulalongkorn University.

Yooyen, J., Wijitkosum, S. and Sriburi, T. (2015). Increasing yield of soybean by adding biochar. *Journal of Environmental Research and Development,* 9, pp. 1066–1074.

Zheng, W., Sharma B. K. and Rajagopalan, N. (2010). *Using Biochar as a Soil Amendment for Sustainable Agriculture*. Sustainable Agriculture Grant's Research Report Series, Illinois Department of Agriculture, Champaign, IL.

Index

1 Utama shopping centre, 4, 7, 177
2015 Paris Climate Conference (COP21), 10

activated carbon, 167
Africa, 3, 375
agricultural residues, 241
agriculture
   small scale, 29
Albizia myriophylla, 373
allellochemicals
   sorption, 340
America
   North, 165, 319
   South, 3
ammonia (NH_3), 353
ammonia volatilisation, 358
ammonification, 355
Amygdalus communis, 29
Andisol, 325
Apennine Massif, 146
aquatic crops, 241
arbuscular mycorrhizal fungi (AMF), 336, 338
ASEAN
   Economic Community (AEC), 181
   region, 166
ash
   pyrolytic, 241
Asia
   Southeast, 3, 375
Australia, 2, 283
Austrian Advisory Council for Soil Protection and Soil Fertility, 132
Austrian Bio-Indicator grid, 127
Austrian Forest Soil Inventory, 127
Austrian HOBI study, *see biomass supply potential*
Austrian National Forest Inventory (ANFI), 124
   sample details, 125
autotrophic respiration, 325

Balkans, 146
bamboo, 173, 281
bentonite sulphur, 109
Berlin-Dahlem Botanical Garden, 4, 97
   material cycle, 99
Betula spp., 142
biochar
   acidity, 215, 259
   activation, 102
   adsorbing agent, 98, 321
   aging, 113
   application, 41
   as a byproduct, 47
   as animal feed, 307
   as co-product, 39
   as soil amendment, 9, 39, 228, 280
   ash content, 212
   bespoke, 3
   BET surface area, 216
   carbon content, 255
   cation exchange capacity (CEC), 259
   characterization, 101
   characterization techniques, 259
   compost, 3, 98, 104, 105, 112
   costs, 306
   definition, 27, 316, 352, 368
   demand, 176
   designer, 3
   economics, 219
   ecosystem services, 353
   effect on nitrous oxide, 360
   feedstock materials in Thailand, 281
   filter, 115
   fit for purpose, 4
   fixed carbon, 211, 214
   from bark, 280
   greenhouse gas (GHG) mitigation, 10
   H/C and O/C ratios, 214
   hydrophobicity, 317

biochar (*cont.*)
  IBI definition, 1, 253
  large-scale facilities, 319
  life cycle stages, 53
  liming effect, 15, 212, 299, 303, 361
  logistics, 55
  manufacture process, *see biochar production*
  mean residence time, 352
  microsites, 346
  monitoring projects, 283
  mulching, 321
  negative priming effect, 292
  nutrient release, 113
  nutrient supply, 84
  organic farming amendment, 352
  organic matter stabilization, 105
  organic pollutants, 101
  particle behaviour, 15
  pellets, 320
  physical protection of soil biota, 346
  physicochemical behaviour, 258
  porosity, 216, 258, 380
  post-processing, 13, 82
  priming effect, 325
  production, 8, 25, 34, 53f. 3.3., 100, 254, 317
    by retort, 372
    feedstock effect, 254
    heating rate effect, 255
    pyrolysis temperature effect, 254
  prolonged water retention, 308
  properties, 81
    biological, 322
    chemical, 322
  prospects in agriculture, 309
  proximate analysis, 211
  public interest, 2
  recalcitrance, *see biochar stability*
  regulation, 73, 75, 116
  sanitation, 4, 96, 114, 115
  small-scale facilities, 319
  stability, 11, 75, 215, 337, 352
  supply chain, 27, 29
  surface area, 255
  surface chemistry, 258
  sustainability, 31
  system, 4, 6, 30, 31, 116
  system scenario, 79
  systems analysis, 71
  systems approach, 12, 70
  toxicity, 76
  ultimate composition, 214
  volatile matter, 211
bioeconomy, 7, 71, 94, 308
BioMaCon, 100
biomass
  byproduct, 35
  certification, 179
  co-combustion, 88
  co-location of conversion systems, 38
  conversion, 38, 173, 175, 178
  co-product, 35
  definition, 240
  energy
    potential in Turkey, 191
  feedstock accessibility, 179
  from forests, 241
  investment, 179
  low-value utilization, 179
  mathematical gasifier model, 232
  mixed woodchips, 294
  moisture content, 201
  pellets, 155, 165, 166, 176, 329
  perennial crops, 86
    circular model, 87
    directional model, 87
    value chain logistics, 88
  plantations, *see forestry, short rotation*
  production of woody, 153
  resource in Malaysia, 163
  rice-based, 164
  source of energy, 189
  stability, 178
  storage, 38
  supply potential, 132
  supply potential in Austria, 124
  sustainable, 6, 46
  Sustainable Production Initiative (Biomass-SP), 172
  technical standards, 178
  thermal decomposition behaviour, 242
  waste, 32, 35
  waste incineration, 164
  wood based, 123
  woodchips, 15, 35, 101, 155
Biomass Industry Action Plan 2020, 169
Biomass Industry Strategic Action Plan, 172
bio-oil, 39, 175, 200, 217, 247
  acidity, 251
  characterization, 252
  chemical composition, 247
  effect of pyrolysis temperature, 248
  organic compounds, 251
  oxygen content, 251
  properties, 250
  upgrading, 217
  viscosity, 251
  yields, 247
biopolymers, 167
Bio-XCell Biotechnology Industrial Park, 170
Blachia siamensis, 381
Brunauer-Emmett-Teller (BET), 216, 294
Brunei, 181
bypass flow, 359

Index

Calathea insignis, 111
Calathea rotundifola, 110
Cambisol, 293
Cambodia, 181, 266
Canada, 2, 323f. 15.2.
carbon
   below-ground sequestration, 339
   black, 317, 352
   dioxide (CO_2), 240, 336
   permanent sequestration, 63
   sequestration, 11, 42, 56, 292, 293, 340
   net, 189
   permanent, 63
   stabilization, 78
carbon capture and storage (CCS), 78
carbon footprint, 32, 169
carbonisation plant, 100
   thermal efficiency, 101
Carica papaya, 106
cation exchange capacity (CEC), 280, 316, 377
   measurement, 296
cellulose, 241
char, 3
   drawings, 3
   primary, 203
   secondary, 203
charcoal, 3, 26, 174, 258, 268
   from mangrove feedstock, 275
   health risks, 274
   pellets, 283
   production
      residuals, 280
      temperature, 270
   production and trade, 268
   properties, 269
Chauvert cave, 3
chemicals
   bio-, 167
   bio-based, 175
Chernozem, 293
China, 166, 175, 276
Citrus aurantium, 341
Citrus aurantium L., 341
Clean Development Mechanism (CDM), 166
   projects, 169
climate change
   adaptation, 11, 152
   mitigation, 8, 11, 33, 292, 337
   monetizing mitigation, 43
Coffea arabica, 106
column chromatography, 253
combined heat and power (CHP), 32
Combretum punctatum, 374
compost production, 103, 166
composting, 111, 117, 164, 176
   mineralization rate, 104
coppice
   ancient silvicultural system, 150
   annual coupes, 143
   conservation, 152
   conversion, 191
   ecology, 151
   extent and conversion, 151
   genetic resource, 152
   high, 145
   history, 150
   low, *see coppice, simple*
   overmatured, 140
   rotation age, 144
   selection system (CSS), 146
   short rotation (SRC), *see forestry, short rotation (SRF)*
   simple, 141, 150
   soil carbon, 152
   standards, 148
   underwood, 149
   with reserves, 153
   with standards, 140, 146, 150, 153
   woodland, *see forest, coppice*
cost of logistics, 38
cotton seed, 243
cradle-to-gate, 47
   environmental outputs, 60
cradle-to-grave, 47
Cratoxylum formosum, 269
crop residue, 29, 372
crop yields, 376
cultivation, 36
cumulative energy consumption, 60

Danube river, 146
decomposition, 337
denitrification, 358
depolymerisation rate determination, 358
Dipterocarpus obtusifolius, 269
dissolved organic nitrogen (DON), 355
distributed activation energies, 230
distribution logistics, 40

empirical correlations, 231
empty fruit bunches (EFB), 164
energy, 239
   bio-, 8, 143, 174
      subsidies, 88
   crisis, 143
   crop, 35, 241
   forest, 191
   renewable, 2, 123, 133, 164, 169, 240
   thermal, 3, 140, 174
ENVANIS, 188
environmental impact, 51
environmental supply chain management (ESCM), 32
EU Fertiliser Regulation, 76
EU Forest Strategy, 72

Euphorbia x lomi, 111
EUROCOPPICE, 141
European Biochar Certificate, 101, 116
European Biochar Research Network (EBRN), 4
European Chemicals Agency (ECHA), 77
European Union, 165, 172
exhaust smoke color, 271
exudates
　from mycorrhizae, 339
　from roots, 339

Fagus sylvatica, 142
feed-in-tariff (FiT), 171
feedstock logistics, 36, 54
felling cycle, 146
fertilizer
　bio-, 170
　efficiency, 280
　organic, 376
field experiments, 296
fixed-bed gasifier, 100
flash carbonization, 208
FOREBIOM, 4, 13, 15, 18
forest
　annual increment in Austria, 132
　biomass stock, 190
　carbon sequestration, 185
　coppice, 139, 140
　degradation, 267, 281
　high, 132, 140, 150
　management, 73, 141, 185
　mangrove, 269
　of North America, 315
　plantation, 267
　policy, 282
　rehabilitation, *see forest restoration*
　residues, 315
　restoration, 316, 319
　stress, 315
forest biomass
　Adaptation of Standing Volume (ASV)
　　scenario, 129
　Constant Standing Volume (CSV) scenario, 128
　definition, 124
　mobilisation scenario, 128
　nutrient pools, 127
　price scenario, 128
　Silviculture (SC) scenario, 129
　theoretical supply potential in Austria, 124
Forest Research Institute Malaysia (FRIM), 173
forestry
　agro, 73
　differentiating plantation, 72
　short-rotation (SRF), 7, 83, 140, 143
　　circular model, 84
　　directional model, 85
　　productivity, 144
　　scenarios, 83
　　value chain logistics, 85
　traditional, 7
　urban, 88
　　circular deployment, 89
　　directional deployment, 90
　　value chain logistics, 90
Fourier transform infrared spectroscopy (FT-IR), 243, 252
free-living nitrogen-fixation, 361
fuelwood, 268
functional unit, 49

gas chromatography-mass spectroscopy
　　(GC/MS), 252
gasification, 77, 87, 199, 201, 210
gasifier
　downdraft, 210
　updraft, 210
General Directorate of Forestry (GDF), 186
Geranium maderense, 106
German Compost Quality Assurance
　　Organisation, 105
German Federal Soil Protection and Contamination
　　Ordinance, 116
German Fertiliser Ordinance, 116
Global Forest Resources Assessment, 72
global rate constant, 231
glomalin, 339
Glycine max, 340, 380
goods
　final, 32
　intermediate, 33
green supply chain management, *see environmental
　　supply chain management (ESCM)*
greenhouse gas (GHG)
　abatement costs, 13
　emission, 9
　emission reduction, 42, 359
　emissions, 325
　　non CO_2, 111
　flux measurement, 297
　mitigation effects, 124
　mitigation potential, 14t. 1.1., 111
　N_2O emissions, 302
　National Inventory Report (NIR), 184
　net emission reduction of biochar, 66
Grewia paniculata, 269
groundwater quality, 298
growing media, 78

habitat heterogeneity, 152
harvest potential
　theoretical, 130
harvesting constraints
　ecological, 126, 131, 132
　economical, 126

technical, 126
harvesting potential
    realisable, 130, 132
hemicellulose, 241
heterotrophic respiration, 337
Hevea brasiliensis, 163, 166, 173
    plantation, 269
Hordeum vulgare, 295
horizontal grinder, 28
horizontal integration, 29
HTC char, *see hydrochar*
hydrochar, 210
hydrothermal carbonization (HTC), 90, 202, 210
hydrothermal liquefaction (HTL), 202
hyperaccumulator, *see phyto-extraction*

illegal logging, 267, 281
Inceptisol, 292, 326
incubation experiment, 300
India, 2, 182
Indonesia, 181
industrial wood, 189
industry
    biomass, 165
        in Malaysia, 165
    brick stone, 163
    charcoal, 175
    furniture, 173
    green technology, 166
    palm oil, 164
    pellets, 175
inoculant
    microbial, 81
    mycorrhizal, 339, 341
Integrated Pollution Prevention and Control (IPPC), 76
Intergovernmental Panel on Climate Change (IPCC), 185
International Biochar Initiative (IBI), 4, 253
    biochar standards, 259
invasive species, 327
Irvingia malayana, 269
isotope, 354
    dilution technique, 356
    enrichment studies, 355
    natural abundance studies, 355
    stable, 355
isotope ratio mass spectrometry (IRMS), 355

Japan, 166, 175, 352

kiln, 246
    advanced technology, 275
    beehive, 205
    brick beehive, 275
    drum, 277
    emissions of GHGs, 281
    ground pit, 204, *see kiln, heap*
    heap, 273
    Hereshoff multiple hearth furnace, 206
    Iwate, 276
    Missouri, 205
    mound, 204
    mud beehive, 275
    New Hampshire, 205
    permanent, 272
    Schottdorf, 206
    Tao Op, 271
    Tao Phi, 272
    Temerity charcoal production, 273
    Thai-Iwate, 275, 276
    traditional technology, 271
    Yoshimura, 275
kinetic models, 229
    global, 229
Korea
    South, 4, 166, 175
Kyoto Protocol, 74, 141, 166, 184

land
    conversion, 267
    development, 89
    historic use, 127
    tenure, 282
Land use, Land-use Change and Forestry (LULUCF), 185
    mitigation potential, 190
Lao Institute for Renewable Energy (LIRE), 276
Lao PDR, *see Laos*
Laos, 181, 266
leaching, 127, 170, 175
    of nitrate, 353, 359
    of nutrients, 113, 127, 298
Leucaena leucocephala, 374
life cycle analysis (LCA), 6, 47, 77, 174
    attributional (ALCA), 51
    consequential (CLCA), 52
    cut-off criteria, 49
    goal and scope definition, 49
    interpretation phase, 51
    material extraction stage, 54
    phases, 47
    primary data, 49
    product system boundary, 49
    secondary data, 49
life cycle impact assessment (LCIA), 50
life cycle inventory analysis (LCI), 49
    flows, 50
life cycle assessment, *see life cycle analysis*
lignin, 241
litter layer removal, 192
local communities participation, 282
Lorentz and Parade's Handbook of Silviculture, 151

maize, 374
Malaysia, 162, 172, 181
   Agriculture Research and Development Institute (MARDI), 175
   Biomass Industry Confederation (MBIC), 172
   Biotechnology Corporation, 170
   Industry-Government Group for High Technology (MIGHT), 175
   National Biomass Strategy 2020, 169, 172
   National Biotechnology Policy, 170
   National Green Technology Policy, 170
   National Renewable Energy Policy and Action Plan, 171
   Timber Industry Board (MTIB), 174
mass spectrometer (MS), 243
mesocarp fibers, 164
microlysimeter, 294
Middle East, 175
mine reclamation, 28
mine tailings restoration, 320
mineralization-immobilization turnover (MIT), 357
mixed organic waste stream, 34
Mollisol, 325
Morvan Massif, 146
mulching, 164
Myanmar, 181
mycelia network, 337
mycorrhizae, 10, 337
   abundance, 346
   colonization, 340

natural abundance approaches, 361
Nerium oleander, 106
net sequestration impact, 12
Netherlands Development Organisation (SNV), 276
nitrate (NO_3), 353
   maximum contaminant level (MCL), 359
nitric oxide (NO), 353
nitrification, 355
nitrogen, 353
   atmospheric deposition, 127
   compounds in soil, 302
   cycle, 9, 352
      biogeochemical processes, 353
   gross transformation rates, 356
   immobilization, 356
   of microbial biomass, 360
   plant uptake, 362
nitrous oxide (N_2O), 353
   emission pathways, 360
nuclear magnetic resonance spectroscopy (NMR), 252
nutrients
   balance approach, 127
   bioavailability, 303
   buffer, 176
   cycle, 165
   extraction of, 127
   primary, 377
   retention, 84

oil palm fronds (OPF), 164
old palm trunks (OPT), 164
organic extractives, 241
Oryza sativa L., 375
Other Wood Land (OWL)
   FAO definition, 187
overwood, 147
oxidation of labile organic matter, 358

palm
   kernel cake, 164
   kernel shells (PKS), 164, 167
   oil mill effluent (POME), 164
   waste disposal, 164
Panicum virgatum, 35
paper mill bio-sludge (PMS), 176
pellets, 167
persistent organic pollutants (POPs), 80
Philippines, 181
phospholipid fatty acids (PLFAs)
   analytical procedure, 297
phosphorus, 12, 99, 298, 377
   reduced leaching, 113
photosynthesis, 337
phyto-extraction, 79
Picea abies, 15, 346
Planosol, 293
plant substrate, 106, 109
pollarding, *see coppice, high*
polycyclic aromatic hydrocarbons (PAH), 251
Populus spp., 35
pot experiment, 293, 294, 304, 341
primary products, 35
PROGNAUS, 125, 127, 129
protective forest, 124
pruning, 374
pulp and paper, 174
Pyrenees, 146
pyroligneous acid, 200, 278
   application, 279
   characteristics, 279
   constituents, 278
   production capacity, 276
   risks, 279
pyrolysis, 199, 227
   ash, 280
   atmosphere, 248
   chemical kinetics, 230
   chemistry, 202
   conditions, 3, 8
   definition, 245
   economics, 236
   effects of conditions, 228

elemental balance, 232
energetics, 236
energy balance, 232
energy requirement, 235
fast, 200, 203, 217, 246, 270
    catalytic treatment, 248
    heating rate, 246
    temperature, 246
feedstock preparation, 101
microwave, 208
of urban food waste, 90
primary reactions, 202
process, 369
products, 8, 246
secondary reactions, 203
slow, 201, 203, 217, 227, 245, 270, 370
    heating rate, 245
    temperature range, 245
steps, 271
pyrolysis reactor
    auger, *see pyrolysis, reactor, screw*
    batch type, 205
    continuous type, 206
    controlled temperature biochar retort, 369
    fluidized bed, 209
    particulate emissions, 271
    retort, 206
    retort construction materials, 369
    rotating cone, 210
    rotating drum, 207
    screw, 207
    slow, 204

Quercus-Carpinus CWS, 153
Quercus ilex, 142

randomized complete block design (RCBD), 376
rattan, 173
remediation, 28
    circular model, 80
    directional model, 81
    land, 74, 79
    scenarios, 79
    value chain logistics, 82
Renewable Energy Act, 171
resprouting, 7, 140
Rhizobium japonicum, 382
Rhizophora spp., 174
Rhododendron simsii, 109
rice
    charred husk, 280
    husk, 164, 177, 272
    Nasarn, 376
    photosensitive, 376
    straw, 164
    upland, *see Oryza sativa*
road removal, 321

root respiration, *see autotrophic respiration*
root sucker, 139, 152
Royal French Ordinance of 1545, 143
rubberwood, *see Hevea brasiliensis*

saline soils restoration, 115
Salix spp., 35, 144
Samanea saman, 374
scanning electron microscopy (SEM), 259
second-generation biofuel, 175
Shorea obtusa, 269
silviculture, 36
Sinapis alba, 295
Singapore, 177, 181
single-stem volume limits, 126
slash, 38
    burning, 346
    logging, 47
soil
    acidity, 377
    aggregates, 338
    agricultural, 293, 319
    amelioration, 368, 377
    available potassium (K), 377
    biota, 10
    bulk density, 321, 322t. 15.2.
    compaction, 133
    degradation, 369
    disturbance, 185
    electric conductivity (EC), 377
    enzyme activity, 322
    exchangeable phosphorus (P), 377
    fertility, 306, 352
    gross nitrification rate, 358
    microbial biomass (SMB), 355
    nutrient availability, 345
    organic carbon (SOC), 337
        contents of amended Amazonian soils, 292
    organic matter (SOM), 165, 337, 355
    pollution, 282
    productivity, 362
    properties, 32
    protein turnover, 358
    weathering, 175
Sorghum bicolor, 340, 345
South America
    Central Amazonia, 97
soybean, *see Glycine max*
stable isotope techniques, 353
stool, *see stump*
Streblus ilicifolius, 373
stump, 139, 140, 143
substitute energy products, 32
supply chain, 25
    framework, 31
    management (SCM), 25
    model, 27, 30, 31

sustainability
　feedstock, 74
　nutritional, 127, 130
Sustainable Energy Development Authority
　　(SEDA), 171
syngas, 59, 173, 227

Terra Preta, 97, 98, 114, 291
Terra Preta do Indio, *see Terra Preta*
Terra Preta Sanitation (TPS), 114
TerraBoGa, 97
terrestrial laser scanning, 17
testaste amoebae (TA), 346
Thailand, 266, 369
thermochemical conversion, 54, 199
thermochemical technologies, 241
thermogravimetric analysis (TGA), 242
torrefaction, 201
tree growth, 326
Trifolium alexandrinum, 339
Trifolium pratense, 295
triplet tracer experiments (TTEs), 361
Triticum aestivum, 294
Tucker Renewable Natural Gas (RNG), 57
Turkey, 185
　forest carbon stocks, 186
　National Forest Inventory (NFI), 188

Universiti Malaya (UM), 174
UN Framework Convention on Climate Change
　　(UNFCCC), 72, 184, 186
underwood, 147
United Kingdom, 2, 89
United States, 2, 51, 56

Universiti Putra Malaysia (UPM), 174, 175
Universiti Sains Malaysia (USM), 174
Universiti Teknologi Mara Malaysia (UiTM), 174
University Kuala Lumpur (UniKL), 175
urban water drainage, 89

van Krevelen diagram, 214
vegetative propagation, 139
vermicomposting, 115
vertical integration, 29
Vitis vinifera, 294

waste
　animal, 241
　industrial, 241
Waste Incineration Directive (WID), 76
wastewater nutrients, 115
water-holding capacity (WHC), 105, 165, 297
water retention characteristics, 297
whole-tree harvesting (WTH), 127, 130, 132
wildfire hazard, 316
wildfire prevention, 47
wood
　hard, 373
　soft, 373
wood vinegar, *see pyroligneous acid*
wood waste, 163, 319

Xylia spp., 269

yield components, 376

Zea mays, *see maize*
zero order kinetics, 356